T0322054

Exploring Ethical Problems in Today's Technological World

Tamara Phillips Fudge
Purdue University Global, USA

A volume in the Advances in Human and Social Aspects of Technology (AHSAT) Book Series

Published in the United States of America by
IGI Global
Information Science Reference (an imprint of IGI Global)
701 E. Chocolate Avenue
Hershey PA, USA 17033
Tel: 717-533-8845
Fax: 717-533-8661
E-mail: cust@igi-global.com
Web site: http://www.igi-global.com

Library of Congress Cataloging-in-Publication Data

Names: Fudge, Tamara Phillips, 1957- editor.
Title: Exploring ethical problems in today's technological world / Tamara Fudge, editor.
Description: Hershey PA : Engineering Science Reference, [2023] | Includes bibliographical references. | Summary: "This volume will focus on ethical dilemmas created by today's ever-changing technologies and how these issues have affected individuals, companies, and society to include policies, responsibilities, abuses, consequences, whistle-blowing, and other factors in a wide variety of technology areas"-- Provided by publisher.
Identifiers: LCCN 2022027514 (print) | LCCN 2022027515 (ebook) | ISBN 9781668458921 (hardcover) | ISBN 9781668458938 (paperback) | ISBN 9781668458945 (ebook)
Subjects: LCSH: Technology--Moral and ethical aspects.
Classification: LCC BJ59 .E97 2023 (print) | LCC BJ59 (ebook) | DDC 174/.96--dc23/eng/20220810
LC record available at https://lccn.loc.gov/2022027514
LC ebook record available at https://lccn.loc.gov/2022027515

This book is published in the IGI Global book series Advances in Human and Social Aspects of Technology (AHSAT) (ISSN: 2328-1316; eISSN: 2328-1324)

British Cataloguing in Publication Data
A Cataloguing in Publication record for this book is available from the British Library.

For electronic access to this publication, please contact: eresources@igi-global.com.

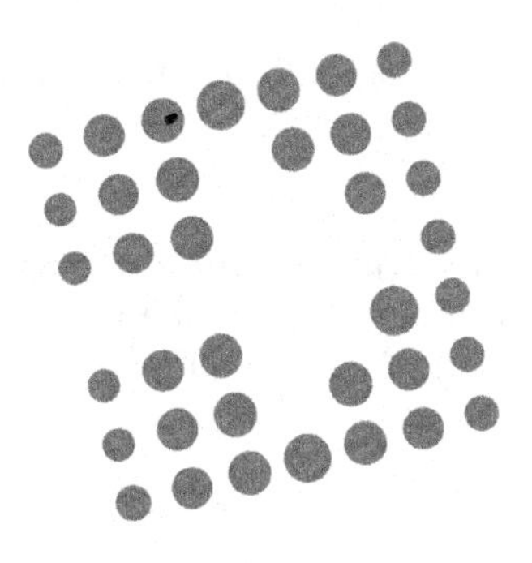

Advances in Human and Social Aspects of Technology (AHSAT) Book Series

Mehdi Khosrow-Pour, D.B.A.
Information Resources Management Association, USA

ISSN:2328-1316
EISSN:2328-1324

Mission

In recent years, the societal impact of technology has been noted as we become increasingly more connected and are presented with more digital tools and devices. With the popularity of digital devices such as cell phones and tablets, it is crucial to consider the implications of our digital dependence and the presence of technology in our everyday lives.

The **Advances in Human and Social Aspects of Technology (AHSAT) Book Series** seeks to explore the ways in which society and human beings have been affected by technology and how the technological revolution has changed the way we conduct our lives as well as our behavior. The AHSAT book series aims to publish the most cutting-edge research on human behavior and interaction with technology and the ways in which the digital age is changing society.

Coverage

- Cyber Bullying
- Technoself
- Information ethics
- Technology and Social Change
- Activism and ICTs
- ICTs and human empowerment
- Technology Adoption
- Technology Dependence
- Digital Identity
- Human Development and Technology

IGI Global is currently accepting manuscripts for publication within this series. To submit a proposal for a volume in this series, please contact our Acquisition Editors at Acquisitions@igi-global.com or visit: http://www.igi-global.com/publish/.

The Advances in Human and Social Aspects of Technology (AHSAT) Book Series (ISSN 2328-1316) is published by IGI Global, 701 E. Chocolate Avenue, Hershey, PA 17033-1240, USA, www.igi-global.com. This series is composed of titles available for purchase individually; each title is edited to be contextually exclusive from any other title within the series. For pricing and ordering information please visit http://www.igi-global.com/book-series/advances-human-social-aspects-technology/37145. Postmaster: Send all address changes to above address.

Titles in this Series

For a list of additional titles in this series, please visit: www.igi-global.com/book-series/advances-human-social-aspects-technology/37145

Handbook of Research on Implementing Digital Reality and Interactive Technologies to Achieve Society 5.0
Francesca Maria Ugliotti (Politecnico di Torino, Italy) and Anna Osello (Politecnico di Torino, Italy)
Information Science Reference • © 2022 • 731pp • H/C (ISBN: 9781668448540) • US $295.00

Technological Influences on Creativity and User Experience
Joshua Fairchild (Creighton University, USA)
Information Science Reference • © 2022 • 305pp • H/C (ISBN: 9781799843542) • US $195.00

Machine Learning for Societal Improvement, Modernization, and Progress
Vishnu S. Pendyala (San Jose State University, USA)
Engineering Science Reference • © 2022 • 290pp • H/C (ISBN: 9781668440452) • US $270.00

The Digital Folklore of Cyberculture and Digital Humanities
Stamatis Papadakis (University of Crete, Greece) and Alexandros Kapaniaris (Hellenic Open University, Greece)
Information Science Reference • © 2022 • 361pp • H/C (ISBN: 9781668444610) • US $215.00

Multidisciplinary Perspectives Towards Building a Digitally Competent Society
Sanjeev Bansal (Amity University, Noida, India) Vandana Ahuja (Jaipuria Institute of Management, Ghaziabad, India) Vijit Chaturvedi (Amity University, Noida, India) and Vinamra Jain (Amity University, Noida, India)
Engineering Science Reference • © 2022 • 311pp • H/C (ISBN: 9781668452745) • US $270.00

Analyzing Multidisciplinary Uses and Impact of Innovative Technologies
Emiliano Marchisio (Giustino Fortunato University, Italy)
Information Science Reference • © 2022 • 275pp • H/C (ISBN: 9781668460153) • US $240.00

Handbook of Research on Applying Emerging Technologies Across Multiple Disciplines
Emiliano Marchisio (Giustino Fortunato University, Italy)
Information Science Reference • © 2022 • 548pp • H/C (ISBN: 9781799884767) • US $270.00

Handbook of Research on Digital Violence and Discrimination Studies
Fahri Özsungur (Mersin University, Turkey)
Information Science Reference • © 2022 • 837pp • H/C (ISBN: 9781799891871) • US $270.00

701 East Chocolate Avenue, Hershey, PA 17033, USA
Tel: 717-533-8845 x100 • Fax: 717-533-8661
E-Mail: cust@igi-global.com • www.igi-global.com

Table of Contents

Detailed Table of Contents

Understanding the various types of innovation technology that are currently disrupting organizations around the globe is an important endeavor. Various innovative technologies are defined, and related ethical challenges are explored that organizations face in today's technology climate. Additionally, key methods are examined for an organization to manage employees when they refuse to change, including unfreezing, transition, and refreezing. The importance of alignment of innovation strategy to organizational resources is reinforced, with the best methods to align organizations with innovation success, including when it is okay to say no to innovation within an organization.

All aspects of human life are affected by technology. It has changed the way individuals communicate with one another and conduct their everyday routines and business transactions. The world is undergoing a Fourth Industrial Revolution, which is radically altering the world of leadership. Artificial intelligence, machine learning, robots, and other cutting-edge technologies, in particular, are increasing efficiency and reshaping leadership and operations. Leaders in today's most successful firms must keep up with technology advancements and successfully leverage them to achieve laudable outcomes. The motivating element behind the new leadership style is technology and its ethical use. Literature gaps indicate that there is much more to learn about this issue and that it is critical to address it in order for enterprises to remain competitive.

Today's world is increasingly based on digital access to information. People conduct essential aspects of life online through their web browsers and mobile applications: education, healthcare, banking, shopping, entertainment, and even jobs are conducted through the internet. To be cut off from the digital world is to miss these essential connections; this is exactly what happens to people with disabilities when the websites and content they try to use have accessibility barriers. People and organizations creating web content need to understand the elements of accessibility, important laws and regulations that guide accessibility efforts, and ways to improve the accessibility of web content. Eliminating these barriers is an important step towards a more inclusive society.

The exponential growth of AI and its applications in different areas of society, such as the financial, agricultural, telecommunications, or health sectors, poses new challenges for the government's public sector, mainly in regulating these systems. Governments and entities in general address these challenges by formulating soft laws such as manuals or guidelines. They seek full transparency, privacy, and bias reduction when implementing an AI-based system, including its life cycle and respective data management or governance. These tools and documents aim to develop an ethical AI that addresses or solves the aforementioned ethical implications. The revision of 22 documents within frameworks, guides, articles, toolkits, and manuals proposed by different governments and entities are examined in detail. Analyses include a general summary, the main objective, characteristics to be highlighted, advantages and disadvantages if any, and possible improvements.

Preface: The Attempt to Define Limits

INTRODUCTION TO THIS VOLUME

The pervasiveness of technology today has brought with it a bevy of ethical questions, many of which are difficult to answer. Average consumers place themselves at risk financially, professionally, and personally by everyday activities executed on computers and smart phones. Therein lies the responsibility of technologists and decision-makers to devise logical and ethical solutions that impact the individual, companies, and society at large.

HUMAN THOUGHT PROCESS

Consequences are unfortunately not always at the forefront during the development of exciting new technologies. Hofmann (2020) explains because new inventions are developed and rolled out to consumers at an incredibly fast pace, the benefits and drawbacks – including the potential for disruption – might not be known until after the technology has already been available to the public for a period of time. The automobile industry provides an example of this: if seatbelts had been required in the early days, surely some fatalities could have been avoided (Ammanath, 2021). Quick distribution to the masses was also experienced with the switch to virtual work and schooling during the pandemic, although this premise had been practiced for quite some time prior (Richter & Sinha, 2020).

Fast rollouts might mean a focus on monetary gain instead of societal benefit, inadequate testing, or some other lack of foresight as to consequences. The timing is not the only problem, though, as even long-standing technologies can still straddle a line between what society may think is acceptable and what is not. Limits could be imposed by legal mandates, industry standards, and/or companies' choices to make socially acceptable ethical decisions; the latter is where the true problems lie, as what might be acceptable to one group of people could be fully unacceptable to others. Ethics is, after all, a subjective construct.

This brings to mind three quotes which aptly (and entertainingly) describe choices that must be made. Note each of these includes a particularly critical keyword: "should."

- "Just because you can doesn't mean you should." – author Sherrilyn Kenyon in the 2008 book *Acheron* (Kenyon, n.d.).
- "Just because you can afford it doesn't mean you should buy it." – financier Suze Orman (Orman, 2015).

Table 1. Biases

Bias	Explanation
Authority	The highest-ranking person in the company has the most appropriate opinion. This is often related to an Overconfidence bias (Glaveski, n.d.).
Bells and whistles	More features must mean it works better (Hofmann, 2020).
Confirmation	It is good if the result backs initial presumptions (Glaveski, n.d.)
Connectivity	It is best when you can connect a new innovation to other technologies (Hofmann, 2020).
Courtesy	The result is good if it does not offend anyone (Glaveski, n.d.).
Human substitution	Technology is better than what humans can do (Hofmann, 2020).
Normalcy	The problem has not occurred in the past and so mitigation plans do not need to be developed (Glaveski, n.d.).
Pro-innovation	If it is new, it must be better (Hofmann, 2020).
Subjunctivization	It is good because of what it could possibly do, not what it actually does (Hofmann, 2020).
Triviality	Focus is given to minor concepts as a means to avoid more difficult ones (Glaveski, n.d.).

- "Your scientists were so preoccupied with whether or not they could that they didn't stop to think if they should." – line delivered by the movie character Ian Malcolm as portrayed by Jeff Goldblum in 1993's *Jurassic Park* (Koepp, 1992, p. 30).

Social scientists are quick to point out specific theories which help guide choices, including *consequentialism*, where the end results matter most; *utilitarianism*, in which the purpose is to focus on the good and minimize unhappiness; and *deontology*, where actions themselves are labeled as fundamentally right or wrong, no matter the consequences ("Moral Theories," n.d.). These theories are complicated and contentious, but the basic tenets help to frame answers to whether or not a new technology "should."

Even with technologies perceived as ethical and beneficial, there can be great resistance. The so-called Digital Divide gives a partial answer to why certain users might be more apt to adopt new innovations; those without adequate computer skills may struggle with adapting to new methods. There is also cost, company culture, and the need for management to adapt in order to more effectively use new systems (Richter & Sinha, 2020). Hofman (2020) likens resistance to change, especially when there is unfamiliarity with the new technology, as asking someone to change their loyalty from one sports team to another. Richter and Sinha (2020) also mention "legacy approaches to innovation and problem-solving" as a company challenge (para. 5). Another consideration is that humans cannot always predict accurate consequences ("Moral Theories," n.d.). If people are fallible, then it might follow computers could be fallible, too.

Theories cannot cover all possible human responses. Glaveski (n.d.) claims over one hundred thought *biases* affecting our decision-making. While there can be a *status quo bias* wherein changes are strongly resisted, a *progress bias* on the other end of the spectrum may also exist, where change may be fervently desired even if not warranted (Hofmann, 2020). It can be difficult to find middle ground between these extremes, and other fallacious predispositions such as those in Table 1 further confound decision-making.

Hofmann (2020) additionally describes an *endowment effect*, wherein people place value on what they possess rather than contemplating true judgements. The familiar Mac versus PC dispute and iPhone versus Android disagreement can be examples of this phenomenon.

Each of these biases can distort decision-making as new technologies are in development and made available to the public. Thus, not all human decisions are rational.

ATTEMPTS TO CONTROL

It can be argued that technology cannot make any decisions on its own; it is controlled by humans, whether they are programmers, end-users, or other stakeholders. For controlling purposes, there are workplace codes of ethics and various laws to consider.

Acceptable Use Policies and Monitoring

Brandt (2022) defines ethics as a set of principles for a group, whereas morals are societal and values are what is espoused by the individual. Within the context of a group is a company's Acceptable Use Policy (AUP). This is a document outlining the proper use of company-owned technology, is similar to a Terms of Service declaration or Code of Conduct (Kostadinov, 2014), and is both a semi-legal assertion to protect the company's assets and an educational tool for employees (Martins, 2022).

The main premises of the AUP are to explain which systems, networks, and devices are covered, as well as outlining required actions (such as logging out when leaving a workstation), prohibited activities (such as sharing access to those who are not authorized), punitive actions taken for violating rules, and mandating training (Sheil, 2020). Although focused on behavioral security, this document also contains logical security such as how passwords are managed (Kostadinov, 2014) and must include employee-owned device use, any remote work restrictions, and guidelines for social media use that could reflect on the company's reputation (Martins, 2022). The AUP is typically acknowledged by signature in order to enforce the rules therein; lastly, it is also important to make revisions when business circumstances change (Martins, 2022).

One area of ethical concern covered in an AUP is the employees' privacy and workplace monitoring. Sheil (2020) indicates as much as 40% of employee time is used for non-work internet time. This can lead to "cyber-slacking" and monitoring might include filtering software, checking server logs, and/or email scans (Kostadinov, 2014). Being watched when using company technology is to be expected, but methods should be described (Kostadinov, 2014; Martins, 2022). Martins (2022) recommends to set reasonable rules, as "hyper control" tends to be counter-productive.

Regulations and Legalities

There are many laws and regulations that attempt to control technology innovation and use. In the U.S., for example, the Health Insurance Portability and Accountability Act (HIPAA) regulates electronically-stored medical data, the Payment Card Industry Data Security Standard (PCI DSS) protects financial data such as credit and debit cards, and the Federal Information Security Management Act (FISMA) controls how federal agencies use IT systems, and the National Institute of Standards and Technology (NIST) publishes a myriad of technology standards.

There are also international "recommendations" such as those from the World Wide Web Consortium (W3C) which assist web and software developers in developing appropriate interfaces.

In France, a law was developed in 2019 forbidding the publishing of court decision data analysis (Langworth, 2019). Sweden's laws cover both hard copy and electronically preserved data and like many other countries, specifies what can and cannot be collected, how data is used, and what is allowed with marketing ("Data Protected," 2020). The European Union has mandated certain rules for all EU residents since May 2018 through the General Data Protection Regular (GDPR); this includes websites based anywhere, not necessarily in the EU (Frankenfield, 2020).

African countries are also writing laws to protect data as well. Kenya, Nigeria, Togo, Uganda both enacted laws in 2019 to protect data privacy, followed by South Africa in 2020 and Rwanda in 2021. Morocco and Ghana passed laws much earlier, in 2009 and 2012 respectively (Schneidman et al., 2021). Obviously, countries worldwide are becoming aware of the need for control.

Despite these restrictions (or perhaps because of them), Venkataramakrishnan (2021) explains a major problem is the pervasive belief in the business world that laws and regulations impede new technologies and innovation. This may lead to unethical research methods or concealed development.

Enforcing laws is thus another ethical issue. According to Wozniak (2021), the Digital Millennium Copyright Act in the U.S. deals with online piracy and requires notification of infringement, but unfortunately, "70% of takedown notices do not lead to the removal of materials" (Online Piracy section). Since the internet is a worldwide phenomenon, it is also difficult if not impossible to reach all parties and meet each country's requirements.

Court cases provide the opportunity for clarification, enforcement, and sometimes review of laws themselves. In mid-2021, the U.S. Supreme Court decided a case wherein a law enforcement employee was accused of using work access (a patrol car computer) to check a license plate against his department's rules, as there was allegedly an exchange of funds for the action; his activity was caught as part of a Federal Bureau of Investigation sting (*Van Buren v. United States*, 2021). The Computer Fraud and Abuse Act of 1986 (CFAA) was called into question for the wording "entitled so to obtain" and the case surprisingly resulted in the ruling that while Van Buren violated department policy, his action did not violate the CFAA (Knight, 2021). Knight (2021) states the Court had concerns about "criminalizing commonplace computer activity" (para. 7).

Digital accessibility in the U.S. is covered partially by the Americans with Disabilities Act (ADA) which deals with physical place/location and the Communications and Video Accessibility Act (CVAA) which deals with the actual technologies needed (Reid, 2022). In one case, a person with sight impairment sued the grocery chain Winn-Dixie because the company's website did not allow for the complainant's screen reader to activate online coupons or renew prescriptions; a federal appellate court, however, determined ADA did not cover the technology and vacated the suit (Reid, 2022).

CONCLUSION

The development and use of 21st-century technologies will always be judged in part by a societal code of ethics. The rather subjective construct of right versus wrong (alternatively, good versus evil) and all of the points between these opposite poles can make decision-making for the individual, company, and government difficult at best. Policies and industry regulations are a good starting point. While laws are designed to back the unwritten rules of ethics, different countries have different laws, complicating global economies, communication, and other connections between human counterparts worldwide.

The interpretation of laws and whether or not policies and laws are followed will continue to further complicate decision-making.

It is hoped this volume's in-depth chapters (as well as the topic snippets in this preface) will provide the reader with ideas to contemplate and possibly spur some to action in keeping technology usable, safe, and ethically suitable for today and tomorrow.

ORGANIZATION OF THIS BOOK

There are six sections with a total of 16 chapters. Each chapter's contribution is summarized below.

Section 1: Technology Adoption, Business Leadership, and Strategies

Chapter 1 introduces the concept of today's disruptive technology innovations and the strategies that may be employed by businesses, including "unfreezing" and "refreezing" as well as making transitions. Benefits and drawbacks t deploying technological advances must be considered. Importantly, this chapter explains that it is perfectly acceptable to refuse some innovations when they do not align with company goals and objectives.

Chapter 2 delves into how leadership can be enabled by technology, and how ethics plays a role. Ten major types of technology are defined, as are the characteristics of today's responsive leadership style. E-leadership had become more prevalent since the pandemic, and additional recommendations and various digital tools are explored. It is a balance between old and new styles of leading companies and people.

Chapter 3 describes Radio Frequency Identification (RFID) and many of its applications in the business world, including implementation in marketing, supply chain management, logistics, retail situations, agriculture, medicine, and other areas. Potential problems are noted such as technology failures, data mining, environmental impact, etc., and two case studies give additional insight into the ethics of RFID use.

Chapter 4 explains digital "nudging," a pressure put on online consumers during a decision-making process that may or may not have the consumer's best interest in mind. The study conducted for this chapter is chronicled in detail and respondent commentary provides a unique view into the kinds of nudges that are used, where they can be employed and for what purpose, and consumer awareness and learning as to the risks.

Section 2: Teaching and Learning

Chapter 5 offers a warning about corporate corruption and how schools can address learning technology ethics for business leadership. Internet use, security, biometrics, data ethics, and cryptocurrency issues are examined. Solutions include incorporating ethics in curriculum in ways that it connects to students' lives and creates a personal connection to what is good for society – and thus good for a company meeting business objectives.

Chapter 6 examines the technology world of people who must work or learn in an English-speaking environment but are not native speakers. An understanding of user tracking and where data is collected, plus a students' data rights, responsibilities and digital citizenship are outlined and digital tools are also reviewed. The unique needs and experiences of English as a Second Language students are addressed.

Chapter 7 focuses on a positive use for social media: teaching empathy, civil discussion, and moral development while guiding students to think critically through a unique curriculum design. The experiment conducted on Facebook is detailed with learning objectives, a variety of writing prompts and purposes (including links and a variety of online resource material) plus thoughtful instructions that faculty may provide to students.

Section 3: Design as a Factor

Chapter 8 presents "technoethics" and how it can be assessed. Environmental, liability, and other issues are explored, along with topics such as orbital space debris and autonomous driving. Societal efficacy must be determined and potential alternatives considered. A lack of ethics training provides a large focus of the chapter, with National Institutes of Health and other associations' guidance for situational analysis.

Chapter 9 describes why design accessibility is a crucial and necessary factor in providing a fair and equitable access to today's technology. In addition to individual country's laws, the World Wide Web Consortium and Web Content Accessibility Guidelines are explained and recommendations given for dealing with text, color, alternative text for images and multimedia, hyperlinks, keyboard compatibility, and other facets that can be controlled by web and application developers.

Section 4: Public Policy Development

Chapter 10 studies the ethical implementation of Artificial Intelligence systems for public policy in development and deployment. Nearly two dozen documents are summarized and analyzed and compared for advantages, disadvantages, and improvements. These include manuals, frameworks and guides, toolkits, and articles and white papers. Recommendations are provided for the acquisition of various AI-based solutions.

Section 5: When Technology Gets Personal

Chapter 11 investigates the uses and potential abuses of Facial Recognition Technology (FRT). Security, privacy, and ethe potential for discrimination are among the subtopics inspected, and various ethics theories are applied. There are several laws explained that can help govern this kind of technology, but there are loopholes and a dearth of clear legal guidance. Future research is suggested as well.

Chapter 12 provides a look into how the collection and use of DNA and biometrics could straddle the ethical compass, and how GPS (the Global Positioning System) is beneficial in various ways but could be used for nefarious purposes as well. This combination of *who* and *where* people are affects privacy and security, but at the same time, such tracking may be useful in major events such as pandemics and war. It all relates to human rights.

Chapter 13 guides the reader through Radio frequency identification (RFID) technology: how it works, and when, where, and why it might be employed. The "microchipping" of pets then takes a step towards microchipping people, where the ethical line is not as clearly delineated. While useful in situations such as medical emergencies, there are physical concerns about inserting technology into the body and other facets to consider.

Chapter 14 scrutinizes the Smart Home. Advantages include Ambient Assisted Living for the elderly and those with any of a variety of impairments, whereas limitations include issues such as phishing attacks and eavesdropping. Specialized companies providing management and security services are explored. The perceived benefits and costs make the choice of installing this technology a more difficult decision. Solutions and best practices are provided.

Section 6: Moving Further Out of the Comfort Zone

Chapter 15 reports about the world of Autonomous Vehicles. An explanation of the technology and how decisions are made are included, and the Six Levels of Automation plus the ISO Standard 26262 are reviewed for understanding the complexity of these machines. Specific cases best chronicle the perils, and major risks described include economic and environmental impacts, individual and company responsibilities, and problems with the lack of clear legislative oversight.

Chapter 16 investigates the Metaverse, which is the term used for virtual and physical worlds when combined. This ecosystem includes avatars, intellectual property and other content, and the potential for censorship. Social acceptability, privacy, trust, and other areas are appraised, and Blockchain's participation is explained along with issues related to governance, design, and specific platforms for further exploration.

LAGNIAPPE: ADDITIONAL TOPICS

The brief segments below are offered as a means to begin classroom discussion or to spark further exploration.

Assistive Technologies

Individuals who have disabilities can be greatly supported by assistive technologies. Along with the benefits comes the potential for errors, opening up some ethical challenges. Among these technologies are the following:

- Screen reader devices are often used by people with sight impairments, but Das et al. (2022) report that these could be better designed to allow for teamwork in ability-diverse work groups. For an ethical inclusion of those who have disabilities, they suggest using non-speech sounds such as earcons and "contextual presentation" in which the person editing is identified (Das et al., 2022).
- Reyes Leiva et al. (2021) report inertial measurement sensors used in some systems designed for people with visual impairments may be able to perform tasks such as identifying moving objects, although accuracy is imperfect and the sight-impaired community has not yet readily accepted the technology.
- The Teletypewriter, known as TTY, provides a text alternative for those with hearing impairment. At first, this only allowed messaging between TTY devices, but since 1990, the FCC funds and controls a system in which any kind of phone can interact with the TTY, and a human assistant creates or corrects the text and can even transcribe into sign language (Reid, 2022). Some users may not want to use the system, citing privacy concerns.
- Assistive technologies specifically for patients with dementia are also available, including electronics worn on a pendant, devices that tell staff about falls or other changes, GPS to monitor

patient location, etc. There can be problems if these technologies fail to adequately assess the individual's circumstances, however (Curnow et al., 2021). These patients may or may not legally have the mental capacity to approve use of these technologies.

- GPS may be incorporated into various devices to assist people who must use wheelchairs; this mobility enhancement is called Geospatial Assistive Technology (GAT). Prémont et al. (2020) tested human reactions to GAT in smart phone applications, smart watches, virtual reality headsets, and augmented reality goggles. The access to location and other data was appreciated, but user concerns included the expense, weight of some devices, difficulty in adjusting smart watches, smart phone apps are not hands-free, and the social acceptability of wearing large devices (Prémont et al., 2020).
- Assistive robots in long-term care facilities help to reduce costs and mitigate caregiver fatigue. Duties may include monitoring, cleaning, managing logistics, and even a degree of social interaction – but acceptance of this kind of technology is often resisted by both patient and nursing staff (Franke et al., 2021).

Stramondo (2020) identifies several problems overall, including determining how much assistance is appropriate, who should cover the costs, and what conditions might be covered through medical insurance. These technologies exist, but determination and distribution of access to it, plus reliance on potentially inaccurate results present ethical dilemmas.

CGI Effects

When an actor dies during movie production, producers have a dilemma: if the project is scrapped, all of the hard work and financing would be lost. In the past there was some trickery such as "body doubles and a cardboard cut-out" which sufficed for Bruce Lee in the 1978 movie *Game of Death* or reusing previously-filmed scenes and covering a body double in facial bandages for Peter Sellers in 1982's *The Trail of the Pink Panther* ("13 actors," 2021).

Thanks to technology, there is an alternative in Computer Generated Imagery (CGI) which can recreate the actors and finish what had been started in a more realistic fashion. Some examples are Brandon Lee's role in 1994's *The Crow* and Oliver Reed's scenes in 2000's *Gladiator* ("13 actors," 2021). Campbell et al. (2022), however, express concern with "synthetic advertising," which is the use of deepfake techniques to recreate personas – including a realistic face and plausible-sounding voice of someone who is well known – and in doing may be considered unfairly manipulative.

Completing what has been started is one issue, but entirely using someone's likeness and persona well after their death could be considered a separate ethical problem. In 1987, Humphrey Bogart and Marilyn Monroe were digitally recreated for a short festival film (Stein, 2021) as perhaps a curiosity. Laurence Olivier's entire "participation" in 2004's *Sky Captain and the World of Tomorrow* was a recreation; the actor had died 15 years prior ("13 actors," 2021). Quite famously, Peter Cushing, who died in 1994, appeared in 2016's *Rogue One: A Star Wars Story* (Haridy, 2016).

Whereas the average movie-goer might enjoy seeing an actor re-animated, film actors' legacies are in the preservation of their work. The question arises as to whether or not this can be considered the work of the actor and how the person's estate can be licensed and protected (Fudge & Williams, 2020). According to Haridy (2016), Cushing's family gave permission. Robin Williams reportedly stipulated in his will that his likeness was not to be used for 25 years after his passing (Haridy, 2016). It is not clear how many actors or their heirs have had clear control, however.

CGI can be used for purposes other than movies. In 2020, a Japanese television network developed and broadcast an encounter between a bereaved mother and her deceased child, the latter recreated from photographs and presented as a 3-D animation. Public reactions ranged from empathy for the mother to fear and even disgust (Stein, 2021).

These recreations can be done surreptitiously and without the knowledge of the person whose face, voice, and mannerisms have been copied. Revenge deepfake pornography can ruin a career or family relationships (Fudge & Williams, 2020), and there is the potential for psychological harm along with emotional and even financial exploitation.

War Technologies and Human Rights

Mpinga et al. (2022) agree Artificial Intelligence and human rights are inextricably connected. Concerns identified in their research include but are not limited to geolocation privacy, self-driving car safety, robot "personality," insurance/financial institution assessment and decision-making based on algorithms, and the development and use of autonomous weaponry.

There is an advantage to war technology in that deployment can be quick and fewer people are needed to commit a strike on an enemy target. However, if a drone or other automatic war device malfunctions, there would be a "responsibility gap" for any errors or war crimes committed by misdirected machinery – and programming a robot to determine when to use lethal force is surely an ethical violation (Rosendorf et al., 2022). When weapons of war are controlled in part with AI, there is also an increased risk of biological and chemical attacks and the threat of unwarranted pre-emptive strikes against perceived enemies (Dawes, 2021).

The European Union has discussed technology ethics at length, and Pfaff (2020) claims a group of researchers and organizations created and signed "Lethal Autonomous Weapons Pledge" in the hope machines that can make life-and-death decisions will not be developed. This does not mean technology is missing from the battlefield; Russian AI-equipped drones have been shot down in the war Ukraine, and the fear of "killer robots" doing exactly what the Pledge wished to stop is still alive (Kahn, 2022). Prior to the start of this war, the United Nations met in December 2021 to discuss banning automatic weapons, but no consensus had been reached (Dawes, 2021).

For all of the laws, regulations, pledges, and promises, the world and technology are still controlled by the will of human beings, some of whom do not have the best interest of others in mind. Even when human desire is what would be considered honorable, the connection between people and technology is still imperfect. As Pfaff (2020) says, people might make decisions based on emotions, but machines are still capable of making mistakes.

Intellectual Property and Cybercrime

Intellectual Property (IP) has various definitions depending on each country's laws, but the wording typically includes the protection of inventions, processes, designs, software, and other creative expressions (Chapman, 2021), all of which are used in innovative technology projects.

Cyberattacks can be perpetrated by outsiders, but "rogue" employees may also cause a threat and share proprietary information with competitors (Chapman, 2021). Methods may include physical break-ins, blackmail, reverse engineering, and fraud perpetrated from inside the company (Chapman, 2021), plus counterfeits/illicit copies (Fernández-Márquez et al., 2020), outright extortion, and holding the IP for ransom (Peck, 2022). As projects involve a collaborative effort of many individuals, access to trade secrets and other IP elements can be extensive.

Not all business decisions set well with consumers. In the last few years, for example, there has been a trend toward monthly cloud subscription software such as with Adobe products for which legal program downloads are no longer available. Sartain (2019) explains companies may be able to pay subscription fees for employees but individual users may be unhappy paying for a suite of programs where only one or two are wanted, and might only be used occasionally. The fees the user is "now obligated to pay in perpetuity" also could rise to where it becomes unaffordable (Sartain, 2019, p. 103). Alternative software, both downloads and free programs, are often substituted (Sartain, 2019).

Other users might resort to piracy. Each instance of copied software is a sales loss which may impact future sales offerings to customers who want legal programs. Wozniak (2021) claims $46 billion is lost every year to software piracy. Fernández-Márquez et al. (2020) explains the pirate's cost includes time and learning how to make copies; social media has unfortunately made it easier to gather that knowledge.

Music piracy is similar. Oad et al. (2021) performed a study in Pakistan and indicated social group pressures, the desire to avoid normal purchase cost, and a belief no one is harmed when music is pirated are prevalent reasons for theft, which is done even at the risk of also downloading a virus. The lack of harm concept is clearly a fallacy. Quain (2019) maintains royalties are lost, but jobs can also be in peril: producers, sound technicians and engineers, talent scouts, and others rely the recording company making money so they can be paid for their hard work and expertise. Streaming services may have hoped to curtail piracy, but it has not been proven to do so (Oad et al., 2021). The FBI reports "intellectual property theft costs U.S. businesses billions of dollars a year and robs the nation of jobs and tax revenues" (Federal Bureau of Investigation, n.d., para. 2).

The term "cybercrime" obviously has legal implications and various applications; by default it is construed as unethical. It is clear that laws, regulations, Acceptable Use Policies, and industry standards cannot control criminal human behavior. The ethical issue, then, is to find the best ways to protect IP and customers and to have mitigation plans for when a criminal strikes. As for piracy, both protection and education may help users understand how the consequences of their actions affect others.

Tamara Phillips Fudge
Purdue University Global, USA

REFERENCES

13Actors who were brought back to life with special effects in movies. (2021, July 26). Business Insider. https://www.businessinsider.in/entertainment/movies/13-actors-who-were-brought-back-to-life-with-special-effects-in-movies/slidelist/33082956.cms#slideid=33082974

Ammanath, B. (2021, November 9). *Thinking through the ethics of new tech ... before there's a problem.* Harvard Business Review. https://hbr.org/2021/11/thinking-through-the-ethics-of-new-techbefore-theres-a-problem

Brandt, J. (2022, February). Courage under fire. *ISACA Industry News*, 1–3.

Campbell, C., Plangger, K., Sands, S., & Kietzmann, J. (2022, January-March). Preparing for an era of deepfakes and AI-generated ads: A framework for understanding responses to manipulated advertising. *Journal of Advertising*, *51*(1), 22–38. https://doi.org/10.1080/00913367.2021.1909515

Chapman, R. J. (2021, February). Theft of intellectual property from advanced technology projects. *PM World Journal*, *10*(2), 1–9.

Curnow, E., Rush, R., Gorska, S., & Forsyth, K. (2021). Differences in assistive technology installed for people with dementia living at home who have wandering and safety risks. *BMC Geriatrics*, *21*(1), 613. https://doi.org/10.1186/s12877-021-02546-7

Das, M., Piper, A. M., & Gergle, D. (2022, March). Design and evaluation of accessible collaborative writing techniques for people with vision impairments. *ACM Transactions on Computer-Human Interaction*, *29*(2), 1–42. https://doi.org/10.1145/3480169

Data protected – Sweden. (2020, April). Linklaters. https://www.linklaters.com/en-us/insights/data-protected/data-protected---sweden

Dawes, J. (2021, December 20). *UN fails to agree on 'killer robot' ban as nations pour billions into autonomous weapons research.* The Conversation. https://theconversation.com/un-fails-to-agree-on-killer-robot-ban-as-nations-pour-billions-into-autonomous-weapons-research-173616

Federal Bureau of Investigation. (n.d.). *Intellectual Property theft/piracy.* https://www.fbi.gov/investigate/white-collar-crime/piracy-ip-theft

Fernández-Márquez, C. M., Vázquez, F. J., & Watt, R. (2020, October). Social influence on software piracy. *Managerial and Decision Economics*, *41*(7), 1211–1224. doi:10.1002/mde.3167

Franke, A., Nass, E., Piereth, A.-K., Zettl, A., & Heidl, C. (2021, August 26). Implementation of Assistive Technologies and robotics in long-term care facilities: A three-stage assessment based on acceptance, ethics, and emotions. *Frontiers in Psychology, 12*. doi:10.3389/fpsyg.2021.694297

Frankenfield, J. (2020, November 11). *General Data Protection Regulation.* Investopedia. https://www.investopedia.com/terms/g/general-data-protection-regulation-gdpr.asp

Fudge, T., & Williams, L. (2020, August 18). Zoom in (but fasten your seatbelt). *International Conference for Media in Education (iCoME) 2020: Diversity Education in ICT Advanced Society*, Kobe, Japan (online), pp. 195-201. https://icome.education/wp-content/uploads/2020/08/icome2020_programproceedings_final_v2.pdf

Glaveski, S. (n.d.). *36 cognitive biases that inhibit innovation.* Collective Campus. https://www.collectivecampus.io/blog/36-cognitive-biases-that-inhibit-innovation

Haridy, R. (2016, December 28). *The ethics of digitally resurrecting actors.* New Atlas. https://newatlas.com/star-wars-ethics-digital-actors-cg/47123

Hofmann, B. (2020, March). Progress bias versus status quo bias in the ethics of emerging science and technology. *Bioethics*, *34*(3), 252–263. https://doi.org/10.1111/bioe.12622

Kahn, J. (2022, April/May). A.I. goes to war. *Fortune*, *185*(2), 28–30.

Kenyon, S. (n.d.). *Acheron (Character Profile).* https://www.sherrilynkenyon.com/character/acheron

Knight, J. (2021, June 3). *The Supreme Court narrows the scope of the Computer Fraud and Abuse Act.* Alston & Bird Privacy, Cyber & Data Strategy Blog. https://www.alstonprivacy.com/the-supreme-court-narrows-the-scope-of-the-computer-fraud-and-abuse-act/?cn-reloaded=1

Koepp, D. (1992, December 11). *Jurassic Park.* Screenplay based on the novel by M. Crichton. http://www.dailyscript.com/scripts/jurassicpark_script_final_12_92.html

Kostadinov, D. (2014, September 23). *The essentials of an acceptable use policy.* InfoSec. https://resources.infosecinstitute.com/topic/essentials-acceptable-use-policy

Langworth, H. (2019, November 19). *French market remains cautious on legal technology.* Law.com International. https://www.law.com/international-edition/2019/11/12/french-market-remains-cautious-on-legal-technology

Martins, A. (2022, February 14). *Why you need an Acceptable User Policy and how to create one.* Business.com. https://www.business.com/articles/acceptable-use-policy

Moral theories. (n.d.). Seven Pillars Institute. https://sevenpillarsinstitute.org/ethics-101/moral-traditions

Mpinga, E. K., Bukonda, N. K. Z., Qailouli, S., & Chastonay, P. (2022, February 2). Artificial Intelligence and human rights: Are there signs of an emerging discipline? A systematic review. *Journal of Multidisciplinary Healthcare, 2022*(15), 235–246. https://doi.org/10.2147/JMDH.S315314

Oad, S., Qu, J., Dai, J., & Rehman Abro, M. ur, & Oad, R. (2021, December). Streaming media: Safeguarding music industry against piracy? *International Journal of Information & Management Sciences, 32*(4), 287–310. doi:10.6186/IJIMS.202112_32(4).0002

Orman, S. (2015, June 7). *An important reminder: just because you can afford it, doesn't mean you should buy it!* [Status Update]. Facebook. https://www.facebook.com/suzeorman/posts/10153392669529551:0

Peck, G. A. (2022, April). Cyber threats to media companies are on the rise. *Editor & Publisher, 155*(4), 36–40.

Pfaff, C. A. (2020, January). The ethics of acquiring disruptive technologies: Artificial Intelligence, autonomous weapons, and decision support systems. *Prism: A Journal of the Center for Complex Operations, 8*(3), 128-145.

Prémont, É., Vincent, C., & Mostafavi, M. A. (2020, November). Geospatial assistive technologies: Potential usability criteria identified from manual wheelchair users. *Disability and Rehabilitation. Assistive Technology, 15*(8), 844–855. https://doi.org/10.1080/17483107.2019.1620351

Reid, B. (2022, May). Law and technology, two paths for Digital Disability Law: Understanding the legal drivers of efforts to make technology accessible. *Communications of the ACM, 65*(5), 36–38. https://doi.org/10.1145/3527201

Reyes Leiva, K. M., Jaén-Vargas, M., Codina, B., & Serrano Olmedo, J. J. (2021, July 13). Inertial measurement unit sensors in assistive technologies for visually impaired people: A review. *Sensors (Basel), 21*(14), 4767. https://doi.org/10.3390/s21144767

Richter, F.-J., & Sinha, G. (2020, August 21). Why do your employees resist new tech? *Harvard Business Review.* https://hbr.org/2020/08/why-do-your-employees-resist-new-tech

Rosendorf, O., Smetana, M., & Vranka, M. (2022, March 21). Autonomous weapons and ethical judgments: Experimental evidence on attitudes toward the military use of "killer robots." *Peace and Conflict: Journal of Peace Psychology.* doi:10.1037/pac0000601

Sartain, J. D. (2019, October). Alternatives for Adobe Acrobat, Photoshop, and more. *PCWorld, 37*(10), 98–105.

Schneidman, W., Cooper, D., Mkhize, M., & Naidoo, S. (2021, December 14). *Tech regulation in Africa: Recently enacted data protection laws.* Covington. https://www.globalpolicywatch.com/2021/12/tech-regulation-in-africa-recently-enacted-data-protection-laws

Sheil, J. (2020, October 13). *What is an Acceptable Use Policy?* Electric. https://www.electric.ai/blog/what-is-an-acceptable-use-policy

Stein, J.-P. (2021, October). Conjuring up the departed in virtual reality: The good, the bad, and the potentially ugly. *Psychology of Popular Media, 10*(4), 505-510. doi:10.1037/ppm0000315

Stramondo, J. A. (2020, October 6). The right to assistive technology. *Theoretical Medicine and Bioethics, 41*(5-6), 247–271. https://doi.org/10.1007/s11017-020-09527-8

Van Buren v. United States, 593 US (2021). https://supreme.justia.com/cases/federal/us/593/19-783

Venkataramakrishnan, S. (2021, December 5). Why ethics must be built into tech development. *Financial Times.* https://www.ft.com/content/43a6a1ab-3f52-43c5-af67-b4e7a1251bde

Wozniak, M. (2021, October 7). *How much is software piracy costing your business and how can you prevent the loss?* Software Key. https://www.softwarekey.com/blog/about-software-piracy/software-piracy-costing-your-business

Acknowledgment

I would like to thank my family for their support, the chapter authors for their hard work and dedication to their topics, the editorial board even though they wish to remain anonymous, Susan Shepherd Ferebee and Linnea Hall for their reassuring advice whenever it was requested, and Kristina Setzekorn for her constant encouragement. This volume is a testament to the fervor we all must share in dealing with the ethics of our technological world.

Section 1

Technology Adoption, Business Leadership, and Strategies

Chapter 1
Technology Innovation and Adoption in the Modern Workplace:
Resistance, Ethics, Reassurances, When to Say "No"

Kelly Wibbenmeyer
Purdue University Global, USA

ABSTRACT

Understanding the various types of innovation technology that are currently disrupting organizations around the globe is an important endeavor. Various innovative technologies are defined, and related ethical challenges are explored that organizations face in today's technology climate. Additionally, key methods are examined for an organization to manage employees when they refuse to change, including unfreezing, transition, and refreezing. The importance of alignment of innovation strategy to organizational resources is reinforced, with the best methods to align organizations with innovation success, including when it is okay to say no to innovation within an organization.

INTRODUCTION

Before an examination of the various ethical concepts within innovation, it is important to define what innovation is (and what it is not). Per Yermak and Lisnichenko (2016), innovation considers the aspects of intensity and complexity, effectively improve profitability, and organizational competitiveness through technology means. Pinheiro, Merino, and Gontijo (2016) mention that innovation is the socially acceptable alternatives of current technology to enhance current organizational concerns and risks. Rejon-Parrilla, Espin, and Epstein (2022) define innovation as a step-change improvement, which causes additional convenience through evidenced based efficiencies. Regardless, in the various ways that innovation is defined, a consistency in the definitions is that innovation allows for greater organizational efficiencies

DOI: 10.4018/978-1-6684-5892-1.ch001

through modifications of current process which results in process effectiveness that presents positive monetary gain for the organization.

Ethical concepts within the innovation landscape are an important concept. Organizations are faced with making moral decisions based on their needs and these ethical decisions are being made quickly due to the rapid advancements in the current technology landscape (Özsungur, 2019). Lei, Ha, and Le (2019), note that ethical leadership is imperative to successful incremental innovation and is achieved through tacit and explicit sharing of knowledge. There are multiple ethical aspects that need to be considered when dealing with implementing innovation technology in any organization. For example, cyber concerns are constantly changing, so the way people think about innovation must also change. Various concepts in innovation are impacting how consumers (of products) think about innovation and how it will impact them. One major aspect to ponder is the actual innovation technology that organizations are considering implementing. The specific technology must be understood and defined as the ethical concerns that are present are unique with each specific technology.

Various environmental aspects must also be considered, such as industry that can be greatly influenced (both negatively and positively) by innovation. Strate (2020) notes that just because people can perform certain innovations does not mean they should. Strate (2020) also notes the environment in which innovation is an important aspect to answer the question of whether or not innovation is warranted. Yaghmaei and van de Poel (2021) mention companies must be responsible with innovation. The assessment of responsibility includes reviewing benefits, costs, and risks. Additionally, there are various factors to consider such as social, environmental, scientific, and economic impacts when performing the assessment (Yaghmaei & van de Poel, 2021). The environmental factors (both internal and external) that can influence organizational decisions to implement an innovation technology must also be considered. Once the environmental factors are defined, ethical considerations for the internal and external forces that impact the success of the innovation technology must be explored.

How organizational culture influences the effectiveness of the innovation implementation and how ethical leadership behaviors drive strategic alignment can affect successful innovation implementation. Also significant for exploration are innovation in the workplace, innovation technology occurring in the workplace and their ethical factors, reasons for resistance, and the ethical concerns when there is a resistance to change, and when to say "no" to innovation. Specific innovation tools (such as an Acceptable Use Policy) and technologies that organizations implement can lead to potential ethical concerns. Tools and technologies must be recognized and explored for ethical consideration. These considerations help to lead to a final answer as when to say no to innovation and the ethical implications of doing so.

ALIGNMENT OF CORPORATE OBJECTIVES AND RESOURCES

Before analyzing the various innovation technologies and tools that organizations are considering and the ethical implications of each tool, a company must first address strategy alignment and the impact to ethical practices. Ethical leadership transforms organizations to think differently (Le et al., 2019). Organizations must implement a culture of change to be successful in advancing innovation technology (Stauffer & Maxwell, 2020). Organizational culture is governed by the concepts of organizational identity and how it transforms over time (McDevitt, 2020). McDevitt (2020) states leadership drives the behavior of employees as employees typically shadow the actions of leadership. When leaders are seen as unethical or make unethical choices when it comes to strategy, employees replicate this behavior.

Strategic alignment of innovation activities is imperative to organizational success of the implementation of an innovation culture. The overall strategy of an organization must match the innovation technology adoptions, else the organization will not be successful in implementing its mission and vision. McDevitt (2020) states that lack of vision and mission can lead to an organizational identity crisis that results in uncertainty and misalignment of overall goals. This leads to organizational disorientation and therefore a lack of purpose across the ministry. The organizational strategy is a path for a business to reach its goals and objectives. Without a clear alignment of strategy and how it links to innovation, an organization can be headed for trouble.

Strategy also defines where the organization should be heading in the future. If an organization does not implement projects that meet strategic objectives, then the organization will most likely not meet its strategic goals. Sjödin et al. (2020) notes that value creation and value capture must be aligned within an organizational strategy to meet overall goals. Additionally, Sjödin et al. (2020) mentions incremental value capture should be implemented with new innovation models. This means when implementing innovation projects, goals must be set for various phases, lessons learned documented, and those lessons built into the next phase of implementation. Due to various risks with implementing innovation projects, a shorter cycle for review is recommended to ensure overall project success.

From a leadership perspective, it is vital to follow the strategic goals of an organization (Perry, 2020). If a leader goes rogue and focuses on their own priorities and not those of an organization the organization can run into a sense of misalignment, as noted above. Leaders must exhibit ethical behavior from a strategic perspective, which will fundamentally lead to a cohesive strategic alignment but also allow for a synergy to implement a culture of change. Ethical leadership also drives ethical employee behaviors which will result in an ethical organizational culture (McDevitt, 2020). Ensuring the fundamental strategy is being followed by all employees is a basis for transforming an organization to a culture of change.

In many cases, change is required for successful innovation alignment to meet organizational strategy. Iskat and Liebowitz (2021) note when organizations go through major change, organizations need to be proactive in managing change. A positive change attitude is required to ensure the best outcomes for innovation projects. Leaders also must find methods to ensure the organization is all in with an innovation mindset. Leaders must foster an innovation culture. When implementing innovation within an organization, employees must be challenged to think differently. This means when implementing an innovation program, organizations must minimize the red tape associated with implementing something new. This could be something as simple as not implementing the innovation on the organizations network or having a pilot project off network. An organization can perform benchmarking with other organizations and understand the benefits and risks of implementing the new technology before implementing within the organization. A company must assess its innovation appetite and what will be accepted within the organization prior to implementation.

The organization must also have an "okay to fail" attitude when implementing innovation. Bales and van Rensburg (2019) note that innovation is not about the ideas, but instead about the people and culture aspect of innovation. Understanding that failing is learning is also crucial. This means if the organization is failing quickly and understanding when to pause or transition to some other innovation effort, the team is still learning and therefore it is seen as a positive movement forward. The only way an okay to fail culture is seen as permitted is if leadership creates and nurtures this change.

Ethical leaders ensure they have the appropriate training and skill to navigate an innovation culture. Leaders should promote and model innovative behaviors, which includes leading innovation projects, instilling the okay to fail attitude, and guiding teams when to transition to new technology. Leadership action is significant at this stage. For example, an ethical leader must not state that it is okay to fail and then fire someone for not meeting goals.

The goals in an innovation culture are very different. The goals are about learning what is possible, failing quickly, and learning. If a positive result happens, celebrate it but also celebrate failing quickly. Leaders are encouraged to reward innovation habits, which do not always correlate to positive project results. These concepts may be challenging for a leader to implement. Leaders that excel in this new way of innovative thinking, excel at building trust within an innovation team.

When employees have a supportive leader that has been educated on the methods of effective innovation management, it results in a positive change that influences successful innovation culture. When employees feel supported, even when negative outcomes occur, and leaders encourage failing fast over positive innovation outcomes, employee's attitudes change which creates a positive organizational culture towards innovation. When employees identify innovation projects, are excited about innovation and have a communication feedback loop, the level of innovation maturity increases within an organization. As the level of innovation maturity increases in an organization, employee engagement in innovation efforts are not demanded but expected and flourish within an organization.

INNOVATION IN THE WORKPLACE

The concept of change within an organization can cause angst, nervousness, and fear for impacted employees. How an organization manages employees with a refusal to change must be examined, as well as implementing positive changes through a good change culture, all why aligning to corporate objectives and resources.

What makes an organization innovative must be understood. It may be easier to determine what actions an organization performs when it is not ready to accept innovation. For instance, performing random acts of innovation is not in strategic alignment with advancing an organizations innovation strategy. When attempting to advance the innovation culture, an organization must perform the following:

1. Create specific strategic innovation mission and vision for the organization.
2. Once the mission and vision are created, then create specific objectives and measurable goals to meet those objectives.
3. Understand how the innovation goals and objectives tie into the existing strategy and understand how the existing strategy must change to meet the new innovation goals.
4. Determine how the organizational resources and structure must be changed to meet the new goals.
5. Evaluate a change and communication plan for the entire organization. This includes change and communication strategies for all levels of the organization.

Another question is to determine why the alignment between business strategy and innovation breaks down. There are multiple reasons for this misalignment. Abdulkader et al. (2020) mentions that misalignment between the value capture and creation can cause a misalignment. Value creation is the total benefit created when transforming the inputs of a process into the end resultant output. Value capture is the ability of an organization to capture or retain the value as a measure of profit or success. When thinking about value capture and creation in terms of innovation, it may be difficult to identify the actual value due to the inherent fact that the product being evaluated is new and there are few methods to capture value from other organizations due to the newness of the technology. The measurement of risks is also something that should be considered in the value creation, and again it will be difficult to identify the risks due to the newness of the technology. Misalignment can be caused strictly due to the lack of understanding or knowledge present about the innovation itself.

Additional rationale regarding the misalignment could be due to losing site of the end goal of the innovation implementation. For example, if the team is working on a specific goal and becomes fixated on an innovation that is not ready instead of one that is ready for implementation, the team could get caught up in the idea of being first to market instead of addressing the organizational need. Goals must be well-defined goals and have measurable objectives. A specific example of this breakdown could be an organizational goal to find an innovation technology that allows for transport from building A to building B within the next year. The measure of success is that it reduces the human involvement of transportation. Another measure of success is that it can be developed, tested, and implemented within the year timeframe. If the team is determined to implement drone technology, which at the time of writing, is not allowed for out of line of sight in the United States, the team can get fixated on the actual technology (drone implementation) instead of the goal of filling the gap with other available technology.

Changes in leadership can also cause a breakdown in strategic alignment. For example, one leader's goals may influence the overall organizations direction for innovation. If a new leader enters the organization and has different beliefs regarding the innovation, that leader could completely change the strategic direction of innovation and cause misalignment with existing innovation technology or corporate strategy.

One of the larger issues when thinking about organizational misalignment is poor communication of the corporate vision, mission, and overall innovation strategy. When the overall innovation strategy is not clear, especially to teams working on innovation projects, it can cause major disconnections. Communication issues from senior leaders can also cause team members to feel very disconnected to the overall strategy, which can lead to transparency issues and result in an overall lack of trust of leadership within the organization.

To reduce overall misalignment with innovation strategy, organizations should understand both customer and employee concerns with innovation. Having a shared purpose and goal with leadership, customers, and employees will ensure all innovation decisions are transparent and are clearly understood by all parties. A feedback loop should be implemented. For example, the business should inform, and influence strategy and strategy should inform and influence the business. Both the leaders and employees should have a clear line of communication and feedback and use continual improvement to continue learning and shifting to meet the overall innovation strategy goals. The communication of the vision, mission, and strategy should be communicated clearly and using the right media and frequency to ensure everyone is comfortable and understands the overall innovation strategy goals. Education is a vital aspect of the communication strategy and should take as much time as needed to ensure the concepts are explained thoroughly. Understanding the availability of the innovation is also an important aspect of reducing misalignment. If a technology is not ready for implementation, it does not mean it will not be ready for implementation forever. It is best to create a mechanism for managing the innovation technologies and understand the triggers in the environment or within innovation that will allow the innovation technology to be revisited.

Ethical leadership practices include, setting an attainable mission and vision, reducing the confusion of strategic alignment through transparent communication, and being open to change. Ethical leaders take steps to drive organizational programs to ensure the organization is ready to implement change (Stauffer & Maxwell, 2020). Leaders are charged with implementing programs and policies to enhance a culture of change. Ethical leaders not only drive strategic alignment, but also give the appropriate training and advice to employees to better accept change within the organization. Successful implementation of innovation projects, starts with a solid strategic vision that includes innovation implementation and is achieved through ethical leadership practices which drive an organizational culture of change.

Innovation: How Does an Organization Manage Refusal to Change?

There are many reasons why people resist change. Some believe if something is working properly, then it should be left alone. This is a very old way of thinking that can cause major issues with successful implementation. Iskat and Liebowitz (2021) mention that if there is too much change occurring in an organization, employees tend to resist change, as it is too much change at once. Employees that resist change usually resist change for a reason. Some of these reasons include fear of doing something wrong and being reprimanded, not understanding how they will be impacted by the change (especially if they will lose their job), and the fear of the unknown (Iskat & Liebowitz, 2021). Refusal to accept change can also be seen as an unethical practice and can lead to negative employee working relationships and even termination. There are ways to implement successful change practices as well as various reasons for resisting change.

Implementing change is not an easy task, especially when the organization change culture is not positive. Understanding the tools that allow for a culture of change is a great way to be successful in implementing innovation projects. Lewin (1947) states that there are three stages of successfully implementing change in an organization: unfreezing, making the transition, and refreezing.

Unfreezing Stage

In the unfreezing stage, leaders prepare the employees for change by noting better methods to perform work. To ensure the success of the unfreezing phase, an explanation of why the change is occurring must be formed. Some employees only need the explanation of why the change is important to the organization. Others will need more time and communication. Other key aspects that occur in this stage are to link the change to the organization strategy, note how the change will allow the organization to be more competitive in the marketplace, and show past examples of other organizations (use cases) and how they implemented the change successfully to include the results of the organization implementing the change. Being empathic regarding how the change will impact the employee is another great aspect of the unfreezing state. If the employees are shown that they are appreciated, understand how the change will impact their lives, and that the change was thoroughly thought through, they would be more accepting of the change. The largest aspect of the unfreezing stage is effective and transparent communication. In many instances, employees want to hear the change, then reciprocate how the change will impact them, and have time to process the change and ensure that their thoughts and concerns have been expressed as part of the solution.

Transition Stage

During the transition phase, the benefits of the change must be shown (Iskat & Liebowitz, 2021). When innovation is imminent, it must be understood how the change will impact employees' work. Explaining how the change will make their life better is another key aspect of the transition stage. Ensuring the employees are aware of awards occurring with the change is also appropriate (monetary, a reduction of repetitive tasks, or recognition). Identifying a change champion also occurs this stage. If the organization can find someone that is highly influential or respected in the organization, these traits are important in finding the best champions as they have the charisma to influence the change. Another method to have a successful transition is to have direct employee input. This process deals with working directly with the employees to influence the change. Also, timing is key. Having an innovation take place at the wrong time can be disastrous. Working with the employees who complete the work daily will also surface any potential issues with implementing a new process or system at a certain time. The employees that work the process will know the best time to implement due to their participation in daily operations. If the change will allow additional job security, this is something to note when explaining the change.

In all instances, innovation may not afford job security. Other opportunities within the organization can be offered, if possible, or assistance for employees to attend job fairs, or even a bonus for staying on until the innovation is implemented. Once the employees are agreeable to changes, the company must provide training so employees feel comfortable with the changes.

Refreezing Stage

The last phase of change per Lewin (1947) is refreezing. In this stage, top management should be vocal about their support of the new innovative process. Employees tend to shadow the behavior of senior leaders and therefore this step is crucial in the adoption process. Another aspect of the refreezing process is to publicize success and to constantly improve the process as potential concerns arise. Being flexible to changing the new process to ensure employee satisfaction is a great way to ensure continued support of the process. One last aspect of refreezing is to provide human resource services for those who simply do not want to change. In some cases, employees just do not want to change and the best option for both them and the organization is for them to leave. This could be in the form of early retirement packages or employee assistance programs.

Offering employees the time to reflect on the changes that are occurring is helpful. Ethical leaders will ensure employees are communicated with about the change and that they are part of the change process. Lewin (1947) notes that ethical implementation of change includes employee engagement. A feedback loop to ensure employees have an active voice in the change, having an open-door policy to address any concerns with the change, and allowing a non-retribution policy regarding voicing concerns are just a few methods that leaders and influence a culture of change. Ensuring employees feel comfortable with changes is part of the ethical responsibility of a leader.

Reasons for Resistance

Rahaman (2019) notes that every organization requires change. If change is such an integral aspect of organizational life, then employees who resist become a problem. There are also no quick fixes to change, this means change takes time, and it is a process. Rahaman (2019) states that employees crave routine practices. In many instances employees are fearful of learning a new skill and fear of performing like they are less than competent at the job they have been performing for a long time.

Richter and Sinha (2020) note that employees resist change due to a lack of employee skills, lack of senior leadership awareness of the change, lack of remote working opportunities, organizational culture, increasing complexity in work processes, major costs and risks for both the organization and employees, and inadequate infrastructure. Older and larger companies run into more complications due to mass amounts of legacy systems, the difficulties in transitioning from the older legacy systems due to things like data inaccuracies, history requirements on data, and other major barriers to increase success in innovation activities (Richter & Sinha, 2020).

The reasons for resistance of change must be managed across the organization. Senior leaders must drive a culture of change throughout the organization when attempting to implement innovation. Masumba (2019) notes that leaders must review the overall organizational landscape of change. Innovation must be the responsibility of every employee for the change to be successful (Masumba, 2019). A major challenge when driving innovation is ensuring there is an all-in attitude towards innovation. Employee-driving innovation does not just occur, it is influenced by leaders and a requirement of employment (Masumba, 2019).

Senior leaders must both model and promote workplace creativity. There are a few methods to create and instill a culture of innovation within an organization noted below:

- Employees participate in online challenges
- Organization sponsored hack-a-thons to solve problems using innovative ideas
- Offer something different to employees such as new innovation projects

Some key activities organizations can implement to enhance an employee culture of change are to incentivize the new technology use, invest in infrastructure to reduce cumbersome processes, and include reskilling and learning as part of the plan to implement the innovation. Rahman (2019) also notes that leaders and managers have a major role to play in implementing successful change. If employees and leaders start from a shared dissatisfaction for the current situation, then work together to build a shared vision of the future this is the best starting point for implementing successful change within the organization. Schein and Schein (2018) note that there are three categories of organization culture that are crucial to understand for a successful change: artifacts, shared beliefs and values, and underlying assumptions. Rahman (2019) states that emotions have a lot to do with why employees resist change. The best way to understand these emotions is to ask about the assumptions for hiring, understand their social contact, since teams and working relationships may change, and to understand their assumptions about the organization, specifically what it does and how it treats its employees. Breaking through resistance of change must be an open conversation with employees and leadership. A method to complete this conversation is to engage in two-way communications, quash rumors, and create psychological safety (Rahman, 2019). If the organization performs these functions, then they should be able to build the trust that is needed for a successful innovation change.

Ethical leaders transform the organization by implementing programs that resist change (Rahman, 2019). Employees engage in ethical organizational behavior by working with their leaders to discuss any concerns, offering honest feedback, and not engaging in rumors but only discussing the facts of the implementation (Rahman, 2019). Ethical employees, at all areas of the organization should have transparent communication that offers a safe space of non-retribution interactions to reduce the resistance of change within the organization.

VARIOUS INNOVATIVE TOOLS AND TECHNOLOGIES IN THE WORKPLACE AND ASSOCIATED ETHICAL CONCERNS

In today's world, there are many opportunities to implement innovation technologies within an organization. To be successful, organizations must ensure they have a positive culture of change. A positive change attitude is required to be productive and stay competitive in the market. The purpose of this examination is to give an overview of the various technologies that many organizations and employees are seeing along with the potential ethical concerns with implementing said technologies.

Acceptable Use Policies (AUPs)

An Acceptable Use Policy governs how employees and other system users, utilize the applications or hardware within an organization. It covers a wide range of issues including rights, responsibilities and privileges, and sanctions with technology. This document allows employees and users of systems to understand how the organization is implementing due diligence to secure data within its walls. Steps must be taken to ensure the IT network is secure, protects sensitive data, and is a founding document for any breech or required audits.

This policy is a tool to drive governance on how to best implement new technology. The reason why companies start with the AUP is that in many cases, it governs how innovation is implemented within an organization. This documentation is not new for some companies, but for others it is very new. Some employees see the policy as another rule they need to follow, instead of the positive aspects. Organizations must explain the positive benefits of the policy and note what can occur if employees do not follow the policy. Many organizations have an AUP to explain how to use organizational systems to reduce the impact to security breaches. In many cases organizations not only have a policy, but additional training and testing to ensure employees understand the expectations of computer usage.

Ethical Concerns With Acceptable Use Policies

An AUP does have some ethical and legal concerns. If the expectations of the policy are not clearly documented, the organization can run into some legal concerns with implementing sanctions in the policy. Privacy expectations must be clearly documented and communicated to all employees. Employees must be informed that audits can take place, if not informed, there could be a lack of trust with employees which could also lead to employees leaving the organization. Based on the type of data, additional fines could be implemented. For example, if a patient's health information goes public, the organization could be sued and in turn the organization could file a civil suit on an employee for the cost of damages. Intellectual rights and employee responsibilities could cause an organization to file a civil suit on employees claiming that the organizations intellectual property was compromised by the data breach that an employee caused. An AUP is an important aspect of an organization's security framework and employees must be educated on the ramifications of not following rules therein.

Data Breaches

Data breaches and remediation of data breaches are noted within an AUP, but due to the importance of data breaches, it is presented here as a separate inquiry. The first step is an appropriate response when a breach occurs. Lee and Choi (2021) note that data breaches are inevitable. Financial implications of data breaches are also growing exponentially. Gwebu, Wang, and Wang (2018) note that the firm's reputation is a very valuable asset and therefore the methods and tools to identify and respond to data breaches are vital. The major response strategies are: apology (apologizing for the occurrence of the breach) with remedial actions (noting the steps to mitigate and control the damage), ingratiation (the goal is to make the stakeholders like the organization), justification (seeking to reduce the perceived damage of the breach), denial (noting that the breach did not occur), excuses (goal is to reduce the organizations responsibilities for the breach), correction commitment (ensuring the stakeholders that the organization will take whatever specific steps necessary to reduce the same breach from occurring in the future), stakeholder commitment (reassuring the stakeholders that the organization is committed to providing the best service), and value commitment (informing stakeholders that the firm is committed to its core values (Gwebu et al., 2018).

Regardless of the steps or actions that an organization takes, the organization should complete some form of communication to the impacted parties. The preferred approach is apologizing for the breach, noting what steps are being taken to resolve the current issue, and note what additional safeguards are being implemented to reduce future concerns. Stating how the breach has impacted or has the ability to impact system users is also a good trust-building method. Any additional monitoring, such as a free year to a credit reporting bureau is another way to rebuild any negative organizational backlash from the breach.

Ethical Concerns With Data Breaches

Many times, data breaches occur due to employees not following the Acceptable Use Policy. There are multiple legal and ethical concerns with data breaches. Victims of a data breach can have trouble trying to meet the requirements of keeping data safe and secure. The victims may not have the resources (time, understanding, or funding) available to ensure their data is secure. Corporations can also experience data breaches if they fail to update their systems with the latest security to reduce these issues from occurring (Spinello, 2021). Also, many organizations may think they are too small for someone to breach their organization, until it is too late, and a breach occurs. The cost of compliance may be too much for a small business to implement. An organization can be morally at fault with a data breach if they choose not to inform the users of a potential concern as well.

Artificial Intelligence (AI), Machine Learning (ML), Natural Language Processing (NLP), and Virtual Assistants

Artificial Intelligence (AI) is an appropriation of various technologies working in tandem to enable machines to sense, comprehend, act, and learn with human-life levels of intelligence. There are many definitions present regarding what AI is and what it is not. There are two types of AI: narrow and strong. In narrow AI, the application performs a single task or a set of closely related tasks. An example of this technology is weather apps or digital assistants. Strong AI applications are multi-functional and usually have multi-layered infrastructure (Liu, 2021).

Machine Learning (ML) and Natural Language Processing (NLP) are two specific technologies that are present within the AI umbrella. When using strong AI, sentiment machines emulate human intelligence. ML and/ or NLP is required for this advanced collaboration. Some machines can perform tasks more efficiently (such as data processing), human-machine collaboration is still crucial to extend these tools to be more effective as machines are not creative, they cannot think strategically, or in an abstract manner like humans can.

A virtual assistant is a tool many organizations use to "assist" employees with low-end, highly standard tasks, which allow employees to focus on higher cognitive tasks. Li and Yang (2021) note that virtual assistants can manage a variety of tasks and services such as order processing and production execution in the manufacturing industry. In other organizations, these assistants are located on an employee's computer and can be accessed when certain pre-programmed functions must be completed. The great thing about virtual assistants is the bot will only do what it was programmed to perform. This technology lends itself to higher quality and more standard output then if multiple humans were performing the task. Virtual assistants are a subset of AI, ML, and NLP as the developer must code and manage the technology. The same ethical concerns exist between these technologies.

Ethical Concerns With AI, ML, NLP, and Virtual Assistants

There are multiple ethical concerns within implementing AL, ML, and NLP. Informed consent, safety, transparency, and patient privacy are just a few ethical concerns in the healthcare industry (Katznelson & Gerke, 2021). Due to the increasing number of changes in the landscape, big data, and other external forces, it will be imperative for students to gain an understanding of the implications of AI and ethical concerns prior to entering the workforce industry (Katznelson & Gerke, 2021). Due to the large amounts of data that is consumed and processed, there are ethical concerns around overall privacy of the data. Data is collected in multiple formats within organizations. When a user navigates online, the click path they take can be recorded. The algorithms required to process the machine learning function must also be vetted by multiple groups for accuracy. There are ethical implications if an organization only has a few people vetting algorithms for high-risk implementations. For example, there is a new technology for detecting abnormal chest x-rays that is actively being developed. This machine learning tool will allow patients to receive results quicker and more effectively if the algorithm works correctly. An ethical and moral aspect of machine learning overall is allowing enough time to vet the algorithm with human professionals and practitioners in the field to ensure the algorithm is working as expected.

As the world continues to advance and consider new automation technologies there is an ethical concern around validation of AI. Doris, Wagner, and Winokur (2022) note the importance of utilizing corporate boards to vet AI and ensure AI is used responsibility within the organization. Multiple other devices fall within the AI, ML, and NLP area of interest as the technology that drives the device utilizes AI, ML, and NLP to work appropriately. For example, Siri (on an iPhone) or Alexa (from Amazon) are considered cognitive assistants also known as voice activated personal assistants (VAPAs). VAPAs are present in homes, banks, healthcare, education, and many more industries. Hamilton, Swart, and Stokes (2021) recommend a rating system on VAPAs for the ethical. legal, and social content that can be created and established by various organizations. The purpose of this rating scale is to inform customers of the implications of using VAPAs within their organizations.

Cryptocurrency and Bitcoin

Cryptocurrency is a method to decentralize digital money over the Internet. The transaction method is called blockchain. Blockchain is the tool used to send the value transaction. Blockchain, specifically, is a distributed database that is shared amongst nodes of a network. The network stores information electronically in digital format. This technique allows for a secure method to send money as it reduces the ability for hackers to decode the transmission of the data since it is being sent over in many packets. Without the key at the end, it is virtually impossible to decrypt the message. Bitcoin is the first cryptocurrency that has been developed. Cryptocurrency makes it possible to transfer value without the need of a bank or payment processor; and like bitcoin, is not issued or controlled by the government.

Carvalho, Sambhara, and Young (2020) note that blockchain will have more applicability than just sending cryptocurrency in the future. Cryptocurrency will become more diverse to include both the audience who uses the technology and the underlying technology (Carvalho et al., 2020). This means Blockchain will be available for other types of secure data transactions in the future.

Ethical Concerns With Cryptocurrency and Bitcoin

Cryptocurrency is a fairly new technology. Bitcoin has been present since 2009 but there are still some major ethical issues surrounding this technology (Billah & Atbani, 2019). There is a moral concern of cryptocurrency being used exclusively, as the banking industry is exposed to threat of disappearing in totality (BIllah & Atbani, 2019). There are challenges to the current systematic laws and regulations, as cryptocurrency users have great uncertainty regarding the state of their funds. Cryptocurrency also faces many threats for viability and has weaknesses due to a lack of standardization and rules that govern this technology (Billah & Atbani, 2019). Blockchain technology is seen as a standard platform to manage cryptocurrency, but there are no laws or regulations regarding this technology. Billah and Atbani (2019) recommend utilizing the halal alternative model which is becoming the standard model for Islamic finance. Bayram (2020) notes that there are other concerns with blockchain technology, especially across country boarders due to the lack of governance. Overall, even though this technology has been around for over a decade, the lack of standardization and rules causes major ethical and legal concerns for cryptocurrency.

Chatbots

A chatbot is a tool used for digital customer service. Chatbots have become common in various industries. Crolic, Thomaz, Hadi, and Stephen (2022) note that chatbots can cause angry end users as the technology still has a long way to go before taking over the human aspect of customer service. Before deploying, chatbots must be carefully designed and tested. There should also be a method to handover the chat to a live agent quickly to reduce negative response a human representative must manage. Chatbots are going to be around for years to come (Crolic et al., 2022). Organizations and users must become used to engaging with a bot in the near future as many organizations are moving towards this technology as a gateway to solve level one support concerns (before contact with a human for more complicated issues). The push towards chatbot technology is occurring due to the organizational requirements to reduce O&M. If 50 percent or more of customer service interactions with chatbots solve problems, the human-to-human engagement can be decreased by at least 50%. This reduces the amount of call time to address simple issues and allows the customer service representatives manage the more complex service calls.

Ethical Concerns With Chatbots

Ethical concerns also plague chatbot technology. Gender bias can be created during chatbots development, and therefore a company must avoid any gender bias, racism, sexism, or abusive language. during the development stage. A good way to reduce any potential concerns is to have multiple testers and reviewers test the technology. For example, Microsoft's Tay chatbot, was created for Twitter, as users interacted with Tay and began posting offensive tweets, Tay reciprocated by emulating the same language in reply.

Additionally, chatbots that are implemented within an organization should have additional security controls in place such as end-to-end encryption, biometric verification, and access controls for certain material. If an employee does not have access to certain material, the chatbot should perform analysis to ensure that the content or answers it is providing is limited to the current employee access.

There are moral concerns regarding this technology. Even though chatbots are seen as assistants, sometimes the information they find is not valid. There must be a feedback loop to ensure wrong answers are remediated. Chatbots can also be seen as replacing human positions, so when implemented within an organization, a change management and communication management strategy must be created to reduce concerns and resistance. Chatbots from an end user's perspective can also cause great strife. As noted above, Crolic et al. (2022) mention that there can be some negative impacts to chatbots that do not respond appropriately. This can cause consumers to get frustrated and could result in a loss of business. Chatbots can be a great tool, if they are appropriately vetted, tested, and have a feedback loop to fix any concerns. They should be secure to reduce any potential security or moral concerns.

Global Positioning System (GPS)

GPS is a navigation system using satellites, receivers, and algorithms to determine location, velocity, and time data. There are multiple industries using GPS technology, such as bike rental companies, electric utilities, agricultural, and many more. This technology will allow hospitals to assist patients in knowing where to go once they are in the facility and can even offer directions on a mobile device. Research was conducted for detection and analysis of COVID-19 cases via GPS technology. This technology would allow the user to know if they were near someone that has come down with the disease (Schmidt et al., 2020). GPS technology is also used in mapping. The application, Zillow, uses GPS to present street views of potential neighborhoods. Google Earth uses mapping from satellites to allow users to zoom in on specific areas around the globe. This technology also allows for very specific coordinates to be presented to get from one place to another. For example, MapQuest also uses this technology to show the best route to get from one place to another.

Ethical Concerns With GPS

GPS technology presents some ethical and moral concerns. GPS technology can be found in cellphones, wearable technology, and vehicles. GPS technology is becoming more prominent in the healthcare industry. It has been utilized in the monitoring of infectious diseases, identification of patterns of health-related transmissions and can also link movement patterns of humans based on their health behavior (Apte et al., 2019). Tracking the movements of people can cause major ethical concerns. Certain countries have geolocation privacy legislation to prohibit the use of GPS technology (Apte et al., 2019). With the use of this technology, it is difficult to deidentify information before analysis, which can also cause concern for data bias. There are great challenges with obtaining informed consent, especially in low-income countries (Apte et al., 2019). Even though GPS technology can be seen as a great method to track patient care concerns, find places on a map, and even find the house of your dreams and understand the surrounding neighborhood, it is important to understand the ethical and moral implications that this technology presents from a privacy and security perspective.

RFID Technology

Radio Frequency identification (RFID) is a technology where digital data is encoded on labels or other devices and is captured by a reader via radio waves. RFID is similar to barcoding, as there is data in the barcode. The difference is that RFID can be read outside the line of sight, whereas barcodes must be read by an optical scanner. Many manufacturers are automatically placing smart or RFID labels on their packaging.

RFID is implemented in many industries in the form of inventory management, asset tracking, personnel tracking, access control, badging, supply chain management, and counterfeit prevention (in pharmaceutical areas). In the healthcare industry, more advanced RFID technology is being discussed. The healthcare industry is working on an implantable RFID device to analyze patients with cellular conditions to enhance the care that these patients receive based on real-time cellular data. There are some issues with technology usage and process around labeling that is troublesome for many supply chain departments within organizations due to the inconsistent practice of labeling. RFID thus can cause additional burden to staff if they need to relabel or check for labeling standards when a new shipment of products enters their organization (Yang et al., 2021).

Ethical Concerns With RFID

The implementation of RFID has created great improvements in the service delivery industry (Abugabah et al., 2020). RFID can also be used to track assets within an organization (Abugabah et al., 2020). For example, IV poles can be tracked with an RFID device; when a patient needs an IV, the RFID trackability can note which poles are in use and which ones are available. The RFID technology reduces the time it takes to perform patient care by easily finding the available technology. Some major concerns with RFID is tracking and tracing individuals within an organization. Obtaining informed consent is difficult when trying to perform patient care as clinical staff are focused on treating the patient and informed consent can be seen as more administrative work, especially if the patient is very ill. Security issues can occur if the system is accessed by unauthorized personnel (Abugabah et al., 2020).

There is also a moral aspect to RFID technologies. Per Soni and Soni (2019), the supply chain operations of an organization can be impacted positively with RFID technology, but human interaction will be reduced due to these new technologies. With the advances in RFID and other automation technologies, a major moral concern is implementing RFID technology and reducing human involvement in the process. As RFID technology usage increases, leadership must ensure the workforce culture is ready to implement these changes (Soni & Soni, 2019).

Self-Driving Vehicles

Self-driving vehicles, also known as autonomous cars, are vehicles that can sense the environment and operate without human intervention. A human passenger is not needed to take control of the vehicle or be in the car at any time. Currently there are challenges with the full-scale implementation of self-driving cars. The lidar and radar technology inherent within this technology, can potentially interfere with other cars driving on the road. The range of radio frequency must be excellent for one of these vehicles to work, therefore cars outside of the range would not be able to function. Poor weather conditions can make it difficult for these vehicles to work. For example, if it is snowing or there is heavy rain, the cameras may not work effectively. Traffic conditions and laws may also cause issues – for example, when emergency vehicles must pass by. There are also state and federal regulations that currently regulate or restrict these cars from being on the roads. Also, there is a morale and legal question of who is responsible if an accident occurs within an autonomous car.

There are a lot of questions surrounding this technology. Elliot, Meng, and Hall (2021) note that self-driving vehicles are possible from a technical standpoint. The authors also believe the technology will be very feasible by 2030 (Elliot et al., 2021). The major issue is the acceptance of this technology. Autonomous vehicles have multiple usage (not just for use on roadways). Doordash has a self-driving bot that transverses sidewalks. College campuses have self-driving vehicles that deliver goods and supplies across the campus.

Ethical Concerns With Self-Driving Vehicles

Self-driving vehicles can offer many advantages, but there are some ethical. legal, and social economic concerns with the implementation of self-driving vehicles. First of all, the safety concern of self-driving vehicles is a real threat to the success of implementing this technology. Per Ryan (2020), there is a major risk of self-driving vehicles being hacked. Self-driving cars can also threaten driving jobs. Liability is of great concern with self-driving vehicles. If a self-driving vehicle is in an accident, the question of liability is a concern. There are also other legal laws to consider such as national differences on road traffic and safety (Ryan, 2020). There is a moral concern with self-driving cars, as humans lose the ability to make decisions or take control of the vehicle which can lead to liability concerns. Also, self-driving vehicles must abide by legal speed limits which remove driving freedoms as well. Privacy of data is another concern with this technology. With the dangers and cautions needed to make self-driving vehicles a reality, a large amount of data must be collected, which some will note is a violation of their privacy. There are major benefits of self-driving vehicles, but it is imperative to consider the major roadblocks that need to be considered before this technology becomes deployed in a wide scale.

Smart Organizations

Smart buildings are starting to emerge. This technology allows the energy system in a business to only be in use when needed, this allows for organizations to cut down on costs. For example, when a meeting room is not in use, the room can shut down and reserve energy. Hospitals are also starting to use more smart technology. An example of this is an AI robot can assist with the cleaning of a dirty room and perform tasks like cleaning the floors and assisting employees with transporting dirty bedding to the laundry room. In the book, Evaluation of Energy Efficiency and Flexibility in Smart Buildings, the author notes that multiple systems can be configured and programmed based on the organizational preference. HVAC and system controls are two areas where there has been a lot of movement in smart design in recent years (“Evaluation of Energy Efficiency,” 2021).

Ethical Concerns With Smart Organizations

Smart organizations and cities are a new concept. With the amount of automation occurring in these cities, it is imperative that people within these organizations understand the potential ethical, legal, and security challenges that can occur within these organizations. In smart organizations, there is a lot of data that is collected. The goal of obtaining the mass amounts of data is to perform analysis to review certain behaviors and urge employees to act or perform actions a certain way (Ranchordás, 2020). In smart organizations there could be security concerns regarding ensuring access control and monitoring. There should be checks to ensure the access control is appropriate based on the individuals current access rights. There are also concerns regarding the use of the Internet of Things (IoT) devices in smart organizations (Ranchordás, 2020). The interoperability of the devices can show how people interact with one another. With IoT technology, there is a risk of profiling specific employees based on their current actions. For example, a smoker may end up taking more breaks than a non-smoker, and therefore there may be a negative bias towards smokers in the organization based on the data. Transparency of data is difficult to manage as the smart organization forces leadership to think about the data concerns involved around data privacy versus transparency concerns. Even though smart organizations can offer many benefits, such as reduction of costs and more efficiency, there are some security, privacy, and moral implications to consider with the implementation of this technology.

Robotics

A robot is a physical machine that resembles a human being or a part of a human being that performs work or replicates human movements and functions. Robotic arms have been used in the car manufacturing industry for decades (Ritchie & Landis, 2021). The arms perform specific functions and removes the potential for safety concerns within certain functions on the assembly line. In many industries, both robots and humans work side-by-side. The human factor is still required to make decisions, whereas the robot is utilized for reducing safety and quality concerns. Robots are used in various industries such as aerospace, beverage, manufacturing, computers, food, electronics, and the healthcare industry. Each robot typically performs a certain task to reduce quality concerns and increase productivity. Robotics will continue to grow in the future. Organizations have more work than employees and this trend will continue to grow. One method to meet the increasing demand within operations is to implement robotics.

Ethical Concerns With Robotics

There are numerous ethical and moral concerns with the implementation of robotics. Organizations that implement robots, must first understand how to implement robotics in the workplace in conjunction with the existing workforce. Many robots cannot perform complex work and are used alongside their human counterparts. For example, at Amazon, robots follow workers around and deliver goods to the correct aisles, which reduces the steps that an employee takes every day and increases the overall productivity. There is a possible safety concern with the robots, as the power system or lidar system that guides the robot may misfunction. Human interaction with robots can cause other issues such as unauthorized access or improper installation of equipment. Robotics programs must contain training and retraining on a frequent basis. When robots are placed in certain environments such as healthcare industries, implementation must be ethically sound. Schoenhofter et al. (2019) mention that understanding patients' needs and adaptability to robots prior to implementing robot-assisted care is important. Each patient will have a specific vulnerability and assessment with robotic support especially when in a vulnerable state; the patients' needs should be considered above all (Schoenhofter et al., 2019). Robotic implementation will continue to grow over the next decade, best implementations must be determined, and it should start with a strategic vision and transparent communication with the workforce. Applying robotics to an organization works well by starting with lower-level robotics and working up to more advanced technology. Understanding the potential ethical and moral aspects of a robotics program will assist in a successful implementation.

The information above is not meant to be an exhaustive list, but describes some major innovation technology considerations that organizations are addressing today.

WHEN IS IT OKAY TO SAY "NO" TO INNOVATION?

It is imperative that the strategic innovation goals are being met, but when an innovation idea falls outside of the current innovation strategy, it must be stated that the project is not being addressed as it does not fit within the current organizational innovation strategy. Transparent and effective communication is imperative in a successful innovation strategy. There will probably be many more innovative ideas than the organization can implement, and therefore it is imperative to prioritize the innovation projects based on strategic alignment and then priority of importance. Ensuring leaders understand why the proposed innovation project is not moving forward will allow them to gain focus and clarity on the strategic innovation vision. If the strategy changes, and the project fits within the organizational vision, it would be appropriate to review the project at that time.

Risk is another consideration. Ko and Kim (2020), mention the term responsible innovation. Responsible innovation includes not performing work if it is not socially acceptable or if it does not fall into the organization's strategy (Ko & Kim, 2020). If the risk profile of the organization is light to moderate, and the innovation is highly risky, it would be a good idea to explain that due to differences in risk acceptance the innovation will not be reviewed at this time. If the innovation becomes more widespread or less risky, and the organization is willing to accept the overall risk of the implementation, it would be appropriate to review the innovation when the risk profile is acceptable to the implementing organization.

If the viability of the product is not appropriate, then it would also be appropriate to decline the innovation idea at this time (Ko & Kim, 2020). If the product is vaporware (not yet to market or not viable) or does not meet the requirements that are being requested for the innovation effort, then it is best to note that the product does not yet meet the requirements, note the requirements, and also note when to re-review the product. An example of this is robotics. An organization wanted a mobile robot that could pick supplies, deliver the supplies picked, and then complete the order in the system. The technology that was being requested was a mobile robot. The robot had an arm, but the robot was not successful in picking the supplies or integrating with the backend system to fulfill the order. Companies must note what the expectations are and if the company had any future intensions of delivering the functionality in future iterations of the product.

All the above scenarios note that the innovation is not right at the present time, but it does not mean that it will never be ready to implement. A standing list of all innovation ideas, the current status, and the current research would be appropriate. If the innovation was placed on a hold, noting why is an important aspect of when the innovation should be reviewed again for consideration. Digitally transforming an organization should be completed at the right time for the organization and its employees (Zhou et al., 2022).

From an ethical perspective, leaders should not implement technology that is not ready for production. Implementing technology that is not ready, can cause issues with customer retention and even safety concerns. The leaders within an organization should be implementing one innovation vision and not solely focus on their own innovation needs. An ethical leader will review the entire innovation strategy, determine how their ideas fit within the strategy, and ensure their department is ready to accept the innovation when it meets the needs of the overall organization.

CONCLUSION

The implementation of innovation technology is a great method for organizations to stay relevant and competitive in the market (Pinheiro et al., 2016) There are many things to consider when implementing any innovation technology. Organizational leaders should survey the innovation landscape and the organization's risk levels to determine the best innovation strategy to pursue (Yaghmaei & van de Poel, 2021). Leaders implementing innovative solutions also need to ensure the organization has a culture accepting of innovation. The climate is just as important as the innovation itself. A negative innovation climate can cause organizations to fail before they have even begun to implement. Implementing the steps for an accepting change culture is pivotal to an organization's success. Leaders must be willing to change the way they think about implementation of innovation as a learning strategy instead of a implementation success strategy. Regardless, innovation strategy must align with the overall goals of the organization for successful innovation management. Ethical leadership ensures the organization is ready to change by having transparent communication with employees and instilling a culture of change.

Leaders must exhibit ethical behaviors, as employees mimic these behaviors. An unethical leader can cause unethical employee behavior which leads to an unaccepting culture of change. Some technologies are viable and in use today, and some are not being used due to either legal, ethical, moral, or socio-economic challenges.

Teaching ethics when teaching innovation is also of great importance when considering any type of innovation (Budd, 2018). Today, all organizations must consider innovation to stay competitive. Students must be taught the importance of ethics and how ethical behavior is key when creating new tools and skills to manage organizational activities (Budd, 2018). Obtaining an understanding of the various innovation technology opportunities and the ethical, legal, and moral aspects to consider before implementing them offers basic concepts to consider as new technology becomes available. Organizations will continue to implement new technology, but ensuring the technology is implemented at the right time with an employee (customer base) that understands why the technology is being implemented and the benefits is pivotal to the success of implementation. Additionally, the ethical ramifications of implementation are key to a successful deployment strategy.

REFERENCES

Abdulkader, B., Magni, D., Cillo, V., Papa, A., & Micera, R. (2020, August 28). Aligning firm's value system and open innovation: A new framework of business process management beyond the business model innovation. *Business Process Management Journal, 26*(5), 999–1020. doi:10.1108/BPMJ-05-2020-0231

Abugabah, A., Nizamuddin, N., & Abuqabbeh, A. (2020). A review of challenges and barriers implementing RFID technology in the Healthcare sector. *Procedia Computer Science, 170*, 1003–1010. doi:10.1016/j.procs.2020.03.094

Apte, A., Ingole, V., Lele, P., Marsh, A., Bhattacharjee, T., Hirve, S., Campbell, H., Nair, H., Chan, S., & Juvekar, S. (2019, June). Ethical considerations in the use of GPS-based movement tracking in health research - lessons from a care-seeking study in rural west India. *Journal of Global Health, 9*(1), 010323. doi:10.7189/jogh.09.010323 PMID:31275566

Artegoni, A. (Ed.). (2021, January). *Evaluation of energy efficiency and flexibility in smart buildings*. Multidisciplinary Digital Publishing Institute. doi:10.3390/books978-3-03943-850-1

Bales, S., & van Rensburg, H. (2019). *Innovation wars: Driving successful corporate innovation programs*. Morgan James Publishing.

Bayram, O. (2020). Importance of Blockchain use in cross-border payments and evaluation of the progress in this area. *Dogus University Journal, 21*(1), 171–189. https://dergipark.org.tr/en/pub/doujournal/issue/66682/1043235

Billah, M. M., & Atbani, F. M. (2019, January-March). SWOT analysis of cryptocurrency an ethical thought. *Journal of Islamic Banking & Finance, 36*(1), 22–27.

Budd, J. M. (2018). Teaching ethics: A framework for thought and action. *Journal of Education for Library and Information Science, 59*(3), 53–66. doi:10.3138/jelis.59.3.2018-0022.06

Carvalho, A., Sambhara, C., & Young, P. (2020, January). What the history of Linux says about the future of cryptocurrencies. *Communications of the Association for Information Systems, 46*, 18–29. doi:10.17705/1CAIS.04602

Crolic, C., Thomaz, F., Hadi, R., & Stephen, A. T. (2022). Blame the bot: Anthropomorphism and anger in customer–chatbot interactions. *Journal of Marketing, 86*(1), 132–148. doi:10.1177/00222429211045687

Doris, B., Wagner, D., & Winokur, H. (2022, Winter). Director's cut: How boards can help ensure the responsible use of AI: The question for boards and management is not whether to use Artificial Intelligence, but how to ensure it is used responsibly. *NACD Directorship, 48*(1), 38–43.

Elliott, K., Meng, J., & Hall, M. (2021, July 14). An integrated approach for predicting consumer acceptance of self-driving vehicles in the United States. *Journal of Marketing Development and Competitiveness, 15*(2), 10–20. doi:10.33423/jmdc.v15i2.4330

Gwebu, K. L., Wang, J., & Wang, L. (2018, May 15). The role of corporate reputation and crisis response strategies in data breach management. *Journal of Management Information Systems*, *35*(2), 683–714. doi:10.1080/07421222.2018.1451962

Hamilton, C., Swart, W., & Stokes, G. M. (2021). Developing a measure of social, ethical, and legal content for intelligent cognitive assistants. *Journal of Strategic Innovation and Sustainability*, *16*(3), 1–37.

Iskat, G. J., & Liebowitz, J. (2021, June). What to do when employees resist change. *Super Vision*, *82*(6), 7–9.

Katznelson, G., & Gerke, S. (2021, March 3). The need for health AI ethics in medical school education. *Advances in Health Sciences Education: Theory and Practice*, *26*(4), 1447–1458. doi:10.100710459-021-10040-3 PMID:33655433

Ko, E., & Kim, Y. (2020, May 20). Why do firms implement responsible innovation? The case of emerging technologies in South Korea. *Science and Engineering Ethics*, *26*(5), 2663–2692. doi:10.100711948-020-00224-2 PMID:32436167

Lee, J., & Choi, S. J. (2021, July). Hospital productivity after data breaches: Difference-in-differences analysis. *Journal of Medical Internet Research*, *23*(7), e26157. doi:10.2196/26157 PMID:34255672

Lewin, K. (1947, June 1). Frontiers in group dynamics: Concept, method, and reality in social science; social equilibria and social change. *Human Relations*, *1*(1), 5–40. doi:10.1177/001872674700100103

Li, C., & Yang, H. J. (2021). Bot-X: An AI-based virtual assistant for intelligent manufacturing. *Multiagent & Grid Systems*, *17*(1), 1–14. doi:10.3233/MGS-210340

Liu, B. (2021). *"Weak AI" is likely to never become "Strong AI", so what is its greatest value for us?* doi:10.48550/arXiv.2103.15294

Masumba, D. (2019). *Leadership for innovation: Three essential skill sets for leading employee-driven innovation*. Morgan James Publishing.

McDevitt, P. (2020). *Anchoring cultural change and organizational change: Case study research evaluation project All Hallows College Dublin 1995-2015*. Information Age Publishing.

Özsungur, F. (2019). The impact of ethical leadership on service innovation behavior: The mediating role of psychological capital. *Asia Pacific Journal of Innovation and Entrepreneurship*, *13*(1), 73–88. doi:10.1108/APJIE-12-2018-0073

Perry, F. (2020). *The tracks we leave: Ethics and management dilemmas in healthcare* (3rd ed.). ACHE Management Series.

Rahaman, A. (2019, June). Address the real reasons employees resist change. *HR News Magazine*, 18–21.

Ranchordás, S. (2020). Nudging citizens through technology in smart cities. *International Review of Law Computers & Technology*, *34*(3), 254–276. doi:10.1080/13600869.2019.1590928

Rejon-Parrilla, J. C., Espin, J., & Epstein, D. (2022, January 3). How innovation can be defined, evaluated and rewarded in health technology assessment. *Health Economics Review*, *12*(1), 1–11. doi:10.118613561-021-00342-y PMID:34981266

Richter, F.-J., & Sinha, G. (2020, August 21). Why do your employees resist new tech? *Harvard Business Review*. https://hbr.org/2020/08/why-do-your-employees-resist-new-tech

Ritchie, E., & Landis, E. A. (2021, June 23). Industrial robotics in manufacturing. *Journal of Leadership, Accountability and Ethics*, *18*(2), 110–116. doi:10.33423/jlae.v18i2.4258

Ryan, M. (2020). The future of transportation: Ethical, legal, social and economic impacts of self-driving vehicles in the year 2025. *Science and Engineering Ethics*, *26*(3), 1185–1208. doi:10.100711948-019-00130-2 PMID:31482471

Schein, E. H., & Schein, P. A. (2018). *Humble leadership: The power of relationships, openness, and trust*. Berrett-Koehler Publishers.

Schmidt, F., Dröge-Rothaar, A., & Rienow, A. (2021, August 28). Development of a web GIS for small-scale detection and analysis of COVID-19 (SARS-CoV-2) cases based on volunteered geographic information for the city of Cologne, Germany, in July/August 2020. *International Journal of Health Geographics*, *20*(1), 40. Advance online publication. doi:10.118612942-021-00290-0 PMID:34454536

Schoenhofer, S. O., van Wynsberghe, A., & Boykin, A. (2019). Engaging robots as nursing partners in caring: Nursing as caring meets care-centered value-sensitive design. *International Journal for Human Caring*, *23*(2), 157–167. doi:10.20467/1091-5710.23.2.157

Sjödin, D., Parida, V., Jovanovic, M., & Visnjic, I. (2020). Value creation and value capture alignment in business model innovation: A process view on outcome-based business models. *Journal of Product Innovation Management*, *37*(2), 158–183. doi:10.1111/jpim.12516

Soni, R. G., & Soni, B. (2019). Evolution of supply chain management: Ethical issues for leaders. *Competition Forum*, *17*(2), 240–247.

Spinello, R. A. (2021, Spring). Corporate data breaches: A moral and legal analysis. *Journal of Information Ethics*, *30*(1), 12–32.

Stauffer, D. C., & Maxwell, D. L. (2020, May 18). Transforming servant leadership, organizational culture, change, sustainability, and courageous leadership. *Journal of Leadership, Accountability and Ethics*, *17*(1), 105–116. doi:10.33423/jlae.v17i1.2793

Strate, L. (2020, July-October). The ethics of innovation. *Etc.; a Review of General Semantics*, *77*(3–4), 182.

Yaghmaei, E., & van de Poel, I. (2021). *Assessment of responsible innovation: Methods and practices*. Routledge.

Yang, M. X., Xiaolin, H., Demir, A., Poon, A., & Wong, P. H.-S. (2021, March 16). Intracellular detection and communication of a wireless chip in cell. *Scientific Reports*, *11*(1), 5967. Advance online publication. doi:10.103841598-021-85268-5 PMID:33727598

Yermak, S. O., & Lisnichenko, O. O. (2016). Studying the aspects of establishing the definition of "innovation activity" and its determining factors. *Bìznes Ìnform*, *3*, 49–55.

Zhou, P., Zhou, S., Zhang, M., & Miao, S. (2022, May 12). Executive overconfidence, digital transformation and environmental innovation: The role of moderated mediator. *International Journal of Environmental Research and Public Health, 19*(10), 5990. Advance online publication. doi:10.3390/ijerph19105990 PMID:35627526

Chapter 2
What Effect Does Technology Have on a Responsive Leadership Style?

Ibidayo Awosola
https://orcid.org/0000-0003-1944-9900
Purdue University, USA

ABSTRACT

All aspects of human life are affected by technology. It has changed the way individuals communicate with one another and conduct their everyday routines and business transactions. The world is undergoing a Fourth Industrial Revolution, which is radically altering the world of leadership. Artificial intelligence, machine learning, robots, and other cutting-edge technologies, in particular, are increasing efficiency and reshaping leadership and operations. Leaders in today's most successful firms must keep up with technology advancements and successfully leverage them to achieve laudable outcomes. The motivating element behind the new leadership style is technology and its ethical use. Literature gaps indicate that there is much more to learn about this issue and that it is critical to address it in order for enterprises to remain competitive.

INTRODUCTION

Globalization ushers in a new social and occupational paradigm. It is possible to communicate and interact anytime, from anywhere, face-to-face or virtually, making the organizational configurations volatile, uncertain, complex, and ambiguous (VUCA). Technology's leadership processes are increasingly mediated by technology, making room for tech-leadership, e-leadership, and virtual teams, whose conceptualization revolves around technology as a facilitator of communication, engagement, development, and work sharing (Machado & Brandão, 2019). Leadership is a crucial determinant of a successful employee, team, and organizational creativity and innovation (Hughes et al., 2018). The world is not in the grip of a technology crisis; instead, it is in the grip of a technological revolution. Humanity will

DOI: 10.4018/978-1-6684-5892-1.ch002

see technological transformations and developments in the near future never before witnessed (Kluz & Nowak, 2016).

Technology's impact on organizations as change drivers and solutions to some of those very change imperatives makes it critical levers for the much-needed transformation of outdated management and leadership models. Unfortunately, most organizations are not entirely using the possibilities of this new ecosystem while it is still in its early phases. The connection between leadership practices and the emerging role of technology is underappreciated (Deiser & Newton, 2015). The efficacy of technology in various leadership styles has not been extensively examined in the literature.

Leadership experts have spent the last two decades attempting to track the consequences of digitization processes. Part of the scholarly discussion has centered on leaders' abilities to incorporate digital transformation into their organizations while also inspiring staff to embrace change, which is frequently regarded as a challenge to the present status quo (Cortellazzo et al., 2019). Technology advances at a rapid pace, and firms must continually adopt cutting-edge and bleeding-edge technologies in order to remain competitive. This necessitates flexible leadership, and businesses must have the proper leadership structure in place to allow for this change. It is possible that a hierarchical structure is not the optimal solution. Rather than developing a small group of individuals, management duties will most likely be distributed throughout a company. In today's increasingly digital environment, leadership models must be able to grasp the new road forward (Miller, 2021). This study aims to conduct a theoretical assessment of the impact of technology on responsive leadership style and provide theoretical clarity.

TECHNOLOGY OVERVIEW

The definition of "technology" involves applying scientific knowledge to achieve human goals, both for practicality and in industry. The mechanisms can be exceedingly simple or tremendously complicated.

Ten Types of Technology

Technology is a very broad term that refers to the use and comprehension of electronic tools and crafts, as well as how these goods and talents influence the ability to control and adapt to surroundings. Science and engineering have resulted in today's technology ("What is technology," 2019).

Today, various types of technology are available and induce advancements in multiple ways (Crook, n.d.). There are ten key forms of technology thus to explore, including how they assist their sphere of operation.

Entertainment Technology

Entertainment technology makes use of numerous manufactured or created pieces to increase or create the possibilities of any entertainment activity. This includes entertainment and activities such as video games, television shows, and so forth. Some persons are involved in the production, sale, design, creation, or use of such items and services in order to give greater entertainment experiences. The breadth of entertainment technology encompasses every sphere of human activity, including movies, music, literature, and so on, that allows people to entertain themselves. Due to increasing demand in the market, the education, scientific, and technology sectors are seeing significant expansion these days (Hayden,

2018). Many schools, colleges, and universities offer entertainment technology courses, and are viewed as an excellent career option by many. Furthermore, numerous businesses and commercial organizations have formed alliances with various educational institutions and colleges in order to educate students for a long and successful career in this sector. As a result, the quality of the employment market for persons working in the entertainment technology business has dramatically improved. This has intensified competitiveness among those working in this field.

Operational Technology

The science of measuring, controlling, detecting, and diagnosing systems, devices, people, and events is known as operational technology. The scientific study of how diverse systems interact is known as operations research. The operations science may also be applied to human behavior. Operational technology includes computer systems, optical devices, electronics, and electrical engineering. Monitoring and management of industrial process assets and manufacturing/industrial equipment are done by operational technology ("Operational technology," n.d.). It also comprises employing mathematical concepts to devise and implement solutions for specific industrial processes. Electronic devices are one of the most essential areas of operational technology since they directly affect how business is done in recent times. All parts of production and the economy rely on operational technology. An airline, for example, utilizes flight data from instruments to deliver more efficient services. Weather and climatic information are essential aspects of commerce, and operational technology developed to detect possible threats determines an airplane's safety.

Assistive Technology

Help with movement, ambulation, turning or other bodily action, bathing, eating, or ascending stairs are examples of technological systems that enhance the way people do things. To define a specific application, the equipment utilized may be mechanical devices such as crutches, walkers, canes, wheelchairs, stairlifts, or an artificial arm, hand, or leg. Aids, in other words, are goods or systems that increase a person's capacity to perform tasks on their own at home or within their community. The most prevalent use of assistive technology is for mobility. Mobility gadgets enable persons with disabilities to move about safely and freely. Mobility chairs, wheelchairs, walkers, canes, scooters, and other equipment may be used to define the application. Bathing and eating, climbing, turning or other bodily actions, walking, and opening and closing doors are additional uses and needs (Assistive Technology Industry Association, n.d.).

Agriculture Technology

Agriculture now routinely employs technology such as temperature and moisture alarms, various types of machinery, instruments, cameras, aerial imaging, and global positioning (GPS). Agriculture is a diverse business that necessitates substantial use of modern technologies and gear. This technology is concerned with the production of food, feed, feedstock, animals, fruits, and vegetables, as well as with the handling, storage, and distribution of agricultural goods such as cereals, dairy, livestock, poultry, sugarcane, and so on. Modern agriculture technology has become a key component of agricultural production and processing, supporting the agricultural industry's productivity and competitiveness on the global market. This field has contributed chiefly to the development of modern farming and boosted agricultural production in a very short period of time. Using contemporary technologies, feed processing may be made more cost-effective, and more crops can be cultivated in the same area at a lower cost. Because of the reduced need for fertilizer, insecticides, and herbicides, farming has become more accessible and lucrative (National Institute of Food and Agriculture, n.d.).

Superintelligence

Super Intelligence technology is neither a robot nor a computer system; rather, it is a tool that a person may utilize to accomplish particular tasks that no human can perform. There has indeed been much debate concerning Super Intelligence, and any various explanations have been offered. Still, it is a means of using computers to help humans in a variety of situations, and the benefits are already being realized. This technology field enhances humans' decision-making abilities since they will be able to think faster and make better judgments based on the real intelligence of their brains as compared to a computer. There are unfounded concerns that artificial intelligence could eventually replace humans. The potential to employ artificial intelligence to improve elements of life can be beneficial. People require assistance, much as pilots in flight require other humans to assist them in flying and landing, and robots aid mankind in industrial work and other industries. Humans would not have to rely only on themselves or rely totally on an artificial intelligence system; instead, there can be a hybrid of the two (Kanade, 2022). As a consequence, a new human being with athletic ability and artificial intelligence can potentially be created.

Artificial Intelligence

Artificial intelligence is a discipline of computer science concerned with the research and creation of artificially intelligent systems. The concern is with how an artificial system acts rather than how it responds to real-life events. Human and animal intelligence demonstrate emotions and social consciousness, whereas AI exhibits behavior controlled by a computer or software. The phrase "artificial intelligence" itself exemplifies the distinction between the two forms of intelligence. As worries about machine learning spread to applications outside of computer science, artificial intelligence researchers are always seeking new ways to greatly improve the technology (Wigmore, 2018). To execute daily chores, artificial intelligence may make judgments based on patterns and learning behavior, as well as use previously set rules. It has already transformed several fields of research, including as health, transportation, and industry. Apprehensions about artificial intelligence are rising as more businesses turn to automation for better efficiency. This has heightened interest in computer security, software patents, and other areas of research. As artificial intelligence becomes increasingly widespread in everyday life, researchers will continue to investigate these many concepts and their societal implications. Experts have recently begun to investigate the influence of artificial intelligence on society, specifically the impact on future artificially intelligent devices (West & Allen, 2018). Some academics are concerned about individuals' future rights to govern the behavior of such gadgets, while others are troubled by the future of business and employment automation.

Information Technology

IT, or information technology, collects, retrieves, organizes, and transmits data using computers and related equipment. Information technology is often used for corporate activities rather than for personal or recreational purposes and remains a major component of communications. This includes any technology that makes information more accessible, such as email, instant messaging, desktop publishing software, spreadsheets, multimedia, web page design, music, movies, and so on. The IT sector, which includes computer software, engineering services, network and storage systems, information technology support, networking goods, and peripherals, accounts for the bulk of the worldwide economy. The speed of technology and globalization is growing knowledge and use of IT technologies ("What is Information Technology," n.d.). As information technology becomes a more important element of humans' lives, the demand for information technology workers will rise.

Blockchain Technology

Blockchain technology established the foundation for a new class of web-based monetary systems by allowing information to be encrypted yet not corrupted. Blockchains, in their most basic form, are networks of ledgers or other virtual data storage that users may exchange and edit. Originally developed for the digital asset BitUSD, the technology is currently being utilized to develop apps for anything from social networking sites to online stock markets. Blockchains are the most recent and significant invention to emerge from this group. This technology has opened up whole new avenues for everyone who wants to construct an internet application by combining the functionality of various earlier technologies ("What is blockhain technology," 2022). One of the key goals of Blockchain technology is to provide consumers with a centralized location for storing, managing, and securing their private transaction ledger. Transactions are quick and may be changed as soon as they occur. Furthermore, because of the increased speed with which transactions occur, the average transaction time is substantially less than in the past. Because blockchains are available to everyone who wants to join, anyone may develop an internet application that allows users to send money and other assets.

Educational Technology

A systematic approach to all educational resources and systems, including computer-based teaching and learning systems, is known as educational technology, or EdTech. It relates to the many demands placed on individuals, the use of technological solutions to academic education, and the assessment of student academic achievement. It is used to improve educational quality by making it more relevant and engaging for pupils. Educational Technology is a crucial resource and application of information technology that positively affects formal learning. It covers a wide range of topics, from classic learning techniques like print and visual media to more modern systems such as interactive learning tools, digital audio-visual systems, and internet learning (Koss, 2022). This technology has made substantial contributions to the fields of education and research. It is used for gathering, organizing, storing, distributing, transferring, assisting, and supporting individuals' education and information. Students using technology have greater access, can learn more quickly, tend to perform better, and experience a more enriching learning environment through standard and developing processes, materials, practices, and methods. Educational technology has a wide range of applications. It is critical in the teaching and learning processes of many of today's disciplines (Duff, 2021).

Medical Technology

Medical technologies are medical items and procedures used in health care. This technology field may be viewed as a significant way of exchanging ideas and experiences concerning health care commodities in order to improve, simplify, and secure the lives of patients. Technological advancements have resulted in more efficient methods of diagnoses and treatment, reducing the complexity of formerly complex operations and requiring fewer healthcare staff to accomplish these goals. Medical technologies are focused on providing great service, high-quality goods, and exceptional customer service (David et al., 2020). To stay positioned at the forefront of medical technology advancements and preserve the very nature of patient care, healthcare professionals must remain up to date on the latest scientific discoveries and technology trends. To safeguard the healthcare professionals who apply these new procedures and products, as well as the patients who utilize them, strict regulatory restrictions must follow these technical advancements. Scientific societies must also take the lead in educating the public about the need of improving healthcare systems via the implementation of innovative medical technology. Medical technology has the potential to significantly alter how medical facilities and healthcare personnel address both the typical and emergency problems that affect huge populations ("What is Medical Technoloyg," n.d.). The impact of such advancements have a favorable impact on public safety and health.

Technology as an Enabler

Globalization, technological progress, and the increasing number of connections between individuals, activities, and events have accelerated the change rate in today's world. People worldwide are mixing and mingling as never before, forming new alliances and fostering more culturally diverse workplaces.

Companies are increasingly working together to investigate potential opportunities and enter new markets. Every day, new technological surprises emerge, some of which have the potential to upend whole markets in a matter of months. Consumers, the media, and other players now wield immense power due to the internet and the rise of social media, which they freely use to praise or chastise companies at a rapid speed (Stephenson, 2011).

As an enabler, technology assists leaders in achieving their objectives. Technology can help business or political leaders be more sensitive to change as it arises, providing them with a wealth of real-time data and insight into humanity's many encounters with nature and its consequences.

Digital technology has the potential to both empower and disrupt business leaders. Working out the most efficient ways to use a new application or moving offline processes online, for example, may not be every leader's strength, but they must be willing to take up the challenge. Business leaders must recognize the profound shift in the business landscape and gain a competitive advantage by seizing new opportunities before their competitors.

Corporate leaders must create a transformative digital vision, empower followers through collaboration, concentrate on digital governance, and establish technical leadership (Della Corte et al., 2019). In addition, technological advancements require leaders to be more responsible and responsive. Leaders must act and react wisely due to increased awareness of the impact of technological advancements (Craig, 2017).

Figure 1. Leadership capabilities and digitalization: Key factors and business consequences
Source: (Della Corte et al., 2019).
Image used by kind courtesy of the Creative Commons Attribution 3.0 license.

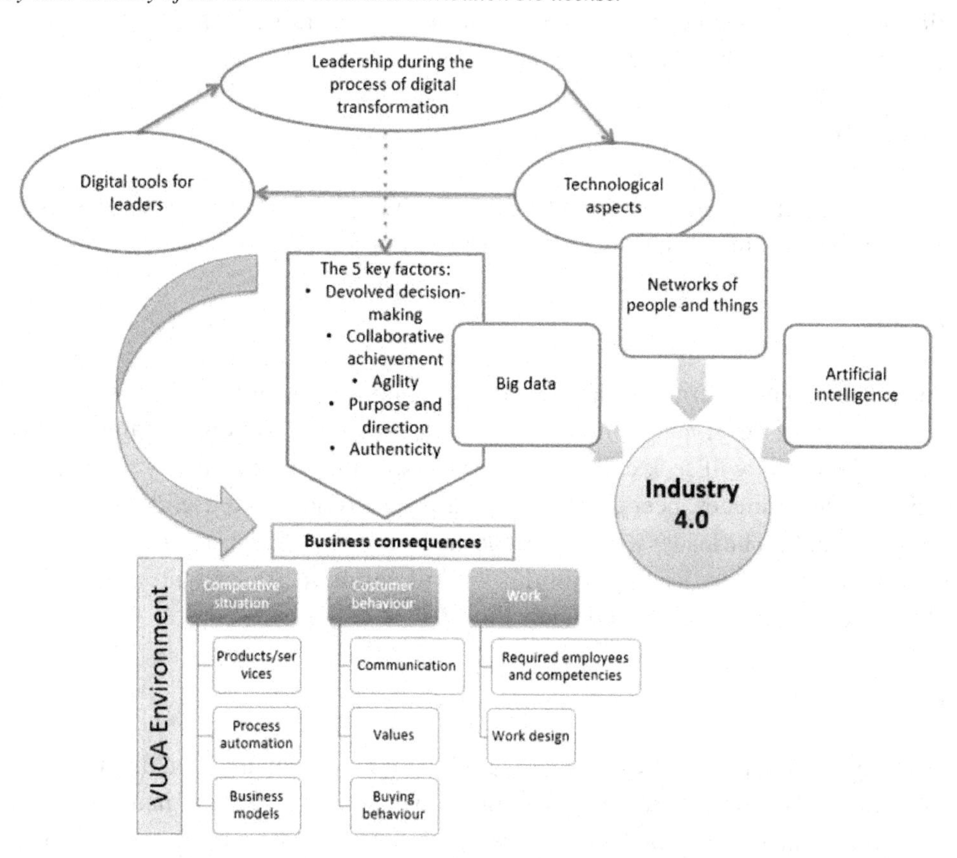

LEADERSHIP OVERVIEW

The term "leadership" refers to a social influence strategy that uses the efforts of others to achieve a goal (Kruse, 2013).

There are key components in the above definition. Instead of authority or power, leadership is generated via social influence. This also implies the *presence* of others, indicating that they are not required to be "direct reports." There is no description of personality traits, attributes, or even a title; there are several styles and methods to effective leadership. It must have an objective, as opposed to influence with no intended result (Kruse, 2013).

Thus, the fulfillment of a goal via the guidance of human assistance is referred to as leadership. A leader skillfully rallies his human partners to achieve certain goals. A strong leader can maintain this relationship day after day, year after year, in a variety of situations (Prentice, 2004).

Many Types of Leadership

Leadership styles, like personality types, are not easily classified. Instead, they may be classed broadly based on characteristics such as autonomy and flexibility. In practice, leadership styles will combine components from each of these categories and adapt to the needs of the organization (Hogarty, 2021). Great leaders have developed throughout history with distinct leadership styles in offering direction, implementing strategies, and encouraging others (International Institute for Management Development, 2022). The focus here is solely to investigate the responsive leadership style.

Leadership Attributes

The top four attributes of a leader are curiosity, empathy, humility, and resilience. These are integral to responsive leadership, and responsive leaders are needed today more than ever (Jenkins-Scott, 2020).

Curiosity

This is the urge to continued learning (Merriam-Webster, n.d.). Curiosity will be essential in driving an organizational transition, developing a new product, embracing innovation, or gaining a deeper grasp of competitors. The leader will seek expertise and understanding from a variety of sources, including outside specialists, subordinates, peers, experts, and trusted advisers, and use ideas with an inquisitive mind (DiGeronimo, n.d.). The leader will inspire constant learning, respect, and regard for a greater grasp of the organization's culture by leading with this kind of curiosity. As a result, a continuous learning culture will boost company creativity and innovation.

Empathy

This is the ability to feel and appreciate other people ("What is empathy," n.d.). Understanding the feelings of others will keep people in touch with their own emotions as the organization addresses difficulties and develops answers. Empathy is seen as a prerequisite for workplace cooperation and fruitful collaboration. People must collaborate with others to flourish in various professional contexts. A leader's own empathy will instill awareness of the impact of the drastic changes on the people around them and within the company. Regardless of leadership style, many CEOs would agree that empathy is a basic and necessary trait of a successful leader. It must be displayed during a crisis or difficult period and will also serve to create trust and confidence in the leader and their judgments (Cherry, 2020).

Humility

Humility fosters curiosity about people as well as physical surroundings ("Humility," n.d.). Humility allows leaders to have the courage to surround themselves with the brightest, most skilled, and potentially even more inventive individuals than themselves. People know that they may learn from others with humility, fully conscious that the individual does not have "all the answers." Employees will typically respond most successfully to a humble boss, which is another important factor in establishing trust and confidence in the leader and the firm (Stith-Flood, 2018).

Resilience

This is the ability to recover and persevere in the face of hardship ("Resilience," n.d.). At some time in their careers, all leaders confront difficulties. Recovering from a setback is one of the most difficult tasks for resilient leaders. An essential aspect for the resilient leader is "not taking it personally," which can be difficult. Recovering quickly from what is viewed as a failure, and maybe what everyone else considers as a failure, might hasten the rehabilitation process. It can give the opportunity for faster organizational growth and continuing development as needed (Kohlrieser et al., n.d.).

Responsive Leadership Style

A responsive leader is defined as using healthy, proven behaviors to react in a complex environment (Smith, 2012). A responsive leader understands the ever-changing nature of business and can adjust rapidly to new problems and circumstances. A responsive leader looks to foresee issues and takes proactive actions to address them. There must also be motivation to learn about people and the environment in which they work. Responsive leaders attempts- to comprehend what is occurring rather than what they want to see. They recognize when they lack the required tools/skills to handle a problem and work to develop those skills, no matter how difficult that growth may be. There must be commitment to ongoing development, both personally and organizationally (Jenkins-Scott, 2020).

Leadership analysis dates to the 1950s, when theorists looked for traits in a good leader. Behavioral and style theorists focused on action and style rather than personal characteristics to describe a good leader. Various leadership hypotheses have been established, and scientists have defined leadership types (Gençer & Samur, 2016).

Leaders face a significant challenge in these fast-paced, ever-changing, sometimes chaotic, and often challenging times: reducing cynicism and creating greater trust and confidence in their position and organization. (Jenkins-Scott, 2020). There have been cases wherein an organization has failed due to a misunderstanding or miscalculation about how inventions and technology should be viewed and used. Mercedes-Benz and Sony suffered losses because of new dynamic systems and a failure to anticipate the consumer market's technical demands. Scholars have labeled these conditions "disruptive technology" and "innovation" (Wadood et al., 2016).

Using safe, validated behaviors to respond in a dynamic environment creates a sensitive and responsive leader. The ever-changing nature of business must be recognized and quick adjustments made to unforeseen problems and new circumstances. This includes prevention: recognizing problems and taking constructive steps to resolve them before they occur (Smith, 2012).

ETHICS OVERVIEW

In its most simple form, ethics is a set of moral principles. They have an impact on the way individuals make decisions and conduct their lives. Ethics is often known as moral philosophy since it is concerned with what is best for people and society.

The phrase comes from the Greek word *ethos*, which can indicate custom, habit, personality, or temperament. Ethics is the process of examining, identifying, and defending beliefs, ideals, and purpose. It is about individuals discovering who they are and remaining loyal to their identity in the face of temptations, obstacles, and uncertainty. It is not always enjoyable and never simple, but if leaders commit to it, they create an atmosphere in which they can make decisions they justify, a career that is genuinely theirs, and a future that others wish to share ("What is ethics," n.d.).

Ethics is centered on well-established norms of right and wrong that dictate what humans should do, generally in terms of rights, duties, societal advantages, fairness, or special qualities (Velasquez et al., 2010). Being ethical is not the same as obeying the law. The law frequently involves ethical norms to which most persons adhere.

Technology and Ethics

Technology ethics is a significant field that is concerned with the interaction between technology and humans in terms of well-being, values, how technical improvements occur, and what societal consequences of technological development may be. Ethics should apply to all fields of technology. Comprehension comes first, followed by identifying and addressing moral dilemmas relating to technology, its design, development, and implementation. Essentially, technology ethics is a set of often unwritten rules that ensures technology is produced and used for the common good and benefit of humanity (Ethical AI Advisory, 2020).

With the introduction of new technology comes an ethical obligation, and it is critical that the leaders who will be taking on these future tasks understand their ethical duties (Lovegrove, 2020). Because new technologies provide more capacity to act, there may be new decisions to make. The best approaches to dealing with the ethics of emerging technology will not be one-size-fits-all. A wide range of potential consequences may need to be investigated, as well as a diverse set of potential dangers. However, most businesses would benefit from assigning these tasks to a single person who can specialize in these kinds of tasks. This is why firms should consider appointing a chief ethics officer — or a chief technology ethics officer — who would be tasked with both the obligation and power to mobilize essential resources (Ammanath, 2021).

Leadership and Ethics

Ethical leadership is guided by a regard for ethical views and ideals, as well as for the dignity and rights of others. It focuses on moral growth and virtuous action. Ethical leadership entails company executives acting appropriately both within and outside of the workplace. Good values are demonstrated by ethical leaders via their words and deeds. According to an online article authored by the Harvard Business Review, ethical leaders will not tolerate dishonesty, even if it benefits their companies (Prentice, 2004). Being an ethical leader is acting with integrity and doing what is right. Ethical leaders serve as role models for the rest of the organization (Kuligowski, 2022).

Ethics and leadership are inextricably linked. While some people believe that they must choose between being profitable and successful and acting morally, this is simply not the case. Ethical individuals and leaders are critical success drivers. A leader motivates and encourages subordinates and followers to work together to achieve a shared objective, whether it be teamwork, an organizational mission, or any endeavor ("Leadership Ethics," n.d.).

Leadership's Influence on Technology Ethics

Technology is the catalyst for transformative change and the force that propels most modern businesses toward profitability. Technology is no longer just inherent to explicitly tech-based conglomerates. Technology fuels everything from banks and financial services to food-delivery services and traditional retail as it facilitates the quick migration online to meet the needs of a world spending more time at home than ever before. Due to the widespread use of technology to facilitate goods and services, executives who design and implement it as well as consumers who interact with it are thinking more and more about ethics. As a crucial fundamental component of their digital transformation, the smartest businesses must discover how to comprehend and embrace the subtleties at the nexus of technology and human values (Frankfurt, 2021).

An organization cannot establish an ethical standard without strong leadership. Being just technologically savvy is insufficient for a leader. The savviness needs to extend to the moral principles used to support and advance an organization's technological foundation. As a company's technological footprint grows, leaders must always prioritize ethics, taking into account the potential ethical implications that a new function, piece of software, or company-wide implementation may have for the company, its customers, and its employees (Frankfurt, 2021).

Modern society faces challenging ethical issues, not the least because of the potential benefits and risks of Information and Communications Technology (ICT). Business leaders need to be able to solve or manage a variety of moral issues that may arise during the course of their professional activity. Failure to properly address moral issues has a significant impact on how a business operates, affecting revenue, public perception, internal organizational procedures, and the wellbeing of the CEO and staff (Kavathatzopoulos, 2017). The increasingly technology-centered world will be a much more comfortable place for everyone if the company promotes trust, celebrates cultural diversity, and is open to learning how to harness new tools and use them ethically.

INTERSECTION POINT: BRIDGING TECHNOLOGY AND LEADERSHIP

There is a growing agreement these days about the changes that technology has brought about in how people live, work, communicate, and arrange activities. Organizations require structures that differ from traditional hierarchical systems as their flexibility grows, their borders blur, and their job content changes. It has, to a large extent, created a demand for transitory systems whose pieces (people and technology) are erected and dismantled in response to changing conditions connected with specific tasks mediated by technology. Thus, cognitive aspects supplied via technologies such as email, knowledge management, and organizational information systems describe the interdependent dynamics between leadership and technology. There are several sources involved, both organizational and inter-organizational (Machado & Brandão, 2019).

From administration to senior management, technology has invaded every level of companies. Whereas technology was once an unknown threat to many, it is now an automatic component of business, pushing change and innovation. Today's leaders make effective use of technology, which was impossible even five years ago. This increased grasp of technology's seemingly infinite benefits is profoundly altering the corporate environment. Technology's continual evolution has left no business unaffected. Technology is increasingly influencing all aspects of life, transcending all boundaries. Technology has systematically enabled employees to work from anywhere, as well as managers to manage from a distance. Leadership is evolving as a result of technological advancements, altering market realities, new company models, and unforeseen external factors (Prichard, 2021). Traditional management techniques are being supplanted by a new leadership philosophy as a new market, and workforce needs emerge. This distinctive management style is responsive, and leaders must grasp how – and when – to employ various leadership styles

(Zorzoli, 2018). Leaders may view the amount of tasks completed by team members in real-time using technology. It also allows them to convey any changes in their functions regardless of where they are. It has resulted in significant changes in how leaders communicate and interact with their staff. Leaders' social media presence also propels them to new heights of popularity. Technology has given rise to a distinct leadership style that enables and encourages employees to grab the most compelling business possibilities by leveraging the greatest technology available (Patel, 2022).

To establish a truly engaged culture and workforce, as well as to harness new ideas and technology advancements, the enablers and drivers should meet along a well-defined route and via constant leadership initiatives. This is accomplished by emphasizing the changes that leadership anticipates with their new vision and how they plan to engage staff with it. The new vision must be presented in such a way that every employee understands their part in attaining the larger goals connected with it, as well as how it will affect their own development. Leaders must bridge the enabler-driver divide by explaining the intellectual and emotional implications of the whole change. Fostering the right vision and culture should go hand in hand with technological improvements (Jupudi, n.d.).

E-Leadership

In the last century, technological advancement has fueled huge growth in organizations and society. This so-called "Fourth Industrial Revolution" necessitates fresh approaches to leaders and leadership. A paradigm change is occurring that prioritizes participation, relationships, inclusiveness, independence, and involvement toward both employees and consumers. To be relevant to employees and consumers, individuals must be responsive. To adapt to this changing world, such an individual need to be the responsive leader (Østergaard, 2018). This has led to the emergence of "e-leadership" and digital or virtual teams (Machado & Brandão, 2019).

E-leadership refers to this new period of the information age, which is marked by rapid technological growth. Businesses routinely cross boundaries to earn a profit in today's global economy. Many of the difficulties caused by the information era require leadership to be resolved (Wang, 2009).

According to Torre and Sarti (2020), e-leadership was considered to be:

- a multidimensional notion distinguished by an individual and organizational focus, as well as the capacity to concentrate on both the big picture and the details;
- indicates, as is customary, a social process in which a leader inspires a follower (thus proving its inherent constitutional character);
- mediated by technologies, the function of which is expanding (and so reflecting the unavoidable change that everyone is experiencing in all working and non-working contexts);
- a process involving the effective and adaptive management of both electronic and traditional methods of communicating (underpinning both the coexistence of the two relationship levels and the need to preserve the human dimension) while keeping in mind that e-leadership is part of the broader domain of the science and practice of leadership and must be examined coherently. (pp. 4-5)

To handle conflict and complexity, a good e-Leader aligns team members with the proper communication technology for the team's objectives, communicates expectations and goals, and links all team members utilizing a cross-cultural viewpoint (Dalhousie University, n.d.). Leading people through digital media necessitates a blend of multiple leadership and management roles. As a result, leaders confronted with new technology must design effective working solutions and management procedures, as well as manage people by building and preserving team identity through supporting the company goal, vision, and values. (Torre & Sarti, 2020).

Below listed, are ten characteristics of a technologically inclined leader, which may also be termed the leader of the future (Apffel et al., 2020):

1. Creating a culture of continuous learning by actively listening to stakeholders, acquiring an "outside-in" viewpoint in order to grasp the customer/client experience.
2. Empathy and understanding in communication; setting a clear aim, providing vision, and encouraging organizational engagement
3. Developing teams with excellent subject expertise and mentoring them through change, with a focus on outcomes, not physical presence.
4. Recruiting from a range of technological domains for superior staffing.
5. Collaborating with other parts of the company, including and bridging the gap between business departments and information technology personnel.
6. Adaptability and the ability to make swift judgments when warranted.
7. Thinking about and recognizing the long-term consequences of technological projects.
8. Acting as a valued partnership, gaining inspiration from challenges to normality.
9. Nurturing a culture of innovation in which employees are not afraid to make mistakes and take risks.
10. Using influence and persuasion rather than employing authoritarian methods.

According to a new study, digital workforces demand digital transformation to better represent their concerns and beliefs, rather than simply increasing corporate capabilities and possibilities. Leaders must pay special attention to how their leadership is seen in the present context and assess if digital tools, strategies, and technology make important stakeholders like workers, consumers, and investors feel more appreciated (Schrage et al., 2021).

With technology driving leadership transformation, the long-term sustainable value will only be created by unifying business and technology strategies to cocreate exponential value for companies (Kark et al., 2019).

Technology Tools for Leaders

In today's fast-paced global market, business executives confront ongoing problems. Every day, critical decisions must be taken. As technology advances, business processes must also change ("4 Tech Tools," 2020 February 4).

Leaders that are open to incorporating cutting-edge technology to empower their organization will be more successful in their roles. With so much responsibility in a fast-paced world, leaders frequently use a variety of technologies to manage their work and personal lives. Services, resources, or gear that assist people in managing their workloads, keeping up with the news, staying connected, or even managing their health are examples of such items ("15 Tools," 2021).

To stay competitive in the industry and feel secure making the best decision for their organization, leaders must use some of the major technology resources are enumerated below:

Workplace Communication Tools

One of the most critical aspects of every successful organization is effective workplace communication. Leaders that are technologically savvy are constantly seeking for new and improved methods to expand and enhance their internal communications efforts in order to reach workers in the most efficient manner (Duncan, 2022). Technology is very crucial in workplace communication. Coworkers can exchange fast messages to one another using apps like Slack and Google Chat. This keeps teams informed and encourages teamwork for collaborative initiatives. These chat applications are great for rapid texting, and email may, of course, be used for longer-form conversation. Both of these platforms enable coworkers to communicate with one another in real time. Furthermore, virtual platforms enable employees to stay engaged even when peers are working flex hours or remotely.

Human Capital Management Tools

The phrase "human capital management" (HCM) can be used to describe both a company strategy and a collection of contemporary IT tools and other technology that are employed to carry out that plan ("What is human capital management," n.d.). Organizations require a wide range of assets to run smoothly and effectively, from a physical office to the staples in the desk drawer. One of the most crucial strands of any business model is talent acquisition; finding workers with the talents that an organization need is critical to its success. Acquiring excellent people is critical for the health and success of the organization. Without it, businesses lose money in the long run by having to train staff for longer periods of time or simply not matching the appropriate individual to the correct position. The greater the talent when acquired, the faster an organization can move an employee from the training stage to the lucrative stage.

Customer Relationship Management Tools

Maintaining a great client relationship is frequently the key to a lucrative and successful company strategy. Businesses that use a Client Relationship Management (CRM) system will benefit from the cutting-edge advantage that comes with the centralized collecting of customer data. CRM software coordinates all business contacts with present and future customers. It is especially important in marketing, sales, and customer service (Hansen, 2022). CRM software saves data including as leads, customer transaction histories, product returns, account numbers, procurement team information, and other details.

Cloud Storage Tools

Cloud storage refers to a cloud computing approach in which data is systematically reposited on the internet by a cloud computing service provider who manages and administers data storage as a service. It is offered on demand with just-in-time capacity and pricing, and it eliminates the need for an organization's data storage infrastructure to be purchased and managed. With "anytime, everywhere" data access, firms gain agility, global scalability, and durability ("Cloud storage," n.d.). With customer and staff management systems migrating to the cloud, it stands to reason that information storage would follow suit. Cloud storage may save an organization both time and money. All organizations require information storage, but the chore of keeping all that data frequently necessitates enormous server rooms, expensive IT employees, complex regulatory compliance, and pricey equipment updates. Business leaders can save these resources by migrating to a cloud service. The burden of keeping data on-site may be eliminated by employing a cloud storage service. The charges are passed on to the service provider, and a company's information is significantly more secure. On-site storage also confines personnel to a physical place. With more remote employees and flexible work hours, storing information in the cloud provides for greater mobility. The ability to view corporate information from any location will also provide organizations greater flexibility when it comes to hiring personnel.

FURTHER DISCUSSION

Businesses are continually altering as the globe transitions to an ultra-digital future. Technology has ushered in this new era by facilitating every shift. This shows that there is a significant link between leadership, business change, and technology. Despite this apparent association, simply technological transformations frequently fail to achieve long-term corporate goals. The causes for this range from overinvesting in technology while disregarding the soft aspects of business transformation to confusing successful technology deployment with successful business transformation without analyzing its durability. This difficulty is exacerbated when executives fail to link technological change with transformation drivers. These forces are intangible, yet they are essential to every change endeavor (Jupudi, n.d.).

The technology department has become an essential component of day-to-day business operations since the onset of the digital era, holding authority at all levels of decision-making. Technology is no longer an afterthought when it comes to simplifying and improving company operations. Technology has taken a prominent position, fundamentally altering the ideas and practices of executive leadership (Page, n.d.). Because of technological advancements, the balance of power is shifting. The leaders' task is to implement emerging technology to increase human imagination, ingenuity, and judgment while gaining new efficiencies. The leaders' ability to solve obstacles and achieve true future success will significantly improve if technology is paired with leadership (Jonk et al., 2018). Recent technology advances will simplify much of what is known as administration, giving leaders more time to lead. It is crucial when digitalization is disrupting at a rapid rate. Company leaders face the task of effectively forecasting changes in the business world and transforming their companies into more flexible organizations. Leaders must be liberated from tedious duties to make an effort toward strategic transformation (Jonk et al., 2018).

Many facets of human life are being influenced by technology. It has revolutionized how people interact, go about their everyday lives, and conduct business. A leader who masters this paradigm shift will gravitate toward the new paradigm while understanding how to blend and dose the old and new worlds efficiently. A responsive leader shapes the contemporary workplace; it is where people desire to come to work (Østergaard, 2016). For those aspiring to the highest management leadership levels, technology can be a handy tool. There is no doubt that technology is the driving force behind today's leadership style, which values transformation, learning, collaboration, and diversity (Arians, 2017). Leaders must influence, encourage, empower, and develop a clear and convincing vision for the workforce and the organization's progress as technology and process complexity increase.

According to research, five characteristics describe the modern leaders; leaders who balance the old-school strategy execution with the new-school responsive leadership:

- **People come first:** A responsive leader prioritizes people and treats them with respect, understanding, opportunity, and decency. It is very important to note that it is not staff or consumers who come first (Hougaard, 2019). It is people, as well as society, community, and rivals. The contemporary leader aspires to build an organization in which everyone feels like they belong. Outstanding leadership is measured by social capital.

- **Purpose, meaning, sense-making, and value creation:** Instead of focusing on products and services, the responsive leader attempts to address issues. This is completely consistent with the trend toward purpose-driven leadership, meaningfulness, and sense-making (Dieffenbacher, 2022). Going to work needs to make sense. Modern corporations attempt to address challenges for consumers and society, as shown by the United Nations' 17 Sustainable Development Goals. Success is assessed not just in money but also in value creation.
- **Continuous innovation and experimentation:** The responsive leader consistently encourages everyone to experiment with procedures, methods, thinking patterns, and technology. This need a culture that avoids alienating innovation and instead supports it in everything from daily improvement to big, major changes and pivoting. When one experiment, the responsive leader acknowledges that things will change (Amirat & Reeps, 2018). It might be difficult for some leaders to let go of old-school predictability and management.
- **An unquenchable desire for results:** The responsive leader is concerned with the human aspects of the business and experimentation and purpose (Østergaard, n.d.). Most responsive leaders prosper in this are laser-focused on delivering outcomes, acquiring customers, increasing market share, and increasing revenue. It requires talent, ability, and work.
- **Everyone has the opportunity to take the initiative:** Finally, the responsive leader understands that the only way to do this is through distributed leadership ("15 Ways," 2020). Both from a business standpoint, to engage the intelligence and cleverness of every employee in shaping the future (clearly, there are people in the organization who are smarter in specific expert fields and have the right project fit for execution of this particular task), and from a "people first" perspective, to ensure that everyone feels acknowledged, seen, in development, challenged, and safe.

CONCLUSION

Globalization (connectivity and integration), mass immigration, and multiculturalism have all prompted a rethinking of leadership approaches in a digital world that is constant, diverse, constructive, and nuanced due to digital technology (Bowen, 2021). The current technological change rate results if sophistication that outpaces leaders' capacity to make sound decisions. The ability to make fast decisions in the face of ambiguity and confusion is a quality that leaders must now possess. The lines used to divide sectors and companies are blurring in this age of globalization and technological advancement. Competition in real-time is a fact. Long strategy horizons are no longer available to leaders (Stephenson, 2011).

Technology has had a significant impact on leadership. Leaders of technology in an organization are developers (specify and create systems), commercializers (determine how to earn money), and stewards (get systems developed and used). The role of organizational leaders with technological expertise is evolving (The Strategy Institute, 2020). They recognize the technology life cycle, understand consumer personas, and serve as a vital connection between technology and strategy. Technological advancements are evaluated, managed, and forecasted, and technology transfers are facilitated; they act as a bridge between people and procedures. They are willing to work across functional lines to find innovative ways to employ technology and data, and consider technology to be a vital component of business at all levels, not simply the products research and development activities of the organization.

Although there are several challenges and opportunities, the biggest challenge and opportunity for leadership is to ensure that decisions are continually guided by their organization's mission and values when using technology. A good organization relies on its most vital and robust component: leadership. Organizations will work effectively in both positive and negative markets and market disruptions if their leaders have the requisite resources and skills to empower and influence their followers. The twenty-first century has proved to be a time of unprecedented technological development. As a result, businesses must integrate technology into their long-term leadership strategies. Any organization's present and prospective leaders must be prepared to face the threats posed by technical change (Draper, 2019).

Clearly, technology is a primary, driving factor in today's new leadership style. Leadership must value change, learning, communication, and diversity. The most effective leaders on a global scale realize how much technology can help them manage processes and people—assisting them in forming teams and keeping track of work across all channels and in any place across the world. There is no ready-made solution for leaders to cope with technological ethical concerns. There is a need, however, for leaders' responses to unique situations and relationships to follow a societal norm for ethical behavior and action for maintaining strong purpose, principles, and company values.

REFERENCES

15 tools and resources top tech leaders use every day. (2021, May 28). Forbes. https://www.forbes.com/sites/forbestechcouncil/2021/05/28/15-tools-and-resources-top-tech-leaders-use-every-day

15 ways leaders can encourage employees to take initiative. (2020, June 11). Small Biz Trends. https://smallbiztrends.com/2020/06/15-wasy-leaders-can-encourage-employees-to-take-initiative.html

4 tech tools every business leader should know about. (2020, February 4). Tanveer Naseer Leadership. https://tanveernaseer.com/4-tech-tools-critical-for-organizational-success/

Amirat, C., & Reeps, R. (2018, June-July). Continuous innovation through experimentation. *2018 IEEE Technology and Engineering Management Conference (TEMSCON)*. doi:10.1109/TEMSCON.2018.8488399

Ammanath, B. (2021, November 9). Thinking through the ethics of new tech… before there's a problem. *Harvard Business Review*. https://hbr.org/2021/11/thinking-through-the-ethics-of-new-techbefore-theres-a-problem

Apffel, C., Bernad, P., Gollenia, L. A., Lupo, C., Mijnarends, H., & Westland, J. (2020, December). *The future of technology leadership*. Spencer Stuart. https://www.spencerstuart.com/research-and-insight/the-future-of-technology-leadership

Arians, H. (2017, September 10). *The impact of technology on leadership*. The People Development Magazine. https://peopledevelopmentmagazine.com/2017/09/10/technology-leadership

Assistive Technology Industry Association (ATIA). (n.d.). *What is AT?* https://www.atia.org/home/at-resources/what-is-at

Bowen, G. (2021, January). Digital leadership, ethics, and challenges. In H. Jahankhani, L. M. O'Dell, G. Bowen, D. Hagan, & A. Jamal (Eds.), Strategy (pp. 23–29). Leadership, and AI in the Cyber Ecosystem. https://doi.org/10.1016/B978-0-12-821442-8.00013-6

Cherry, K. (2020, July 21). *What is empathy?* Very Well Mind. https://www.verywellmind.com/what-is-empathy-2795562

Cloud Storage. (n.d.). *Amazon Web Services*. https://aws.amazon.com/what-is-cloud-storage

Cortellazzo, L., Bruni, E., & Zampieri, R. (2019, August 29). The role of leadership in a digitalized world: A review. *Frontiers in Psychology, 10*(1938). doi:10.3389/fpsyg.2019.01938

Craig, D. (2017, April 11). *How technology is making leaders more responsive and responsible*. Thomson Reuters blogs. https://blogs.thomsonreuters.com/answerson/disruptive-leadership-technology-making-leaders-more-responsive-responsible

Crook, A. (n.d.). *What are the different types of technology?* https://digitalizetrends.com/types-of-technology

Dalhousie University. (n.d.). *What is e-Leadership?* https://www.dal.ca/sites/celnet/about/eleadership.html

David, Y., Judd, T. M., & Zambuto, R. P. (2020). Introduction to medical technology management practices. In Clinical Engineering Handbook (2nd ed., pp. 166–177). Academic Press. https://doi.org/10.1016/B978-0-12-813467-2.00028-6

Deiser, R., & Newton, S. (2015, January 15). *Social technology and the changing context of leadership*. The Wharton School. https://leadershipcenter.wharton.upenn.edu/research/social-technology-changing-context-leadership

Della Corte, V., Del Gaudio, G., & Sepe, F. (2019, November 20). Leadership in the digital realm: What are the main challenges? In M. Franco (Ed.), *A New Leadership Style for the 21st Century*. IntechOpen. doi:10.5772/intechopen.89856

Dieffenbacher, S. F. (2022, March 4). *Value creation definition, model, principles, importance & steps*. Digital Leadership. https://digitalleadership.com/blog/value-creation

DiGeronimo, J. J. (n.d.). *How to be a curious leader*. https://jjdigeronimo.com/how-to-be-a-curious-leader

Draper, A. (2019, January 19). *3 ways technology has affected today's leaders*. https://www.business-2community.com/leadership/3-ways-technology-has-affected-todays-leaders-02160570

Duff, C. (2021, November 9). *Everything you need to know about education technology "EdTech."* Owl Labs. https://resources.owllabs.com/blog/education-technology

Duncan, C. (2022, January 18). Communication tools in the workplace. https://www.alert-software.com/blog/internal-communication-tools

Ethical A. I. Advisory. (2020, June 5). *What are technology ethics?* https://aiadvisory.ai/2020/06/05/what-are-technology-ethics/

Frankfurt, T. (2021, December 13). Why all companies must explore the role of ethics in technology. *Forbes*. https://www.forbes.com/sites/forbestechcouncil/2021/12/13/why-all-companies-must-explore-the-role-of-ethics-in-technology

Gençer, M. S., & Samur, Y. (2016, August 19). Leadership styles and technology: Leadership competency level of educational leaders. *Procedia: Social and Behavioral Sciences*, *229*, 226–233. https://doi.org/10.1016/j.sbspro.2016.07.132

Hansen, L. (2022, July 20). *How CRM and ERP integration can benefit your business*. CIO Insight. https://www.cioinsight.com/enterprise-apps/crm-erp-integration

Hayden, C. (2018, January 11). *Entertainment technologies*. International Studies Association and Oxford University Press. doi:10.1093/acrefore/9780190846626.013.386

Hogarty, S. (2021, November 1). *Five common leadership styles, and how to find your own*. We Work Ideas. https://www.wework.com/ideas/professional-development/management-leadership/five-common-leadership-styles-and-how-to-find-your-own

Hougaard, R. (2019, March 5). The power of putting people first. *Forbes*. https://www.forbes.com/sites/rasmushougaard/2019/03/05/the-power-of-putting-people-first

Hughes, D. J., Lee, A., Tian, A. W., Newman, A., & Legood, A. (2018, October). Leadership, creativity, and innovation: A critical review and practical recommendations. *The Leadership Quarterly*, *29*(5), 549–569. https://doi.org/10.1016/j.leaqua.2018.03.001

Humility. (n.d.). https://www.dictionary.com/browse/humility

International Institute for Management Development. (2022, August). *The 5 leadership styles you can use*. https://www.imd.org/imd-reflections/reflection-page/leadership-styles

Jenkins-Scott, J. (2020, January 29). *Responsive leadership: Needed now more than ever*. Leadership Now. https://www.leadershipnow.com/leadingblog/2020/01/responsive_leadership_needed_n.html

Jonk, G., Anscombe, J., & Aurik, J. C. (2018, March 29). *How technology can transform leadership – for the good of employees*. World Economic Forum. https://www.weforum.org/agenda/2018/03/how-technology-can-transform-business-performance-for-human-good

Jupudi, S. (n.d.). *Technology is the enabler, not the driver, for business transformation*. Dallas Business Journal. https://www.bizjournals.com/dallas/news/2021/01/05/technology-is-the-enabler-not-the-driver-for-business-transformation.html

Kanade, V. (2022, March 11). *What Is super Artificial Intelligence (AI)? Definition, threats, and trends*. Spice Works. https://www.toolbox.com/tech/artificial-intelligence/articles/super-artificial-intelligence

Kark, K., Briggs, B., & Tweardy, J. (2019, May 13). *Reimagining the role of technology*. Deloitte. https://www2.deloitte.com/us/en/insights/focus/cio-insider-business-insights/reimagining-role-of-technology-business-strategies.html

Kavathatzopoulos, I. (2017). Ethical leadership in business: The significance of Information and Communication Technology. In *Keynote speech at Japan Society for Information and Management 75th Annual Conference* (pp. 1-4). Tokyo: Japan Society for Information and Management.

Kluz, A., & Nowak, B. E. (2016, December 23). *The impact of technology on leadership*. The Oxford University Politics Blog. https://blog.politics.ox.ac.uk/impact-pressures-technology-leadership

Kohlrieser, G., Orlick, A. L., Perrinjaquet, M., & Rossi, R. L. (n.d.). *Resilient leadership: Navigating the pressures of modern working life*. https://www.imd.org/research-knowledge/articles/resilient-leadership-navigating-the-pressures-of-modern-working-life

Koss, E. (2022, August 8). *Education technology: What is Edtech? A Guide*. Built In. https://builtin.com/edtech

Kruse, K. (2013, April 9). What is leadership? *Forbes*. https://www.forbes.com/sites/kevinkruse/2013/04/09/what-is-leadership/?sh=52a9c3345b90

Kuligowski, K. (2022, June 29). How to be an ethical leader: 7 tips for success. *Business News Daily*. https://www.businessnewsdaily.com/5537-how-to-be-ethical-leader.html

Leadership ethics - Traits of an ethical leader. (n.d.). *Management Study Guide*. https://www.managementstudyguide.com/leadership-ethics.htm

Lovegrove, M. (2020) *Why we need to talk about ethics in technology*. Hello World. https://helloworld.raspberrypi.org/articles/HW06-why-we-need-to-talk-about-ethics-in-technology

Machado, A. M., & Brandão, C. (2019). Leadership and technology: Concepts and questions. In *New Knowledge in Information Systems and Technologies. World CIST'19: Advances in Intelligent Systems and Computing* (pp. 764-773). Springer. doi:10.1007/978-3-030-16184-2_73

Merriam-Webster. (n.d.). *Curiosity*. https://www.merriam-webster.com/dictionary/curiosity

Miller, B. E. (2021, July 23). *Leadership skills in the age of technology*. Inc. https://www.inc.com/inc-masters/leadership-skills-in-the-age-of-technology.html

National Institute of Food and Agriculture. (n.d.). *Agriculture technology*. U.S. Department of Agriculture. https://www.nifa.usda.gov/topics/agriculture-technology

Operational technology (OT) – definitions and differences with IT. (n.d.). *i-SCOOP*. https://www.i-scoop.eu/industry-4-0/operational-technology-ot

Østergaard, E. K. (2016, March 27). *Responsive leadership - A guide*. Slide Share. https://www.slideshare.net/ErikKorsvikstergaard/responsive-leadership-a-guide

Østergaard, E. K. (2018, February 21). *The responsive leader: How to be a fantastic leader in a constantly changing world*. LID Publishing.

Østergaard, E. K. (n.d.). *What is the place for modern, responsive leadership in 2020?* https://www.vunela.com/what-is-the-place-for-modern-responsive-leadership-in-2020

Page, M. (n.d.). *The impact of technology on executive leadership*. Michael Page. https://www.michaelpage.com/advice/management-advice/development-and-retention/impact-technology-executive-leadership

Patel, A. (2022). *How technology can be used to empower leadership*. About Leaders. https://aboutleaders.com/technology-empower-leadership

Prentice, W. (2004, January). Understanding leadership. *Harvard Business Review*. https://hbr.org/2004/01/understanding-leadership

Prichard, S. (2021, April 12). *Master the four fields of leadership*. Skip Prichard. https://www.skipprichard.com/master-the-four-fields-of-leadership/

Resilience. (n.d.). *American Psychological Association*. https://www.apa.org/topics/resilience

Schrage, M., Pring, B., Kiron, D., & Dickerson, D. (2021, January 26). Leadership's digital transformation: Leading purposefully in an era of context collapse. *MIT Sloan Management Review*. https://sloanreview.mit.edu/projects/leaderships-digital-transformation

Smith, D. R. (2012, May 30). *What does it mean to be a responsive leader*. Smart Business. https://sbnonline.com/article/what-does-it-mean-to-be-a-responsive-leader

Stephenson, C. (2011, July/August). How leadership has changed. *IVEY Business Journal*. https://iveybusinessjournal.com/publication/how-leadership-has-changed

Stith-Flood, C. (2018, May/June). It's not hard to be humble: The role of humility in leadership. *Family Practice Management*, *25*(3), 25–27. https://www.aafp.org/fpm/2018/0500/p25.html

The Strategy Institute. (2020, November 20). *How technology leadership can accelerate disruption in your business strategy*. https://www.thestrategyinstitute.org/insights/how-technology-leadership-can-accelerate-disruption-in-your-business-strategy

Torre, T., & Sarti, D. (2020, November 11). The "way" toward e-leadership: Some evidence from the field. *Frontiers in Psychology*, *11*, 1–14. https://doi.org/10.3389/fpsyg.2020.554253

Velasquez, M., Andre, C., Shanks, T. J. S., & Meyer, M. J. (2010, January 1). *What is ethics?* Markkula Center for Applied Ethics at Santa Clara University. https://www.scu.edu/ethics/ethics-resources/ethical-decision-making/what-is-ethics

Wadood, S., Gharleghi, B., & Samadia, B. (2016). Influence of change in management in technological enterprises. *Procedia Economics and Finance*, *37*, 129–136. https://doi.org/10.1016/S2212-5671(16)30103-4

Wang, V. (2009). Traditional leadership in light of E-HRMS. In T. Torres-Coronas & M. Arias-Oliva (Eds.), Encyclopedia of Human Resources Information Systems: Challenges in e-HRM (pp. 849–854). IGI-Global. https://doi.org/10.4018/978-1-59904-883-3.ch125

West, D. M., & Allen, J. R. (2018, April 24). *How artificial intelligence is transforming the world*. Brookings. https://www.brookings.edu/research/how-artificial-intelligence-is-transforming-the-world

What is blockchain technology? How does blockchain work? (2022, August 9). *Simplilearn*. https://www.simplilearn.com/tutorials/blockchain-tutorial/blockchain-technology

What is empathy? (n.d.). *Greater Good Magazine, Science Center.* https://greatergood.berkeley.edu/topic/empathy/definition

What is ethics? (n.d.). *The Ethics Centre.* https://ethics.org.au/about/what-is-ethics

What is human capital management (HCM)? (n.d.). *Oracle.* https://www.oracle.com/human-capital-management/what-is-hcm

What is information technology? (n.d.). *CompTIA.* https://www.comptia.org/content/articles/what-is-information-technology

What is medical technology? (n.d.). *MedTech Europe.* https://www.medtecheurope.org/about-the-industry/what-is-medical-technology

What is technology? (2019, September 4). *4HL Net.* https://4hlnet.extension.org/what-is-technology

Wigmore, I. (2018, February). *Artificial superintelligence (ASI).* https://www.techtarget.com/searchenterpriseai/definition/artificial-superintelligence-ASI

Zorzoli, E. (2018, October 19). *Why a responsive management style is the future of leadership.* https://www.wearebeem.com/why-a-responsive-management-style-is-the-future-of-leadership/

ADDITIONAL READING

Diaz, J. B. B., & Young, S. F. (2022). The future is here: A benchmark study of digitally enabled assessment and development tools. *Consulting Psychology Journal, 74*(1), 40–79. https://doi.org/ doi:10.1037/cpb0000201

Ellis, B. (2018, May 15). *Tech tools for leaders.* TCEA. https://blog.tcea.org/tech-tools-for-leaders

Fitzhugh-Craig, M. (2021). Simple tech tools to help your business soar in 2021. *Information Today, 38*(1), 34–35.

Nica, I. (2016, July 1). *10 Essential productivity tools for tech-savvy leaders.* Hubgets. https://www.hubgets.com/blog/productivity-tools-for-leaders

Chapter 3
An Examination of RFID Ethical Issues Supports the Need for Improved Business and Legal Strategies

Ellen Marie Raineri
Pennsylvania State University, USA

Lori S. Elias Reno
Pennsylvania State University, USA

Pauline S. Cho
Pennsylvania State University, USA

Gina M. Dignazio
Pennsylvania State University, USA

ABSTRACT

RFID (Radio Frequency Identification) is an important and frequently utilized technology in our modern society. It gives its users the ability to capture, store, and access vast amounts of data with ease and efficiency. These abilities make RFID both an asset and a potential hazard. If used in the wrong hands, RFID could be appropriated for nefarious reasons such as data theft and invasion of privacy. Although RFID technology has been around for over 70 years, few laws directly address RFID or define illegal uses of it. Some of the ways that RFID is used in the following industries are thus explored: marketing, transportation, travel, shopping, supply chain management, agriculture, and hospitals. Some of the benefits, drawbacks, ethical concerns, and legal implications of using RFID technology are included, as well as recommendations for addressing ethical concerns.

DOI: 10.4018/978-1-6684-5892-1.ch003

INTRODUCTION

RFID (Radio Frequency Identification) was utilized as a method for aircraft radar as early as WWII (Violino & Roberti, n.d.). Each subsequent decade produced new and innovative uses for RFID. By the late 1960s, electronic article surveillance (EAS) had been developed to prevent merchandise theft (Landt, 2005). By the 1980s, RFID had been incorporated into applications as diverse as animal tracking, personnel access, and automated toll roads (Landt, 2005). Today, RFID is found in everyday items such as credit cards, metro cards, passports, and car keys. Hospitals, hotels, government buildings, warehouses, amusement parks, concert venues, sporting events, libraries, museums, and airports all rely on RFID technology. Much of its appeal comes from the fact that it is small, holds more information than a barcode, and can be powered remotely. However, each novel amenity that technology affords creates new points of access for malicious activity or hacking. In 2003, the ACLU sued a statement arguing that if RFID is improperly implemented, it has the potential to compromise consumer privacy and violate civil liberties (ACLU, 2003). Ethical concerns regarding RFID tracking and data collection are still highly debated today. It is important to examine the uses, benefits, drawbacks, and legal implications of using RFID technology in various industries.

RFID BACKGROUND INFORMATION

How RFID Works

RFID is a technology that uses electromagnetic fields to automatically capture the digital data encoded in RFID tags. It is a simple system that consists of only three components that identifies, collects, and receives data. The devices that identify and collect data act as a RFID reader, which can be either fixed or mobile. The network-connected reader uses radio waves to transmit signals that activate the tag. Once activated, the tag sends a wave back to the antenna, where it is translated into data (Amsler, 2021). The transponder, which receives data, can have a longer read range based on its power source and type of tag. Two main types of RFID tags are active and passive. The active RFID tag has its own power source, and the passive RFID tag receives its power from the reading antenna, whose electromagnetic wave induces a current in the RFID tag's antenna (Amsler, 2021). The data that is received by the tags is transferred to the main computer system, where the data is stored in the database to be analyzed.

Businesses can use the data that is collected to increase efficiencies and create cost saving solutions for the organization. Various features of RFIDs -- read range, frequency, and interference -- are tailored to each organization's processes and needs. RFID technology extends to many industries to perform tasks such as supply chain management, asset tracking, personnel tracking, controlling access to restricted areas, ID badging, supply chain management and counterfeit prevention ("What is RFID," n.d.). Large companies use it to check in and track their extensive employee bases for safety reasons during travel and in the event of catastrophes at any of their many sites.

Table 1. U.S. Federal Laws relating to RFID

Federal Law	Content
Privacy Act of 1974	Pertains to PII maintained by government
Section 208 of the E-Government Act of 2002	Pertains to PII maintained by government as well as proper privacy education for employees
Section 522 of the Consolidated Appropriations Act of 2005	Privacy mandates for the Department of Transportation and Treasury
Federal Information Security Management Act (FISMA)	Government framework pertaining to tools, products, and controls
Office of Management and Budget (OMB) memoranda on the implementation of privacy requirements	OMB suggestions for implementing privacy laws
Health Insurance Portability and Accountability Act (HIPAA) of 1996	Rules to protect personal health information (PHI)

RFID Legal Perspectives and Standards

Some of the existing privacy federal laws include aspects of RFID pertaining to privacy of individuals' personal identifiable information (PII) regarding accessing, using, and storing such information. Table 1 summarized some of the privacy acts that impact RFID (Karygiannis et al., 2007).

Despite the ethical concerns about RFID technology, laws are not as prevalent. Within the U.S., some laws have been passed through the years. In 2007, Michigan cattle owners passed a law requiring cattle to have RFID tags. The outcome is that owners are able to track their cattle; additionally, if a foodborne illness occurs, the source of the disease can be traced to its herd (Walker, 2007). In New Hampshire, vendors are required to include a warning label for any items that contain RFID tags, which is an amendment to the Consumer Protection Act ("NH bill regulates RFID," 2006). In Washington, a law prohibits others from spying on individuals to obtain PII (personally identifiable information) (Gaudlin, 2008). California, North Dakota, Missouri, Oklahoma, and Wisconsin have laws that prohibit individuals from being required or compelled to have RFID implants – most often used for tracking individuals (Fowler, 2019; Rein, 2007).

Outside of the U.S., countries have varying RFID laws. Canada does not have specific RFID laws, but the privacy laws within Canada's Personal Information Protection and Electronic Documents Act (PIPEDA) within the federal private sector or the Privacy Act within the federal public sector may be relevant when personal information is collected and used (Office of Privacy Commissioner of Canada, n.d.). China has enacted a law requiring all cars to have RFID tags such that the government can track and monitor the travel and location of citizens (Fowler, 2019). Following this initiative of China, Egypt is planning the roll-out of RFID tracking of 10,000,000 vehicles. In addition to tracking vehicles, the tags convey information on drivers such as speeding or talking on the phone while driving ("The new Egyptian traffic," 2021; Moody, 2019). Sweden does not have national RFID laws. Furthermore, thousands of Swedes have opted to have RFID chips inserted under their skin, citing the benefits of access to door entrances, phones, and transportation (Savage, 2018). The European Union requires that products containing RFID must have a label to alert customers as well as a communication to customers about what data is collected, how the data will be used, and how the RFID tags can be removed (Sullivan, 2005).

Various associations have developed standards that foster interoperability, consistency, and reliability. A summary of such associations is listed in table 2.

Table 2. Association standards relating to RFID

Standard	Description
GS1	Barcodes
International Electrotechnical Commission (IEC)	Electrical and electronic technology
International Standards Organization (ISO)	167 national standard bodies
Joint Technical Committee (JTC 1)	ISO and IEC
Federal Communication Commission (FCC)	Interstate and international communications
Association of American Railroads (AAR)	Freight railroads
Automotive Industry Standards Group (AIAG)	Automotive supply chain
American Trucking Associations (ATA)	Trucking trade association
International Air Transport Association (IATA)	Airline trade association

Sources: (American Trucking Associations, n.d.; Association of American Railroads, n.d.; Automotive Industry Standards Group, n.d.; Federal Communication Commission, n.d.; GS1 US, n.d.; Impinji Inc, n.d.; International Air Transport Association, n.d.; International Electrotechnical Commission, n.d.; International Standards Organization, n.d.; Joint Technical Committee, n.d.).

DISCUSSION OF SELECT RFID BUSINESS SECTORS AND FUNCTIONS

RFID technology is used in many different business sectors including but not limited to healthcare facilities, agriculture, retailing, entertainment, manufacturing, non-profit, government, etc. Additionally, RFID can also be found in broader business functions that touch or include several sectors like marketing, supply chain management, and finance. The following highlights the use of RFID technology and its ethical implications in marketing, supply chain management, agriculture, and healthcare.

Marketing

In viewing the aforementioned uses, RFID plays a significant role in business. Remembering that marketing decisions include those made for the core functions of marketing–promotion, product, price, and place (supply and distribution) --the examples cover the gamut of these decisions. It is thus helpful to examine examples based on the four core functions of business transactions across a variety of industries.

Promotion

Promotion includes both traditional and digital forms of advertising, PR, sales promotion, direct marketing, and personal selling, and RFID technology is used in all of them. For example, when customers pay with RFID debit, credit, or rewards cards, purchasing behavior provides direction in targeted promotional activities and allows for merchandise layouts to be rearranged in meaningful ways (Mack, 2017). This is a form of sales promotion called merchandising. Another example (and quite early use) of RFID as a sales promotion is the “Heineken Moments” event at the 2012 Shanghai ATP Tennis Masters tournament. Visitors registered their Weibo (a social media platform in China) accounts with their RFID bracelets, and as they journeyed through displays, the technology uploaded photos via their bracelets to their personal Weibo accounts. The RFID bracelets also entitled them to free beers at the event (Li, 2012).

The possibilities of interactive advertising with RFID seem endless. One early example is Budweiser Brazil's "Buddy Cup" for two RFID encoded cups to clink together in "cheers" and subsequently add the holders to each other's Facebook friends list (Thrasher, 2013). It simply requires linking one's cup to their Facebook account using a QR code on the cup. When new drinking buddies "clink" their activated cups, they instantly become Facebook friends. More recently, companies create even more interactivity by adding RFID chips to their printed materials. This keeps ads, business cards, and direct mail promotions short and eye-catching, but by simply raising their phones to the promotion, customers can have immediate and dynamic access to videos, e-commerce sites, web or social media pages, and even coupons and discounts (Kalany, 2018).

Product

When it comes to product decisions, marketers provide input regarding design and style, features and attributes, and quality. To improve the ease and quality of border crossings at key US cities, four states already allow for an Enhanced (with RFID technology) Driver's License and four other states intend to implement them. The RFID links your license to Homeland Security and takes the place of your passport for entry at 95% of the 39 most used crossing locations (U.S. Department of Homeland Security, n.d.).

Regarding features, design, and style, brands like Keurig and McKinsey and Co. with Kendra Scott use RFID technology to provide customization according to consumers' personal preferences. Keurig's MyBrew once used RFID technology to read the pod that is inserted in the brewer and pull up the proper temperature and other brew settings (Thrasher, 2013). Keurig has since updated their technology, now called BrewID, to recognize the roast of the K-Cup® pod and then change the brew settings to make each cup to the distinct taste of the consumer ("Keurig debuts," n.d.). Similarly, McKinsey and Co. created an interactive Kendra Scott jewelry color bar in the Mall of America at which consumers can pick up RFID-activated colored stones to view various configurations of the stones and eventually create a personalized bracelet (Taylor, 2019). In this case, the Color Bar satisfies both product personalization and shopping experience needs of consumers.

Price

Most consumers associate price with cost, but marketers look at price from a broader perspective, including how consumers pay. Thus, many of the price-related uses of RFID technology take care to make the paying experience secure and easy. A common example is RFID technology in credit and debit cards. This technology is widely recognized for its identity-theft benefit, but it is also used to ease the payment process in traffic toll booths and key-fob gasoline purchase payments (Laczniak & Murphy, 2006). Similarly, many commuters who use public forms of transportation like the subway, train, buses, or public bike rentals use "smartcards" with RFID chips that no longer require cash or even card-swiping to use this form of transportation (Schino, 2019).

RFID checkout technology is on the cusp of revolutionizing the retail checkout experience. Barcode walkout technologies exist in stores like Sam's Club, but it still requires that consumers scan each barcode on each item before leaving the store. RFID checkout technology checks everything out at once with one reader, and that means a quicker and less stressful checkout experience for consumers ("Is Deploying," 2020). An even more sophisticated Amazon/Whole Foods "Just Walk Out" technology combines

RFID with Artificial Intelligence (Palmer, 2021) and thus takes the whole checkout experience to new and easier levels.

Place

Marketing decisions regarding place include supply and distribution of products. Distribution includes how and where consumers access products and some crossover exists with how RFID technology is used in Supply Chain Management, which is examined in its own section below. Therefore, this Marketing section addresses only the function of Place that applies to the shopping experience, and RFID technology is used extensively during in-store shopping experiences. Chanel, for example, creates smart fitting rooms using RFID tags, enabling customers to view the latest style trends, sizes, and colors. It also makes personal outfit recommendations for each consumer (Adhi, et. al. 2021). Another boutique retail store, Burberry, uses product RFID tags to show the craftsmanship of each piece brought into the fitting room (Thrasher, 2013). Yet another clothing store in Singapore attaches RFIDs to each article of clothing. When taken into a dressing room, the RFID tag on each piece of clothing initiates music associated with the style of clothing. Additionally, consumers receive a text message offer for a free download of the song (Thrasher, 2013).

Another engaging shopping experience is a partnership between a manufacturer and one of its retail distribution channels. Hellman's piloted an interactive shopping experience in São Paulo, Brazil, where RFID tags were selectively positioned around the store and customers used shopping carts armed with RFID tags to access videos and recipes when near products that could be used with Hellman's mayonnaise (Thrasher, 2013).

Supply Chain Management

The flow of goods or services and the movement from raw materials to finished goods can benefit from the RFID technology. With automatic tracking and uploading data through RFID tag points, it is a low-cost and efficient way to keep track of product movement and operations. Companies such as Amazon are integrating RFID to improve their fulfillment system to further improve supply chain efficiencies (Roberti, 2015). RFID technology can help companies speed up deliveries, lower costs, and improve the identification of products. RFID tags are the data foundation for large-scale retail and wholesale product movements. Every step of the supply chain can be improved and optimized to efficiently track the process.

Production

On the production and raw materials stage of the supply chain, the movement of goods can be tracked with tags to efficiently track production steps. This leads to an improvement in the quality of the supply chains with aims to enable visibility of supply chains and reduce operations costs of supply chains (Matičević et al., 2011). The tags aid manufacturers in product aggregation, tracking the movement of goods, and inventory management to semi-finished and finished goods. Many companies are using RFID to gain a competitive advantage by improving their operations. Companies are solving production material demand problems in the manufacturing environment (Poon et al., 2011). The technology can also contribute to better quality control in production by contributing through greater effectiveness in various processes such as asset tracking, flow of materials, control of production, and follow-up long supply chain ("RFID for Quality Control," n.d.).

Warehouse

Companies that operate large warehouses have implemented RFID along with robotics to efficiently move goods. By implementing a RFID warehouse tracking system, items can be scanned and cataloged from anywhere, even when they are hidden behind boxes or pallets (Prologis, n.d.). The tags increase stock tracking accuracy, real-time inventory visibility, and on-shelf availability ("RFID in Supply Chain," n.d.). Within the warehouse, the technology is able to monitor vehicles and transporting devices to make sure collision is avoided. By automating these processes, RFID was able to decrease human error and streamline warehouse operations.

Logistics and Shipping

Along with warehouse storage, RFID can be used alongside the warehouse tracking system to get real-time data of shipments. By allowing automation, companies are able to run shipments continuously through the calendar year with less error. The tags allow for communication with the shipment center's software and automatically document their arrival and exit from ports. It automatically tracks goods being received and loaded on transportation vehicles. This creates an efficient cargo or shipment tracking system that eliminates human error.

Distribution and Retail

In the distribution of products, the use of RFID extends from logistics and shipping to automatically pre-allocate goods to send to designated locations. Oftentimes, the goods are sent to fulfillment centers where the packages are tracked to send to retail or end-users. Once the package reaches retail, RFID is used in other ways to improve operations and reduce costs. RFID has the power to unlock up to 5 percent top-line growth from better stockout management and shrinkage reduction as well as up to 15 percent reduction in inventory-related labor hours (Adhi et al., 2021). In a store, it can be used to find items, automatically reorder products to set amounts, and reduce manual inventory counting. On the end-user side, stakeholders and customers are able to use RFID tags to authenticate products and check the origin of products from the beginning stage at manufacturing ("RFID in Supply Chain," n.d.).

RFID has opened doors for businesses to explore and create new and innovative checkout processes and overall supply chain. Amazon has introduced a retail marketplace, Amazon Go, that allows customers to automatically check out without counters or clerks. This invisible cashier is possible due to the RFID tags that were placed on every product in the grocery stores. With their "Just Walk Out" cashierless technology that combines AI and RFID, automation has become possible at the very end of the supply chain process (Hedgepeth, 2021). Amazon has partnered with Auburn RFID lab to further improve their overall supply chain processes by connecting to their existing software and robotics infrastructure (Roberti, 2015). Further improvements in the supply chain and warehouse management are currently being developed by MIT researchers that are introducing airborne drones with smarter RFID systems in large warehouses.

Agriculture

Outbreaks of foodborne zoonotic diseases (infectious diseases that are transmitted between animals and humans) such as bovine spongiform encephalopathy (mad cow disease) and mycobacterium bovis (bovine tuberculosis) have created an increased demand for food traceability (Yu et al., 2018). RFID tracking is an efficient way to trace animals from birth to market, providing valuable information for swiftly containing any outbreaks. “Lifetime traceability” has become the standard for Japan, Korea, and Europe. Australia’s National Livestock Identification System (NLIS) ranks as one of the largest and most sophisticated databases (Hossain et al., 2017). The United States Department of Agriculture (USDA) encourages the use of RFID tracking but has yet to adopt nationwide mandates despite its profitability in international markets.

Tagging

method of tagging the animal differs depending on the species. Commonly used are ear tags which can be read with a handheld scanner or a fixed scanner that is placed in a strategic location, such as a gateway or feeding trough. Traditional farming requires tags to be read manually, which is time consuming and prone to error (Gough, 2021).

Feeding

RFID can monitor precision feeding of animals, therefore, reducing the risk of overfeeding or underfeeding, resulting in healthier animals and better profit margins for farmers (Maida, 2019; “HerdX,” 2022). Precision feeding in cattle may also help reduce methane emissions from cows by up to 30% (Maida, 2019).

Veterinary Care

RFID tags can store a wealth of information about each individual animal, such as sex, date of birth, weight, medical records, and offspring (Gough, 2021). Veterinarians can simply scan the animal’s tag to view its information. HerdX recently released a new app that notifies users “when an animal does not go to food or water, a potential sign of illness” (“HerdX,” 2022, para. 6). This feature gives ranchers the ability to respond rapidly with medical intervention for animals whose health may be dwindling.

RFID Within Hospitals

RFID technology has also been implemented within hospitals producing benefits of improved efficiency, tracking, and safety in the areas of asset management, patient care, and internal processes. Additionally, by interfacing with vendors, RFID has been instrumental in managing vendor product updates and recalls.

Asset Management

Hospitals have numerous common assets such as medical devices and instruments. Prior to RFID technology, hospitals tracked assets using manual methods of bar codes (Cheng & Kuo, 2016). Such time-consuming efforts totaled approximately 6,000 hours per month of nurses’ time to locate assets

("MGM solutions," 2018). With the advance of RFID, hospitals can easily help identify the location of such assets, if they are misplaced, or if they are missing (Yao et al., 2011).

Patients

Bracelets with RFID tags have been used for patients. Depending upon what is stored, information may save patients' lives, may allow for quicker patient treatment, and may help locate a patient. Additionally at Pratt Regional Medical Center, both mother and baby wear RFID bracelets to ensure that newborns are not switched (Wong, 2021). In some other patient uses, RFID tags have been used on laundry and clothing to support tracking when needed (Cao et. al, 2014). Last, RFID tags have been used to assist patients with medication such that nurses easily know the amount of medication to be given to reduce medication errors (Cheng & Kuo, 2016).

Processes

Numerous processes within hospitals have benefitted from RFID technology. For example, a patient's sample can be tagged and monitored as it goes through the testing process. Additionally, RFID technology can be used to track the number of supplies or drugs used and the reorder point, thereby improving order accuracy and inventory management. Last, RFID can be used to track the expiration dates of drugs (Lai et al., 2014).

Recalls From Medical Vendors' Products

Since recalls do not occur frequently, hospitals do not always invest in personnel and systems to document products, regularly check vendor and government databases, and remove and replace such products. Yet, recalls occur due to a variety of reasons such as discovered adverse reactions in patients or damaged products (i.e. due to cleaning methods or distribution methods). To better assist hospitals, vendors can tag their equipment such that hospitals can easily know if a recall was issued as well as when an upgrade is needed (Barlow, 2019).

TECHNOLOGICAL AND ETHICAL CONCERNS

Many industries rely on RFID and are continuing to find new inventive applications for its use. Unfortunately, data stored on RFID tags can become compromised in multiple ways.

Technology Problems

Since RFID is a technology, numerous cyber security problems can occur. The main components of RFID technology include stored information that is read and decoded by a reader RFID. Collision occurs when several readers are used, and the signals from more than one reader interfere with each other (Hsu et al., 2009; "Problems with RFID," n.d.). In RFID skimming, perpetrators use a device to intercept RFID data such as credit card information from a credit card. In addition to using stolen data, perpetrators may also duplicate such information for future use or to even make it available for sale (Shaw, 2018; "Problems with RFID," n.d). Another cyberattack is the introduction of viruses to a network or devices (Kumar et al., 2021). A man-in-the-middle attack or relay attack occurs when a perpetrator intercepts and changes the sign between the reader and tag, while the reader is believed to be legitimate (Kumar et al., 2021). Active jamming occurs when a perpetrator introduces another signal in the same transmission range for the reader that is trying to read tags (Mitrokotsa et al., 2010). In Denial of Service (DoS) attacks, many packs of data are transmitted to the tag, which is eventually removed (Kumar et al., 2021). Any of these attacks can have consequences that impact upon the public as well as businesses that may have damaged reputations, lost services, and lost revenue.

Ethical Issues

The ease with which information can be collected and stored through RFID makes it an obvious technology choice for marketing and other fields. Companies can track customers' preferences and spending habits to provide a more customized, personal experience. In some cases, it is difficult to determine the threshold between trying to create a personalized consumer experience and an unwarranted invasion of privacy. The risks become even more critical when considering the medical industry. Health care professionals need access to patients' charts and medical histories. The same RFID technology that can provide efficient data access for doctors and nurses also runs the risk of being intercepted by bad actors with the intent to commit identity theft. The lack of consistent laws and protocols for RFID can make the public feel uneasy about widespread unquestioned use, even in fields such as agriculture, where the benefits of using RFID are abundantly clear.

Marketing Ethical Issues

Ethical considerations and dilemmas grow from the use of RFID technology in marketing. Examples include the fear that RFID technology embedded in products would allow companies to "enter" consumers' homes (Hajewski, 2003), as well as concerns regarding how to handle data mining sensitive consumer data and electronic copyrights (Laczniak & Murphy, 2006).

The use of RFID as part of the customer experience, products tagged with RFID chips allow readers anywhere including in the homes of consumers. Therefore, RFID chips in products must be deactivated before entering the home. In stores this is feasible, but customers have no way of knowing if they are, indeed, deactivated. If the products are ordered online, the ambiguity regarding deactivation is presumably increased.

It gets more complicated when companies use RFID tags to help prevent fraudulent returns. For example, Gap clothing stores add a label to the RFID tags at checkout indicating that the items were, indeed, purchased. If a customer attempts to return an item, the RFID reader detects whether the item was purchased or stolen (Hajewski, 2003). This type of initiative means that the chip is not deactivated.

In order to make decisions about products and promotion, marketers often use RFID technology, chip by chip and in aggregate, to collect and sell consumer information, which has its own set of well-known privacy concerns. Many uses of RFID chips, however, include sensitive information such as banking information as well as personal identifiers like social security and driver's license numbers and all of the personal information associated with these identifiers like birthdays, addresses, and so on. Ethical considerations need to be at the forefront of decisions about what to do with this information.

Data mining also creates an even more specific ethical dilemma regarding copyright law. Several different organizations may use, add to, or even change the date on the chips, so it may not be clear who owns the data in aggregate and on individual chips. Only limited legislation exists to protect this information at all, but not knowing who owns the data makes knowing who has rights under that legislation and about what data the laws apply even more vague. (Smith, 2006). Copyright laws generally do not apply to individual facts, but they do cover compilations of data. The questions are, then, does this legislation cover the information collected while tracking each chip (is that a compilation?) and if so, who owns the copyright? Without specific legislation to guide users and sellers of this information, it is up to each company to interpret existing legislation to make ethically acceptable decisions about use of this information.

Marketers, of course, refute or minimize the issues in contrast to the benefits of efficiency and effectiveness both to business and to consumers. Still, decades after the introduction of RFIDs in marketing, laws and policies lag in addressing the new twists on age-old consumer concerns of privacy, access, and property. Considerable opportunities for discussion exist regarding the ethical considerations of RFIDs in marketing.

Supply Chain Ethical Issues

Supply chain management is increasingly implementing the usage of RFID and substantially benefiting businesses in various ways. Despite the improvements in efficiency, accuracy, and quality of operations, ethical concerns exist. The concerns arise over privacy issues, security issues, green RFID, and coercion of use (Abu-Shanab, 2015). The RFID information is stored in local servers, which are prone to security breaches due to their centralized nature. The security implications caused by such unauthorized access can be critical where sensitive personal identifiable information can be accessed. Specifically, on the retail side of the supply chain, the information that is stored on the tags can be compromised and accessed to monitor individual shopping trends. The tags are inherently insecure in that anyone with a compatible reader can retrieve the information on them. The security issues that pertain to supply chain are similar to other e-commerce security issues, but the sheer scale may present new issues that must be addressed at a system level (Niederman et al., 2007).

However, not only the security risks through external parties can be a threat. The owner of the RFID information can leverage the information for malicious or unethical purposes. The profiling and surveillance of individuals can be used by businesses to create individualized marketing campaigns. It is recommended that in order for RFID to become widely accepted by industry and end-users, security and privacy-preserving authentication protocols are required (Abu-Shanab, 2015). Consequently, data encryption is critical also in the context of RFID.

With the increasing use of RFID in supply chain and other sectors, the environmental impact on manufacturing, designing, and disposal has been questioned. Proper disposal of the tag and antenna is important to make sure that the technology is having the least amount of impact as possible. The adhesives, computer chips, and pieces of metal from the antenna can affect the process of recycling and the manufacturing of new boards from recycled feedstock (Das, 2009). Other pieces of the RFID, such as the copper in the plastic tag casings, are contaminants in steelmaking as well as possible air emissions (Das, 2009). They can become a risk to the environment if the tags continue to use non-biodegradable RFID tags. Some parts of the tags and antennas are hard to recycle. By reducing the environmental damage, RFIDs can increase their net positive contribution to a greener planet.

Another concern with RFIDs in the supply chain is that the tags can often be read after an item leaves a store or supply chain, especially if the tag has a unique serial number (Amsler, 2021). Some are concerned with consumers' privacy by allowing tracking of products beyond the point of sale. RFID tag data can be read by anyone with a compatible reader after leaving the store or supply chain. A survey that measured consumer preferences concluded that the majority of consumers want to kill RFID chips at store exits rather than using any presented complex technical solutions. Customers were willing to forgo the benefits to protect their privacy. Greater privacy risk comes from embedding a tag in the product itself rather than the tag of a product that remains in place. The argument for this case is that it does not break privacy any more than barcodes and credit card uses unless the hackers have access to readers and associate databases.

Agriculture Ethical Issues

The USDA's Animal and Plant Health Inspection Service (APHIS) has been working on implementing mandatory RFID tracking of cattle. The animal disease traceability (ADT) regulations apply to interstate movement of all dairy cattle and sexually intact beef animals over 18 months of age (United States Department of Agriculture, 2021). The first mandate attempt was released April, 2019 with a start date of January 1, 2023. After the Ranchers-Cattlemen Action Legal Fund, United Stockgrowers of America (R-CALF USA) filed a lawsuit, APHIS withdrew the mandate but released a new one the following year ("USDA starts formal rulemaking," 2021). The revised mandate was published July, 2020 with the same 2023 start date. After reviewing public comments of the proposed mandate APHIS decided to once again withdraw its mandate to continue the rulemaking process (United States Department of Agriculture, 2021).

Critics of the ADT regulations have cited civil liberties, financial, and religious concerns. Some U.S. ranchers interpret a national livestock database as an example of government overreach; under U.S. laws, livestock and other animals are considered private property. Suspicions have been brought up that government contracts with private companies to produce smart farming equipment are a cloaked means of funneling money to special interest groups. A statement by R-CALF USA CEO Bill Bullard accused the USDA and APHIS of making "back-room deals with ear tag manufacturing companies'' (R-CALF USA, 2021, para. 8).

Another concern is that the cost of tags, readers, and other operating equipment will render undue financial strain on small producers and family farms, putting them out of business (Ahl, 2021). In a 2008 lawsuit against the USDA, a group of Amish farmers in Michigan stated that administering RFID tags violated their religious beliefs, comparing the tags to the "Mark of the Beast" (Cumming & Farmer, 2008). The case was later thrown out on the basis that it fell under Michigan state law and not national law (Ray, 2009).

Hospital Ethical Issues

The beneficial uses of RIF within hospitals also have associated ethical issues, primarily in the area of privacy. For example, patients' personal information is stored so anyone who has access to the reader can obtain that patient's information (Abugabah, et. al, 2020). A perpetrator can then use patient data in malicious ways such as illegitimately obtaining prescription drugs, processing false insurance claims, or receiving illegitimate medical services/procedures. Additionally, the perpetrator can make a counterfeit patient tag that they may use in the future or may sell to others (Rosenbaum, 2014).

SOLUTIONS AND RECOMMENDATIONS FOR RFID PROBLEMS AND ETHICAL CONCERNS

Some states are proposing upcoming laws to further assist with RFID and privacy; some other states have objected to RFID initiatives that have been adopted. For example, although New York passed the Tenant Data Privacy Act (TDPA), it will not be enforceable until 2023. Although New York has already implemented the Administrative Code that contained biometric regulations for commercial entities, TDPA addressed the use of RFID within smart access commercial structures. Tenants will have to provide consent for their data collection. Commercial landlords will have to provide varied information to tenants elaborating upon how data will be collected, shared, retained, protected, and removed ("New Laws on Biometric," 2021). In a different scenario, the governor of California has objected to RFID initiatives in 2015 because of security concerns (De Looper, 2015). Since that time, California has permitted RFID in ID cards but has been met with serious concerns about privacy (Woodrow, 2018). To address such concerns, amendments to bills would need to be proposed.

Numerous cyber security measures can be implemented to assist with cyber security threats. For example, to prevent an intruder from physically removing RFID tags, surveillance can be increased. Also, a stronger glue or solvent can be utilized. For attacks on the data, a second form of authentication or encryption of data between the tag and reader can be used, though the computational resources are significant. Additionally, RSA blocker tags can be used. They repetitively respond positively to unauthorized readers and create fake tags. For RFID that is part of financial cards, aluminum-lined wallets can be used. For extreme cyber situations, a tag can be destroyed by using the KILL command. In some scenarios, perpetrators may engage in social engineering by sharing false information in order to borrow someone's RFID access. Several security measures can be deployed by a company, such as implementing security measures and employee training to address social engineering. Organizations should also take precautions to avoid loss or theft (Cavoukian, 2006). Last, a risk assessment by an external company could be performed.

Studies have shown that when it comes to RFID and ethical issues, personal privacy was the most cited concern by consumers (Boeck & Durif, 2014; Smith et al., 2014). Surprisingly, suspicions about how personal information (such as spending habits, demographics, and tracking of movement) will be used by businesses and government entities are of higher importance to consumers than issues relating to banking or medical information (Boeck & Durif, 2014). By developing a specific, detailed, yet brief privacy statement or guarantee, companies may be better able to earn trust from their customers (Smith et al., 2014).

RFID ETHICAL CASE STUDIES

Two case studies were selected to investigate the ethical issues faced in the marketing departments and retail supply chain. RFID in marketing used by major corporations such as Disney and Amazon impact many individuals, bringing forth more ethical concerns. Implementing new technology increases efficiency and improves performance but faces privacy issues often not apparent to the users. The following case studies provides a perspective on privacy issues and how RFID can cause ethical concerns.

Company Example 1: Disney

The Walt Disney Company experienced ethical and technical problems from the RFID technology. Their MagicBands, RFID-enabled bracelets that Disney issues to guests to make their experience more convenient and memorable, have been accused of invading the guests' privacy. These bands are part of their MyMagic+ system, allowing guests to make reservations for attractions, shows, and restaurants during their stay at Disney World. It also allows access to their accommodations, the theme park itself, as well as reserved rides, shows, and restaurants (Niles, 2013). Some consumers are concerned about being closely monitored by Disney by tracking their spending habits, movements, and activity interests. Several guests have reported seeing pictures uploaded to their MyMagic+ account from rides where they did not scan their MagicBands (Yee, 2013). The implication was that these RFID enabled bands could be tracked from a longer distance or track using active features.

Although controversies exist and Disney has been accused of harnessing the power of big data, examining their intent on using this technology is essential. Disney's privacy policy on MagicBands states in part that it is

...used to deliver personalized experiences and photos, as well as provide information that helps us improve the overall experience in our parks. Guests can participate in MyMagic+ and visit the resort without using the MagicBand by choosing a card, which cannot be detected by the long-range readers. (Disney, 2022, para. 8)

The amount of information collected through RFID should not violate privacy laws as long as the consumers using them agree beforehand. It would be hard for an organization to improve user experience without collecting information on their target market. These chips can see collectively how many people are going where and how often they are going to certain spots. By analyzing the collected data and mapping out Disney world based on their data sets, they will be able to further improve their attractions and overall experience for their customers. With more consented information, analyzing data and figuring out why certain attractions are more popular will be more effective. The same can be done with certain areas that are not retaining customers. These devices can also improve Disney's operations and save expenses by identifying which areas can more efficiently allocate their resources.

Consumers having digital rights and the option to opt-out should prevent any privacy law violations. This might lead to another question, whether a company should make it inconvenient for customers to not have a Magic Band so they can increase their usage. Similar issues apply to other everyday devices people use. For example, Apple's ecosystem makes it harder for the customers to cross over to other brands since switching costs are high.

Company Example 2: Amazon

The Amazon Go stores permit shoppers to avoid registers, scanning, and lines. Amazon's new "just walk out" system was introduced with the help of artificial intelligence, RFID, sensors, cameras, facial recognition, and machine learning algorithms. With this convenience, shoppers simply scan the Amazon app, shop, and leave (Tillman, 2022). This "just walk out" system permits Amazon to gather large amounts of customer data using their new stores. The RFID system linked to the customer's Amazon account identifies the shopper and their in-store behaviors (Sattel, 2016.). How long a customer looks at a product and which items are picked up or placed back are also monitored. Shoppers' information is then shared in a database in the cloud. This raises privacy issues for customers who are shopping at Amazon Go.

Within the store, recognition algorithms are used to monitor the customer's movements and virtual cart. Motion, item, and human recognition methods identify multiple movements and transactions. The sensors and cameras digitally map the customer's face, hair, and clothing to track the removal and placement of items on shelves. This information is recorded per customer along with their purchase history in their database. By recording products that customers almost bought but did not give Amazon more information than ever with in-store shopping (Strom, 2021). Although how long one scroll and stays on a page is recorded through mobile shopping, facial expressions or body language was not previously recorded. With the real-life decision-making process being recorded, more surveillance than ever is on the shoppers. Most customers are unaware of how much information they give up by shopping at Amazon Go stores. With the data collected through computer vision, Amazon can create buyer profiles by adding another layer of information sets with physical habits (Strom, 2021).

Along with the privacy issues, implementing RFID technology and creating an employee-free system leads to jobs being lost in the largest occupation in the United States. According to the U.S. Bureau of Labor Statistics, around 4 million Americans are working in retail sales. Automation through these technologies is having an impact on human jobs. By increasing the efficiency, accuracy, and speed of checkout lines of traditional, the concept of cashiers is fading.

SIGNIFICANCE

As the information above indicates, RFID technology pervades many organizational functions across industries. Yet the reality is that it is just not a "sexy" technology compared to artificial intelligence, virtual and augmented reality, robotics, and other Internet-of-Things and Metaverse technologies. RFID compares more to brakes on a car rather than a car's driver-less capabilities, but the use of RFIDs is proven and widespread and does not show signs of disappearing any time soon. Still, not much evidence exists that RFID technology is a consistent topic addressed in business classes or that it is known to business entrepreneurs, corporate decision-makers, or even for that matter many of us as consumers. Readers have been shown both historical and opportunistic applications and, importantly, their ethical considerations in operations, human resources, all aspects of marketing, supply chain management, and more. Readers are able to use a pervasive technology for which the ethical considerations are many and for which global and national standards, policy, and law lag.

FUTURE RESEARCH

Future research could occur in numerous initiatives. First, researchers could examine and categorize RFID ethical issues by country. Analysis could be done to determine if the country's culture or political leaders has an impact on the associated RFID ethics. In another initiative, a focus could be on RFID and entrepreneurship. Noticing the benefits of RFID, small business owners may also want to include RFID.

Accordingly, a study can be done to determine if resource centers that help small businesses (i.e. Small Business Development centers) have the skills to provide such assistance. If so, the study could further probe the type of RFID assistance that is offered. Last, a future study could interview businesses that incorporate RFID to catalog what steps they are taking to address consumer concerns such as privacy, security, and so on. Results can be compared by industry, company size, or geography.

CONCLUSION

RFID is an essential technology in today's world that affords both companies and consumers alike accuracy, efficiency, and convenience. Its ability to store and access data makes it invaluable for many different industries but also creates the potential for data theft and privacy issues. Because of this, it is important for companies to be transparent about what information they are gathering from consumers, how it is used, and how that information is being protected. To achieve privacy and security, protocols need to be put in place, such as tags must not compromise the holder's privacy, information must be kept private from unauthorized readers, prevention of long-term tracking connection between tags and holders, and holders should be able to find and disable any tags (Abu-Shanab, 2015). The new technologies should be implemented with thorough testing for vulnerabilities (Zhou & Piramuthu, 2013). Initiatives like two-factor authentication, data encryption, KILL commands, and employee training on social engineering awareness could help prevent the theft of data. Deactivation of RFID chips on merchandise upon purchase can help put customers at ease and reduce the risk of privacy violations. Additionally, legislators need to take the concerns of the public into consideration when addressing RFID with regards to privacy. Consumer awareness, legislation, and ethical business practices are all important aspects of using RFID technology in a manner that is safe, moral, and ethically sound.

REFERENCES

Abu-Shanab, E. (2015, January). Big issues for a small piece: RFID ethical issues. *The 7th International Conference on Information Technology.* http://icit.zuj.edu.jo/ICIT1

Abugabah, A., Nizamuddin, N., & Abuqabbeh, A. (2020). A review of challenges and barriers implementing RFID technology in the healthcare sector. *Procedia Computer Science, 170*, 1003–1010. doi:10.1016/j.procs.2020.03.094

ACLU. (2003, November 14). *RFID position statement of consumer privacy and civil liberties organizations.* https://www.aclu.org/other/rfid-position-statement

Adhi, P., Harris, T., & Hough, G. (2021, May 20). *RFID's renaissance in retail.* McKinsey & Company. https://www.mckinsey.com/industries/retail/our-insights/rfids-renaissance-in-retail

Ahl, J. (2021, March 17). Electronic tracking of livestock has many opponents in the U.S. *NPR.* https://news.stlpublicradio.org/economy-business/2021-03-17/electronic-tracking-of-livestock-has-many-opponents-in-the-u-s

American Trucking Associations. (n.d.). *About ATA.* https://www.trucking.org/about-ata

Amsler, S. (2021, March). *RFID (radio frequency identification).* Tech Target. https://www.techtarget.com/iotagenda/definition/RFID-radio-frequency-identification

Association of American Railroads. (n.d.). *About us*. https://www.aar.org/about-us

Automotive Industry Standards Group. (n.d.). *About us*. https://www.aiag.org/about

Axelrod, D., & Rove, K. (2021, July 8). *Department of Agriculture: What does the USDA do? - 2022.* MasterClass Articles. https://www.masterclass.com/articles/what-does-the-usda-do

Barlow, R. D. (2019, December 19). Future proofing product alerts and recalls. *Healthcare Purchasing News.* https://www.hpnonline.com/sourcing-logistics/article/21118103/futureproofing-product-alerts-and-recalls

Boeck, H., & Durif, F. (2014). An overview of ethical considerations when using RFID with consumers. *International Journal of Cyber Society and Education*, *7*(2), 157–164. doi:10.7903/ijcse.1172

R-CALF USA. (2021, August 26). *Cattle producers appeal mandatory RFID case against USDA*. https://www.r-calfusa.com/cattle-producers-appeal-mandatory-rfid-case-against-usda

Cao, Q., Jones, D. R., & Sheng, H. (2014). Contained nomadic information environment: Technology, organization, and environment influences on adoption of hospital RFID patient tracking. *Information & Management*, *51*(2), 225–239. doi:10.1016/j.im.2013.11.007

Cavoukian, A. (2006). *Privacy guidelines for RFID information systems.* Information and Privacy Commissioner of Ontario. https://www.ipc.on.ca/wp-content/uploads/resources/rfid-guides&tips.pdf

CFI Education Inc. (2022, February 1). *Supply chain*. https://corporatefinanceinstitute.com/resources/knowledge/strategy/supply-chain

Chen, S., Yu, C., & Park, J. H. (2009, April 17). Alleviating reader collision problem in mobile RFID networks. *Personal and Ubiquitous Computing*, *13*(7), 489–497. https://doi.org/10.1007/s00779-009-0224-9

Cheng, C., & Kuo, Y. (2016). RFID analytics for hospital ward management. *Flexible Services and Manufacturing Journal*, *28*(4), 593–616. doi:10.100710696-015-9230-6 PMID:32288935

Copeland, B. J. (2022). *Artificial Intelligence*. Encyclopædia Britannica, Inc. https://www.britannica.com/technology/artificial-intelligence

Cumming, I., & Farmer, O. (2008, December 9). *Michigan farmers protest RFID: Amish farmers there say that using the tags in animals violates their religious beliefs.* Ontario Farmer. https://www.proquest.com/magazines/michigan-farmers-protest-rfid-amish-there-say/docview/2166536047

Cybersecurity and Infrastructure Security Agency. (2019, November 14). *Security Tip (ST04-001) What is Cybersecurity?* https://www.cisa.gov/uscert/ncas/tips/ST04-001

Das, R. (2009, April 22). *How green is RFID?* IDTechEx. https://www.idtechex.com/fr/research-article/how-green-is-rfid/1382

De Looper, C. (2015, October 13). California governor vetoes RFID tags in driver's licenses, chooses security over confidence. *Tech Times*. https://www.techtimes.com/articles/94507/20151013/california-governer-vetoes-rfid-tags-drivers-liceneses-chooses-security-over.htm

Disney. (2022). *Privacy at the Walt Disney World Resort, the Disneyland Resort, and Aulani, a Disney resort & spa: FAQs.* https://disneyworld.disney.go.com/faq/my-disney-experience/my-magic-plus-privacy

Federal Communication Commission. (n.d.). *About the FCC*. https://www.fcc.gov/about/overview

Fowler, M. C. (2019, October 10). Chipping away employee privacy: Legal implications of RFID microchip implants for employees. *The National Law Review*. https://www.natlawreview.com/article/chipping-away-employee-privacy-legal-implications-rfid-microchip-implants-employees

GS1 US. (n.d.). *GS1 Standards*. https://www.gs1.org/standards

Gaudlin, S. (2008, March 27). Washington state passes RFID antispying law. *Computerworld*. https://www.computerworld.com/article/2536199/washington-state-passes-rfid-antispying-law.html

Gough, S. (2021, August 10) How RFID is transforming the livestock-management industry. *RFID Journal Live*. https://rfidjournallive.com/content/blog/how-rfid-is-transforming-the-livestock-management-industry/

Hajewski, D. (2003, February 9). High-tech ID system would revolutionize retail industry. *Milwaukee Journal Sentinel*. https://www.proquest.com/newspapers/high-tech-id-system-would-revolutionize-retail/docview/261768960

Hedgepeth, O. (2021, August 11). *Amazon unveils RFID technology for 'shop and go' consumers*. American Public University. https://apuedge.com/amazon-unveils-rfid-technology-for-shop-and-go-consumers/

HerdX® unveils RFID livestock tracking solution tailored to independent cattle producers. (2022, February 1). *Business Wire*. https://www.businesswire.com/news/home/20220201005430/en/HerdX%C2%AE-Unveils-RFID-Livestock-Tracking-Solution-Tailored-to-Independent-Cattle-Producers

Hossain, M. A., Standing, C., & Chan, C. (2017). The development and validation of a two-staged adoption model of RFID technology in livestock businesses. *Information Technology & People*, *30*(4), 785–808. https://dx.doi.org/10.1108/ITP-06-2016-0133

Impinji Inc. (n.d.). *RFID Standards*. https://www.impinj.com/products/technology/rfid-standards

International Air Transport Association. (n.d.). *About us*. https://www.iata.org/en/about

International Electrotechnical Commission. (n.d.). *Who we are*. https://iec.ch/who-we-are

International Standards Organization. (n.d.). *About us*. https://www.iso.org/about-us.html

Is deploying an RFID self-checkout system affordable and easy? (2020, September 22). Nordic ID Group. https://www.nordicid.com/resources/blog/is-deploying-an-rfid-self-checkout-system-affordable-and-easy

Joint Technical Committee. (n.d.). *About*. https://jtc1info.org/about

Kalany, S. (2018, October 18). *The future of RFID: Making print marketing come alive*. Blue Star Inc. https://blog.bluestarinc.com/the-future-of-rfid-making-print-marketing-come-alive

Karygiannis, T., Eydt, B., Barber, G., Bunn, L., & Phillips, T. (2007). *Guidelines for securing Radio Frequency Identification (RFID) systems*. National Institute of Standards and Technology. https://www.govinfo.gov/content/pkg/GOVPUB-C13-8506922fd706b620b3373caefb73e9f5/pdf/GOVPUB-C13-8506922fd706b620b3373caefb73e9f5.pdf

Keurig debuts BrewID, next-generation technology platform. (2021, July 27). *Beverage Industry*. https://www.bevindustry.com/articles/94300-keurig-debuts-brewid-next-generation-technology-platform

Kumar, A., Jain, A. K., & Mohit, D. (2021). A comprehensive taxonomy of security and privacy issues in RFID. *Complex & Intelligent Systems, 7*(3), 1327-1347. doi:10.1007/s40747-021-00280-6

Laczniak, G. R., & Murphy, P. E. (2006). Marketing, consumers and technology: Perspectives for enhancing ethical transactions. *Business Ethics Quarterly*, *16*(3), 313–321. https://www.jstor.org/stable/3857918

Lai, H., Lin, I., & Tseng, L. (2014). High-level managers' considerations for RFID adoption in hospitals: An empirical study in Taiwan. *Journal of Medical Systems*, *38*(2), 1–3. https://doi.org/10.1007/s10916-013-0003-z

Landt, J. (2005, October-November). The history of RFID. *IEEE Potentials*, *24*(4), 8–11.

Li, B. (2012, October 18). *Case study: Sharing "Heineken moments" at ATP Tennis Masters: Digital*. Campaign Asia. https://www.campaignasia.com/article/case-study-sharing-heineken-moments-at-atp-tennis-masters/319684

Mack, S. (2017, November 21). *Benefits of RFID in retail marketing*. Chron. https://smallbusiness.chron.com/benefits-rfid-retail-marketing-57549.html

Maida, J. (2019, Sep 10). Global RFID tags market for livestock management 2019-2023 | Government regulations mandating transition to RFID tags to boost growth. *Bloomberg Business Wire*. https://www.bloomberg.com/press-releases/2019-09-10/global-rfid-tags-market-for-livestock-management-2019-2023-government-regulations-mandating-transition-to-rfid-tags-to-boost

Matičević, G., Čičak, M., & Lovrić, T. (2011). RFID and Supply Chain Management for manufacturing enterprise. In *Supply Chain Management - New Perspectives*. IntechOpen. doi:10.5772/18625

MGM Solutions: 6000 hours per month wasted on nurses finding lost equipment. (2018, February 13). Healthcare Facilities Today. https://www.healthcarefacilitiestoday.com/posts/MGM-Solutions-6000-Hours-Per-Month-Wasted-on-Nurses-Finding-Lost-Equipment--17611

Mitrokotsa, A., Rieback, M. R., & Tanenbaum, A. S. (2010). Classifying RFID attacks and defenses. *Information Systems Frontiers*, *12*(5), 491–505. https://doi.org/10.1007/s10796-009-9210-z

Moody, G. (2019, December 2). *Egyptian government plans to track the movement of 10 million vehicles with low-cost RFID stickers*. Techdirt. https://www.techdirt.com/2019/12/02/egyptian-government-plans-to-track-movement-10-million-vehicles-with-low-cost-rfid-stickers

National Institute of Standards and Technology. (n.d.). *Privacy*. CSRC. https://csrc.nist.gov/glossary/term/privacy

New laws on biometric, RFID, and other "sensitive" data collection and use. (2021, June 15). The National Law Review. https://www.natlawreview.com/article/new-laws-biometric-rfd-and-other-sensitive-data-collection-and-use

NH bill regulates RFID. (2006, March 31). *Packaging World.* https://www.packworld.com/machinery/coding-printing-labeling/news/13341063/nh-bill-regulates-rfid

Niederman, F., Mathieu, R. G., Morley, R., & Kwon, I. K. (2007, July). Examining RFID applications in supply chain management. *Communications of the ACM, 50*(7). https://www.researchgate.net/profile/Ik-Whan-Kwon/publication/220426733_Examining_RFID_application_in_supply_chain_management/links/0fcfd5136245268dc9000000/Examining-RFID-application-in-supply-chain-management.pdf

Niles, R. (2013, December). *Are Disney MagicBand privacy concerns legit?* Theme Park Insider. https://www.themeparkinsider.com/flume/201312/3801

Office of Privacy Commissioner of Canada. (n.d.). *RFID in the workplace: A consultation paper on recommendations for good practices.* https://www.priv.gc.ca/en/about-the-opc/what-we-do/consultations/completed-consultations/rfid

Palmer, A. (2021, September 8). Amazon brings its cashierless tech to two Whole Foods stores. *CNBC.* https://www.cnbc.com/2021/09/08/amazon-brings-its-cashierless-tech-to-two-whole-foods-stores.html

Problems with RFID. (n.d.). Technovelgy LLC. http://www.technovelgy.com/ct/Technology-Article.asp

Ray, B. (2009, July 31). Amish farmers lose court battle against RFID. *The Register.* https://www.theregister.com/2009/07/31/rfid_cows

RFID for quality control in production processes. (n.d.). TraceID. https://www.trace-id.com/rfid-for-quality-control-in-production-processes/

RFID in supply chain. (n.d.) Zetes. https://www.zetes.com/en/technologies-consumables/rfid-in-supply-chain

Roberti, M. (2015, May 29). *Amazon announces program with Auburn RFID Lab.* RFID Journal. https://www.rfidjournal.com/amazon-announces-program-with-auburn-rfid-lab

Rosenbaum, B. P. (2014, February). Radio Frequency Identification (RFID) in healthcare privacy and security concerns limiting adoption. *Journal of Medical Systems, 38*(3), 1–6.

Sattel, S. (2016). *Amazon Go $ RFID - The automation dilemma.* Autodesk. https://www.autodesk.com/products/eagle/blog/amazon-go-rfid-automation-dilemma

Savage, M. (2018, October 22). Thousands of Swedes are inserting microchips under their skin. *NPR.* https://www.npr.org/2018/10/22/658808705/thousands-of-swedes-are-inserting-microchips-under-their-skin

Shaw, E. (2018, February 22). *What is RFID skimming.* Tripwire. https://www.tripwire.com/state-of-security/featured/what-rfid-skimming

Smith, J. S., Gleim, M. R., Robinson, S. G., Kettinger, W. J., & Park, S. (2014). Using an old dog for new tricks: A regulatory focus perspective on consumer acceptance of RFID applications. *Journal of Service Research, 17*(1), 85–101. https://doi.org/10.1177/1094670513501 39

Smith, L. S. (2006). RFID and other embedded technologies: Who owns the data? *Santa Clara Computer and High-Technology Law Journal, 22*(4), 695–755. https://digitalcommons.law.scu.edu/chtlj/vol22/iss4/2

Standard. (n.d.). *Encyclopædia Britannica, Inc*. https://www.britannica.com/dictionary/standard

Strom, D. (2021, June 23). *Amazon's "just walk out" tech: Should you just walk away?* Avast. https://blog.avast.com/amazon-go-security-risks-avast

Sullivan, L. (2005, March 4). Privacy laws: Europe protects against RFID abuses. *Information Week*. https://www.informationweek.com/it-life/privacy-laws-europe-protects-against-rfid-abuses

Taylor, G. (2019, October 22). *McKinsey & Co. debuts combination store-learning lab in Mall of America*. Retail TouchPoints. https://www.retailtouchpoints.com/topics/store-operations/mckinsey-co-debuts-combination-store-learning-lab-in-mall-of-america

The new Egyptian traffic police law regulations, technologies, and RFID chips for electronic surveillance. (2021, March 9). *Egypt United*. https://egyptunitedvoice.com/2021/03/09/egypt-traffic-police-technology-electronic-surveillance

Thrasher, J. (2013, July 15). *9 Examples of RFID & NFC in marketing*. Atlas RFID Store. https://www.atlasrfidstore.com/rfid-insider/examples-of-rfid-nfc-marketing

Tillman, M. (2022, March 11). *Amazon Go and Amazon Fresh: How the 'Just walk out' tech works*. Pocketlint Limited. https://www.pocket-lint.com/gadgets/news/amazon/160266-amazon-closing-shut-down-four-star-books-pop-up-stores

Understanding official USDA 840 ear tags. (n.d.). CCK Outfitters. https://www.cckoutfitters.com/blogs/blog-posts/understanding-official-usda-840-ear-tags

U.S. Department of Homeland Security. (n.d.). *Enhanced drivers licenses: What are they?* https://www.dhs.gov/enhanced-drivers-licenses-what-are-they

U.S. Food and Drug Administration. (2021, March 23). *USDA announces intent to pursue rulemaking on radio frequency identification (RFID) use in animal disease traceability*. https://www.aphis.usda.gov/aphis/newsroom/news/sa_by_date/sa-2021/rfid-traceability-rulemaking

U.S. Food and Drug Administration. (n.d.). *Radio frequency identification RFID*. https://www.fda.gov/radiation-emitting-products/electromagnetic-compatibility-emc/radio-frequency-identification-rfid

USDA starts formal rulemaking process for RFID ear tags. (2021, Mar. 26). *Food safety news*. https://www.foodsafetynews.com/2021/03/usda-starts-formal-rulemaking-process-for-rfid-ear-tags

Violino, B., & Roberti, M. (2005, January 16). *The history of RFID technology*. RFID Journal. https://www.rfidjournal.com/the-history-of-rfid-technology

W. (2007, November 2). *New California law forbids forced RFID implantation in humans*. Lexology. https://www.lexology.com/library/detail.aspx?g=d6e8cb95-3911-4a89-b52f-c4c80ed7bb43

Walker, C. (2007). *Tagging cattle: Mandatory RFID tags in Michigan.* Michigan Senate. https://www.senate.michigan.gov/SFA/Publications/Notes/2007Notes/NotesMarApr07cw.pdf

What is Digital Marketing? (2022, February 25). American Marketing Association. https://www.ama.org/pages/what-is-digital-marketing

What is RFID and how does RFID work? (n.d.). AB&R. https://www.abr.com/what-is-rfid-how-does-rfid-work/

Wong, W. (2021, January 5). *How RFID solutions improve patient safety and hospital workflow.* Health Tech. https://healthtechmagazine.net/article/2021/01/how-rfid-solutions-improve-patient-safety-and-hospital-workflow

Woodrow, M. (2018, January 22). Real ID cards available in California come with controversy. *ABC News.* https://abc7news.com/california-real-id-in-flying-for/2978718

Yao, W., Chu, C., & Li, Z. (2011). Leveraging complex event processing for smart hospitals using RFID. *Journal of Network and Computer Applications, 34*(3), 799–810. https://doi.org/10.1016/j.jnca.2010.04.020

Yee, K. (2013, December 3). *Privacy at Disney World waning? RFID and tracking updates.* MiceChat. http://www.micechat.com/50252-rfid

Yu, X., Liu, P., Ren, W., Zhang, C., Wang, J., & Zheng, Y. (2018). Safety traceability system of livestock and poultry industrial chain. In *Cloud Computing and Security. ICCCS 2018. Lecture Notes in Computer Science, 11068*. Springer. doi:10.1007/978-3-030-00021-9_1

Zhou, W., & Piramuthu, S. (2013, August). Technology regulation policy for business ethics: An example of RFID in supply chain management. *Journal of Business Ethics, 116*(2). https://doi.org/10.1007/s10551-012-1474-4

ADDITIONAL READING

Atkins, R., Sener, A., & Russo, J. (2021). A simulation for managing retail inventory flow using RFID and bar code technology. *Decision Sciences Journal of Innovative Education, 19*(3), 214–223. https://doi.org/ doi:10.1111/dsji.12232

Bunduchi, R., Weisshaar, C., & Smart, A. U. (2011). Mapping the benefits and costs associated with process innovation: The case of RFID adoption. *Technovation, 31*(9), 505–521. https://doi.org/ doi:10.1016/j.technovation.2011.04.001

Chrysochou, P., Chryssochoidis, G., & Kehagia, O. (2009). Traceability information carriers. the technology backgrounds and consumers' perceptions of the technological solutions. *Appetite, 53*(3), 322–331. https://doi.org/ doi:10.1016/j.appet.2009.07.011 PMID:19631704

Coisel, I., & Martin, T. (2013). Untangling RFID privacy models. *Journal of Computer Networks and Communications, 2013*, 1–26. https://doi.org/ doi:10.1155/2013/710275

Dimitriou, T. (2016). Key evolving RFID systems: Forward/backward privacy and ownership transfer of RFID tags. *Ad Hoc Networks*, *37*, 195–208. https://doi.org/ doi:10.1016/j.adhoc.2015.08.019

Felt, U., & Öchsner, S. (2019). Reordering the "World of Things": The sociotechnical imaginary of RFID tagging and new geographies of responsibility. *Science and Engineering Ethics*, *25*(5), 1425–1446. https://doi.org/ doi:10.100711948-018-0071-z PMID:30357561

Irani, Z., Gunasekaran, A., & Dwivedi, Y. K. (2010). Radio frequency identification (RFID): Research trends and framework. *International Journal of Production Research*, *48*(9), 2485–2511. https://doi.org/ doi:10.1080/00207540903564900

Levine, M., Adida, B., Mandl, K., Kohane, I., & Halamka, J. (2007). What are the benefits and risks of fitting patients with radiofrequency identification devices. *PLoS Medicine*, *4*(11), e322. https://doi.org/ doi:10.1371/journal.pmed.0040322 PMID:18044979

Liukkonen, M. (2015). RFID technology in manufacturing and supply chain. *International Journal of Computer Integrated Manufacturing*, *28*(8), 861–880. https://doi.org/ doi:10.1080/0951192X.2014.941406

Lockton, V., & Rosenberg, R. S. (2005). RFID: The next serious threat to privacy. *Ethics and Information Technology*, *7*(4), 221–231. https://doi.org/ doi:10.100710676-006-0014-2

Yazici, H. J. (2014). An exploratory analysis of hospital perspectives on real time information requirements and perceived benefits of RFID technology for future adoption. *International Journal of Information Management*, *34*(5), 603–621. https://doi.org/ doi:10.1016/j.ijinfomgt.2014.04.010

KEY TERMS AND DEFINITIONS

840 RFID tags.: "Official ear tags that are a source of animal identification that, with accurate movement records, provides the ability to trace animals to their origin" ("Understanding official USDA 840," n.d., para. 2).

Artificial intelligence (AI).: "The ability of a digital computer or computer-controlled robot to perform tasks commonly associated with intelligent beings" (Copeland, 2022, para. 1).

Cyber security.: "The art of protecting networks, devices, and data from unauthorized access or criminal use and the practice of ensuring confidentiality, integrity, and availability of information" (Cybersecurity and Infrastructure Security Agency, 2019, para. 1).

Digital marketing.: "Any marketing methods conducted through electronic devices" ("What is Digital Marketing," 2022, para. 4).

GS1.: "Creates a common foundation for business by uniquely identifying, accurately capturing and automatically sharing vital information about products, locations, assets and more" (GS1 US, n.d., para. 1).

Privacy.: "Assurance that the confidentiality of, and access to, certain information about an entity is protected" (National Institute of Standards and Technology, n.d., para. 1).

RFID receiver or reader.: "A device that has one or more antennas that emit radio waves and receive signals back from the RFID tag" (U.S. Food and Drug Administration, n.d., para. 1).

RFID tags.: Small objects that "use radio waves to communicate their identity and other information to nearby readers" (U.S. Food and Drug Administration, n.d., para. 1).

Standards.: "A level of quality, achievement, etc., that is considered acceptable or desirable" ("Standard," n.d., para. 1).

Supply chain.: "An entire system of producing and delivering a product or service, from sourcing the raw materials to the final delivery of the product or service to end users" (CFI Education Inc., 2022, para. 1).

United States Department of Agriculture (USDA).: "A cabinet-level agency in the federal government responsible for matters involving farming, food, forestry, rural development, and nutrition programs" (Axelrod & Rove, 2021, para. 2).

Chapter 4
Is Digital Nudging an Ethical Marketing Strategy?
A Case Study from the Consumers' Perspectives

Po Man Tse
The Hong Kong Polytechnic University, Hong Kong

ABSTRACT

Consumption through mobile applications is becoming popularized especially after consumers' adaption of the "new normal" due to the COVID-19 pandemic. The same as other marketing strategies, nudges are often applied digitally through the design of smartphone applications to nudge consumers into performing the marketer's preferred actions subconsciously. The study aims to contribute to the existing literature on the acceptance of digital nudges by exploring its ethicalness as a marketing strategy as seen through the consumers' perspective. Eighteen semi-structured interviews were conducted, and thematic analysis was applied to analyze the data.

INTRODUCTION

Rapid advancements in technologies today have significantly altered the delivery processes that businesses employ. Consumers can get quick and accurate information, filter options, and complete transactions through a wide variety of mobile applications. It is now common for people to order groceries via the internet for delivery, paying for their food through yet other applications on their smartphones. Similar to other marketing materials, internet marketers promote their clients' products to potential customers through carefully crafted online means, including through social media, client websites, and advertisements. Marketers have been applying different "nudges" in digital platforms, such as the automatic opt-in for the sharing of personal data when completing the registration of an online membership. Consumers may feel overwhelmed, however, with promotional messages such as "last minute deals," "best offers of the day," "70% off today only," and other similar messages. Since the world is still in the early stages for

DOI: 10.4018/978-1-6684-5892-1.ch004

this phenomenon, there is a lack of literature regarding consumers' perceptions on the ethics of nudging strategy applied through digital venues. Besides, recent research has been focusing on the potential benefits from these ploys and consumers' technology acceptance but may be neglecting the importance of exploring consumers' post purchase experiences of consuming through these platforms.

LITERATURE REVIEW

Nudges

Businesses have long been accused of "priming" customers to make decisions based on the business' suggestions. Dennis et al. (2020) indicate that this can be in the form of promoting product value, image, and of course pricing, but consumers can also be influenced by the website or application's attractiveness and other factors. This priming – now more commonly called "nudging" – can be done through pricing and commentary about related or even unrelated items in an attempt to make prices more attractive (Dennis et al., 2020). A clear definition of this phenomenon is found in White (2013):

Nudges are designed to change people's behavior in someone else's interest by relying on unconscious anomalies in decision making. They do so not by force to threat, and not by persuasion to get voluntary compliance, but by relying on the same cognitive biases and heuristics that justified their use in the first place. (p. 95)

According to the behavioral economists Thaler and Sunstein (2008) – who identified these strategies as "nudging" – there is an assumption that nudging is a useful tool to assist human beings in choosing their purchasing and other technical-related options especially when they are in situations in which making decisions is difficult.

Heuristic "rule of thumbs" may assist humans to simplify complex choices (White, 2013), and nudges are most effective when people are relying on "quick and intuitive judgement" (Egebark & Ekström, 2016). Researchers Thaler and Sunstein (2008) explained that humans may lack the ability to pick the best choice for themselves in some circumstances; for example, everyone knows smoking may cause cancer and is bad for one's health, but all over the world, you still see people smoking. No matter what the reason is for the person to smoke, it can be said that undoubtedly smokers are not making the best choices for themselves. In view of this, nudges may assist the decision-maker during periods of human frailties such as weakness of will (White, 2013).

Governments in different countries apply their own nudges by requesting or requiring tobacco companies to place worded and visual warnings on cigarette packages to remind smokers of the negative consequences of smoking. Ploug et al. (2014) also mention company policies that limit smoking to specific areas as another means of nudging the smoker to quit.

Figure 1 shows one of the package designs by the Hong Kong Government; the warning is visually disturbing and the message "Smoking takes away my voice" can be easily seen and is a strong reminder. These package designs were expected to have an effect in assisting Hong Kong citizens to reduce daily consumption of cigarettes.

The next figure shows a similar example from Malaysia and this photo is equally distressing. It must be said that the message is convincing, especially for pregnant persons who still smoke.

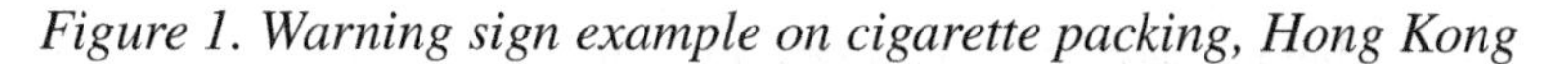

Figure 1. Warning sign example on cigarette packing, Hong Kong

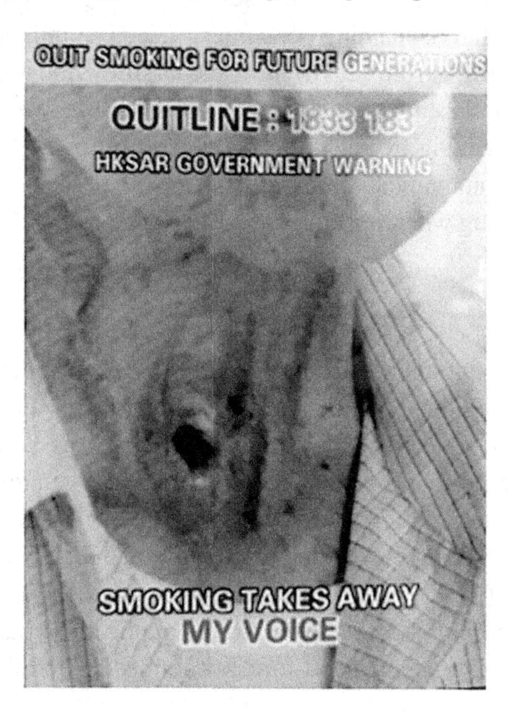

Nudges can thus be based on societal expectations and cultural patterns (Kussainova et al., 2018). They can also be based on economic and security factors, including fears of making expensive mistakes or divulging personal data that cannot be retracted. Sullivan (2018) also indicates a concern about improper nudging as a potential legal issue for companies who may violate security and anti-fraud laws.

Apart from behavioral economists, governmental and company-based researchers in different disciplines have also been exploring the application of nudges to assist humans into performing socially

Figure 2. Warning sign example on cigarette packages in Malaysia

desirable actions (Münscher et al., 2016). In the business field, nudges were applied to assist enterprises in satisfying the fast-changing and demanding consumers' needs (Bammert et al., 2020).

As an example, Egebark and Ekström (2016) have successfully nudged participants into improving resource efficiency in the workplace when making copies. The directive is meant to decrease simplex printing and increase duplex printing by changing the default settings of printers. This is paired with the application of persuasive communication; email is also sent to encourage cutting back on printing in general and by using duplexing when printing is really necessary.

Digital Nudges

Digital nudges are those applied through electronic platforms, again to guide human behavior (Weinmann et al., 2016). Like nudges applied in an offline context, digital nudges aim to assist people in choosing options that result in what they hope will be optimal experience (Tussyadiah & Miller, 2019). Digital nudging has transferred the same offline mechanisms of sometimes gentle and sometimes not-so-subtle nudging to the digital world. These strategies have been proven by researchers Jesse, Jannach, and Gula (2021) to be as effective as offline nudging in altering human behavior towards socially desirable actions. According to Schneider et al. (2018), nudges that appear on any user interface (UI) are all classified as digital nudges; it can be a pop-up message on a mobile applications, an advertisement on a website or in social media, a field inserted in a payment form, or simply a symbol appearing on the company's official website.

Despite the similarities between these offline and online marketing strategies, researchers have also spotted some noticeable differences. Lembke et al. (2019). found that humans tend to be more vulnerable in following their cognitive biases to pick their choices in digital platforms, which means digital nudges may have a greater effect in altering human behavior than nudges in the non-digital world. Besides, researchers Schneider et al. (2018) see great potential in digital nudges since these features can be added and changed easily by either modifying the UI design or the website, social media, or mobile application content.

Figure 3 shows a pop-up screen from an online shop's website. Consumers may not know that 10% will truly be rewarded by signing up as a member, but the information shown in this pop-up may assist in obtaining what would be considered an optimal experience by nudging them to sign up as a member and subsequently getting the desired discount for their purchase.

The next figure shows another example that consumers may frequently see when creating an account on a digital platform; the user intends to create a password to complete the activity but is nudged into creating one that is automatically generated. The nudge is that this is supposed to be the "right" way of creating a strong password.

Similar to research about nudging in the offline context, most research on digital nudging tend to be focused on the effectiveness. As an example, Sharma, Zhan, Nah, Siau, and Cheng (2021) have explored the nudging effectiveness in regard to information security behavior while Pescher, Reichhart, and Spann (2014) previously have examined the impact of the spread of digital information through viral marketing campaigns. While there are such studies, there is still a lack of focused research studying the ethics of digital nudging.

Figure 3. An example of a pop up message shown on an online shop's website

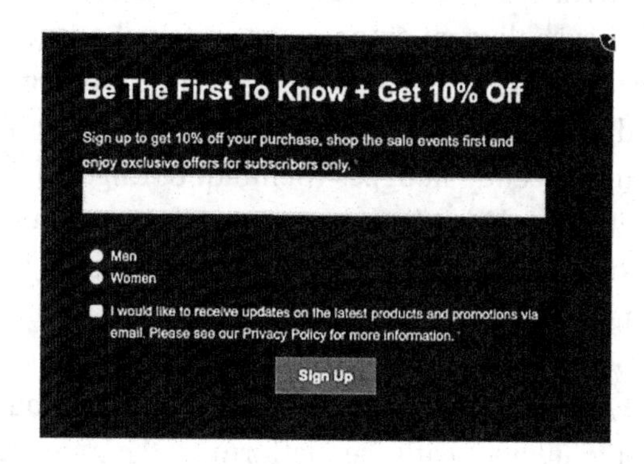

Debate on Nudging Ethics

Thaler and Sunstein (2008) claimed that nudges are typically designed to assist consumers in making decisions that are supposed to be the best for them. These mechanisms are supposed to be ethical, as they modify rational behavior without limiting the choices given to the human being nor constraining their freedom in making decisions that feel appropriate to their situations. However, the same researchers were aware that some marketing strategies have attempted to change consumers' perceived value of alternatives and can powerfully interfere with customers' choices.

Some researchers in the past such as Hausman and Welch (2010) as well as White (2013) also had doubts on how or even if nudging strategies are applied ethically by corporations. White (2013) argued that nudges are coercive, that "they are designed with the express purpose of manipulating people's decision-making processes to change their behavior in pursuit of interests that are not their own" (p. 91). Ideally, it is said that nudging should be paternalistic rather than beneficial. The goal of being paternalistic is to benefit human beings by directing them towards making decisions for their own good, but unfortunately, far too many companies are trying to nudge consumer behavior to make decisions that mostly benefit the company instead. For example, some companies take advantage of humans' cognitive biases and nudge their customers to purchase items and services with the purpose of earning more profits for the company. Other companies may try to nudge customers into selecting fewer desirable items in order to get rid of inventory. In these cases, the strategies may simply be seen as manipulating the customers for the company's profit instead of focusing on customer benefit and satisfaction.

Schmidt and Engelen (2020) have questioned the autonomy of nudges in terms of limiting the consumer's "freedom of choice." This may be perceived as "psychological autonomy as volitional autonomy." While it is clear the ethical path would be to encourage decisions that reflect the individual's actual desire (wherein the "absence of domination" may be cited), critics are concerned that politicians

Figure 4. An example of an automatic message shown when creating an account

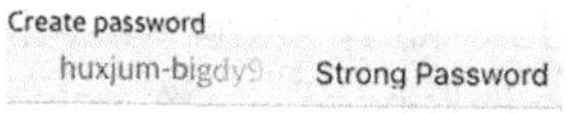

Create password
huxjum-bigdy Strong Password

may make use of the nudging mechanisms to exercise an unjustifiable and problematic control over the local community. In addition, "psychological autonomy as rational agency" can be explored, in which nudging mechanisms are applied in respect that humans are rational beings. By design, nudges are supposed to interfere with the nudgee's irrational heuristics and biases and therefore these strategies can "condescend and infantilize" unsuspecting nudgees (Schmidt & Engelen, 2020).

Zimmerman and Renaud (2021) studied the intersection of cybersecurity and nudging; they found that a "hybrid" approach of providing a nudge and including additional information along with it was far effective than simply relying on the nudge by itself. Thus, desired cybersecurity behaviors could be promoted more easily. Social media is a major area of interest in terms of data privacy today, as millions of computer users post regularly on Facebook and other forums; as Kroll and Steiglitz (2021) warn, "the more that people are motivated to interact with the platform … the more they disclose" (p. 1).

Lu et al. (2020) describes nudging as either overt (noticeable to the consumer) or covert (in which the intent may be masked from the consumer. One effect of constant nudging is that the consumer may develop a predictable purchasing behavior. While this may be profitable for a company, it teaches the customer subconsciously to remove critical thinking from the purchasing process (Lu et al., 2020). Stryja and Satzger (2019) point out that as many as half of all new products and services are destined to fail in the marketplace; there is resistance to new things, and nudging is very often used to keep companies from experiencing financial disaster when they attempt to roll out innovative offerings.

Thornhill et al. (2019) clarify that nudging can also be executed in news stories ("fake news" and misleading content) and can affect political and other beliefs. These can be observed in *presentation*, such as the location of an advertisement, and/or *information*, wherein an awareness is raised about something that may or may not be true (Thornhill et al., 2019). Jung et al. (2022) include that clickbait– another form of nudging – has the potential in social media to controversially manipulate purchases and beliefs and credits Thaler with coining the word "sludge" for when a nudge has such nefarious purpose.

White (2013) also stressed that marketers often are not nudging human behavior ethically and suggested these nudging mechanisms are applying hard paternalism. The problem is that "this is hard paternalism disguised as soft paternalism: pretending to judge the voluntariness of people's decision-making processes but really questioning their choices and their interest" (White, 2013, p. 90). Gerald Dworkin (cited in White, 2013) defined paternalism as "usurpation of decision-making, either by preventing people from doing what they have decided or by interfering with the way in which they arrive at their decisions" (p. 91). Hard paternalism is regarding the interference of human choices with what should be seen clearly as voluntary action. White (2013) questioned its ethicalness as these nudges invoke the problem of value substitution. Other critics are also suspicious of the degree of paternalism of digital nudging, and believe they are an "affront to human dignity" (Waldon, 2014 as cited in Schmidt & Engelen, 2020, p. 7).

Each person's personal mores and principles may also be in peril by these marketing ploys. White (2013) strongly insists that human beings should have the freedom to choose freely following their own principles. Besides, their choices do not necessarily mean they have to benefit the person's well-being, but fulfill a kind of personal satisfaction. Other critics are also concerned about excessive paternalism and related illicit means. Some are questioning how the nudgers are actually qualified to be choice architects and how their trustworthiness can be measured or assessed (Renaud & Zimmermann, 2018). Others cast doubt on how nudgers could possibly know what the best choices are for consumers. Last but not least, some researchers are concerned on the ethical behavior of governments or policymakers in imposing their goals and values on local citizens, especially knowing that different parties in the society may hold very different perspectives of what is *good* (Schmidt & Engelen, 2020).

According to White (2013), it may be unethical for a government or the tobacco companies to display those disturbing photos on the packages seen in the first two figures here, as the best choice for smokers themselves may not have any linkage with their personal health. Smoking is only an undesirable action defined by most people but not by the smokers themselves, and therefore, the altering of the smokers' options by imposing the principles of socially desirable actions can be construed as unethical.

In regard to Figure 3 in which a pop-up message appeared when entering a retail website, some researchers – notably Acquisti et al. (2017) view the digital nudge as an unethical act as it is an intervention in which "exploit decision hurdles to nudge users to disclose more information or behave detrimentally to their interests" (p. 25). This debate between marketers and researchers may never be truly resolved, so to better define the ethicalness of digital nudges, it is important to explore more about the consumer's point of view. They are, after all, the end-users of the online website, social media, or smartphone application, and their perceptions have direct impact on how digital platforms plan to perform. Their purchase behavior and loyalty to the brand can also be at stake.

Nowadays, the use of computers, tablets, and mobile phones have become an essential part of life. Use of these devices helps to sustain connection with friends and family, provide a variety of entertainment options, and the sharing of information to the world through social media platforms, email, etc. (Okeke et al., 2018). Due to this rather recent change in social behavior, people have been spending more and more time on digital devices to work from home and for leisure time. Statistics have shown that many consumers have changed purchasing habits and now do a large part of their shopping through digital platforms. There are increasing concerns that the privacy of citizens and their individual autonomy may be endangered by the application of digital nudges (Ranchordás, 2020), so the consumers perspective and whether or not nudging is performed ethically is crucial.

Thus, data security and political/economic manipulation aside, the focus here is directed more towards consumer behavior towards purchases of goods and/or services.

THE STUDY'S METHODOLOGY

To explore consumers' awareness of being nudged and their perception of its ethicalness as a marketing strategy, a non-probability sampling method was adopted to select the respondents, in which the probability that a sample is being selected is unknown (Acharya et al., 2013). A semi-structured interview method has been proved to be successful in allowing reciprocity between the interviewee and interviewers and enables interviewee verbal expression to be assessed as well (Kallio et al., 2016).

Semi-structured interviews were conducted among 18 consumers who have made a purchase online through a company's official website or a mobile application within one month of the interviews. The interviews took place in April 2022, and due to the social distancing regulations imposed by the interviewer's government, all discussions were conducted through Microsoft Teams. The average duration for the interviews was approximately 45 minutes, and questions about awareness of digital nudges and perceived ethicalness of these strategies were asked. The interviews were recorded and transcribed into textual data, and then thematic analysis was applied to analyze the collected data.

FINDINGS AND DISCUSSION

Three themes were identified: "awareness of digital nudges," "perceived risk of manipulation by digital nudges," and "perceived ethicalness of digital nudges." Each is explored in depth below, with examples of respondent reactions.

Awareness of Digital Nudges

According to Thaler and Sunstein (2008) and White (2013), nudgees are being nudged to perform an action unconsciously, so nudgees may not be able to identify the nudges or remember being nudged to perform the final actions or make the final decisions. Unlike what was assumed by researchers, when respondents were being asked to recall their current online shopping experiences, all the respondents of this study claimed they are well aware that the official websites and smartphone applications have added different cues to give them extra information about the products.

I am aware that they use different colors, fonts, highlights, symbols to attract my attention when I am scrolling through the website (Respondent 1)

Smartphone apps these days have been using a lot of different interesting layouts and pop-ups (Respondent 12)

Respondents described that they are being overloaded with information on digital platforms, and so they may not have a clear idea on what to focus to an extent that they are worried they may have missed some important information. Therefore, most believe that the digital nudges may highlight the information that they need to know and help them in making better choices. Some admit that they were proactively looking for these cues when they review the websites or smartphone applications.

There is just too much information provided on the websites, and sometimes I really don't know where to start, so I will look out for any highlighted messages, especially with the ones signaling discounts. (Respondent 3)

I know those are marketing strategies, but I believe the companies are giving us a heads up on the latest promotions, so we won't miss it. (Respondent 4)

Most respondents classified themselves as technologically savvy and they are familiar with the marketing tactics applied both online and offline. This means most view the application of nudges as "common" or "normal." Although respondents may not be able to identify all nudges applied by the websites or the mobile applications as they thought, they are aware of them and will sometimes rely on them to assist in making what they feel will be better choices for themselves. In general, all have a neutral to positive attitude towards nudges they have encountered in their online shopping experiences.

Perceived Risk of Manipulation by Digital Nudges

Respondents were asked "do they feel the nudges applied are really for the good of the customers" and "do they feel being manipulated to act towards the benefit of the company." For the first question, respondents showed an understanding that companies need to gain profits to survive, so they think that nudges are one of the marketing tactics to help the company promote and increase sales. It was pointed out that it is helpful to highlight the information that companies want customers to know, so none of the respondents are particularly feeling the nudges applied are focused on benefiting either only the customers or only the company.

For the second question, all participants believe that they are rational human beings, and they are intelligent enough to identify what they feel is right and wrong for themselves. Besides, all claimed that they did pick all the choices themselves and they believed they have the freedom of choice - no one has forced them to choose something against their own will. Therefore, none think that they have encountered experiences of being manipulated in their recent online purchase.

I only focused on the ones I am interested and ignored their pop-ups or messages on items that are irrelevant to me. (Respondent 9)

I believe I am the one to make the final decision, so I don't think I was being manipulated by any marketing strategies implemented in the apps (Respondent 6)

I did check everything before I confirmed the choices, I am quite sure I did not consume something that I don't want due to the promotional strategies applied on the website. (Respondent 1)

Respondent 5 made mention of the password suggestion (similar to that shown in Figure 4) during creation of a digital user account with the auto-generation function. She stated that even if a strong password was generated and advised to the user, she will only take it as advice, and she strongly believed that most users will not adopt the suggestion for actual usage as the "strong passwords" tend to be overly complicated to remember in general. She further explained that she thinks such digital nudges serve as useful reminders for consumers rather than tools for the company to manipulate consumers. Even though consumers may not have adopted the suggestions, to a certain extent they will be aware of the strength of the final password that they will be creating themselves due to the nudge.

However, even when none of the respondents claimed that they have encountered any unwanted purchases due to the different nudges on the websites or smartphone applications, all expressed their worry that it is possible that they will misread some terms and conditions in the future and become manipulated to make choices that marketers want them to do. Their worries are mainly due to the fact that most have heard of others "being tricked" when purchasing online through WOM (Word-of-Mouth) or eWOM before. Therefore, respondents in general perceived there is a risk of being manipulated by digital nudges.

There might be days that I am in a rush or not cautious enough, so I think it is possible that I will be tricked by the cues to buy something that I do not need one day. (Respondent 17)

My friend has experiences added in travel insurance that she does not want in her booking accidentally, so I do think some tactics are trying to trick consumers to consume products that benefiting the company more than the consumer themselves and I am worried this will happen to me as well. (Respondent 14)

Perceived Ethicalness of Digital Nudges

Respondents were also asked about their perception of the ethicalness of applying the variety of nudges on official websites and smartphone applications. The responses differed and appeared to depend highly on the situation in which each of these nudges were applied. Some of the respondents have a comparatively neutral attitude towards the digital nudging mechanism as they see it as inevitable. Similar to promotions in the offline context, they believe the digital platforms have to present the choices in a certain way, so it is unavoidable for the company to maximize their opportunities to make more profits by placing it in a certain order.

Just like in any physical stores, there will definitely be items placing in a more eye-catching place and some don't on the websites and apps. (Respondent 18)

Few interviewees mentioned that they could not control but would not mind digital data being collected on a continual basis, which is considered as unethical practice by some researchers (Ranchordás, 2020). However, when the nudges are directly affecting the benefit they receive, especially when it may affect the total amount of money they are paying, they will be really quite concerned and think that the company is not ethical in its actions. Some respondents expressed that especially some default options can be misleading, and the company should not choose something for their customers, especially when the options are benefiting the company more than the customer.

Sometimes we cannot really control our digital footprints being collected as big data is the trend now. As long as the data collected was not being used illegally like credit card fraud, or causing me some troubles or losing money somehow, I am fine with it. (Respondent 2)

I think the boxes are tricky, it is unethical for the company to pre check all for their customers. There might be someone mistakenly accept the options and thought they have clicked it themselves when scrolling down the page. (Respondent 12)

Apart from the "pre checked" default options that may result in consumers paying more than desired, most respondents feel comparatively negative towards particular nudges that signify suggestions made by the company. Figure 5 shows an example raised by Respondent 7, although it is a suggestion about reducing the use of cutlery, which is beneficial for the environment rather than the delivery company itself.

Similarly, in the non-digital nudging world, Goldstein et al. (2008) mention that hotel guests are strongly nudged towards certain conservation/cost-saving efforts, as they "are almost invariably informed that reusing one's towels will conserve natural resources and help save the environment from further depletion, disruption, and corruption" (p. 472).

Respondent 7 felt that the nudge in this instance is giving the customers some form of "peer pressure" and he felt guilty to turn the button "on" when he was ordering take out from work since he had no reusable cutlery with him there. Respondent 7 expressed that this is unethical too for the delivery

Figure 5. An example of cutlery options provided in a delivery app

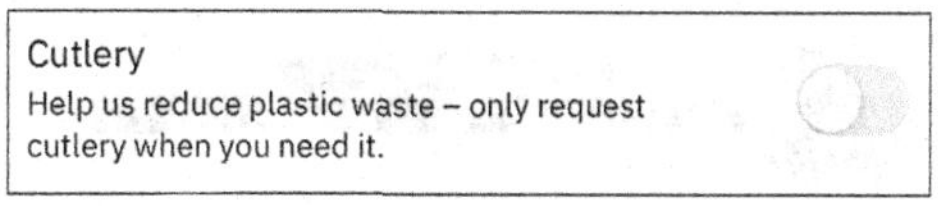

Cutler
Help us reduce plastic waste – only request cutlery when you need it.

company to impose this needless guilt on their customers while they did not purposely make the best decision for the world but for themselves.

RECOMMENDATIONS FOR ENHANCING DIGITAL NUDGING MANAGEMENT

The popularization of online shopping has both positive and negative consequences for marketers and consumers. Using websites and mobile applications for marketing strategies can certainly create great business potential for any company. On the other hand, it also places the marketers in a position in which their reputation can become vulnerable to threats. Underestimating the importance of the ethical use of nudges may damage the image of the company and its marketing departments, and result in the loss of loyal customers. A few recommendations are thus made based on the research findings.

Be Cautious With Nudging Situations

Marketers should be aware of the different situations in which nudges are being applied and should be exceptionally cautious with nudges that are directly related to creating extra cost or losses for the customer. These costs are not just monetary costs, as the effects can be emotional and create distrust as well. Although none of the respondents from the study truly indicated strong negative experiences, all of them have heard of others being nudged to purchase things that they actually do not want, and all have expressed negative attitudes towards those marketing tactics. Companies should not forget that digital nudges should assist customers in selecting the best option for the customers themselves, not the company. Customers view misleading nudges as interference in their choices, and this is unethical. Even if the nudges are successful in encouraging preferred customer behavior in the short run, it may create long term consequences for the reputation of the company that marketers should not neglect.

Another consideration is the technological prowess of the consumers themselves. Sobolev (2021) warns that not all users of technology are savvy in its use, and this digital divide can make a difference in the effects of marketing ploys such as nudging.

Be Cautious With the Selection of Nudges

Apart from costs related to nudges, marketers should also be aware that different nudges can be applied to consumer behavior, so some strategies need to be avoided if at all possible. Respondents of this study have highlighted nudges like buttons in Figure 4, suggesting customers to take the default choices or even making them feel tricked into accepting those default choices.

Figure 6. An example of voluntary donation options provided in a delivery application

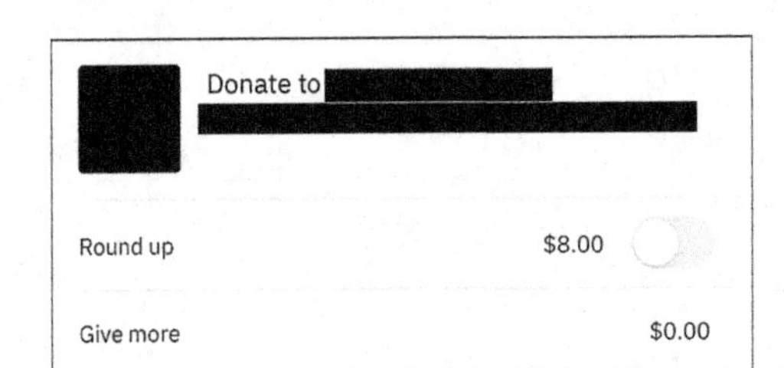

The perceived ethical nature of the nudges was similar for buttons which is not suggesting customers to perform the socially desirable actions with the default setting as in Figure 6, in which the button rounding up a total donation amount for a charity is turned off.

In such scenarios, the respondents suggested that they would prefer the use of other designs such as the checking of blank boxes. Ethical designs will allow customers a higher degree of freedom in choosing their own options, so that the company's default does not make decisions for them and thus interfere with the customer's intent.

It is fundamental for marketers and UI designers to focus on creating attractive designs and layouts for websites and smartphone applications, but a similar effort should also be put into exploring the perception of any nudges. The perceived ethical nature of the company may be far more important than the visual attractiveness of the graphical user interface in retaining customers and providing satisfaction.

Implementation of Post-Purchase Services

Marketers are recommended to explore the implementation of post purchase strategies, as research findings suggested it is effective in minimizing the possible negative consequences of nudging mechanisms. All of the respondents stated that they have some worries to a certain extent when purchasing through digital platforms, and some are afraid that marketers have purposely or accidentally hidden some important information which led them to make wrong choices. Some are simply not confident in finding the best deals for themselves through browsing unfamiliar websites or smartphone applications; and some are concerned of being tricked into paying for unwanted items. Respondents suggested that post purchase services may assist the companies in gaining their trust or even loyalty. Customers appreciate features such as a designated hotline to call with inquiries after purchases, and this would increase their confidence of making choices in the first place. Additionally, being allowed to have free cancellation or assistance in correcting decisions made within a reasonable period of time would negate problems caused by undue nudging and reinforce confidence in company ethics. Respondents claimed that they do not regret their choices most of the time and some had never taken advantage of free cancellation offered by some companies, but these post purchase features would prove that the digital nudges are ethically applied, as the company is not trying to "trick" their customers.

LIMITATIONS AND FUTURE RESEARCH

There are several limitations within the current research. The study has only invited 18 participants for the qualitative interview and the current research has focused only on consumers' perception of nudges applied on mobile applications for shopping purposes, which may not be representative of digital nudges that could appear on the many mobile applications available to the consumer. Future research would be prudent to develop methods for measurement based on the three identified themes. Quantitative research should be conducted to test and validate the results of this study. In addition, researchers Schmidt and Engelen (2020) believe that ethical concerns differ from practice to practice and from nudge to nudge, and as different techniques are employed, different heuristics may be triggered. There are also a wide variety of different objectives in the digital nudging mechanisms applied, and so a comparative study among different digital nudges applied in various online applications would be valuable in testing and validating the results.

CONCLUSION

The currently available literature focused mainly on the effectiveness of digital nudges and may have neglected the perception of the users and the potential consequences of nudging when it has affected consumer decision-making. This study has attempted to provide a beginning insight to the current research gap. Findings suggested even when the digital nudges are successful in shifting consumption behavior, most consumers are aware that different types of nudges were applied in the mobile applications. In general, most consumers find it unethical for companies to surreptitiously manipulate consumers in order to increase profits, and this could well also cause damage to the brand's image. There could also be a decrease in the use of these mobile applications if consumers experience post purchase dissonance due to the digital nudges applied. This, along with potential legal issues for companies themselves, signals that there is an urgent need for marketers to explore the ethics in applying digital nudges and to differentiate between manipulation and assistance for making better decisions for their customers.

REFERENCES

Acharya, A. S., Prakash, A., Saxena, P., & Nigam, A. (2013, July-December). Sampling: Why and how of it. *Indian Journal of Medical Specialties*, *4*(2), 330–333.

Acquisti, A., Adjerid, I., Balebako, R., Brandimarte, L., Cranor, L. F., Komanduri, S., & Wilson, S. (2017, August). Nudges for privacy and security: Understanding and assisting users' choices online. [CSUR]. *ACM Computing Surveys*, *50*(3), 1–41. https://doi.org/ doi:10.1145/3054926

Bammert, S., König, U. M., Roeglinger, M., & Wruck, T. (2020). Exploring potentials of digital nudging for business processes. *Business Process Management Journal*, *26*(6), 1329–1347. https://doi.org/ doi:10.1108/BPMJ-07-2019-0281

Dennis, A. R., Yuan, L., Feng, X., Webb, E., & Hsieh, C. J. (2020, March 1). Digital nudging: Numeric and semantic priming in e-commerce. *Journal of Management Information Systems*, *37*(1), 39–65. https://doi.org/ doi:10.1080/07421222.2019.1705505

Egebark, J., & Ekström, M. (2016, March). Can indifference make the world greener? *Journal of Environmental Economics and Management*, *76*, 1–13. https://doi.org/ doi:10.1016/j.jeem.2015.11.004

Goldstein, N. J., Cialdini, R. B., & Griskevicius, V. (2008, October). A room with a viewpoint: Using social norms to motivate environmental conservation in hotels. *The Journal of Consumer Research*, *35*(3), 472–482. https://doi.org/ doi:10.1086/586910

Hausman, D. M., & Welch, B. (2010, January 8). Debate: To nudge or not to nudge. *Journal of Political Philosophy*, *18*(1), 123–136. https://doi.org/ doi:10.1111/j.1467-9760.2009.00351.x

Jesse, M., Jannach, D., & Gula, B. (2021, December 20). Explorations in digital nudging for online food choices. *Frontiers in Psychology*, *12*, 1–12. https://doi.org/ doi:10.3389/fpsyg.2021.729589 PMID:34987443

Jung, A.-K., Stieglitz, S., Kissmer, T., Mirbabaie, M., & Kroll, T. (2022, June 29). Click me …! The influence of clickbait on user engagement in social media and the role of digital nudging. *PLoS One*, *17*(6), 1–22. https://doi.org/ doi:10.1371/journal.pone.0266743 PMID:35767538

Kallio, H., Pietilä, A.-M., Johnson, M., & Kangasniemi, M. (2016, January). Systematic methodological review: Developing a framework for a qualitative semi-structured interview guide. *Journal of Advanced Nursing*, *72*(12), 2954–2965. https://doi.org/ doi:10.1111/jan.13031 PMID:27221824

Kroll, T., & Stieglitz, S. (2021, January). Digital nudging and privacy: Improving decisions about self-disclosure in social networks. *Behaviour & Information Technology*, *40*(1), 1–19. https://doi.org/ doi:10.1080/0144929X.2019.1584644

Kussainova, A., Rakhimberdinova, M., Denissova, O., Taspenova, G., & Konyrbekov, M. (2018, Winter). Improvement of technological modernization using behavioral economics. *Journal of Environmental Management and Tourism*, *7*(31), 1470–1478. https://doi.org/ doi:10.14505/jemt.v9.7(31).11

Lembke, T.-B., Engelbrecht, N., Brendel, A. B., & Kolbe, L. (2019, June). To nudge or not to nudge: Ethical considerations of digital nudging based on its behavioral economics roots. *European Conference on Information Systems.* Stoholm & Uppsala, Sweden. https://www.researchgate.net/publication/333421600_To_Nudge_or_Not_To_Nudge_Ethical_Considerations_of_Digital_Nudging_Based_on_Its_Behavioral_Economics_Roots

Lu, S., Chen, G., & Wang, K. (2020, November 23). Overt or covert? Effect of different digital nudging on consumers' customization choices. *Nankai Business Review International*, *12*(1), 56–74. doi:10.1108/NBRI-12-2019-0073

Münscher, R., Vetter, M., & Scheuerle, T. (2016, December). A review and taxonomy of choice architecture techniques. *Journal of Behavioral Decision Making*, *29*(5), 511–524. doi:10.1002/bdm.1897

Okeke, F., Sobolev, M., Bell, N., & Estrin, D. (2018, September). Good vibrations: Can a digital nudge reduce digital overload? *Proceedings of the 20th International Conference on Human-Computer Interaction with Mobile Devices and Services, 4*, 1-12. https://dl.acm.org/doi/10.1145/3229434.3229463

Pescher, C., Reichhart, P., & Spann, M. (2014, February 1). Consumer decision-making processes in mobile viral marketing campaigns. *Journal of Interactive Marketing, 28*(1), 43–54. doi:10.1016/j.intmar.2013.08.001

Ploug, T., Holm, S., & Brodersen, J. (2014). Scientific second-order 'nudging' or lobbying by interest groups: The battle over abdominal aortic aneurysm screening programmes. *Medicine, Health Care, and Philosophy, 17*(4), 641–650. doi:10.100711019-014-9566-9 PMID:24807744

Ranchordás, S. (2020). Nudging citizens through technology in smart cities. *International Review of Law Computers & Technology, 34*(3), 254–276. doi:10.1080/13600869.2019.1590928

Renaud, K., & Zimmermann, V. (2018, December). Ethical guidelines for nudging in information security & privacy. *International Journal of Human-Computer Studies, 120*, 22–35. doi:10.1016/j.ijhcs.2018.05.011

Schmidt, A. T., & Engelen, B. (2020, February 27). The ethics of nudging: An overview. *Philosophy Compass, 15*(4). Advance online publication. doi:10.1111/phc3.12658

Schneider, C., Weinmann, M., & Vom Brocke, J. (2018, June 25). Digital nudging. *Communications of the ACM, 61*(7), 67–73. doi:10.1145/3213765

Sharma, K., Zhan, X., Nah, F. F.-H., Siau, K., & Cheng, M. X. (2021, October). Impact of digital nudging on information security behavior: An experimental study on framing and priming in cybersecurity. *Organizational Cybersecurity Journal: Practice, Process and People, 1*(1), 69–91. doi:10.1108/OCJ-03-2021-0009

SobolevM. (2021, March 24). Digital nudging: Using technology to nudge for good. doi:10.2139/ssrn.3889831

Stryja, C., & Satzger, G. (2019, November/December). Digital nudging to overcome cognitive resistance in innovation adoption decisions. *Service Industries Journal, 39*(15/16), 1123–1139. doi:10.1080/02642069.2018.1534960

Sullivan, J. (2021, October 18). *Fintech under fire: Benartzi responds to SEC's 'Digital Nudge' warning.* 401k Specialist Magazine. https://401kspecialistmag.com/fintech-under-fire-benartzi-responds-to-secs-digital-nudge-warning

Thaler, R. H., & Sunstein, C. R. (2008). *Nudge: Improving decisions about health, wealth, and happiness*. Penguin.

Thornhill, C., Meeus, Q., Peperkamp, J., & Berendt, B. (2019, June 6). A digital nudge to counter confirmation bias. *Workshop Proceedings of the 13th International AAAI Conference on Web and Social Media.* 10.3389/fdata.2019.00011

Tussyadiah, I., & Miller, G. (2019, September). Nudged by a robot: Responses to agency and feedback. *Annals of Tourism Research, 78*, 102752. doi:10.1016/j.annals.2019.102752

Weinmann, M., Schneider, C., & vom Brocke, J. (2016, December). Digital nudging. *Business & Information Systems Engineering, 58*(6), 433–436. doi:10.100712599-016-0453-1

White, M. (2013). *The manipulation of choice: Ethics and libertarian paternalism* (1st ed.). Palgrave Macmillan. doi:10.1057/9781137313577

Zimmermann, V., & Renaud, K. (2021, February). The nudge puzzle: Matching nudge interventions to cybersecurity decisions. *ACM Transactions on Computer-Human Interaction, 28*(1), 1–45. doi:10.1145/3429888

ADDITIONAL READING

Batta, S. (2020, April 19). *Digital nudging in our personal and professional lives.* Digital Literacy for Decision Makers @ Columbia B-School, via Medium. https://medium.com/digital-literacy-for-decision-makers-columbia-b/digital-nudging-in-our-personal-and-professional-lives-ed06866f0598

Dhar, J., Bailey, A., Mingardon, S., & Tankersley, J. (2017, May 17). *The persuasive power of the digital nudge.* Boston Consulting Group. https://www.bcg.com/publications/2017/people-organization-operations-persuasive-power-digital-nudge

Luger-Bazinger, C., Marquez, R. M., Harms, C., Loidl, M., Kaziyeva, D., & Hornung-Prähauser, V. (2022, June). Ethics of digital, data-based nudges: The need for responsible innovation. *Proceedings of ISPIM Conferences*, 1–14.

Meske, C. (2020, January). Ethical guidelines for the construction of digital nudges. *Proceedings of the 53rd Hawaii International Conference on System Sciences,* pp. 3928-3937. https://arxiv.org/ftp/arxiv/papers/2003/2003.05249.pdf

Mirosa, M., Munro, H., Mangan-Walker, E., & Pearson, D. (2016, September 5). Reducing waste of food left on plates: Interventions based on means-end chain analysis of customers in foodservice sector. *British Food Journal, 118*(9), 2326–2343. https://doi.org/10.1108/BFJ-12-2015-0460

Omerhodzic, A. (2021, October 21). *Digital nudging – 4 common techniques for an e-commerce.* Flexiana. https://flexiana.com/2021/10/digital-nudging-4-common-techniques-for-an-e-commerce

Sela, A. (2019). E-nudging justice: The role of digital choice architecture in online courts. *Journal of Dispute Resolution, 2019*(2), 127–164.

Section 2

Teaching and Learning

Chapter 5
Leadership and Business Ethics for Technology Students

Jennie Lee Khun
https://orcid.org/0000-0001-6076-1782
Purdue University Global, USA

ABSTRACT

Technology ethics is a sub-field of ethics education addressing the dilemmas that are specific to the information technology age and encompasses a societal shift as technological devices provide a more efficient transfer of information. The use of technology poses unique ethical dilemmas and is an important topic to explore. Corruption in society is a problem and is further segmented by focusing on the technology industry. An overview of ethics in academia is discussed to impact a person's ethical values before entering the workforce. A review of business and technology ethics provides the background on the differences between topics. Ethical concerns about the use of the internet, security, biometrics, data, cryptocurrency, and elevated privileges for technology professionals are also explored. An overview of leadership ethics is included and the impact it has on culture. A curriculum approach to teaching ethics to technology students through practical ethics education in the program through simulation, case study, and real-world approaches is also provided.

INTRODUCTION

Corporate corruption is a persistent problem in a global society. The news consistently presents evidence that academia is failing to implement methods that provide ethical education for students to successfully adapt to their professional careers. These failures are evident in the numerous scandals surrounding government, financial and risk management, the personal conduct of senior leaders, and quality assurance issues in products being produced (Prisacariu & Shah, 2016). These industries employ well-educated graduates from some of the most prestigious academic institutions yet continue to fail in teaching practical applications of ethics. In a business setting, the answer to preventing unethical behavior is to adopt codes of conduct, compliance training, and whistleblowing programs as part of a multifaceted approach. Codes of conduct define the compliance framework, training helps employees understand and apply the

DOI: 10.4018/978-1-6684-5892-1.ch005

framework, and whistleblowing is used as a tool to detect code violations and is designed to influence desired behavior by increasing the detection of wrongdoing (Stöber et al., 2019). Education from a top school, a moral upbringing, and protective business processes do little to prevent corporate corruption by individuals who quite clearly, should know better.

The adverse impacts of such treacherous events have rippling effects that can permanently disrupt business. These unethical business practices have detrimental and long-lasting consequences, including financial disaster and loss of trust (Lilly et al., 2021). Lilly et al. (2021) further describes how leaders at the top do not act alone; they must have the complicit support of others within the organization. Ethics refers to the behavior used in decision-making and conduct of an individual, leading individuals in their dealings with each other, determining the moral responses to situations for which the best course of action is unclear, and directing managers in decisions on what to do in various business situations (Ezenwakwelu et al., 2020). The first step in remediating this issue is to understand why individuals who have been taught ethics in their academic journey idly standby or participate in such adverse behaviors. The second step is to determine how academia can better develop and implement a curriculum that reinforces ethical decision-making to help students understand the impact their future decisions have on themselves, their company, their community, their industry, and society on a holistic level.

APPLICATION OF ETHICS EARLY IN ACADEMIA

Educators are challenged with teaching students the broad and various topics within their program while also providing guidance ethics to individuals whose morals have largely been developed during their younger years. Because the development of moral reasoning continues in adulthood (Kohlberg, 1984) and advances in logical and socio-cognitive capabilities occur during late adolescence and early adulthood, there is some reason to expect further normative developmental changes in prosocial moral reasoning into the 20s (Eisenberg et al. 2014). At a time when many young adults are discovering themselves, it only makes sense to ask if this is a crucial moment in development that is under-leveraged. Schwitzgebel et al. (2020) ask if university ethics classes influence students' real-world moral choices and conclude the question is important but difficult to answer. This question is important to consider when citing curricular reasons and applying the practical value of teaching ethics because real-world moral behavior is, in general, difficult to measure accurately and systematically (Schwitzgebel et al., 2020). Eisenberg et al. (2014) suggest the college experience supports the improved growth of moral reasoning because the academic environment provides opportunities for engaging in complex discussions, alternative perspectives, and abstract thinking that lend well toward maturity; the ability to improve moral reasoning tapers off in early adulthood.

As a society, it is recognized that college is often the last opportunity to substantially inform and expand the role of an individual's ethical behaviors. Kidd et al. (2020) say universities seek to teach ethics across a range of contexts - through a selection of humanities courses, across professional schools, and within general education programs or other core curriculum requirements. It is vital that all training and education develop students' critical analysis skills framed by ethical judgment, considering that ethics must be the guiding light that allows discernment of choice for the future and reflects upon the present (Monteiro et al., 2019). The beginning of an academic program traditionally focuses on core or general education courses. A core class with a syllabus packed full of the material makes it difficult to incorporate teaching ethics intentionally. Faculty contend with the question of not only how to include ethics in the conversation but how often to teach ethics. Departments might offer an entire course dedicated to the concepts surrounding ethics, but for many faculty members, they are on their own to interpret the inclusion of ethics into the curriculum. Monteiro et al. (2019) suggest the goal for faculty is to promote and deepen ethical judgment for their students, framed by a broad and well-grounded philosophical perspective; therefore, it is necessary to refocus on a multidisciplinary perspective. It is insufficient for students to simply have a passive role that exposes and presents facts to foster ethical development; it is necessary that students observe and be intellectually and emotionally involved (Perrenet et al., 2010). Vygotsky's zone of proximal development (ZPD) has been especially influential, explaining that learners who are situated in historical, social, and cultural contexts develop gradually through interpersonal learning relationships (Wald & Harland, 2022). While ZPD does not account for the precise depiction of learning needs or motivational influences at a person's current level and does not explain the process of developing, the approach does offer a starting point for exploring a more practical approach. It is significant to consider how faculty can develop a curriculum that actively matures the students' ethical growth process through these relationships as part of the evolution of their college experience as a means of fostering the purposeful creation of their personality and identity. Data science skills are becoming a necessity, and institutions are developing data science curricula to teach programming and computational skills that are needed to build and maintain data infrastructures and maximize the use of data; however, few of these courses include an explicit ethics component (Bezuidenhout et al., 2020). Faculty must thoroughly demonstrate and create opportunities for actively immersed learning that imparts students with a solid adoption of business ethics as a fundamental personality attribute. This role is further strained for those in the technology field; variations and additional complexities present a challenge for information technology (IT) educators and students alike. First, educators must fully examine IT ethics and how this unique subset parallels and differs from traditional business ethics teaching.

BUSINESS ETHICS VS. IT ETHICS

It is crucial to begin by clarifying the subtleties between business and personal ethics and by discussing ethics, morals, and values to create a baseline understanding of the concepts. The study of ethics in business leadership presents great potential for company growth. Meanwhile, there is an inherent challenge in studying a concept that lies at the intersection of three other broad and ambiguous concepts: business, ethics, and leadership (Palanski et al., 2020). Business ethics are governing rules of conduct levied on an employee or profession; personal ethics are the internal guidelines that govern an individual's approach to how they navigate through life. Ethics, morals, and values are terms that have been used interchangeably; nevertheless, each term has distinct differences. Some would suggest ethics is merely the merger of morals, values, and culturally accepted behaviors. Muhammed-Shittu (2021) explains ethics can be specifically interpreted and described in three dimensions; it can be used synonymously as morality, the definition of globally accepted standards, and the values of conduct that every logical individual expects to receive from other individuals. Ethics is a conventional branch in the field of philosophy that exemplifies the foundations of human standards, ideals, principles, and struggles to pinpoint them in human theories and communal conditions; ethical theory is not universal as it reflects the general codes of conduct followed by varying social groups (Muhammed-Shittu, 2021). Values are more objective in the way society dictates desirable attitudes and behaviors. Morals are often rooted in socialization from an early age. The behaviors that individuals adopt come from their families, communities, schools, and organized religion. Morals are more subjective as individuals navigate what they define as right or wrong, good, or bad, and acceptable or unacceptable.

Business ethics is a branch of ethics that focuses on applying a framework to the way organizations conduct business. Business ethics encompasses the organizational principles, sets of values, standards, and norms that influence the actions and behavior of all individuals in the organization (Ezenwakwelu et al., 2020). Business is fundamentally a cooperative effort for organizing material and social dimensions of common life; identifying the business as a source of collectivity may seem naïve when repeatedly presented with news of corporate malfeasance, mistreatment, and environmental manipulation (Islam, 2020). Such explanations often come down to how people perceive their worlds, make decisions, and interact with each other; business is ultimately something people do to themselves that they make for themselves (Islam, 2020).

A subsection of business ethics involves ethical considerations related specifically to technology. IT ethics is the study of concepts arising from the development and use of various electronic technologies, including the individuals who work to support those technologies. The rapid development of science and technology has significantly impacted how closely connected technology is with people's daily life. While progress is made by society, it is also necessary to think about how to face various problems resulting from technologically advancing activities, such as privacy protection, big data technology, the ethical boundaries of the application of gene-editing technology, and the responsibilities associated with autonomous vehicles (Yan, 2017). It is critical to recognize the value of studying IT ethics as a distinct and separate concept under the comprehensive ethical category. Ethics in IT allows an organization to create a culture of trust, accountability, integrity, and excellence in the use of resources while also bolstering the privacy, accessibility, and confidentiality of information systems.

ETHICS IN INFORMATION TECHNOLOGY

There is a growing consensus that, as an industry, the conversation around ethics is of increasing importance and that this dialog must become part of the socially accepted language that technologists use in daily life. Technology is linked to social behaviors. IT ethics include reviewing issues that arise from the development, usage, and advance of various electronic technologies in society. Ethical issues such as privacy, data collection and handling, biometrics, artificial intelligence, security, software piracy, property rights, acceptable use, and levels of access can cause confounding contradictions for those employed in the industry. Philosophers traditionally approach ethics from actions and decision making, which are described in examples bereft of context. Works in the social sciences consider how ethical considerations play out from a more personal and social perspective (Murukannaiah et al., 2020). The misappropriation of technology resources has become exceedingly commonplace, further supporting the need to increase the emphasis on ethical behavior.

The goal of IT ethical studies is to create and recognize moral responses to the basic individual responsibilities and actions in varying situations that rely on personal judgment. In computer ethics, various sets of issues have to do with privacy, computer programs obtain and store information about the online activities of users, and concerns about the morality of the use of these programs are that it can violate a person's right to privacy (Price, 2020). Another cause for concern is autonomy and enhanced level of access IT professionals need to complete their work. IT professionals have access to user data, systems, restricted access areas of buildings, and the logs that provide accountability. IT professionals can benefit from an industry-accepted code of ethics that is established to help guide individual behavior during conflicting situations. Professional responsibility is the ability to recognize, interpret, and act on ethical principles with values based on the standards considered acceptable in the industry (Afifi et al., 2020). It is beneficial to take a more in-depth look at some of the types of ethical considerations IT professionals may face to better understand how instructors can adapt these concepts into the program of study.

Ethics covers a broad range of topics in various technology fields. The need to demonstrate teaching ethics to students and the benefits it has for the student's professional development can best be explored through breaking down ethics into appropriate subtopics. The following subsections describe existing ethical considerations in various technological areas. Career paths within the technology industry often include and overlap many of these subtopics. This is not an exhaustive list but rather a brief introduction to the concepts most technology professionals are likely to encounter.

Internet Ethics

Internet ethics is a broad and generalized concept that encompasses how the internet is accessed and the impact its use has on society. Internet ethics is an essential issue for academics and professional practitioners alike. Wang et al. (2020) describe internet ethics, sometimes called cyberethics, as the education, discussion, or analysis of the legal, ethical, and moral issues raised by the emergence of cyber technology. General principles regarding internet ethics provide technology professionals with a set of guidelines that provide direction on conduct on the internet. Murukannaiah et al. (2020) declare the elements of the internet ethics that are relevant to people include users, developers, administrators, machines such as smart or Internet of Things (IoT) devices, and data resources. Internet ethics principles include contributing to the overall well-being of humanity, avoiding harm or destruction, encouraging honest and trustworthy conduct, non-discrimination, respect for others and their work, respecting privacy, and honoring confidentiality. IT professionals must know and respect these principles while utilizing the internet personally and professionally. The industry must promote public awareness that professionals understand computing technologies and their consequences and are actively designing and implementing systems and standards that are more secure to protect the user community.

Security Ethics

Cybersecurity and security ethics aim to safeguard people, data, computer systems, networks, software, and hardware. There is much discussion around technical solutions to cybersecurity issues, but there is far less focus on the ethical issues raised by cybersecurity (Formosa et al., 2021). Cybersecurity practices focus on helping protect the integrity, availability, confidentiality, functionality, and reliability of technologies from human influences, good or bad. The basis of security threats will inevitability be rooted in the human factor. Most threats require some form of human interaction to execute; this can occur when someone clicks a malicious email link or opens an attachment from an unknown source. As a result, attackers often target the easiest and most likely to succeed vector, people. The implementation of security practices allows security professionals to protect the lives, well-being, and happiness of the people who depend on the systems. Formosa et al. (2021) explain that cybersecurity raises important ethical trade-offs and complex moral issues, such as whether to pay hackers to access data encrypted by ransomware or to intentionally deceive people through social engineering while undertaking penetration testing. At the core of security are practices aimed to secure data, systems, and networks from human practices that might negatively interfere.

Biometrics Ethics

Figure 1. Fingerprint
(Prentza, 2020)

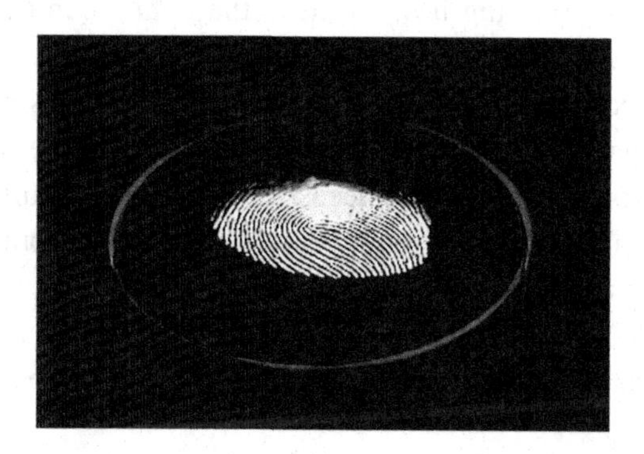

Biometrics represent an advantageous shift from the use of documentation such as identification cards (ID), birth certificates, passports, and visas, providing a rather convenient form of personal identity. Biometric technology concerns the use of the physiological and behavioral characteristics of individuals, and biometric data is used primarily for identity management or authentication purposes (North-Samardzic, 2020). Biometric technology has become widely accepted and will continue to be adopted due to society's preference for convenience. A person may forget their password or leave their wallet with their ID at home, but they are never without their fingertips or face (Hayes, 2019). Unlike other technological innovations over the decades, biometrics lead to additional ethical concerns due to the collection of biometric data being described as an intrusive process where the individual metaphorically gives up a piece of themself (North-Samardzic, 2020). Facial recognition technology has been used to evaluate the responses of students learning in the classroom (North-Samardzic, 2020). Additionally, industry-recognized technology certifications use similar processes for authenticating applicants taking examinations either in their facility or online. The International Information System Security Certification Consortium $(ISC)^2$ is an industry leader nonprofit organization concentrating on the development and certification of security professionals. $(ISC)^2$ requires applicants to show two forms of paper identification, have a photo taken, and have a palm vein scan (unless prohibited by law) before taking an examination (ISC^2, n.d.).

Data Ethics

Information is a powerful tool and knowing how to use data within ethical and legal boundaries is indispensable for any industry. The hype surrounding big data does not seem to subside, regardless of the scandals that persist following privacy breaches in the collection, use, and sharing of data, even from major tech players such as Facebook, Google, Apple, and Uber (Chen & Quan-Haase, 2020). Responsible and ethical data handling means applying principles of transparency, impartiality, and respect to how data are gathered, stored, used, secured, and disposed of. The benefit of data ethics for business includes following government compliance regulations, promoting fair practices, reducing risk exposure, building trust, and creating a positive public perception. Modern life is exponentially generating increasing amounts of data; to effectively navigate this data deluge, are IT professionals educated in using tools to manage, curate, and examine data trends (Bezuidenhout et al., 2020). These analyses contain personally identifiable information that includes data points that can deanonymize a single person. As a result, organizations must be intentional in the data life cycle process by only collecting data that needs to be collected or processed, determining how to make the data meaningful, how does data analysis drive decision-making, how is data disseminated, when is data no longer useful, and how to archive or dispose of it properly. Thus, decisions about what data to use, how to design algorithms, how to design databases and dissemination pathways, while seemingly harmless to the professional, can lead to significant downstream ethical crises such as the unnecessary inclusion of cultural biases within algorithm design that lead to gender disparities within Google searches (Bezuidenhout et al., 2020). The practice of implementing ethics for data users includes concepts such as the development of a comprehensive and formalized data program that considers data throughout the lifecycle and implements codes of conduct, policies, and strategies that are designed to create responsible innovation and sustainable progress.

Cryptocurrency and Business Ethics

Figure 2. Representation of cryptocurrency "coins"
(Kanchanara, 2021).

Cryptocurrency makes use of cryptography to enable financial exchange in digital form. Cryptography translates legible information into digitized algorithms that cannot be broken using what is called block-chaining. Blockchains record individual transactions and ownership of cryptocurrencies, and this decentralized system is managed by blockchain miners who update the accuracy of the information (Milutinović, 2018). The miners use cryptography to gather the data, and it passes through blockchains. Cryptocurrencies are difficult for the government to control. The government cannot tangibly affect cryptocurrency without regulation as the agencies have no power in making decisions. No one can control the flow of this currency as it belongs to no government or state, but also, it belongs to all the people who have decided to risk investing in it (Milutinović, 2018). The theoretical application appears to be brilliant on the surface, but underneath the unregulated world of digital currency lies an ethical dilemma for organizations.

The good versus evil struggle of cryptocurrencies can be argued both ways. Those who support and oppose agree that due to their digital nature and global dissemination, cryptocurrencies have the potential to be much more ubiquitous than any currently established form of money (Dierksmeier & Seele, 2018). The argument for both sides present a moral dilemma that is not easily solved. For example, digital currency can help reduce poverty because it removes associated banking fees for transactions, thereby keeping money in the hands of those who desperately need it. Alternatively, a lack of regulation for digital currency is used by criminals to evade government detection (Dierksmeier & Seele, 2018). Cryptocurrency has become the currency of choice for drug dealers and extortionists to commit criminal activities such as tax evasion, money laundering, Ponzi schemes, and kidnapping for ransom due to anonymity and presumed privacy of the monetary system (Kethineni & Cao, 2020). The benefits of cryptocurrency, such as granting financial sovereignty to civil society, the enhancement of global economic participation, and value creation, could produce enough benefits to outweigh the ethical risks and moral dilemmas that currently exist in an unregulated state (Dierksmeier & Seele, 2018). Thus, businesses must learn to navigate digital currency in an ethical way that aligns with the organization's business practices and objectives. Consumers could view the use of cryptocurrency negatively due to the support of nefarious activity and choose to take their business to competitors. There are virtually no transaction costs for using cryptocurrencies, which provides a business with an advantageous cost-saving opportunity that is difficult to overlook. The existing business models that are utilized for the transfer of money are being confronted by new and revolutionary technologies. As a result, IT professionals must be prepared to understand the impact of their decisions on a situation that is unlikely to easily resolve itself.

Elevated Privileges for IT Professionals

A technology professional must provide the highest quality of integrity by always acquiring and maintaining professional competence. IT professionals have permissions the average user does not have, and as a result, the industry is subject to additional ethical concerns. With the increased volume of and access to electronically held assets, the ethical behavior of IT professionals is critical (Aasheim et al., 2021). These technology professionals are trusted with elevated access, including adjusting safeguards designed to prevent others from circumventing processes. The question to consider is how the industry provides the means necessary for the professionals to perform their job duties while also maintaining the assurance of confidentiality, integrity, and accessibility. The industry adopts best practices, code of ethics documentation, checks and balances, logging, and least privilege measures to ensure the technology professional adheres to the highest ethical standards. The concept of least privilege is an industry-accepted best practice of restricting access rights for users, accounts, and computing processes to only those essential functions required to perform legitimate activities needed to do the job. Corporate codes of ethics are intended to guide employee behavior and are expected to be effective tools as violations carry penalties up to and including termination of employment, but their effectiveness is uncertain (Aasheim et al., 2021). Yet, with all these safeguards in place, there remains an opportunity, and as a result, the individual must only have the motivation to carry out an unscrupulous act.

LEADERSHIP ETHICS

Ethical leadership in business focuses on the organization's shared values, morals, and beliefs of all individuals that create and develop the culture of the company. The omnipresence of leadership in business indicates that the study of leadership is likely to have implications for business; it could be argued that most leadership studies are relevant to the business field even if they do not take place in a business context (Palanski et al., 2021). Leaders, including those in business, must uphold, demonstrate, and embody a company's ethical values. Given this situation, one could argue that just about any leadership study can overlap with business and especially ethics, particularly when the leadership constructs itself has an ethical component (Palanski et al., 2021). It is only sensible to conclude that business leadership ethics requires further deliberation.

Leadership in the context of business ethics is approached from an assumption that ethical leadership has the connotation and perception of positively influencing the environment. Roque et al. (2020) ask what can be done to develop ethical cultures that may be less permeable and more resilient to changes in leadership from an ethical point of view. As the foundation of the organization, these leaders do more than simply manage staff, make business decisions, and attend meetings. Leaders independently perform behaviors that actively encourage ethical behaviors in their followers, and leaders who behave in ways that are perceived as honest, trustworthy, fair, and caring are viewed as righteous, reliable role models by their followers (Kang, 2019). Thus, this behavior has a cascading effect on the entire organizational structure. The leadership of the organization sets the tone for what is to become normative behavior.

When a business acts ethically, the organization attracts customers to the company through loyalty to the brand's ethics, thereby increasing product sales in the market and increasing revenues. Ethical businesses reduce turnover and increase productivity because employees feel more comfortable when their personal ethics align with the organization's ethical culture. Further compounding the benefits of ethical behavior is the way investors feel that the company is acting responsibly, which raises the share price and reputation of the organization. This cycle of behavior continuously offers advantages to the company, employees, and consumers. The opposite is also true; unethical behaviors can damage the company's image, create a toxic culture, increase turnover, reduce profits, and decrease the organization's sustainability. Attempts to articulate ethical leadership tend to rely heavily on terms and concepts. For example, leadership scholars often appeal to notions of authenticity, integrity, and responsibility, ideas that bring with them the same kind of contentiousness associated with ethics itself (Price, 2018). It would be remiss not to further deliberate specific ethical behaviors to fully define the topic of leadership in ethics.

Ethical leaders are the optimal role model for employees by creating, inspiring, and maintaining a culture supported by the foundational aspects of trust and respect. Ethical theories on leadership approach the topic from two main concepts. The first includes the actions and behavior of leaders, and the second describes the character of leaders. Leadership behaviors and character are inherently intertwined. Ask someone to describe what they consider to be advantageous qualities of a good leader, and the responses often include honesty, fairness, integrity, accountability, empathy, effective communication, and possesses strong morals and values. People fundamentally want to work for someone they consider having characteristics that align with their beliefs.

Developing the Ethical Leaders of Tomorrow

An ethical leader is one who strongly believes in following their personal set of values and ideals in all areas of decision-making, actions, and behavior. Among the other attributes of an ethical leader, one needs to be responsible for making unbiased decisions for the benefit and overall interest of people and the organization. Professionals in the technology industry are guided by professional organizations, their morals and values, leadership, business culture, and education to provide situational guidance on ethical dilemmas they will inevitably face. It is not a situation of if but when they will face these challenges. As students prepare to enter the workplace in their chosen profession, they must learn the skills required to thrive in a professional environment. However, there is a need for academia to better teach and assess how tomorrow's professionals are learning ethical concepts as it relates to the world they are about to enter. Ethics is not simply a concept to explore on merely theoretical terms. It is crucial to remember that the practical application of ethics is entwined in the future leadership of all areas of society and is the most impactful right within local communities. These areas include academia, law, politics, religion, health care, and businesses of all types; no industry or business is exempt from ethical considerations. The fundamental values of a personality, linked with ethics, form the personal identity that may be resistant to change. The ethical and moral development of professional leaders must be rooted in more early personality development stages within school education (Hermannsdottir et al., 2018). It is abundantly clear that given the impact ethical considerations have on the outcomes impacting society that a shift in how ethics is developed and taught to technology students is of growing concern.

CURRICULUM APPROACH TO TEACH ETHICS TO TECHNOLOGY STUDENTS

After the high-level overview of the content on the multifaceted approach needed for ethics previously explored, it is reasonably concluded that academia must take a more thorough and intentional approach to teaching ethics to technology students. This ethical education must take place through a purposeful curriculum-wide approach. The goal of education is not just to pass a test or regurgitate facts. The goal of education is to develop future generations of thoughtful, intelligent individuals who have the aptitude and skills to discern the best choices for a given situation.

References to the knowledge society and information age are commonplace characteristics associated with a time when society is using technology to enhance decision-making. It is understood that individuals need a variety of skill sets to be successful, which must include a practical approach to ethics. Despite an explosion of ethics courses and programs over the past half-century, the understanding of the current landscape of ethics instruction in higher education is inadequate (Kidd et al., 2020). Higher education must better educate and assess ethical education outcomes to fully measure student comprehension as a preventative strategy against ethics violations once students leave academia. The comprehension of instructional goals and assessment methods for ethics education is insufficient for assessing comprehension and application at a time when there is greater demand for more robust ethics preparation for students moving into the workforce (Kidd et al., 2020). The problem to be solved is how education can be used to increase students' awareness and reflection on ethical concerns before the individual enters the workforce.

Ethics must be a topic of conversation in all programs, but it is extremely important for those in the IT program. This need for adjustment is evident because public ethical failures remain an unrelenting flaw in society. Educational institutions must be intentional in fully shaping the future leaders of communities, schools, and organizations. Educators must be mindful of developing moral adjustments in college-age young adults as this is the last chance for impactful corrections. A single ethics course is not enough to make a lasting impact on anyone. Additionally, these courses are often high-level theoretical papers and not an immersive learning experience. The intersection of leadership and business ethics for technology students must occur across the program curriculum. Kidd et al. (2020) conclude that understanding ethics education in the context of cross-curriculum coursework requires at least some references to ethically relevant content in each offering, acknowledging it is possible to embed ethical lessons in all content. In cultivating consciousness of ethics in technology, college courses should emphasize a combination of moral autonomy and heteronomy and public and personal morality to ensure the cultural condition of when traditional norms and values become unyielding are deeply embedded in socially acceptable ethical behaviors (Wang & Yan, 2019). This is done to train the student to use practical wisdom when dealing with the problems of ethics of science and technology, as well as to form the conscience and judgment of workers in science and technology careers (Wang & Yan, 2019). The remainder of this section offers suggestions on varying approaches that can be beneficial for the student.

Incorporating Ethics Education

Figure 3. Graduation ceremony image
(Hoehne, 2020).

Teaching ethics is necessary to prepare students to handle ethical dilemmas they will face in their working careers. The benefit of exposing students to ethical concepts and applications is to better prepare them so that they are not surprised by the challenges they may face in the future. Vygotsky's ZPD mentions the main difference is between what a learner working on their own can achieve compared with working alongside someone who is more capable; an experience is seen to support advanced understanding, which is foundational to higher education (Wald & Harland, 2022). Faculty must focus on teaching foundational building blocks that last well outside of the classroom setting. Instructors have an opportunity early on to address internet ethics in the syllabus by including plagiarism, piracy, dishonesty, falsification, and misuse of academic materials (Wang et al., 2020). Internet ethics curriculum covers topics like ethical problems associated with using the Internet, conducted using assigned reading, homework, assignments, and discussions; these are designed to motivate students' conversations about Internet ethical issues and to establish a culture of proper use (Wang et al., 2020). The program can benefit from consistently challenging students to ask themselves whether a specific action is good or bad, right, or wrong, and to apply the principles early and often. This learning experience is designed to allow students to openly question a situation or decision, prompting them to solicit good advice. This creates an opportunity to encourage reflection by considering the implications, taking a breath, and thinking fully before acting on a decision. Many instances of what society deems a white-collar crime start innocently or small, and time evolves into something much more damaging.

Harvard University's Computer Science undergraduate program does not mention ethics in the overview of the program or as a course requirement. However, other programs such as Electrical Engineering and Engineering Sciences specifically define the need to study ethics as part of the program curriculum. The handbook states "An ability to recognize ethical and professional responsibilities in engineering situations and make informed judgments, which must consider the impact of engineering solutions in global, economic, environmental, and societal contexts" ("Fields of Concentration," 2021, p. 77). It is interesting to note that the guide defines similar requirements for ethical studies for other programs including Anthropology, Chemistry, Philosophy, African and African American Studies, and Comparative Literature programs. Surprisingly, the History and Science program, and more specifically the Technology, Information, and Society concentration define "Four courses in the history of science and technology [are] designed to study the larger historical, ethical, and social implications of technology, engineering, and information in the modern world" ("Fields of Concentration," 2021, p. 77). It appears that ethics is being considered for some programs but the importance of exploring ethics in the Computer Science program have not been fully developed at Harvard University. The importance of incorporating ethics into all academic programs cannot be overstated. Higher education must explore ways to more competently incorporate ethical education for all students, but more specifically for technology students.

There are multiple ways to incorporate ethics into a meaningful and transformational learning experience. The easiest and most important thing to consider is finding a way to connect ethics to students' lives. Professors should introduce the topic in a way students can relate to and forces them to contemplate their own decisions. An accounting professor at California State University, Fullerton's Irvine campus, asks students if they have ever taken something seemingly inconsequential from their employer, such as office supplies—like a pen or notepad—that has not been explicitly given to them (Wotapka, 2016). Students discuss whether such trinkets are perks or if the employer would welcome the free marketing from having a logo seen by the public (Wotapka, 2016). The ensuing argument is designed to allow students to think independently and understand how they might consider the implications of their actions if the student was in that same situation. "One way to build an emotional connection through the case method is to require students to take on the roles of various stakeholders in the case" (Kimbell & Dos Santos, 2020, p. 42). Kimbell and Dos Santos (2020) suggest this method addresses the primary goal of creating an emotional connection, better retention of the information and addresses the contemporary issue by allowing students to consider the idealism and relativism of each case study in a specific context.

Situational awareness as a concept aims to allow the distinction between a person and the environment they are experiencing as a means of perceiving information, comprehending the variables, reflecting, and evaluating. In an age where much of the average person's time is spent immersed in an activity, time for reflection is pertinent. This style of teaching can also incorporate case studies as an example of real-world situations. Business ethics case competitions and broad ethics case competitions offer students opportunities to hone communications skills and could also be opportunities to teach business ethics (Kimbell & Dos Santos, 2020). Case studies can make ethics more concrete to students as they reveal the implications of unethical behavior (Wotapka, 2016). The use of scenarios in competitions offers a platform for applying the practical application of a case method by teaching business ethics to students. These competitions incorporated into a curriculum structure could become effective settings for influencing the personal adoption of ethics about contemporary business matters. For technology students, case studies should attempt to incorporate concepts directly related to the IT profession.

Modern instruction should aim to address awareness of how influential others can be in decision-making. Potential is a key concept in Vygotsky's argument that a learner has a certain amount of potential, and the help of a more capable person helps them realize it. However, if the peer does not have the required capability to do so, the learner's potential may not be reached (Walt & Harland, 2022). Another tactic that can be leveraged during the building of the curriculum approach is for faculty to have an open-door policy and remain available to students with questions even after graduation. This requires a more personal connection and a stronger bond with students. It is crucial for students to feel comfortable approaching difficult conversations with honesty. As a society, it needs to become more commonplace to discuss the dilemmas that many people face throughout their life. Encourage students to seek out advice after the term or graduation from faculty, a trusted mentor, close friends, peers, or their own family. These individuals provide a sounding board if the student is struggling with a decision surrounding ethical behavior. If someone is questioning a decision, the need to reflect, discuss, and consider the outcomes before determining a course of action is vital. For ethics instruction to be effective and add value for students, the topics must be practical in their delivery.

There is a generally accepted agreement that teaching technology ethics is crucial but little agreement on what is the best way to accomplish it. One unmet challenge is increasing the capacity of technology students to make decisions about the ethical challenges embedded in their technical work. Computer ethics instruction must persuade students that ethics is relevant to their professional ambitions. Academia must create an environment that provides technology students experience with practicing reasoning while refuting ideologies that portray technical work as only mechanical and unpolitical. Simulation or gaming is a framework that can help to teach ethics and can include "strategies such as role-taking and role-playing, storytelling, deliberation, civil discourse, collaboration, choices and consequences, [and] application to real-world issues" (Shilton et al., 2020, p. 2913). Shilton et al. (2020) further explain joining values into video games through three steps: discovery, translation, and verification; for example, privacy as a discovery topic, translating privacy into a policy development task, and verification by deploying the policy in a simulation environment.

CONCLUSION

Academia continues to struggle with identifying and measuring the outcomes of ethics education. Personal ethical foundations fill the gaps that exist in-between laws and policies that fail to cover all scenarios leaving room for interpretation. Students engaged in higher learning have one last influential opportunity to better understand the ethical dilemmas that will be faced during their careers. The reason technology ethics as a subset of ethics is growing in importance is the fact that new technologies give individuals more power to act. These individuals must make more deliberate and informed decisions because the choices are more impactful than ever before. The need for ethical IT professionals is paramount. These changes in the application of technologies present powerful risks that require reflection and purposeful intent. The use of technology laws, policies, and regulations are not enough to be effective as the sole approach. Ethical IT professionals are needed to ensure the morality and appropriate application of technology in the future. Academia must provide instructional practices aligned with the technology used to create a solid foundational ethics platform that provides comfort and familiarity in navigating complex challenges. Research is encouraged through observational and measurable studies to clarify the outcomes of these suggested concepts. Although presented separately above, it is acknowledged that the dimensions of ethics, leadership, and academia blend into a comprehensive approach. If standards represent a concept designed to balance and preserve societal powers, then the implementation of ethical instructional technology processes must be a larger part of the process. The dependency society has on technology and the professionals working in the industry only continue to grow. IT professionals must be capable of making good decisions and reaching informed choices in those ethical situations. Ethical IT professionals are greatly needed to ensure businesses realize all the benefits of using the technology while also protecting the customer, data, and business objectives.

REFERENCES

Aasheim, C., Kaleta, J., & Rutner, P. (2021, March 13). Assessing IT students' intentions to commit unethical actions. *Journal of Computer Information Systems*, *61*(3), 219–228. doi:10.1080/08874417.2019.1584544

Afifi, M., Kalra, D., Ghazal, T., & Mago, B. (2020, January). Information technology ethics and professional responsibilities. *International Journal of Advanced Science and Technology*, *29*(4), 11336–11343.

Bezuidenhout, L., Quick, R., & Shanahan, H. (2020, February 17). "Ethics when you least expect it": A modular approach to short course data ethics instruction. *Science and Engineering Ethics*, *26*(4), 2189–2213. doi:10.100711948-020-00197-2 PMID:32067185

Chen, W., & Quan-Haase, A. (2020). Big data ethics and politics: Toward new understandings. *Social Science Computer Review*, *38*(1), 3–9. doi:10.1177/0894439318810734

Dierksmeier, C., & Seele, P. (2018). Cryptocurrencies and business ethics. *Journal of Business Ethics*, *152*(1), 1–14. doi:10.100710551-016-3298-0 PMID:30930508

Eisenberg, N., Hofer, C., Sulik, M. J., & Liew, J. (2014). The development of prosocial moral reasoning and a prosocial orientation in young adulthood: Concurrent and longitudinal correlates. *Developmental Psychology*, *50*(1), 58–70. doi:10.1037/a0032990 PMID:23731289

Ezenwakwelu, C. A., Nwakoby, I. C., Egbo, O. P., Nwanmuoh, E. E., Duruzo, C. E., & Ihegboro, I. M. (2020, August). Business ethics and organizational sustainability. *International Journal of Entrepreneurship*, *24*(3), 1–14.

Fields of concentration - 2021-22 Harvard College Student Handbook. (2021). https://handbook.college.harvard.edu/files/collegehandbook/files/fields_of_concentration_2021_2022.pdf

Formosa, P., Wilson, M., & Richards, D. (2021, October). A principlist framework for cybersecurity ethics. *Computers & Security*, *109*, 102382. doi:10.1016/j.cose.2021.102382

Hayes, N. (2019, March 18). *Ethics and biometric identity - security info watch.* Security Info Watch. https://www.securityinfowatch.com/access-identity/biometrics/article/21072152/ethics-and-biometric-identity

Hermannsdottir, A., Štangej, O., & Kristinsson, K. (2018). When being good is not enough: Towards contextual education of business leadership ethics. *Management, 23*(2), 1–13. doi:10.30924/mjcmi/2018.23.2.1

Hoehne, J. (2020, May 20). *Untitled.* [Photograph]. https://unsplash.com/photos/iggWDxHTAUQ

(ISC)^2. (n.d.). *Prepare for your (ISC)^2 exam day.* https://www.isc2.org/Exams/Exam-Day

Islam, G. (2020). Psychology and business ethics: A multi-level research agenda. *Journal of Business Ethics, 165*(1), 1–13. https://doi.org/10.1007/s10551-019-04107-w

Kanchanara. (2021, September 14). *All crypto coins are together in the dark* [Photograph]. https://unsplash.com/photos/fsSGgTBoX9Y

Kang, S. (2019, June). Sustainable influence of ethical leadership on work performance: Empirical study of multinational enterprise in South Korea. *Sustainability, 11*(11), 3101. https://doi.org/10.3390/su11113101

Kethineni, S., & Cao, Y. (2020). The rise in popularity of cryptocurrency and associated criminal activity. *International Criminal Justice Review, 30*(3), 325–344. https://doi.org/10.1177/1057567719827051

Kidd, D., Miner, J., Schein, M., Blauw, M., & Allen, D. (2020, December). Ethics across the curriculum: Detecting and describing emergent trends in ethics education. *Studies in Educational Evaluation, 67*, 100914. https://doi.org/10.1016/j.stueduc.2020.100914

Kimbell, J. P., & Dos Santos, P. L. (2020). Business ethics case competitions: A fresh opportunity to teach business ethics. *Southern Journal of Business and Ethics, 12*, 40–62.

Kohlberg, L. (1984). Essays on moral development: Vol. II. *The psychology of moral development.* Harper and Row.

Lilly, J., Durr, D., Grogan, A., & Super, J. F. (2021, September-October). Wells Fargo: Administrative evil and the pressure to conform. *Business Horizons, 64*(5), 587–597. https://doi.org/10.1016/j.bushor.2021.02.028

Milutinović, M. (2018). Cryptocurrency. *Economics, 64*(1), 105–122. doi:10.5937/ekonomika1801105M

Muhammed-Shittu, A.-R. B. (2021, September 30). A study of philosophical theory and educational science of insights on ethics, values, characters, and morals rooted into the Islamic and contemporary western perspectives. *Tarih Kültür ve Sanat Arastirmalari Dergisi, 10*(3), 47–58. https://doi.org/10.7596/taksad.v10i3.3090

Murukannaiah, P. K., Singh, M. P., Singh, M. P., & Murukannaiah, P. K. (2020, July 24). From machine ethics to internet ethics: Broadening the horizon. *IEEE Internet Computing, 24*(3), 51–57. https://doi.org/10.1109/MIC.2020.2989935

North-Samardzic, A. (2020). Biometric technology and ethics: Beyond security applications. *Journal of Business Ethics, 167*(3), 433–450. https://doi.org/10.1007/s10551-019-04143-6

Palanski, M., Newman, A., Leroy, H., Moore, C., Hannah, S., & Den Hartog, D. (2021). Quantitative research on leadership and business ethics: Examining the state of the field and an agenda for future research. *Journal of Business Ethics*, *168*(1), 109–119. https://doi.org/10.1007/s10551-019-04267-9

Perrenet, J. C., Bouhuijs, P. A., & Smits, J. G. (2010). The suitability of problem-based learning for engineering education: Theory and practice. *Teaching in Higher Education*, *5*(3), 345–358.

Prentza, G. (2020, March 24). *Untitled* [Photograph]. https://unsplash.com/photos/SRFG7iwktDk

Price, M. S. (2020, January 20). Internet privacy, technology, and personal information. *Ethics and Information Technology*, *22*(2), 163–173. https://doi.org/10.1007/s10676-019-09525-y

Price, T. L. (2018). A "critical leadership ethics" approach to the ethical leadership construct. *Leadership*, *14*(6), 687–706. https://doi.org/10.1177/1742715017710646

Prisacariu, A., & Shah, M. (2016, August 2). Defining the quality of higher education around ethics and moral values. *Quality in Higher Education*, *22*(2), 152–166. https://doi.org/10.1080/13538322.2016.1201931

Roque, A., Moreira, J. M., Dias Figueiredo, J., Albuquerque, R., & Gonçalves, H. (2020, July). Ethics beyond leadership: Can ethics survive bad leadership? *Journal of Global Responsibility*, *11*(3), 275–294. https://doi.org/10.1108/JGR-06-2019-0065

Schwitzgebel, E., Cokelet, B., & Singer, P. (2020, October). Do ethics classes influence student behavior? Case study: Teaching the ethics of eating meat. *Cognition*, *203*, 104397. https://doi.org/10.1016/j.cognition.2020.104397

Shilton, K., Heidenblad, D., Porter, A., Winter, S., & Kendig, M. (2020, July 1). Role-playing computer ethics: Designing and evaluating the privacy by design (PbD) simulation. *Science and Engineering Ethics*, *26*(6), 2911–2926. https://doi.org/10.1007/s11948-020-00250-0

Stöber, T., Kotzian, P., & Weißenberger, B. E. (2019). Design matters: On the impact of compliance program design on corporate ethics. *Business Research*, *12*(2), 383–424. doi:10.1007/s40685-018-0075-1

Wald, N., & Harland, T. (2022). Reconsidering Vygotsky's 'more capable peer' in terms of both personal and knowledge outcomes. *Teaching in Higher Education*, *27*(3), 417–423. https://doi.org/10.1080/13562517.2021.2007474

Wang, Q., & Yan, P. (2019, November 21). Development of ethics education in science and technology in technical universities in China: Commentary on "Ethics 'upfront': Generating an organizational framework for a new university of technology.". *Science and Engineering Ethics*, *25*(6), 1721–1733. https://doi.org/10.1007/s11948-019-00156-6

Wang, Y.-Y., Wang, Y.-S., & Wang, Y.-M. (2020, October). What drives students' internet ethical behavior: An integrated model of the theory of planned behavior, personality, and internet ethics education. *Behaviour & Information Technology*, *41*(3), 588–610. https://doi.org/10.1080/0144929X.2020.1829053

Wotapka, D. (2016, August 9). *How to teach ethics when your syllabus is packed*. AICPA. https://us.aicpa.org/interestareas/accountingeducation/newsandpublications/how-to-teach-ethics

Yan, P. (2017). *Research on C. Mitcham's thoughts of engineering ethics*. Dalian University of Technology.

Chapter 6
Privacy and Data Rights for Adult English as a Second Language (ESL) Students

Kimberly M. Rehak
Indiana University of Pennsylvania, USA

ABSTRACT

In the USA, instructors need to ensure the user privacy and data rights of their adult English as a second language (ESL) students. The ways in which educational technology (EdTech) companies track user activity and sell user data to third parties raises ethical concerns for student privacy and data rights. ESL students are particularly vulnerable because of the vague language in privacy policies and user agreements, differences in terms of state surveillance, and insufficient user privacy and data protections. In addition to a discussion on the ethical concerns within EdTech and higher education, one method and two tools to help ESL instructors and educators are provided. These assist with ESL or international students in their classrooms as a means to evaluate EdTech tools and make decisions on whether to adopt or require a digital tool.

INTRODUCTION

After downloading a new application to their phone or computer, users will—more often than not—quickly scroll to the bottom of the small-print terms and conditions to hit the "agree" button without a second thought. Similarly, when visiting a new website, there is a propensity to agree to "accept all cookies" and dismiss annoying pop-ups as quickly as possible.

Recent documentary films, such as *Coded Bias* (Kantayya, 2021) and *The Social Dilemma* (Orlowski, 2020), have exposed the American public to issues of data security, public surveillance, and the danger of unregulated tech. While both documentaries raise concerns that modern-day digital citizens (Ribble & Baley, 2007) should consider before engaging with technology, the effects of these stories on user behavior are underexplored. Rather, these documentaries highlight the small gains activists have made

DOI: 10.4018/978-1-6684-5892-1.ch006

in policy changes and show the need "to take on the massive amount of work still left to be done" (Han, 2020, para. 13).

Less attention has been given to similar issues in student data and users of educational technology (EdTech). EdTech is a term that encompasses software and applications with activities that allow for practice and lead to learning gains inside and out of the classroom (Lestari & Subriadi, 2021). The EdTech industry has expanded to an estimated market size of over $100 billion per annum ("Education Technology Market Size," 2022). Despite best intentions, EdTech companies perpetuate the educational achievement gap—making already marginalized populations even more so (Macgilchrist, 2019; Reich, 2020). Industry regulation varies significantly from country to country, meaning that research into the ramifications of EdTech's rapid expansion is usually contextualized by location in addition to grade level (i.e., K-12; higher education).

In the context of higher education in the United States of America, student digital records are protected under the Family Educational Rights and Privacy Act of 1974 (FERPA) (Checrallah et al., 2020). FERPA limits access to personal data in student records. More recently the Gramm-Leach-Bliley Act (GLB Act) has extended to protect the financial and financial aid records of post-secondary students from cyber attacks ("Gramm-Leach-Bliley Act," n d.). Additional laws protecting student data vary state by state (Gallagher et al., 2017), meaning student data privacy and protections are negligible at the federal level. Without proper protections and regulations, students' rights may be violated without their knowledge.

The ethics of using digital tools in learning environments with adult students is thus explored, as well as how educators can engage with pedagogical theories that ensure their students' privacy and data rights. While FERPA protection extends to international students studying in the USA, the unique concerns regarding the data rights of adult English as a Second Language (ESL) learners and the reasons why this subset of students is especially vulnerable will also receive consideration.

Some privacy and digital rights issues for tertiary-level ESL or international students include:

- Who is responsible for explaining and ensuring the comprehension of FERPA protections and privacy and digital rights to international students who speak English as a second language (ESL)?
- Which factors should instructors of ESL students consider before adopting an EdTech tool for their classroom?
- What should adult ESL students be taught with regards to their rights in a U.S. context and what does such instruction look like?

For the purposes of this publication, ESL will refer to students who are studying the English language in a location where English is common and used by the majority of the population. While the term has gone out of fashion in the Teaching English to Speakers of Other Languages (TESOL) field, this acronym has been used here to differentiate between ESL and English as a Foreign Language (EFL), referring to language instruction that takes place in locations where English is not routinely used by most of the population. As the context herein is within higher education institutions in the USA, ESL has been used to indicate international students whose native language is anything other than a form of English comparable to American English. This includes both English language learners (ELLs) matriculating to post-secondary institutions in the USA from K-12 settings and full-time international students arriving on an F-1 visa to study at the tertiary level.

Despite a focus on the rights of ESL students, any educator who has their adult learners engage with EdTech and/or has ELLs or international students in their classrooms with a first language incomparable with American English can benefit from this discussion.

EDTECH: ETHICAL CONCERNS, PEDAGOGICAL THEORIES, AND APPLICATIONS

Ethical Concerns Within EdTech and Higher Education

Two main ethical concerns with EdTech in higher education relate to user privacy and the anonymity of user data, or, rather, the lack thereof. User privacy refers to user tracking and encompasses not only activity tracking (e.g., search history, biometrics) but also the tracking of geolocation information and IP addresses. On the other hand, user data issues center around third parties accessing and using data without the knowledge or consent of the user and/or the selling of user data to third parties via data brokers. Myriad user privacy ethical issues within American post-secondary institutions raise questions concerning who is responsible for ensuring student privacy and data protection.

User Tracking

According to *The Social Dilemma* (Orlowski, 2020), gaining a person's attention is one of the biggest motivators for the use of tracking technology. Tech companies want to hook users into spending more time using digital tools, including applications (i.e., apps, for short) and other software, to provide companies with more data. More user tracking allows for better algorithms, which can inform private tech companies on ways to sell more targeted advertisements to feature on the "free" versions of their products.

Despite the ethical gray area of using data to manipulate technology to steal consumer attention, in education, a larger concern deals with the tracking and selling of student user data. The lack of transparency in EdTech regulation is an additional complication. Student data tracking by EdTech companies, academic researchers, and/or universities compromises students' data rights.

Much of the academic research into student digital rights and the ethics of tracking technology centers around wearable fitness apps used in disciplines, such as dietary sciences or physical education.

While one might assume that research subjects wearing tracking devices as part of a research study would have concerns about the tracking of their data, research has shown otherwise. For instance, in a human kinetics research study using student focus groups for data collection, 83% of all respondents acknowledged their understanding that their data would be used for the study—but not by third parties (Clark & Driller, 2020). This suggests that students often fail to consider the long-term ramifications that come with relinquishing their data rights. Clark and Driller (2020) recommend that data privacy be explicitly explained to participants in briefing sessions and that research should only use tracking devices when necessary for data collection. Yet, there is little guidance on the extent that research subjects should be briefed on their data rights or how to determine whether using tracking tools for data collection is necessary.

Tracking Through Learning Management Systems and Learning Analytics

The surveillance of students through their technology use has become commonplace. As a result, students may not consider the larger consequences of universities, researchers, and private EdTech companies tracking their activity and data. Hope (2018) also suggests that the surreptitious nature in which tracking is performed could also contribute to a collective "formal indifference" (p. 67) towards surveillance activities.

Most higher education institutions subscribe to learning management systems (LMSs) that assist instructors in the tracking of student work and progress. In fact, learning analytics is a burgeoning topic in higher education predicted to make a significant impact on the future of the field. The higher education non-profit, EDUCAUSE, published a multidisciplinary study looking into current perceptions on learning analytics. They found less than a third of the student participants reported concerns about the tracking of student demographic data or the equity of tracking identity markers (Brown et al., 2022). Faculty were 10% less concerned than students about demographic tracking, suggesting that most instructors fail to consider the consequences of student surveillance and, consequently, have formal indifference to student data tracking.

Furthermore, LMS tracking data and the field of learning analytics raise the issue of student data ownership. Many LMS vendors offer learning analytics in their systems. These data challenge student autonomy by allowing universities to collect a vast amount of data points on registered students. Jones (2019) argues that higher ed institutions should adopt a model for gaining consent from students to use their data for educational purposes. Additionally, students should be informed of the data practices of their institution through data visualization and be allowed to opt out of sharing their data (Jones, 2019).

Other harmful ways learning management systems and EdTech tools track student users include measuring the time users spend on various webpages and using predictive algorithms for adaptive learning technology. By tracking user behavior and monitoring student activity, adaptive technology makes it easier to predict future user activity. While this technology has the capability to individualize learning for students, this educational innovation also has the tendency to pigeonhole students who either underperform or are slower to show learning gains (Goodman, 2015). While this might not seem like a blatant ethical violation, Regan and Jesse (2019) warn that the "more sophisticated prediction that is built into many big data analytics transforms tracking and surveillance into a more powerful tool that can be wielded in ways that have not yet been identified and understood" (p. 172). The scale of user data collection (i.e., *Big* Data) is what provides EdTech companies with power, influence, and capital gains. Furthermore, not knowing the potential use of data or the end game of EdTech companies raises ethical concerns, particularly when these companies are working in the confines of the law. This tracking of student users raises larger questions about the regulation of the EdTech industry.

Issues With User Data

When students use EdTech tools, their data have a high probability of being sold to third parties or getting into the hands of the companies that create digital tools. Although advocates can speak out or warn against using certain apps or software that collect user data more than others, student users are left with little autonomy if they are expected or required to use digital tools for coursework.

Student user data does not have full protection at the federal level. Regan and Jesse (2019) make a strong case against basing activism only on student privacy issues. They argue that, in the USA, the emphasis on user privacy leads to "Band-aid fixes" at the policy level. In other words, focusing solely on privacy issues fails to address the complexity of the problem with student data collection. Not considering issues like student autonomy or the ramifications of Big Data prevent legislators from taking a more comprehensive approach to protecting student data (Regan & Jesse, 2019). This, in turn, leaves student data vulnerable. Other, more expansive efforts at user data advocacy have also had minimal effect as they have not united with broader social movements. For instance, technology employees formed Tech Workers Coalition (TWC) in 2016 to call for their employers to employ more equitable practices (Costanza-Chock, 2020). However, most of the examples of tech worker advocacy are centered around federal government contracts that might violate user's federal civil rights.

Although they might not be digital rights activists, educators need to be aware of student user data vulnerabilities. Students, instructors, and university administrators are beholden to the procurement practices of their university and the ramifications of agreements with private EdTech companies. Richter et al. (2021) provide review criteria to consider for selecting an appropriate LMS, including a transparent procurement process, adherence to state laws, and "ethical decision-making" (Richter et al., 2021, p. 92). Yet, university procurement can sign contracts with whichever vendors they choose—regardless of the recommendations from the informational technology department or university review committee. EdTech companies are "at least one step removed from the data subject" (Regan & Jesse, 2019, p. 171), meaning that universities or instructors, in particular, need to know how and through what means student user data can be compromised. Without proper legal protections codified within state and federal law, student data can be "owned" by universities (Regan & Jesse, 2019) or at the mercy of EdTech companies' self-regulation.

Although Hewson (2015) delves into ethics related to digital research, much of the ways that researchers approach subject data is comparable to issues with student user data. First, consent becomes problematic when researchers mine "anonymous" student data—even when researchers specifically try to gain consent (e.g., from public discussion forums). Second, it is difficult for subjects to withdraw from digital research, especially if no consent was given or if research subjects have no knowledge of their data being researched. When research subjects or, by extension, students have a username that is completely unrelated to their legal name with no directly identifiable information, confidentially in online environments is regularly disrupted. Not only can users be identified through e-mail or IP addresses; but if researchers print direct quotes from publicly available pages, identifying data sources is as easy as Googling the quote in parentheses. Researchers should perform a risk assessment (Hewson, 2015) before using user data, just like instructors should do before requiring their students use a digital tool. Threat modeling, a tool that assists with performing a risk assessment and can help instructors with their EdTech decision-making processes, is also important to investigate.

Responsibility for User Privacy and Data Rights

In the USA, the EdTech industry is primarily self-regulated, meaning the responsibility of data protection falls on the end user. While there are some concerted efforts for company transparency and a pledged commitment to student data ("Student Privacy Pledge," n.d.), these efforts are focused on K-12 EdTech. EdTech marketed towards post-secondary students has less governmental and industry oversight, which calls accountability and responsibility for these students' data rights into question. For instance, if an end user is assigned to use a digital tool to complete an assignment, it is unclear whether the teacher or the school should be held accountable if student data is collected and used unethically. Accountability issues for ELL and ESL student populations receive even less attention.

The Teaching English to Speakers of Other Languages (TESOL) International Association addresses the responsibility question in their 2008 publication *TESOL Technology Standards Framework*. With goals for both students and adult educators, the technology standards center around the use of technology to enable ESL learners to become responsible digital citizens. The TESOL technology standards put the onus of responsibility for student privacy and data rights on the instructors, as Goal 1, Standard 4, reads, "language teachers use technology in socially and culturally appropriate, legal, and ethical ways" (Healey et al., 2008, p. 31). Although ESL educators might be liable for the misuse of technology, such as the tracking of student user data, none of the student technology standards address privacy or data rights. Perhaps this is because user tracking, Big Data, and user surveillance were less prominent issues when the standards were written in 2008.

The ethical dilemma around student data collection is related to the type of EdTech being used in the classroom. For instance, LMSs are usually purchased through a contract with a private vendor, meaning that any data gathered from the LMS would be effectively "owned" by the university (Regan & Jessie, 2019). What actions universities take with LMS data will most likely be an ethical concern for learning analytics researchers in the coming years. On the other hand, if instructors ask their students to use a publicly available digital tool, such as Flipgrid or Edpuzzle, student user data will be proprietary to private EdTech companies. Rooksby (2020) warns that EdTech companies can meet the spirit of the law rather than the letter of the law (p. 1). If companies are tracking students in ways that are technically legal yet ethically questionable, teachers should be cautious when requiring digital tools in their classrooms.

Pedagogical Theories for the Instruction of Adults in Digital Learning Environments

Beliefs and experiences shapes instructors' attitudes towards utilizing EdTech in their classrooms. Similarly, instructors who adopt EdTech can apply a number of pedagogical theories that shape their practice. These theories guide instructors in their approach to both the instruction and application of digital tools to various learning environments. Two theories for instructors of adult ESL students to consider are digital citizenship (Ribble & Bailey, 2007) and critical digital pedagogy (Stommel, 2014). These theories shape the responsibilities of the instructor and the goals and/or outcomes of student engagement with EdTech as part of their learning.

Digital Citizenship

For their concept of digital citizenship, Ribble and Bailey (2007) outline nine elements, including digital literacy and digital rights and responsibilities, which claim to create responsible student users of digital technology. The main goals in the advocacy for digital citizens are to equip students with 21st century skills and improve their learning outcomes. Ribble and Bailey (2007)'s nine elements aim to meet those goals through a three-tiered model that moves from the personal (i.e., student performance) to the school-level to the life outside of the school (p. 44).

Davis (2017) recommends two approaches for teaching digital citizenship (Ribble & Bailey, 2007): the proactive and the experiential. The former includes the nine elements, like passwords, private and personal information, and professionalism; whereas the latter refers to providing students with the opportunity to apply and test out their learning. For instance, Davis (2017) mentions having students practice recognizing facts and allowing for collaborative learning.

As an extension to digital citizenship, Bhargava et al. (2015) argue that instructors and students gain data literacy. Currently, most training in data literacy focuses on training teachers how to interpret student performance data (Bhargava et al., 2015), missing an opportunity to empower students by showing them how to use data for good, such as engaging with community issues happening right outside of the classroom walls.

Critical Digital Pedagogy

Critical digital pedagogy (Stommel, 2014) has derived to account for critical pedagogy in digital learning environments. Central to critical pedagogy is getting students to think beyond the walls of the classroom and consider how "they, as students, fit into broader social and cultural context" (Young, 2019, para. 2). Furthermore, critical pedagogy implies action—students engaging in their communities, out in "the real world." The *critical* in critical pedagogy implies instructor advocacy and the empowerment of students. Power is distributed equitably in the classroom, which is seen when instructors provide students with the language, tools, and context in which the instructor and students find themselves in.

Bradshaw (2017) investigates the intersection of critical pedagogy and educational technology, stating that culture is key and often missing when EdTech is used by educators who adhere to a positivist philosophy (p. 16). Culture can dictate how technology is created, designed, and used (Bradshaw, 2017, p. 20). Rorabaugh (2012) also recognizes the influence of culture in the classroom centered around critical pedagogy. "If students live in a culture that digitizes and educates them through a screen, they require an education that empowers them in that sphere, teaches them that language, and offers new opportunities of human connectivity" (Rorabaugh, 2012, para. 7). Critical digital pedagogy takes these ideas one step further.

Moreover, critical digital pedagogy engages students beyond their phone or computer screen. Community, collaboration, and communication are used in tandem to unite students "across cultural and political boundaries" (Stommel, 2014, What is Critical Digital Pedagogy section). Critical digital pedagogy calls for instructors to unite student voices in online learning environments and facilitate knowledge construction that is applicable beyond the "screens" of a typical digital learning environment. In sum, critical digital pedagogy asks instructors to connect students with the outside world in a meaningful way, to spark curiosity in learning itself and to get them to ask critical questions within a community of practice (CoP) (Lave & Wenger, 1991)—a tenet of adult education.

Unique Learner Needs for ESL Students

Over one million international students are enrolled in American colleges and universities in any given year (Israel & Batalova, 2021). Universities will set required minimum scores on language proficiency tests (usually the Test of English as a Foreign Language (TOEFL) for institutions in the USA) for international student admission. Individuals with lower proficiency scores are usually placed in language support classes and considered ESL students. These students will then work to improve their English proficiency, with an emphasis on academic English vocabulary and productive skills required for post-secondary studies (i.e., speaking, writing, and notetaking).

Even with a high proficiency in English, many international students are experiencing instruction both in English and within American academic culture for the first time (Bergey et al., 2018). Without proper institutional attention and support programs, these students are often left on their own to navigate their new "education norms, communication habits, and classroom participation structures" (Bergey et al. 2018, p. 4). In addition to orientation sessions, international students require help and support from higher ed institutions to succeed. Unless instructors have training or certification to work with ESL students, educators are oftentimes unaware of the unique circumstances affecting ESL, ELL, and/or international students highly proficient in English. Because of this, it is important for all educators in post-secondary institutions to consider the digital tools and rights for this subset of the student population.

Digital Tools for Adult English Language Learners

The goal of ESL classes is to equip students with the cultural know-how and language skills needed to communicate effectively in a variety of the English language—primarily academic English or English for Specific Purposes (ESP).

Some ESL students might be informed digital citizens in their native language or home country. For instance, they might be active on global platforms, such as Twitter or Instagram, or social media outlets created by technology firms in their home countries. However, ESL instructors should not assume their students have similar competency with digital tools in an educational or American context. Critical digital pedagogy (Stommel, 2014) explains the need for ESL instructors to utilize digital tools to have students interact with one another in digital and non-digital spaces. This means that while being advocates for student engagement with the digital world beyond the classroom, instructors also must ensure the rights and digital privacy of their students.

Few would argue against the use of digital tools for learning and in foreign language learning. Digital tools, such as Snapchat, FlipGrid, or VoiceThread, allow students to record themselves and engage with one another as well as users around the world. Students can practice their productive skills, making audio or video recordings and writing comments, which allow for situated learning (Goodwin-Jones, 2017, p. 6) and interaction in a CoP. Adaptive technology allows for personalized learning that meets the unique needs of learners, using spaced retrieval practice for correcting student errors and encourages learners to learn, as reported by Bourekkache & Kazar (2020), on their own outside of the classroom.

Anecdotally speaking, teachers who are early adopters of tech may not be able to realize or measure the ramifications of having students use a new digital tool. The new use or potential for student engagement might overshadow the risk that the adoption of the new product. For instance, Duolingo is a platform that uses gamification in a mobile-assisted language learning (MALL) context to allow user to study an additional language with an approximate 15-minutes-per-day commitment. Despite the popularity of the platform in the mainstream and as a topic of academic research, the effectiveness of the program and the outcomes of users are rarely measured. In a systematic review of research on the Duolingo platform, Shortt et al. (2021) found that the majority of the 35 research studies focused on the design of the app rather than its effect. This suggests a trend in the EdTech industry to focus more on the making and design of digital tools rather than "the process and outcomes of language learning from using these tools" (Shortt et al., 2021, p. 1). These mirror, in many ways, the actions of instructors who are early adopters of EdTech, who place more attention on the novelty and potential of digital tools instead of the risks and efficacy that come along with them.

DATA RIGHTS FOR ESL STUDENTS

Language in Terms and Conditions

Despite the advantages that digital tools can provide to ESL classes, instructors need to be careful before requiring any for student use. In the American context, most free tools come with hidden costs to students' privacy. As there is "no national standard for how to acquire consent" (Checrallah et al., 2020, p. 138), it is essential for instructors to read and understand the language in privacy policies and end-user license agreements or terms of service. ESL students are particularly vulnerable to privacy rights because of the complicated and intentionally vague language contained in these statements, which are difficult for educated native English speakers to understand (Checrallah et al., 2020). In a study of the transparency of language in privacy and digital security policies in mental health apps for individuals with depression, five out of the 116 apps investigated used transparent language (O'Loughlin et al., 2019). Perhaps even more concerning is that O'Loughlin et al. (2019) also found that only 49% had privacy policies and, of those, the vast majority (nearly 80%) were apps that collected identifiable user data.

According to LePan (2020), the average American, non-ESL reader (at 250 words per minute (wpm) would need ten minutes to read Instagram's terms of service, which is one of the shortest user agreements, while the user agreement from Microsoft is estimated to take an entire day to get through. In addition to the length of privacy and end-user agreements, the language complexity of these agreements also makes it more difficult for ESL students to read and comprehend the entirety of their digital rights. The results of a quantitative research study conducted by Mora et al. (2021) show that reading speed and reading comprehension are influenced by a few factors, such as English level, profession, and gender. While a highly proficient ESL reader can read common, everyday reading materials, such as newspaper or magazine articles, at a rate of around 50 fewer words per minute (wpm) than the average American non-ESL reader, more complex reading materials that contain advanced, specialized vocabulary and more complex grammatic structures, require deeper language processing. This means that the average reading speed is reduced significantly—to 70 fewer wpm. Therefore, it would take a proficient ESL reader an extra four minutes to read Instagram's terms and services and an extra nine hours and 20 minutes to get through Microsoft's policies.

The vague language in the terms and conditions allows the EdTech companies to track and sell user data with user consent and disadvantages ESL learners. Instructors of ESL, ELL, and international students should also be reviewing the terms and conditions of tools used in their classroom. Additionally, individuals at post-secondary institutions responsible for EdTech procurement should be equipped to understand and advocate for student user rights before agreeing to contracts with private vendors.

Privacy and User Tracking

When instructors require students to register for a digital tool, they are asking the students to provide private information (e.g., demographic information, e-mail address) to tech companies that often do not have the best interest of the students in mind. Depending on the language in the user agreement, user information, like the IP address or geolocation coordinates, is provided to companies directly (Checrallah et al., 2020) or sold to external parties through third-party data brokers. Data points can be used to find users' locations, meaning that online anonymity is near impossible (Thompson & Warzel, 2019). For instance, Snapchat users agree to location tracking and facial recognition software is used in their filters.

Companies also sell location data to third parties for a return on investment for their services. This selling of data is the hidden cost of free software and applications. ESL instructors should avoid assigning work in digital tools that make it easy to track ESL students' activity. Additionally, ESL instructors need to consider the unique aspects of their students' lives that might make them particularly susceptible to tracking and surveillance.

Since U.S. government agencies, like Immigration and Customs Enforcement (ICE), have purchased tracking data to surveil individuals at the USA-Mexican border (Molla, 2020), requiring ESL students to use apps that track user data, which, in turn, can be sold to federal government agencies, could be harmful for students whose immigration status is questionable or unknown. Furthermore, instructors should not encourage or require ESL students to do anything that makes their students susceptible to surveillance or internment by a foreign government (in this case, a U.S. federal agency).

Students or teachers who think they "have nothing to hide" are still vulnerable to Big Tech and EdTech companies. Data is a profitable resource for figuring out user behavior to exploit it through targeted ads or using it for more nefarious ends, such was the case of Cambridge Analytical and the 2016 U.S. election. Importantly,

This data is immensely valuable to those who know what to do with it – and that value has a lot to do with scale. The more data that a company or group has to play with, the higher its chances of achieving its goals, either by identifying a larger number of people who might be interested in what it has to say, or by figuring out exactly what they are thinking, and speaking to their views specifically. (Ghosal, 2018, para. 6).

ESL instructors should consider their students' privacy and data rights a top priority, especially as the learning analytics field continues to grow.

Different Experiences With State Surveillance

Depending on their home country, the experiences of ESL students with state surveillance can be an extremely disparate situation than the one in the USA. Students from countries with more state surveillance may be unaware of their digital rights under U.S. federal law (i.e., FERPA) which could also lead to different attitudes towards user tracking and data brokering.

To illustrate such a difference of opinion, Zhu & Yang (2019) used an online survey method to investigate the perceptions of digital ownership and digital rights of American and Chinese students at the tertiary level. While American students were more likely to report a sense of ownership of their digital property, Chinese students were concerned with understanding digital rights, which could be attributed to the difference in China's and the USA's digital cultures. While China has been regulating its digital EdTech markets and their delivery, the American federal government provides less regulation of EdTech companies. This suggests that when Chinese students attend American universities, they may not realize their data rights are different than they are at home and that they might have fewer protections than they realize.

ESL educators can raise this issue outright in their classrooms by having students compare the situation in the USA with that of their home country. Dedicating part of a lesson to have students compare and contrast the level of state surveillance in different countries can raise awareness of the extent that governments protect the privacy of their citizens and regulate their tech companies. Instructors can use Comparitech's ranking of surveillance states at https://www.comparitech.com/blog/vpn-privacy/surveillance-states/ to facilitate this exercise. Additionally, instructors can have their students hypothesize reasons that explain the various rankings of the countries from Comparitech's website. ESL instructors can turn this activity into an exercise for practicing both digital citizenship as well as grammatical structures, such as comparative and superlative adjective forms.

TEACHING PRACTICES AND TOOLS THAT INFORM AND PROTECT ESL STUDENTS

ESL instructors should be sure that they are equipping students with the right tools to facilitate communication while protecting their students' rights. Unfortunately, research has shown that student privacy and data rights are a low priority for educators in multiple disciplines. Lupton (2020)'s survey looking into the considerations and practices of health and physical education instructors in Australia found that educators rarely think about who has access to student data when using digital tools in the classroom. Similarly, only 10% of respondents to an educator survey on data privacy were aware of what happens with student data ("Educator Toolkit," 2018). Marín et al. (2021)'s mixed-methods study of 148 pre-service teachers from 3 different countries (i.e., Germany, Spain, and the USA) also found that avid social media users, so-called "digital natives" (Prensky, 2001), knew little about data privacy from their own social media use. Digital natives are individuals who have grown up with modern technology integrated into their lives, and the priorities of digital citizenship curricula tend to vary by age group. For example, cyberbullying affecting K-12 students more than adult learners. Many adult learners and adult educators are "digital immigrants" (Prensky, 2001) whose experiences with digital technology differ from younger generations of digital natives. The result from Marín et al. (2021) suggest that it is difficult to get "digital native" educators to think about their students' data rights if they are not considering their own.

Instructors should use a reference, like The Common Sense Media evaluation tool at https://privacy.commonsense.org/evaluations/1 to ensure they are assigning digital tools, which are minimally detrimental to students' privacy and digital rights. Taking a few minutes to visit the Common Sense Media resource might have a big impact on the data privacy of ESL students in the USA.

While it is important for ESL students to engage in the world beyond their devices, they must be shown how to do so in a way in which their digital rights are protected. Instructors should dedicate time to explaining to their students what they are agreeing to, what information tech companies are collecting, and what rights they might be relinquishing. As privacy rights differ from country to country, some students might need to learn what rights they have before they agree to consent to an end-user license agreement or terms of service or a privacy policy. Furthermore, instructors should be prepared to provide alternative assignments if students choose not to consent. With the vast majority of educated Americans not reading terms and conditions, it is safe to assume that ESL students are not fully aware of their privacy and user data rights when using digital tools while studying in the USA.

Teaching Students to Protect Their Privacy

The onus of protecting the privacy and digital rights of ESL students is in the hands of instructors or curriculum developers who require their students to use various language learning apps and digital tools. However, digital citizenship (Ribble & Bailey, 2007), data literacy (Bhargava et al., 2015) and critical digital pedagogy (Stommel, 2014) all call for instructors to empower students by helping them help themselves. Threat modeling is one way that instructors can show their students how to protect their data rights. Instructors should consider using this method for themselves as well as their students.

Threat Modeling

As explained by members of the privacy advocate group Electronic Frontier Foundation (EFF), threat modeling is a way to protect what is important to users and which individuals or organizations they need protection from ("How to Protect Your Online Privacy," 2017). Instructors can get students to think about what aspects of their user data they need to keep secure and the individuals or organizations from which they should protect it. The latter should warrant a longer conversation with students who come from countries with a tradition of heavy state surveillance, like China (Bischoff, 2019) or the United Arab Emirates (Mackenzie, 2020). The remaining three steps to threat modeling involve, considering the likelihood of the threat, the consequences if the threat is violated, and the amount of work it would take to avoid negative consequences ("How to Protect Your Online Privacy," 2017). A decision-making tool, which incorporates many of the considerations included in threat modeling, is provided in the appendix.

In addition to threat modeling, instructors can show their ESL students some practical tips to secure their user privacy both in the USA and in general. First, passwords should contain a list of completely randomized words to avoid hacking. Second, free tools or plugins, such as Adblock or DuckDuckGo, can help to block web tracking. Instructors should get in the habit of testing any digital tool or plugin prior to recommending it to students. If there is a teacher or paid version of any app or piece of software, instructors should be sure to check out the privacy and user experience on the free and/or student versions before they have students download them.

Instructors should also educate their students on issues related to school surveillance and encryption. EdTech tools, such as Chromebooks, have lenient default settings that allow for more user data collection by tech firms (e.g., Google, in this case) ("How to Protect Your Online Privacy," 2017). Accessing and changing default settings can become a classroom activity that might even result in students teaching their peers or family members about their own data rights. Finally, students should learn how to protect their data using encryption. One easy way to do this is to inform students to look for the "S" in https:// to know they are using encrypted sites. Students should also avoid using open Wifi networks or learn how to use encryption tools if they must use open networks to prevent others from tracking their data.

Tools for Instructors to Make Informed Decisions About EdTech

In addition to threat modeling, two tools are offered to help the decision-making process of educators who want to use EdTech and also protect their students' privacy and digital rights. It is also recommended that educators limit the amount of EdTech tools that are used in their classrooms. Sometimes the novelty of a new app or platform will take attention away from the actual usefulness of the tool or the learning gains it might facilitate.

Instructors need to be mindful that they and their students have time to investigate and learn the system. Additionally, instructors need to set aside time to explain the benefits and risks that the new tool can have on their students' privacy and data rights. Not only will this time and attention allow for well-informed digital citizens, but these capture the ethic and advocacy considerations engrained in critical digital pedagogy

Should I Use This Edtech Tool? Decision Making Guide for Instructors

To help ESL instructors make more informed decisions on whether to include EdTech tools in their classroom, Figure 1 includes a decision-making tool with ten questions and various ranking scores to determine if a tool is more or less likely to violate user privacy and data rights. This decision-making guide was created using input from Gallagher et al. (2017) and Rooksby (2020) with the intention of nudging educators to think about ways in which they can protect the privacy and data of their students.

Common Sense Media Evaluation Tool

One step embedded in the decision-making tool is the Common Sense Media Evaluation Tool at https://privacy.commonsense.org/evaluations/1, which determines the probability that a tech company will sell user data to third parties. The Common Sense website should be the first stop of any educator concerned for their students data privacy rights.

CONCLUSION

There are many concerns about the privacy and data rights of ESL adult students. When instructors require students to use various digital tools for learning, they should ensure they are protecting and the data rights of their students and properly explaining these rights to them. However, instructors have reported knowing very little about the content of privacy and data user agreements—if they even consider this information at all. Digital tools raise ethical concerns as they can track user behavior, their personal information, their geolocation, and other identity markers. This could make ELL and ESL students vulnerable to tracking, which could unintentionally affect students with questionable immigration status.

Concerns for ESL instructors include the complicated language in terms and conditions and their students' reading speed, proficiency, and comprehension. Additionally, instructors should explicitly teach ESL students about differences in data privacy and state surveillance between the local context and their home countries by using a website that ranks countries on the amount of state surveillance. Finally, using threat modeling and/or decision-making tools, such as the *Common Sense Media Evaluation Tool and the Should I Use This EdTech Tool? Decision-Making Guide for Instructors* found in the appendix can help to guide instructors to using EdTech that will support the learning of their students while protecting their students' privacy and data rights.

REFERENCES

Bergey, R., Movit, M., Baird, A. S., & Faria, A.-M. (2018). *Serving English language learners in higher education: Unlocking the potential.* American Institutes for Research. https://www.air.org/sites/default/files/downloads/report/Serving-English-Language-Learners-in-Higher Education-2018.pdf

Bhargava, R., Deahl, E., Letouze, E., Noonan, A., Sangokoya, D., & Shoup, N. (2015, September). *Beyond data literacy: Reinventing community engagement and empowerment in the age of data.* Data-Pop Alliance (Harvard Humanitarian Initiative, MIT Media Lab and Overseas Development Institute) and Internews. https://dspace.mit.edu/bitstream/handle/1721.1/123471/Beyond%20Data%20Literacy%20 2015.pdf

Bischoff, P. (2019, October 15). *Data privacy laws & government surveillance by country: Which countries best protect their citizens?* Comparitech. https://www.comparitech.com/blog/vpn-privacy/surveillance-states/

Bourekkache, S., & Kazar, O. (2020). Mobile and adaptive learning application for English language learning. *International Journal of Information and Communication Technology Education, 16*(2), 36–46. doi:10.4018/IJICTE.2020040103

Bradshaw, A. C. (2017). Critical pedagogy and educational technology. In A. D. Benson, R. Joseph, & J. L. Moore (Eds.), *Culture, learning, and technology: Research and practice* (pp. 8–27). Routledge. doi:10.4324/9781315681689-2

Brown, A., Croft, B., Dello Stritto, M. E., Heiser, R., McCarty, S., McNally, D., Nyland, R., Quick, J., Thomas, R., & Wilks, M. (2022, February 9). *Learning analytics from a systems perspective: Implications for practice.* EDUCAUSE. https://er.educause.edu/articles/2022/2/learning-analytics-from-a-systems-perspective-implications-for-practice

Checrallah, M., Sonnett, C., & Desgres, J. (2020). Evaluating cost, privacy, and data. In T. Trust (Ed.), *Teaching with Digital Tools and Apps.* EdTech Books. https://edtechbooks.org/digitaltoolsapps/evaluatingcostprivacydata

Clark, M. I., & Driller, M. W. (2020, February). University students' perceptions of self-tracking devices, data privacy, and sharing digital data for research purposes. *Journal for the Measurement of Human Behaviour, 3*(2), 128–134. doi:10.1123/jmpb.2019-0034

Costanza-Chock, S. (2020). *Design justice: Community-led practices to build the worlds we need.* The MIT Press. doi:10.7551/mitpress/12255.001.0001

Davis, V. (2017, November 1). *What your students really need to know about digital citizenship: Ideas on how to guide students to the knowledge and experience they need to act responsibly online.* Edutopia. https://www.edutopia.org/blog/digital-citizenship-need-to-know-vicki-davis

Education technology market size, share & trends analysis report, by sector (preschool, k-12, higher education), by end-user (business, consumer), by type, by deployment, by region, and segment forecasts, 2022 – 2030. (2022, April). Grand View Research Publishers. https://www.marketresearch.com/Grand-View-Research-v4060/Education-Technology-Size-Share-Trends-31517238

Educator toolkit for teacher and student privacy: A practical guide for protecting personal data. (2018, October). Parent Coalition for Student Privacy & the Badass Teachers Association. https://cdn.ymaws.com/www.a4l.org/resource/resmgr/files/sdpc-publicdocs/PCSP_BATS-Educator-Toolkit.pdf

Gallagher, K., Magid, L., & Pruitt, K. (2017, May 4). *The educator's guide to student data privacy.* Connect Safely. https://www.connectsafely.org/wp-content/uploads/2016/05/Educators-Guide-Data-.pdf

Ghosal, A. (2018, April 25). *Why we should collectively worry about Facebook and Google owning our data.* The Next Web. https://thenextweb.com/news/why-should-you-care-if-google-and-facebook-own-your-data

Goodman, E. (2015, April 28). *Privacy in the classroom: What you need to know about educational software.* The International Association of Privacy Professionals. https://iapp.org/news/a/privacy-in-the-classroom-what-you-need-to-know-about-educational-software/

Goodwin-Jones, R. (2017). Smartphones and language learning. *Language Learning & Technology, 21*(2), 3–17.

Gramm-Leach-Bliley Act (GLB Act). (n.d.). *EDUCAUSE.* https://library.educause.edu/topics/policy-and-law/gramm-leach-bliley-act-glb-act

Han, A. (2020, December 30). *Two Sundance docs sound the alarm on the dangers of modern AI: Is the tech industry... bad?* Mashable. https://mashable.com/article/coded-bias-social-dilemma-documentary-review

Healey, D., Hegelheimer, V., Hubbard, P., Ioannou-Georgiou, D., Kessler, G., & Ware, P. (2008). *TESOL technology standards framework.* Teachers of English to Speakers of Other Languages, Inc. https://www.tesol.org/docs/default-source/books/bk_technologystandards_framework_721.pdf

Hewson, C. (2015). Ethics issues in digital methods research. In H. Snee, C. Hine, Y. Morey, S. Roberts, & H. Watson, (Eds.) Digital methods for social science: An interdisciplinary guide to research innovation. Palgrave Macmillan.

Hope, A. (2018, May). Creep: The growing surveillance of students' online activities. *Education and Society, 36*(1), 55–72. doi:10.7459/es/36.1.05

How to protect your online privacy with threat modeling [Video]. (2017, November 15). Above the Noise. https://www.youtube.com/watch?v=VlYjtWg4Thw&ab_channel=AboveTheNoise

Israel, E., & Batalova, J. (2021, January 14). *International students in the United States.* Migration Policy Institute. https://www.migrationpolicy.org/article/international-students-united-states-2020

Jones, K. M. L. (2019, July 2). Learning analytics and higher education: A proposed model for establishing informed consent mechanisms to promote student privacy and autonomy. *International Journal of Educational Technology in Higher Education, 16*(24), 24. Advance online publication. doi:10.118641239-019-0155-0

Kantayya, S. (2021). *Coded bias* [Film; online video]. Independent Lens. https://www.codedbias.com

Lave, J., & Wenger, E. (1991). *Situated learning: Legitimate peripheral participation.* Cambridge University Press. doi:10.1017/CBO9780511815355

LePan, N. (2020, April 18). *Visualizing the length of the fine print, for 14 popular apps.* Visual Capitalist. https://www.visualcapitalist.com/terms-of-service-visualizing-the-length-of-internet-agreements

Lestari, N. D. I., & Subriadi, A. P. (2021, September). EdTech investment: Optimism, pessimism, and uncertainty. *2021 International Conference on Electrical and Information Technology (IEIT)*, 239-245. 10.1109/IEIT53149.2021.9587429

Lupton, D. (2020, March 3). 'Honestly no, I've never looked at it': Teachers' understandings and practices related to students' personal data in digitised health and physical education. *Learning, Media and Technology*, *46*(3), 281–291. doi:10.1080/17439884.2021.1896541

Macgilchrist, F. (2019). Cruel optimism in edtech: When the digital data practices of educational technology providers inadvertently hinder educational equity. *Learning, Media and Technology*, *44*(1), 77–86. doi:10.1080/17439884.2018.1556217

Mackenzie, L. (2020, January 21). *Surveillance state: How Gulf governments keep watch on us.* Wired. https://wired.me/technology/privacy/surveillance-gulf-states

Marín, V. I., Carpenter, J. P., & Tur, G. (2021, September 20). Pre-service teachers' perceptions of social media data privacy policies. *British Journal of Educational Technology*, *52*(2), 519–535. doi:10.1111/bjet.13035

Molla, R. (2020, February 7). *Law enforcement is now buying cellphone location data from marketers.* Vox. https://www.vox.com/recode/2020/2/7/21127911/ice-border-cellphone-data-tracking-department-homeland-security-immigration

Mora, F., Quito, R., & Macías, L. (2021). Reading comprehension and reading speed of university English language learners in Ecuador. *Journal of English Language Teaching and Applied Linguistics*, *3*(11), 11–31. doi:10.32996/jeltal.2021.3.11.3

Orlowski, J. (2020). *The social dilemma* [Film; online video]. Exposure Labs. https://www.thesocialdilemma.com

Prensky, M. (2001, October). Digital natives, digital immigrants. In *On the Horizon, 9* (Vol. 5). MCB University Press.

Regan, P. M., & Jesse, J. (2019, September 15). Ethical challenges of edtech, big data and personalized learning: Twenty-first century student sorting and tracking. *Ethics and Information Technology*, *21*(3), 167–179. doi:10.100710676-018-9492-2

Rehak, K. (2022, May 29). *Should I Use This EdTech Tool? Decision-Making Guide for Instructors* [Image]. Academic Press.

Reich, J. (2020). *Failure to disrupt: Why technology alone can't transform education.* Harvard University Press. doi:10.4159/9780674249684

Ribble, M., & Bailey, G. (2007). *Digital citizenship in schools.* International Society for Technology in Education.

Richter, S., Rhode, J., Arado, T., & Parks, M. (2021, Fall). Principles for conducting a comprehensive LMS review. *The Community College Enterprise*, *27*(2), 89–94.

Rooksby, J. H. (2020, January 13). Consider impact of institution's tracking apps on privacy, best interest of students. *Campus Legal Advisor: Interpreting the Law for Higher Education Administrators*, *20*(66), 1–3. doi:10.1002/cala.40173

Rorabaugh, P. (2012, August 6). *Occupy the digital: Critical pedagogy and new media.* Hybrid Pedagogy. https://hybridpedagogy.org/occupy-the-digital-critical-pedagogy-and-new-media

Shortt, M., Tilak, S., Kuznetcova, I., Martens, B., & Akinkuolie, B. (2021, July 5). Gamification in mobile-assisted language learning: A systematic review of Duolingo literature from public release of 2012 to early 2020. *Computer Assisted Language Learning*, 1–38. Advance online publication. doi:10.1080/09588221.2021.1933540

Stommel, J. (2014, November 17). *Critical digital pedagogy: A definition.* Hybrid Pedagogy. https://hybridpedagogy.org/critical-digital-pedagogy-definition

Student privacy pledge. (n.d.). *Student Privacy Compass*. https://studentprivacycompass.org/audiences/ed-tech

Thompson, S. A., & Warzel, C. (2019, December 19). *Twelve million phones, one dataset, zero privacy.* The New York Times. https://www.nytimes.com/interactive/2019/12/19/opinion/location-tracking-cell-phone.html

Young, J. R. (2019, June 4). *What is critical digital pedagogy, and why does higher ed need it?* EdSurge. https://www.edsurge.com/news/2019-06-04-what-is-critical-digital-pedagogy-and-why-does-higher-ed-need-it

Zhu, X., & Yang, T. (2019, October 18). Do I own it?: US and Chinese college students' digital ownership perceptions. *Proceedings of the Association for Information Science and Technology*, *56*(1), 346–355. doi:10.1002/pra2.28

APPENDIX

Figure 1. Should I Use This EdTech Tool? Decision-Making Guide for Instructors
Source: Rehak, 2022

Should I use this EdTech tool?: Decision-making guide for instructors **1) List how the EdTech tool will serve students in terms of its benefits and drawbacks.**	
Benefits for students (+1 for every benefit)	**Drawbacks (-1 for every major drawback)**
2) Are you able to measure the benefit of the tool? Yes (+1) No (+0)	
List how you are able to measure the benefits of this tool.	
3) Do the benefits of this tool outweigh the drawbacks? Yes (+1) No (+0)	
Explain your answer.	
4) The company or group that makes the EdTech tool has been verified by my school or institution. **Yes (+1) No (+0)**	
5) The Common Sense Media Evaluation Tool shows the company or group that makes the EdTech tool is trustworthy. Pass (+3) Warning (+1) Fail (-1)	
6) If applicable, the company complies with my state's data privacy laws and/or the EU's General Data Protection Regulation? Yes (+1) No (+0)	
7) The EdTech tool forces students to register with an e-mail address. No (+1) Yes (-1)	
8) The EdTech tool will publicly display user content. No (+1) Yes (-1)	
9) The company or group that makes the EdTech tool has explicitly said they will not sell user data. **Yes (+1) No (-1)**	
10) The learning benefits of the EdTech tool cannot be achieved through other means. **Yes (+1) No (+0)**	
TOTAL POINTS=	
10+ points **Yes, definitely!** | **4-10 points** **Maybe, but consider other options.** | **under 4 points** **No! Do not use this tool!**	

Chapter 7
Social Media for Teaching Empathy, Civil and Moral Development, and Critical Thinking

Thomas Huston
https://orcid.org/0000-0002-5820-8010
Purdue University Global, USA

ABSTRACT

There is a growing argumentative, hostile element found in many parts of society today; this growth, combined with increasing anonymity in online communications, suggests a dire need for youth to be harnessed with skills for civil deliberation. Numerous educational scholars have highlighted the importance for educational entities to consider utilizing youths' digital interests for teaching and learning. Additionally, recent research conducted during the COVID-19 pandemic found social media used for online education helped engage students in experiential learning and created positive feelings. Thus, a discussion of advantages to using social media is presented for fostering youths' skills in critical thinking, empathy, community, culture, deliberation, discovery, and more, so they might effectively contribute to their society and have a better, fuller understanding of the world around them. Sample curricula for teaching the deliberative arts via social media is included.

INTRODUCTION

Why It Is Important to Teach Youth Empathy, Civil Deliberation, Critical Thinking, and Moral Development

Today, one might argue Western culture is training youth to be adversarial. Murray State psychology professor Daniel Wann stated, "Civility is going down in our society. Empathy is going down in our society," and the rising, socially abusive behaviors among people "could stem from the culture of ag-

DOI: 10.4018/978-1-6684-5892-1.ch007

gression that exists online" (as cited in Schad, 2022, paras.13 & 25). Recently, the town of Rome, New York, canceled their youth basketball program on account of increasingly aggressive, abusive, and violent behavior by parents. In the cancellation letter, Ryan Hickey, deputy director of Rome's Department of Parks and Recreation wrote, "I think we have lost our way a little. We have to somehow come back together and remember we are in this together" ("Rome youth basketball," 2022, para. 2). Karissa Niehoff, chief executive for the Office of National Federation of State High School Associations, said fans of all ages are more extreme in their confrontational, offensive, and savage behavior (Schad, 2022). Apart from parents and online culture, the hidden and competing "curricula" found in today's media must be considered.

When youth are exposed to cable television's routine shouting of commentators' opposing views, they likely gain a poor understanding of how to think or speak about controversial topics or positions with which they disagree. From these types of programs, youth learn it is better and more logical to meet someone's opposing position with angered, blunt, and quick reaction rather than critical thinking and civil, respectful deliberation. Because too many sources routinely provide poor examples of what it means to civilly deliberate on controversial topics such as abortion, religion, civil rights, equity, politics, and other current events, it is hard to argue against schools promoting curricula with an emphasis on the civil development of youth. American sociologist Deborah Tannen suggested the cost for an argumentative culture was a "price paid in human spirit" (Tannen, 1998, p. 280).

Walter Parker, professor emeritus at University of Washington, wrote important scholarship for the civil development of youth. His work included strategies for youth to engage in skills for civil, meaningful deliberation and critical thinking. He called those skills the "deliberative arts," a concept that included the following skills for successful critical thinking and deliberation: (1) realize and admit any lack of knowledge, (2) listen and only speak in turn, (3) be brave when discussion of disliked topics occurs, (4) aim to fully comprehend someone's different point of view, (5) engage tactics for critical thinking, (6) critique only ideas being discussed rather than any person, (7) be slow to judgment (Parker, 1997; Parker, 2003). Parker (1997) summarized the mindset of a critical thinker in civil deliberation with a quote made by Voltaire: "I disapprove of what you say, but I will defend to the death your right to say it" (p. 19).

Amy Gutmann, academic and current US Ambassador to Germany, and longest serving president of University of Pennsylvania, wrote scholarship with strong support of curricula designed to enhance student empathy and moral deliberation, and believed public education was the appropriate space for such learning. She stated the absence of such curricula resulted in schools becoming "repressive by virtue of what it fails to teach" (Gutmann, 1996, p. 157). Gutmann's (1996) recommended approaches for teaching deliberation properly and ethically included showing youth how to respect other speakers' dispositions, backgrounds, and positions, and for them to understand and practice open mindedness and self-control.

Michael Pritchard, the Willard A. Brown professor emeritus of Philosophy at Western Michigan University, also supported curricula for critical thinking and moral education and believed both could successfully be taught in public education. In his book *Families, Schools, and Moral Development* (1996), Pritchard made a strong claim for moral development within public schools: "Children have a right to be given opportunities to become well-developed, moral persons … Families lacking in this responsibility … strengthen the case for public schools' responsibility" (pp. 92-93). Pritchard suggested schools first begin teaching youth what it means to be an active listener during deliberation because when youth know how to listen, they can begin discerning which questions are important to ask next. As a result, classrooms foster a population of critical thinkers.

Like Gutmann, Pritchard (1996) believed moral education, empathy, and civil deliberation might be achieved if youth first understand what it means to respect someone's viewpoint. To this end, Pritchard (1996) suggested youth must learn to "overcome egocentricity in circumstances calling for moral sensitivity and judgment" (p. 93). To achieve this type of understanding, youth need to understand how a person is influenced by examining what they find important and why. Youth can learn to foster their openness to examining how or why their own values might change over time from new experiences, understanding, or deliberation. Pritchard (1996) stated it is important youth know they have power to openly approach and survey their values and learn how to claim them; claiming ideals through an analysis can make them stronger, opposed to adopting weak one fostered only through societal pressures with no critical examination. When youth engage critical thinking strategies such as surveying their own held ideologies, it can help them to develop a familiarity with how others might adopt similar ideals, and as a result, lead to empathy.

Katherine Simon, author of *Moral Questions in the Classroom: How to Get Kids to Think Deeply About Real Life and Their Schoolwork* (2003), discussed how educators can teach youth to learn empathy and moral development. She shared ideas for teaching civil virtue to youth through a cornerstone understanding of what defines fairness, honesty, self-control, equity, and responsibility. Simon (2003) suggested schools first develop a moral mission, provide workshops to develop the mission, foster a deeper awareness through teaching, and continually challenge youth to investigate all moral aspects of other taught curricula. In *Culture in the Curricula* (2006), regarding diversity and school, academic Elaine Chan stated, "There is a lack of consensus about how best to acknowledge diversity in a school" (p. 349). Curricula developed to correspond with Simon's (2003) idea of a moral mission with themes of empathy, civil deliberation, critical thinking, or moral development can aid the social realm in many ways, including schools' attempts to address diversity.

Academic Michael Pardales (2002) believed youth need a more effective understanding of how they make decisions, from routine choices like careers, to more complex, moral decisions such as dating someone with different beliefs or using someone's online password. To help youth analyze how these types of choices are decided, Pardales (2002) defined a concept known as "moral imagination," the "psychological faculty" that allows people to "explore various possibilities for their lives" (p. 426). Imagining different possibilities through a critical self-examination can also lead people to better empathy through an understanding of others' choices and decisions. Like Pritchard, Pardales (2002) believed morality sometimes changes in one's life because of shifting ideologies, thus it is important for youth to learn how to critically examine their decisions through a lifelong, critical lens.

Learning how to engage in civil deliberations on a variety of controversial topics involving ethics, beliefs, morals, and cultural differences, youth might be better prepared to contribute to a more harmonious intercultural existence. Therefore, apart from curricula development, today's educators must also ask the tough questions, so youth have experience with engaging and critically thinking about events and issues that exist in their world. By using social media for teaching and learning, educators can plan addressing issues that most might have had difficulty within the past because of a lack of student interest or engagement, and the result could be an exciting learning opportunity for everyone.

LEARNING AND TEACHING WITH SOCIAL MEDIA

People worldwide use social media for a variety of purposes. Social media has been found to facilitate and increase self-respect, spirited association and collaboration, originality, creative power, involvement and contact with education, second language literacy and academic performance (Papademetriou et al., 2022). Social media has been used for student learning and teaching, with a steady increase of interest in using platforms like Facebook as a learning environment to reengage youth (Barrot, 2018; Chugh & Ruhi, 2018; Naghdipour, 2017; Rahman et al., 2021; Paliktzoglou et al., 2021; Papademetriou et al., 2022; Tkacová et al., 2022). Recent studies suggested use of social media allowed students to continually be informed, quickly distribute large amounts of information, connect with new people and knowledge, and accomplish productivity alone or through collaboration across time and space (Papademetriou et al., 2022).

When education systems shut down during the COVID-19 pandemic, learning and teaching went entirely online. During the COVID-19 pandemic, Tkacová et al. (2022) studied online education through use of social media and found it helped engage students in experiential learning through social activities, and because that learning created positive feelings, involvement led to a decrease of some negative effects from the pandemic. Tkacová et al. (2022) stated teachers were able to connect formal education with students' interests and current events with "attractive forms of knowledge presentation" (p. 16). Over time, students' understanding of online learning evolved into a natural, social activity, and, as a result, students' in-class demeanor changed from inactive to "online contributor" as information sharing increased for everyone (Tkacová et al., 2022, p. 16).

An Instructor's Experiment With Facebook

A few years ago, a Ph.D. candidate and associate instructor at a Midwest university created a Facebook group to exclusively correspond with an undergraduate course he was teaching on campus. In the past, when teaching the same course, the instructor stated only a few students participated in the online discussion forum via the university's online learning management system (LMS) (e.g., Blackboard, Brightspace, Canvas, etc.). As a result, students were barely engaged in discussion of course content in-between their weekly class. Through use of the Facebook group, the instructor wanted to see if students routinely re-engaged course content in a meaningful manner via written deliberation. The instructor's idea of using Facebook for teaching resulted from watching his own friends deliberate, sometimes passionately, about random topics on Facebook's "wall," a space of content displayed where users with access to a particular Facebook profile communicate multimodally and interact asynchronously. According to the instructor, the Facebook group began slow; only one student responded to a course-related video posted by the instructor on the Facebook wall. However, after a student referenced the Facebook post during a regular class discussion, more students joined and were engaged in online, written deliberation about course-related readings and topics outside of class via the Facebook group.

The instructor stated that even with the lure of participation points, students' use of the university LMS discussion forum was minimal to none, yet student participation in the Facebook group, although voluntary and never associated with course assignments or grades, was much higher and consistent. The instructor mentioned there were two occasions where he privately engaged students who "crossed the line" with their online etiquette and noted these types of teachable moments were no different than what commonly occurred inside the traditional classroom. As a result of the Facebook group, the instructor stated he wanted to know how much more learning occurred outside the traditional class, and if more learning or retention might take place if the student group existed beyond a semester. The instructor believed a Facebook group could remain open for students to continually add their critical reflections and deliberations about shared, new experiences with added examples that hearkened back to course content.

The instructor's Facebook group was likely popular because its asynchronous element provided a space for students to engage other students' deliberation, anytime. Because the interface of the asynchronous platform was always "open," it catered to college students' busy schedules. One main component to successful learning is providing an environment where learners experience a feeling of safety (Rahman et al., 2021; Paliktzoglou et al., 2021; Papademetriou et al., 2022; Sparks, 2013; Tkacová et al., 2022). Because a university LMS is often difficult to navigate or outside students' interests, the Facebook group was an easy choice for engagement because of students' familiarity with the functionality and navigation of the platform. Facebook and other social media sites also provide users with an environment anchored in familiarity of friends and/or family. The type of comfort and safety experienced by students in a space such as Facebook also provided necessary components essential in fostering the skills for civil deliberation and, as a result, self-discovery, empathy, and more (Enslin et al., 2001; Hoffman, 2000; Ito et al., 2010; Pardales, 2002).

Social Media and Learning

The groundbreaking study, *Hanging out, Messing Around, and Geeking Out* (Ito et al., 2010) reported findings from a group of ethnographers who examined how current online and digital trends such as smart phones, video games, social media sites, and multimodal communications engaged youth in online learning. Ito et al.'s (2010) study findings were important because they challenged key elements of curriculum, instruction, and learning; unlike top-down systems in traditional classrooms, youth, on their own, were associating, creating, and participating in various learning activities without sometimes realizing it. This type of learning occurred in informal, online, and digital settings where youth, on their own, sought a variety of interest-based "curricula" for learning and continued to search for higher levels of information for even more learning. Additionally, Ito et al. (2010) reported use of digital media helped youth to engage and develop economic and political literacy, and its asynchronous element prompted youth to communicate more personally because they were less self-conscience or embarrassed.

A considerable amount of recent research suggested social media used for educational purposes provided a variety of advantages for teaching and learning. Use of social media for teaching and learning in higher education suggested an increase of quality communication, a greater ease and interchange with information, and a more powerful experience with colleagues and curriculum (Paliktzoglou et al., 2021). Research also considered use of social media aligned with a constructivist approach to learning because it encouraged a relationship between students and instructors as they worked together toward a "knowledge sharing process, a sense of belonging, and a deeper understanding of class content" (Paliktzoglou et al., 2021, p. 162). Trust played a large role in students' use and engagement with social media for learning (Paliktzoglou et al., 2021; Tkacová et al., 2022). Some elements students associated with trust or credibility included the student's relationship to information, the creator of the content, the media presented, and the technology itself (Paliktzoglou et al., 2021; Tkacová et al., 2022). Students consider online learning credible when there is an opportunity for them to build information, share beliefs, and collaborate in groups (Paliktzoglou et al., 2021; Tkacová et al., 2022). Rahman et al. (2021) found students' perceptions of Social-Media-Based Learning (SMBL) was that it provided positive contributions to their learning. Because SMBL was "more interactive, user-friendly, and considerably easier," it affected students' desires for using it (Rahman et al., 2021, pp. 1341-1342). Additionally, other elements contributing to positive perceptions of SMBL among students and faculty were timely support whenever problems arose, the platform's efficiency, and an expectation of fulfilling encounters with learning (Rahman et al., 2021).

Some educators might be concerned with too much "presentism" in today's curriculum and instruction. Ralph Pinar, American pedagogue, curricula theorist, and founder of *Journal of Curriculum Theorizing*, believed in the following curricular approach: "What is necessary is a fundamental reconceptualization of what curricula is, how it functions, and how it might function in emancipatory ways" (Pinar, 1978, p. 154). Research suggested the overuse or bad design of traditional learning spaces, teaching, and assessment tools, continue their failure to align with students' background, interests, or learning needs (Barrot, 2018; Chugh & Ruhi, 2018; Naghdipour, 2017; Rahman et al., 2021; Paliktzoglou et al., 2021). Theodore Brameld, founder of social reconstructionism, believed future curricula should strive to combine "technology with human compassion" (Haindel, n.d., para. 1). Taken together, these reasons are why interest in using social media for teaching and learning has increased, and why it might be used to teach youth necessary skills for their civil development (Rahman et al., 2021; Tkacová et al., 2022).

Advantages of Using Social Media for Teaching Empathy, Civil Deliberation, Critical Thinking, and Moral Development

Social Media Fosters Literacy and Learning via Student Interest and Deliberative Writing

Youth do not respond well to curricula designed for "everyone" or a "one size fits all" approach because they tend to ignore individual needs, talents, or interests of students (Ivey & Fisher, 2007). Writing scholars have called attention to this idea in writing pedagogy; when students were taught writing under a "one size fits all" approach, they became "passive direction-followers" of those who "rule-governed" with instructions (Yagelski, 2006, p. 535). This top-down mentality of writing instruction is often found in common essay assignments tied to a selection of literature, likely canon, where students are asked to "discuss, prove, or summarize" something within specific numbered paragraphs. Without inclusion of students' interests, background or prior knowledge embedded within the curricular design of the assignment, educators fail to make a connection, resulting in the absence of students' interest or new knowledge (Ivey & Fisher, 2007; Yagelski, 2006; Yagelski, 2009).

To counter poor writing pedagogy, academics like Donald Murray and Peter Elbow proposed important and innovative ideas on the teaching of writing. These ideas included meeting students where their interests lie, free writing, classrooms without teachers, and providing a space for youth to express ideas and beliefs so they might discover their voice and place in the world (Yagelski, 2006; Yagelski, 2009). Equally important, research suggests the best reading material for reading and writing literacy includes a variety of multimodal materials like graphic novels or the online, multimodal texts youth normally seek outside of school (Ivey & Fisher, 2007; Ito et al., 2010). Interestingly, the radical, pedagogical vision of free writing in a teacher-less classroom proposed by Elbow in 1973 is more aligned with how students engage reading and writing today, by themselves online writing and reading while engaged in self-directed, scaled learning of a particular area of personal interest (Ito et al, 2010; Yagelski, 2006).

When youth read, write, hear, talk, or see, it is never a bad thing for development of their reading and writing literacy or gaining new understanding of themselves, others, and the surrounding world. To explain this idea further, consider Yagelski's (2009) discussion on how writing is an ontological act; that is, a writer's experience in the act of writing is as important as the text produced. Yagelski (2009) explained that when people write, they engage the past, present, and future at the same time; when writing, writers reflect on the past while at the same time formulating and writing ideas in the present, and then moving forward with ideas on what to write next. This process of exploring one's mind, world, and others, all while engaging the past, present, and future, is quite extraordinary when considering similar analogies, a likely reason why writing is often used as a therapeutic device. According to Yagelski (2009), elements from this simple act of writing open "awareness, reflection, and inquiry" for the writer and can then lead to a "byproduct" of deliberation and, as a result, potential for individual or collective change (p. 7).

If educators examined positive effects from the just the simple act of writing, they might see how traditional writing pedagogy is more preventative of students' literacy, new knowledge, or self-discovery (Ivey & Fisher, 2007; Yagelski, 2006; Yagelski, 2009). By design, social media's asynchronous element could provide an ontological experience for students through their ongoing, deliberative writing where the interpretations and understandings that manifest through students' ongoing, deliberative writing are strengthened through the medium with which the curricula were delivered. That is, because student interest in social media platforms enhances the process of engagement, any awakened or learned discoveries remain active in the students' willingness to continually engage the platform, the curricula, and deliberation. As a result, students' literacy, understanding of themselves and others, acquired knowledge, or an instructor's learning objectives, might arrive at a higher level of learning and retention, perhaps more so than any learning from a semester's worth of required literature or assigned essays administered from a traditional brick and mortar English classroom.

Online Deliberation Fosters Critical Thinking, Empathy, and Moral Development

At the cusp of the 20th century, perhaps educational theorist John Dewey (1897) made the best argument in support of how deliberation within a collected group of individuals can lead to new understanding:

I believe that the only true education comes through the stimulation of the child's powers by the demands of the social situations in which he finds himself. Through these demands he is stimulated to act as a member of a unity, to emerge from his original narrowness of action and feeling and to conceive of himself from the standpoint of the welfare of the group to which he belongs. Through the responses which others make to his own activities he comes to know what these mean in social terms. (pp. 77-80)

Similarly, the philosopher Socrates stated to truly know wisdom, one must look toward understanding through deliberation. Helbowitsh (2010) explained how this process came to light in the Classical period of Ancient Greece,

Plato taught us that the dialogue did, in fact, make some contributions to the participants understanding ... because, if nothing else, the discussion itself, while leaving the participants somewhat disoriented nevertheless provided an experience in working the answer to the question it raised ... Plato instructed us to judge the dialogue less by an expected or desired outcome than by the process it engaged ... Plato shows us that the dialogue, in effect, comprises the main task of knowing. (Hlebowitsh, 2010, p. 508)

Differences are never in isolation of each other; holding different ideals and beliefs together for review, acceptance, and deliberation is a tense process but captures the above example of how deliberation, acknowledging, and exploring differences eventually leads to understanding by people collectively sharing them over time in the same space. Through ongoing, written deliberation, trying to dig at the "why" behind an instructor-designed question can lead a group of students to new understandings of differences, stances, religions, cultures, and the surrounding world. The discussion method facilitates students' journey of answering his or her own questions via understanding, explicating the validity of other students' written comments, and the practice of rendering an explanation of his or her own thoughts to others.

In his book *Teaching Democracy: Unity and Diversity in Public Life* (2003), Parker offered valuable points on what makes up successful deliberation. According to Parker (2003), successful deliberation is creative, fosters interaction that is ongoing for months. Because of a non-traditional classroom setting, curricula for teaching the deliberative arts in an asynchronous environment supports Parker's (2003) points for successful deliberation. As mentioned, social media platforms' design, such as Facebook's asynchronous wall, can foster student interest, engagement, and ultimately extended deliberation and the platform's availability can foster a longevity of student discovery (Ito et al., 2010). With this reality at hand, educators can present a space where the classroom, interactions, social space, and importance of everyone's contributions are valued. For example, youth, with their diverse backgrounds, can create new knowledge for both student and teacher; the instructor can learn from the discourse and draw from that knowledge in future classes or curriculum. Teaching or learning in an asynchronous format also provides opportunity for discourse between educators, parents, administrators, curricula writers, counselors, and anyone with online access to observe, critique, offer ideas or revisions.

Penny Enslin, author of *Deliberative Democracy, Diversity and the Challenges of Citizenship Education* (2001), believed a social group perpetuates an "understanding of position and how it stands to others … reached not by transcending what divides and differentiates, but by speaking across differences to learn … one's own perspective … by expressing, questioning, and challenging differently stated knowledge" (p. 128). With guidance, social media could be a great space for youth to learn how to get along. Part of discovering one's place in the world includes understanding feelings of others in empathy. Deborah Kerdeman (1998) believed self-discovery and the discovery of others' disposition is hermeneutic consciousness; when awakened, it needs remain aroused; an asynchronous platform that engages student interest is a great way for learning to stay active.

Martin Hoffman, professor emeritus of clinical and developmental psychology at New York University and author of *Empathy and Moral Development: Implications for Caring and Justice* (2000), said using empathy to treat adolescents is a basis for teaching moral education, and that deliberation works in tandem with such goals. Hoffman (2000) stated when juveniles take part in role playing that promotes empathy of caring and other forms of perspective-taking during social scenarios, it can stimulate caring and justice for the person taking part in the action. Because these projections help youth to understand how others think, feel, and value, they aid with discovery of how development of such beliefs and behaviors might first emerge in other people or individually. To aid in these roleplaying exercises, Hoffman (2000) suggested teachers provide youth with common, emotional responses to help them feel understood and what to avoid in terms of common, emotive reactions. Second, Hoffman (2000) believed teachers and youth use multimodal media as tools to learn, and third, provide youth with a large amount of time and space for their responses, something easily accomplished within social media's asynchronous platform.

Michael Pardales' (2002) concept of "moral imagination" can lead one to understand how and what other people think, feel, value, and as a result, foster an understanding of how one develops behaviors to make decisions or judgements, and its use in education could have numerous benefits. To enhance student empathy and moral development, social media platforms offer a variety of means for youth to easily enter specific fictional scenarios to deliberate upon, experience other people via fictional profiles, etc., and as a result, experience a decision made through the moral lens of a real or fictional person. For example, Pardales (2002) stated if a student can explore another possibility for his or her life through a fictional social media profile, it could include projecting his or her personality into various forms of contemplation to enhance that student's empathy, moral development, and ultimately civil deliberation as they might begin expressing how and why they came to a decisions or choice. Some of Pardales' (2002) suggested strategies to achieve such goals included changing certain elements of reality via fictional and non-fictional decision making, assessing multiple possibilities for a person's earlier or future life choices, problem solving with one's own definition of morality or someone else's, and investigating possibilities of what it would be like to be someone else in multiple, different circumstances.

Community and Culture

Providing a community is important for learning socialization and the fundamentals of civil deliberation, critical thinking, empathy, and morality. Peter Hlebowitsch (1999, 2010), curricula theorist and current dean of the College of Education at the University of Alabama, emphasized community when it comes to devising curricula, and placed importance on moral themes not only to enhance moral dimensions of youth, but also the process and purpose of inviting community action. Dewey (1897) also believed ideas of virtue are shown in action (Hostetler, 2005). Tyler (1949) stated in terms of student learning, one of three types of needs for youth is an "integrative need," defined as "the need to relate oneself to something larger and beyond oneself, that is, the need for a philosophy of life" (pp. 71-72). Chan (2006) also emphasized the importance of interplay between youth of different backgrounds because it provided an environment for positive views of differences to begin.

Social media platforms provide an easy interface for engagement where youth learn, communicate, or act for a variety of audiences and purposes. By design, a potentially boundless community, social media platforms also offer a mélange of cultures acting as a genus together in community; the asynchronous element of social media has enormous potential for youth to engage and learn with people of different cultures on a global scale. Social media is a new form of community because of this, and the potential repercussions from youth deliberation can offer new insight into how culture and community develop and change.

SAMPLE CURRICULA

Introduction

The following curricula are designed to prompt youth on issues involving empathy, values, morals, democratic deliberations, civil virtue, and the assessment of each, including evaluating quality in multiple artifacts. It is hopeful these curricula will enhance youths' engagement of conversations with their peers in virtual and physical space and produce and contribute knowledge and ideas to the landscape of digital learning. These curricula might also compel individuals to act in solidarity to address social and economic injustice.

Learning Objectives

To avoid inaccessible links or outdated titles, any update similar in content should suffice for the sample's original curricular intent. It is encouraged instructors engage in democratic deliberation regarding a rubric of assessment for each curricula sample and how literacy of the following ten objectives is represented in each class's archived conversations, multimodal posts, and responses:

1. Youth will understand a conceptualization of themselves as a citizen of a multicultural country.
2. Youth will understand how users wield information acquired through information and communication technologies and new media, both civilly and politically.
3. Youth will explore concepts that enable them to engage in civil deliberation.
4. Youth will understand how opportunities and challenges are presented by the adoption or access to information and communication technologies at the global level.
5. Youth will understand how culture and institutions acquire meaning in the age of information.
6. Youth will theorize strategies that help combat forms of discrimination in classrooms, schools, communities, society, and the world.
7. Youth will understand how to act politically and critically in solidarity to address social and economic injustice.
8. Youth will understand how to foster a sense of community when communicating with others.
9. Youth will reflect upon and assess the characteristics and merits of their work, beliefs, and ideals, and their work and the characteristics and merits of others' work.
10. Youth will understand how individuals can differ in their approaches to learning, communicating, and understanding, specifically with respect to culture, ethnicity, class, and gender.

Three Sets of Sample Curricula

Sample Curricula Set One

The first set of sample curricula provide youth with writing prompts about unique social media events that took place in public education; most youth will likely have a strong opinion but will engage Pardales (2002) "moral imagination" for each scenario to foster empathy and self-discovery (See Figure One for similar types of prompts). If youth engage with multiple forms of character analysis and decision-making, they might begin to empathize with other ideas and the people and backgrounds behind them, and as a result will hopefully enhance how one goes about effective and constructive deliberations enhanced via empathy of viewpoint, etc. Because the stories are controversial in nature, they will also provide initial practice in discussing such controversial themes in hope that the desirable habits of what can be found in civil deliberation will eventually surface through this practice. As time progresses and youth have had more practice at assessment of predisposed opinions or viewpoints, instructors could make the scenarios more controversial and/or difficult to agree upon. The ideas for the content of questions and reflection are also based on Katherine Simon's (2003) *Moral questions in the classroom: How to get kids to think deeply about real life and their schoolwork.*

Titles can be typed into a search engine for easy location of these items. It is also recommended for instructors to find and use their own prompts.

Writing Prompt 1: Georgia Court of Appeals: Parents Can be Held Liable for Kids' Social Media Misdeeds (Ove, 2014)

- https://patch.com/georgia/acworth/georgia-court-appeals-parents-can-be-held-liable-kids-social-media-misdeeds-0

Writing Prompt 2: Former student sues Minnesota district and local police chief after "sarcastic" Tweet leads to seven-week suspension (Charlton, 2014)

- https://www.dailymail.co.uk/news/article-2661496/Honor-student-suspended-tweet-joking-hed-teacher-sues-old-school-district.html

Writing Prompt 3: New Jersey student disciplined for vulgar off-campus Tweet about principal sues district (Beym, 2014)

- https://www.nj.com/camden/2014/03/nj_high_school_student_claims_she_was_punished_for_tweet_about_principal.html

Writing Prompt 4: Pennsylvania student suspended for off-campus Snapchat post sues school district (Myers, 2021)

- https://sharylattkisson.com/2021/06/read-supreme-court-sides-with-cheerleader-suspended-for-off-campus-social-media-tirade

Writing Prompt 5: California district hires company to monitor youth' online activity (Quinn, 2013)

- https://www.washingtonexaminer.com/red-alert-politics/california-school-district-hires-company-to-monitor-students-social-media

Writing Prompt 6: New York senior suspended for creating Twitter hashtag concerning school board's budget (Boyette, 2013)

- https://www.cnn.com/2013/05/26/us/n-y-student-suspended-after-controversial-twitter-hashtag

Writing Prompt Instructions

1. Read each writing prompt.
2. Create a Facebook or Google document and write your opinion of each numbered prompt where you explore the following:
    a. Do you disagree or agree with the problem and/or outcome? Discuss why and anything else that comes to mind.
    b. What would you have done if you were the teacher, administrator, student, or parent featured in each situation?
    c. Can you empathize with any of the actions done by the parent, student, administrator, or teacher? Why or why not?
3. With your selected partner, via Facebook Chat or Google IM, discuss the following:
    a. What do we agree on?
    b. What are we uncertain about?
    c. What other information do we need?
    d. Each paired group create a chart using Google Document that lists "Things we agree on," Things we disagree on," and "Things at least one of us wants to know more about."

4. Individually, express your opinion to the whole group by commenting under the appropriate writing prompt on the Facebook wall. Be sure to state why you hold the view that you do and what ideas inform your opinion.
5. After everyone has shared his or her opinion on one of the writing prompts, find someone else's opinion and ask them the following by responding to their comment:
   a. Is there any information that could convince you to change your mind?
   b. Is there any information that would help you know that your opinion is, in fact sound?
   c. If you are undecided, is there information you need to help you make up your mind?

   Feel free to carry on the conversation per your instructor's guidance.
6. Complete the reflection component below.

Reflection Instructions

Create a Facebook or Google document and answer the following questions where you analyze your experience deliberating with others about the subjects discussed in the writing prompts.

1. How well were you able to share your opinion with others?
2. What feelings arose in you as the conversation proceeded?
3. What suggestions do you have for us as a class for the next time we discuss a controversial issue?

Sample Curricula Set Two

The second set of sample curricula include writing prompts that showcase examples of public deliberation over random issues spanning over sixty years (See Figure Two for similar prompts). These exercises are designed to promote student discussion about effective deliberation over issues controversial in nature. The exercises also continue student immersion in analyzing and discussing controversy. Youth will assess how each speaker "went down the wrong path" in their attempts of civil deliberation. Students are also asked to provide solutions for progress and change based on the provided examples of how to engage in successful civil deliberation. This type of assessment will aid youth with their own future practices of deliberation and/or discussing social issues that may have sparked controversy in the past. Ideas for what makes effective deliberation are from Deborah Tannen's, *The Argument Culture: Stopping America's War of Words* (1998) and Walter Parker's (1997, 2003) discussion of the "deliberative arts," and reflection questions are from Katherine Simon's (2003) *Moral questions in the classroom: How to get kids to think deeply about real life and their schoolwork.*

Again, titles can be typed into a search engine to locate these items. Instructors may like to find sources for developing their own prompts.

Writing Prompt 1: The First Kennedy/Nixon Presidential Debate - Part 1/4 (1960) (Throwback, 2008)
- https://www.youtube.com/watch?v=C6Xn4ipHiwE

Writing Prompt 2: Bill O'Reilly and Geraldo Rivera Debate Immigration (usagreencardcenter, 2007)
- https://www.youtube.com/watch?v=Z3U9ENaTPLY

Writing Prompt 3: National Debate Over Defunding the Police, The View (The View, 2020)
- https://www.youtube.com/watch?v=uMu5UshGUPM

Figure 1. Screenshot of Set One Sample Curricula
Source: Houston, T. (n.d.). From Curricula Instruction. Framework for Phase One Curricula. [Video Attached] [Status update]. Facebook.

Writing Prompt 4: 2021 Presidential Debate between Donald Trump and Joe Biden, Debate One (The Telegraph, 2020)

- https://www.youtube.com/watch?v=CweqW7Pzxz8

Writing Prompt 5: Stephen A., Jalen Rose debate Charles Barkley's Super Teams Comments | Get Up! | ESPN (ESPN, 2018)

- https://www.youtube.com/watch?v=dNh_WMJsJDE

Writing Prompt 6: Who Should Be Held Responsible for COVID? Patrick Bet-David Podcast Episode 132 (Valuetainment Short Clips, 2022)

- https://www.youtube.com/watch?v=GolGGSY4GJ0

Writing Prompt Instructions

The numbered items below are recommendations for people who want to perform well in civil deliberation. After reviewing each of the prompts, discuss how each debate participant did or did not perform each numbered item and be sure to include a time stamp from the video to support each answer. When finished, choose one of your responses to these questions and post it by clicking on the "comment" field under the appropriate prompt. Other members may join in the conversations.

1. Each person debating listened and only spoke in turn.
2. Each person debating listened as if they believed in what the other person was saying.
3. Each person debating admitted any lack of their own knowledge.
4. Each person debating critiqued ideas rather than the people or groups who hold the ideas.
5. Each person debating engaged in elements of critical thinking.
6. Each person debating aimed to fully understand other participants' point of view.
7. Each person debating went beyond dualism and avoided the polarization of two sides by discussing multiple viewpoints.
8. Each person debating avoided any serious conflicts and returned to deliberation as opposed to a monologue or screaming match.
9. Each person debating did not offend or insult other participants.
10. Each person debating was slow to judgement and/or appeared to support each participant's right to their belief.

Reflection Instructions

Create a Facebook or Google document and answer the following questions where you analyze your experience deliberating with others about the subjects discussed in the writing prompts.

1. How well were you able to share your opinion with others?
2. What feelings arose in you as the conversation proceeded?
3. What suggestions do you have for us as a class for the next time we discuss a controversial issue?

Sample Curricula Set Three

The third set of sample curricula provide important elements for youth to help further develop their skills of civil deliberation and foster critical thinking. To gain practice at establishing criteria for assessing the quality of a product, the third set of curricula presents examples of various genres with which students are more familiar. As a collective or in pairs, youth shall deliberate to agree on criteria for measuring the quality of these products. For example, youth might list "lyrics," and "ingenuity" as criterions for assessing the quality of a particular genre of music. Youth then take these criteria to construct a rubric with a numbered system (five being the best representation of the subject matter at hand), and then further develop each criterion to show which elements of the criterion should be present to justify a level five, a level four, and so on. Developing criteria to assess the quality of something is difficult work, and it is hopeful that student interest in the genre examples presented will generate plenty of deliberation and critical thinking. Eventually, youth will assess the "quality" of other items, perhaps even others' deliberations as they come to an agreement as a class about what criteria they will use for their assessment. Ideas for the content of questions and reflection are based on the work of Joy Seybold, PhD, and Katherine Simon's (2003) *Moral questions in the classroom: How to get kids to think deeply about real*

Figure 2. Screenshot of Set Two Sample Curricula
Source: Houston, T. (n.d.). From Curricula Instruction. Framework for Phase Two Curricula. [Video Attached] [Status update]. Facebook.

life and their schoolwork and reflection questions are from Katherine Simon's (2003) *Moral questions in the classroom: How to get kids to think deeply about real life and their schoolwork.*

Original works as well as commentary about these items are easily found using a search engine. As with the previous curricula sets, it is recommended that instructors consider developing their own prompts based on their students' needs and understanding.

Writing Prompt 1: Genre: Music Video
- Prompt 1A – Artist: Mark Morrison; Song: *Return of the Mack*
- Prompt 1B – Artist: Talking Heads; Song: *Burning Down the House*

Writing Prompt 2: Genre: Poetry
- Prompt 2A – Author: Robert Frost; Poem: *After Apple Picking*
- Prompt 2B – Author: Elizabeth Bishop; Poem: *The Fish*

Writing Prompt 3: Genre: Classical Musical Work
- Prompt 3A – Composer: Johann Pachelbel; Work: *Canon in D Major*

- Prompt 3B – Composer: Ludwig van Beethoven; Work: *Piano Concerto No. 5*

Writing Prompt 4: Genre: Television Comedies

- Prompt 4A – Situation Comedy: *Sanford and Son*. Episode: *Blood is Thicker than Junk*
- Prompt 4B – Animated Comedy: *The Simpsons*. Episode: *This Isn't Your Life*

Writing Prompt Instructions

Read, listen, and/or view each writing prompt per the instructor's guidance and then create a Facebook or Google document to answer the following. When finished, choose one of your responses to these questions and post it by clicking on the "comment" field under the appropriate prompt. Other members may join in the conversations.

1. Explain what you believe to be the purpose of this product.
2. Explain who you believe is the audience of this product and why.
3. Examine if you made any personal connections with the product and discuss how and why this connection occurred.
4. Identify any held ideologies within the product and explain how each manifest and why you believe they are there.
5. List any quality attributes of the product.
6. List any less than quality attributes of the product.
7. List three criterions you believe people could use to judge the product's quality.
8. Explain if you believe the product is a quality creation by judging it with your three criterions.

Reflection Instructions

Create a Facebook or Google document and answer the following questions where you analyze your experience deliberating with others about the subjects discussed in the writing prompts.

1. How well were you able to share your opinion with others?
2. What feelings arose in you as the conversation proceeded?
3. What suggestions do you have for us as a class for the next time we discuss a controversial issue?

CONCLUSION

Recent studies presented above suggested several areas for further research, including a recommendation for social media to incorporate "a sound evaluation system in the teaching-learning process" (Paliktzoglou et al., 2021, p. 161). Other areas of further study included suggestions for finding and testing suitable pedagogy, identifying quality social media platforms, and investigating instructor workload, platform security and privacy (Paliktzoglou et al., 2021). Although the sample curricula have not been tested, contributions to the ongoing, scholarly conversation about online learning is valuable here. This curricula are provided with the same themes congruent to the curricula goals themselves, the empowerment of all people involved, regardless of background, to have a stake in deliberating the practices of

Figure 3. Screenshot of Set Three Sample Curricula
Source: Houston, T. (n.d.). From Curricula Instruction. Framework for Phase Three Curricula. [Video Attached] [Status update]. Facebook.

learning, curricular development, and pedagogical practice inside the social media platform, specifically the development and revisions of these curricula for the virtual classroom.

Educators strive to bring about a better individual in school settings, regardless of curricula. Equally important, when devising curricula, educators should attempt to prod the question, "How does acquiring this knowledge help the welfare of people?" Pupils with harnessed skills in civil deliberation, empathy, and critical thinking and moral development will effectively contribute to society and have a better, fuller understanding of the world around them. They will also have more patience and an open mind when communicating; if communication skills are enhanced to a higher degree than before, it is quite possible to achieve resolution and common goals for decisions. Thus, literacy in critical thinking will have been met if youth begin to act as democratic change agents in bettering the world community.

REFERENCES

Barrot, J. S. (2018, August 24). Facebook as a learning environment for language teaching and learning: A critical analysis of the literature from 2010 to 2017. *Journal of Computer Assisted Learning*, *34*(6), 863–875. doi:10.1111/jcal.12295

Beym, J. (2014, March 18). *N.J. High School student claims she was punished for tweet about principal.* https://www.nj.com/camden/2014/03/nj_high_school_student_claims_she_was_punished_for_tweet_about_principal.html

Boyette, C. (2013, May 27). N.Y. student suspended after controversial Twitter hashtag. *CNN*. https://www.cnn.com/2013/05/26/us/n-y-student-suspended-after-controversial-twitter-hashtag

Chan, E. (2006). Teacher experiences of culture in the curricula. *Journal of Curriculum Studies*, *38*(2), 349–360. doi:10.1080/00220270500391605

Charlton, C. (2014, June 18). Honor student who was suspended for tweet joking that he'd made out with teacher sues his old school district. *Daily Mail Online*. https://www.dailymail.co.uk/news/article-2661496/Honor-student-suspended-tweet-joking-hed-teacher-sues-old-school-district.html

Chugh, R., & Ruhi, U. (2018, March). Social media in higher education: A literature review of Facebook. *Education and Information Technologies*, *23*(2), 605–616. doi:10.100710639-017-9621-2

Dewey, J. (1897, January). My pedagogic creed. *School Journal*, *54*, 77–80.

Enslin, P., Pendlebury, S., & Tjiattas, M. (2001). Deliberative democracy, diversity and challenges of citizenship education. *Journal of Philosophy of Education*, *35*(1), 115–130. doi:10.1111/1467-9752.00213

ESPN. (2018, September 18). *Stephen A., Jalen Rose get into heated debate over Barkley's super teams comments | Get Up! | ESPN* [Video] YouTube. https://www.youtube.com/watch?v=dNh_WMJsJDE

Gutmann, A. (1996). Challenges of multiculturalism in democratic education. In R. K. Fullinwider (Ed.), *Public education in a multicultural society: Policy, theory, critique* (pp. 156–179). Cambridge University Press. doi:10.1017/CBO9781139172899.008

Haindel, D. B. (n.d.). *Theodore Burghard Hurt Brameld: The prophet father of the coming world.* Southeastern Louisiana University. https://www2.southeastern.edu/Academics/Faculty/nadams/educ692/Brameld.html

Hlebowwitsh, P. (1999, Autumn). The burdens of the new curricularist. *Curriculum Inquiry*, *29*(3), 343–353. doi:10.1111/0362-6784.00131

Hlebowwitsh, P. (2010, September). Centripetal thinking in curricula studies. *Curriculum Inquiry*, *40*(4), 503–512. doi:10.1111/j.1467-873X.2010.00497.x

Hoffman, M. (2000). *Empathy and moral development: Implications for caring and justice*. Cambridge University Press. doi:10.1017/CBO9780511805851

Hostetler, K. (2005, August 1). What is "good" education research? *Educational Researcher*, *34*(6), 16–21. doi:10.3102/0013189X034006016

Ito, M., Antin, J., Finn, M., Law, A., Manion, A., Mitnick, S., Schlossberg, D., & Yardi, S. (2010). *Hanging out, messing around, and geeking out. Kids living and learning with new media*. The MIT Press.

Ivey, G., & Fisher, D. (2007). *Creating literacy rich schools for adolescents*. Association for Supervision and Curriculum Development (ASCD).

Kerdeman, D. (1998). Between Interlochen and Idaho: Hermeneutics and education for understanding. *Philosophy of Education*, 272-279.

Myers, S. (2021, June 28). *Supreme Court sides with cheerleader suspended for off-campus social media tirade*. Sharyl Attkisson. https://sharylattkisson.com/2021/06/read-supreme-court-sides-with-cheerleader-suspended-for-off-campus-social-media-tirade/

Naghdipour, B. (2017). 'Close your book and open your Facebook': A case for extending classroom collaborative activities online. *The Journal of Asia TEFL*, *14*(1), 130–143. doi:10.18823/asiatefl.2017.14.1.9.130

Ove, J. (2014, October 17). *Georgia Court of Appeals: Parents can be held liable for kids' social media misdeeds*. Patch. https://patch.com/georgia/acworth/georgia-court-appeals-parents-can-be-held-liable-kids-social-media-misdeeds-0

Paliktzoglou, V., Oyelere, S. S., Suhonen, J., & Mramba, N. R. (2021, Summer). Social media: Computing educational perspective in diverse educational contexts. *Journal of Information Systems Education*, *32*(3), 160–165.

Papademetriou, C., Anastasiadou, S., Konetos, G., & Papalexandris, S. (2022, April). COVID-19 pandemic: The impact of the social media technology on higher education. *Education Sciences*, *12*(4), 261. doi:10.3390/educsci12040261

Pardales, M. J. (2002). "So, how did you arrive at that decision?" Connecting moral imagination and moral judgment. *Journal of Moral Education*, *31*(4), 423–437. doi:10.1080/0305724022000029653

Parker, W. C. (1997, February). The art of deliberation. *Educational Leadership*, *54*(5), 18–21.

Parker, W. C. (2003). *Teaching democracy: Unity and diversity in public life*. Teachers College Press.

Pinar, W. F. (1978). The reconceptualization of curricula studies. *Journal of Curriculum Studies*, *3*(10), 150–157.

Pritchard, M. (1996). *Reasonable children: Moral education and moral learning*. University Press of Kansas.

Quinn, M. (2013, September 3). California School District hires company to monitor students' social media. *Washington Examiner*. https://www.washingtonexaminer.com/red-alert-politics/california-school-district-hires-company-to-monitor-students-social-media

Rahman, T., Kim, Y. S., Noh, M., & Lee, C. K. (2021, March). A student on the determinants on social media based learning in higher education. *Educational Technology Research and Development*, *69*(2), 1325–1351. doi:10.100711423-021-09987-2

Rome youth basketball has rest of season canceled. (2022, February 15). *Daily Sentinel*. https://romesentinel.com/stories/rome-youth-basketball-has-rest-of-season-canceled,129459

Schad, T. (2022, March 12). 'It seems to be more extreme': Violent sports fans are causing alarm at every level. *USA Today*. https://www.usatoday.com/story/sports/college/2022/03/12/sports-fans-more-violent-abusive-since-returning-after-worst-covid/6986397001

Simon, K. G. (2003). *Moral questions in the classroom: How to get kids to think deeply about real life and their schoolwork*. Yale University Press.

Sparks, S. D. (2013, January 4). *Social-emotional needs entwined with youth' learning, security*. Education Week. https://www.edweek.org/leadership/social-emotional-needs entwined-with-youth-learning-security/2013/01

Tannen, D. (1998). *The argument culture: Stopping America's war of words*. Random House.

The Telegraph. (2020, September 29). *First presidential debate in full: Trump vs Biden | US Election 2020* [Video]. YouTube. https://www.youtube.com/watch?v=CweqW7Pzxz8

The View. (2020, June 9). *National debate over defunding the police* [Video]. YouTube. https://www.youtube.com/watch?v=uMu5UshGUPM

Throwback. (2008, August 21). *The 1st Kennedy/Nixon Presidential Debate - Part 1/4 (1960)* [Video]. YouTube. https://www.youtube.com/watch?v=C6Xn4ipHiwE

Tkacová, H., Králik, R., Tvrdoň, M., Jenisová, Z., & García Martin, J. (2022, February 27). Credibility and involvement of social media in education: Recommendations for mitigating the negative effects of the pandemic among high school students. *International Journal of Environmental Research and Public Health, 19*(5), 2767. doi:10.3390/ijerph19052767 PMID:35270460

Tyler, R. W. (1948). *Basic principles of curricula and instruction*. University Chicago Press.

Usagreencardcenter. (2007, July 20). *Bill O'Reilly and Geraldo Rivera angry fight Immigration* [Video]. YouTube. https://www.youtube.com/watch?v=Z3U9ENaTPLY

Valuetainment Short Clips. (2022, March 15). *Who should be held responsible for COVID? Patrick Bet-David podcast episode 132* [Video]. YouTube. https://www.youtube.com/watch?v=GolGGSY4GJ0

Yagelski, R. P. (2009, October). A thousand writers writing: Seeking change through the radical practice of writing as a way of being. *English Education, 42*(1), 6–28.

Yagelski, R. P., Elbow, P., Freire, P., & Murray, D. M. (2006, May). Review: "Radical to many in the educational establishment": The writing process movement after the hurricanes. *College English, 68*(5), 531–544. doi:10.2307/25472169

ADDITIONAL READING

Alverman, D., Hagwood, M. C., & Moon, J. S. (1999). *Popular culture in the classroom: Teaching and researching critical media literacy*. Routledge. doi:10.4324/9781315059327

Eisenhart, M. (2001, November 1). Educational ethnography past, present, and future: Ideas to think with. *Educational Researcher*, *30*(8), 16–27. doi:10.3102/0013189X030008016

Westbury, I. (1999). The burdens and the excitement of the "new" curricula research: A response to Hlebowitsh's "The burdens of the new curricularist." *Curriculum Inquiry*, *29*(3), 353–363. doi:10.1111/0362-6784.00132

Section 3
Design as a Factor

Chapter 8
Technoethics:
An Analysis of Tech Assessment and Design Efficacy

Jennifer Fleming
https://orcid.org/0000-0003-2845-7128
Purdue University Global, USA

ABSTRACT

Technoethics is a discipline that seeks to analyze technology's effect on society. This is accomplished by evaluating each proposal from two perspectives: holistic impact and practical application. The first approach looks at how the tool will benefit society and any potential risks from its introduction. The second review evaluates the design used to create the product to understand if this is the best possible construct or whether an alternative would reduce potential harm. The history of technoethics in this setting and the efficacy of educational guidance toward better outcomes are examined and evaluated. Included are recommendations about how institutions could enhance their curriculum to better promote societal well-being.

INTRODUCTION

English poet Alexander Pope (1711) once stated that "to err is human, but to forgive is divine." This statement has been studied and analyzed by thousands of psychologists, sociologists, and theologians over the years, each seeking to explain its meaning towards human behavior, values and moral agency. The basic premise is that the human state is imperfect and fragile and, as a result, will make errors and need correction to establish, assimilate and function within society. Historical research and trend analysis (Gregersen, n.d.) support these perspectives as early drawings and records of technological advancement indicate that fire, wheel, air travel, and space technology all chronicle a series of failures and successes in goal attainment. However, what happens when people continuously repeat their errors in judgment or disregard the past learnings?

DOI: 10.4018/978-1-6684-5892-1.ch008

Pavlov taught that most mammals would conform to conditions when seeking to achieve a goal; this is called respondent conditioning (McLeod, 2021). Social system theory (Gibson, n.d.) taught that interrelationship between people, their beliefs, and society contribute to the ability to adapt to in the environment. Therefore, the presence or absence of some critical factors in neural psyche could contribute to undesired outcomes in the technology journey that create more harm than good. For example, over 3.3 million years ago, cave dwellers developed the first set of technological tools using unshaped stones, hammers, and anvils (Gregersen, n.d.). Now, crewless space ships travel to and from the earth, delivering supplies and collecting samples from other planets. However, that same level of intellectual depth and savvy produces challenges like issues with misuse of technology biomedical fraud (Wells & Farthing, 2009), unethical cloning (Tannert, 2006), cybercrime ("U.S. Consumers," 2022), etc.) and psychological distance with moral agency such as robotic cars with cinema features (Tucker, 2021), space debris cluttering the orbit or space technology falling into the ocean compromising marine life (Hutagalung et al., 2020), or the creation of a massive sub-standard dam in areas where human life can be loss ("Survivors of Laos' Worst Dam," 2020). There was an insufficient amount of concern for potential threats, possible flaws in design elements, and a lack of substantive interest in protecting human life. These incidents occur worldwide, and unfortunately, the frequency seems to be increasing.

One of the most notable Harvard Business Review case studies is the Challenger Space Shuttle Disaster (Prusak, 2011). It humbly reminds society of what can happen when people become too immersed in external factors and forget to provide the fundamental element of care in technological advances. Ethics is the branch of knowledge that governs a person's behavior towards an activity-based upon their values, integrity, choice, conscience, sense of fairness, and accountability. Historians found that these core principles had significant relevance in undergraduate studies but were rarely taught in areas of need (i.e. medicine, law, science, etc.). Douglas Sloan (1979) reported his findings in "The Teaching of Ethics in the American Undergraduate Curriculum, 1876-1976," and Beever et al. (2021) continued this research in their study "Where ethics is taught: an institutional epidemiology," which chronicles 1980 forward. The absence of ethical principles in key areas of technology has serious consequences within society because thoughtful technology assessment (TA) and technoethical design (TED) are not being performed, which could deter the creation of harmful products or services for others or reduce the opportunity for others to create harm to themselves.

BACKGROUND

The term technoethics was developed by Mario Bunge in his 1977 publication "Towards a technoethics." Here, Bunge expounds that the old monarchy structure used to govern society no longer exists and those individuals that are most capable of erecting change and growth within society also bear the responsibility of ensuring its protection, stating,

Nobody recognizes rights without duties, privileges without responsibility...everyone is rightly held responsible for what he does and even for what he fails to do when he ought to act...and the outcome of their labors is well-known: a new kind of society, one that may carry mankind either to a higher evolutionary level or to a quick extinction. (Bunge, 1977, pp. 96-97)

Many authors have interpreted this concept of technoethics: Rocci Luppicini, a Canadian social scientist and professor of technology says, "technoethics offers a theoretical base and set of tools for moving forward in the study of social and ethical aspects embedded within our technologically oriented society" (Luppicini, 2009, p. 7), Billinger (2009) defines it as "the inherent moral codes that compel us to utilize technologically sophisticated methodology for the resolution of an ethical dilemma" (p. 45). In addition, Mahmoud Eid (2014) examines the applicability of these principles in the context of social and theological systems, concluding with the perspective "that teleological ethics underscore the consequences of an act or decision" (p. 189). However, one decides to describe this phenomenon, the underlying themes of care for each other, accountability towards society, and moral agency are consistent. The goal is to identify what elements of any proposed new product, service, or technology presented within a community will interact with its constituents and whether their consumption of that product will cause intentional or unintentional harm.

TECHNOETHICAL ASSESSMENT (TEA)

When considering any given technology's impact or consequence within society, it is essential to consider its technoethical assessment. This means people need to evaluate its intended placement, interaction with new and legacy processes, understand its influence on all potential stakeholders and examine its inter-relations at multiple levels: biological, physical, social, and environmental (Luppicini, 2010). Understanding these answers will enable one to recognize the potential risk and impact proposed by the technology and possible mitigation strategies that could be employed. Methods for conducting the technoethical assessment should be introduced and enforced in all areas of industry that create transformational change.

A technoethic assessment is a broadly described concept that consists of engagement, policy, and governance towards managing the introduction of technology in society. This phenomenon has been studied and interpreted by many scholars. Van Est and Brom (2012) posit it as "an analytic and democratic practice which aims to contribute to the timely formation of public and political opinion on societal aspects of science and technology" (p. 1). Rip (1986) argues that TEA is an instrument to "reduce the social costs of learning by error and to do so my systematic anticipation of potential impacts of new technologies and large projects, and feedback into decision making" (pp. 357-358) and Luppicini (2009) repurposes the scenario as a *technoethical inquiry*, stating that: "technological systems are historically at the heart of contemporary transformation in life and society and have created a multiplicity of new social and ethical dilemmas in the public sphere that did not previously exist creating anti-human consequences" (pp. 15-16). Adopting the following definition will apply: technoethical assessment is an instrument of inquiry used to determine any proposed technology's efficacy within society using public and political venues.

Historical Perspectives

Environmental Injustices

Van Est and Brom (2012) chronicled the formal introduction of TEA in the U.S. by describing its four major tenets: technology development, relevant social actors, potential social impacts, and potential solutions. However, one need not look far for examples of how technoethical assessment was demonstrated before formalizing these principles. Concerned public citizens lay siege on the corporations and the government in pursuit of greater awareness and accountability to protect social well-being. Rachel Carson (1962) wrote of environmental injustices in her publication *Silent Spring*:

Along with the possibility of the extinction of mankind by nuclear war, the central problem of our age has therefore become the contamination of man's total environment with such substances that accumulate in the tissues of plants and animals and even penetrate the germ cells to shatter or alter the very material of heredity upon which the shape of the future depends. (Carson, 1962, p. 8)

This is in response to the overt and callous use of harmful sprays, dusts, and aerosols that were in voluminous use in homes and farmlands and eventually contaminated the waterway systems. These actions resulted in the expulsion of the production of dichloro-diphenyl-trichloroethane (DDT) in the U.S., the creation of the National Environmental Policy, which eventually established the Environmental Protection Agency, and the award of the Presidential Medal of Freedom in 1981.

Product and Agency Liability Issues

Complimenting and contrasting these sentiments, an unbeknownst colleague Ralph Nader championed issues of corporate ethics and more. In his "Legislating Corporate Ethics" speech, Nader (2004) takes a sharp tone toward corporations denouncing their violence and posits,

When it comes to corporate violence, we are talking about preventable death and injury. Every year over 300,000 Americans die from just five of the following preventable conditions. Imagine: 58,000 lose their lives in work-connected diseases and trauma; 65,000 lose their lives due to air pollution; 80,000 lose their lives in hospitals, (not even counting deaths attributable to emergency rooms, due to gross hospital and medical malpractice, incompetence, or neglect) 75,000 people die from preventable hospital caused infections; and about 42,000 loose lives on the highway-each year. Of course, there are many other sources of violence that the law does not recognize in the same way, it recognizes a violent burglary, for example, or any other violent act called a "street crime." As a result, society does not have an understanding of the magnitude of corporate induced violence. It is silent and cumulative, and its victims die alone in hospices rather than in violent crashes of skyscrapers." (Nader, 2004, p. 4)

Nader was most notably known for efforts to promote automotive safety, as noted in the 1965 book, *Unsafe at Any Speed*, which eventually paved the way for legislation to create the National Highway Traffic Safety Administration (NHTSA), which was chartered to promote automotive safety and reduce vehicle- and traffic-related injury and death.

It has been sixty years since the introduction of Silent Spring and its campaign toward environmental protection. The valiant efforts of Carlson and Nader represent the framework of the democratic and analytical approach to technical assessments where elected or private citizens advocate for change through demonstration of the fundamental tenets: the identification of technological development, relevant social actors, potential social impacts, and potential solutions. The following section will highlight the alternate approaches to ethical governance using classical and participative models.

Orbital Space Debris

Recent issues with environmental safety are significantly different from those of Carlson's generation. One such prevalent matter is the concern regarding space debris. For example, the National Aeronautics and Space Administration (NASA) reported that there have been several multinational incidents that have contributed to this phenomenon of space debris. Many decades ago, a French satellite was damaged by remnants from one of their prior rocket explosions, a Russian spacecraft collided with an U.S. system and that created more than 2,300 pieces of traceable debris, and China added 3, 500 additional piece of space junk when they destroyed one of their old weather satellites (Garcia, 2021).

Collectively, these incidents illuminated the impending tragedy known as space debris, or the collection of small and large defunct physical matter orbiting the earth at high rates of speed that hinder space travel schedules and create opportunities for collision with the international space station, government spacecraft, and private space travel entities that conduct scheduled research and discovery missions. Tracking of satellites and other orbital items began around 1950 (with the launch of Sputnik1); these pioneering efforts were primarily conducted by the Naval service (Thomson, n.d.). Since space travel is an international venture, the issues affect all participating countries; as a result, there have been significant discussion, organization, and discovery towards the alleviation of this problem. As a result, using the classical (expert-based policy and consequence analyses) and participative models (grassroots activism and decision analysis towards solutions), several cooperatives have emerged to address the problem (van Est & Brom, 2012). The Department of Defense's global Space Surveillance Network (SSN) sensors all detectable orbital debris (ranging from the size of a softball to the size of a marble or even one millimeter) in a catalog and sends alerts whenever they anticipate potential collision or harmful encounters. This is a tedious process, and verification and notification take time. In 2019, European partners began the ClearSpace project to clear space debris; this project has a 2025 target launch (European Space Agency, 2019). A private Japanese company, Astroscale Inc., has developed commercial spacecraft tasked to declutter space of defunct satellites; this is scheduled to launch in 2024 (Kim, 2021). Finally, the Vice Chief of Space Operations, Gen. David Thompson, indicates that although it is not the mission of the U.S. Space Force, they plan to seek alliances with private industry to identify potential options for the space debris problem because it affects everyone (Erwin, 2022).

Assisted and Autonomous Driving

Automobiles have now reached advanced levels of operations that mimic commercial aircraft, where publicly-owned and privately-owned vehicles can be placed in different modes of operations. Assisted driving (formerly known as cruise control) can be activated to support the driver with short non-human-directed steering and braking periods. In contrast, autonomous driving involves vehicles that have been previously programmed to function all normal operations of driving from a designated departure zone to its intended destination without any human intervention. Assisted driving mechanisms have been in production since 1950 and continue to be refined with scientific discovery and consumer experience (Sears, 2018). Autonomous driving is still evolving, and the standards required for safety are being defined by the National Highway Traffic Safety Administration (National Highway Traffic Safety, n.d.). There are currently five levels of standards: Level 0: Momentary Driver Assistance, Level 1: Driver Assistance, Level 2: Additional Assistance, Level 3: Conditional Automation, Level 4: High Automation, and Level 5: Full Automation. This is an example of the policy analysis for technoethics assessment.

In September 2016, NHTSA and the U.S. Department of Transportation issued the Federal Automated Vehicles Policy, which sets forth a proactive approach to providing safety assurance and facilitating innovation ... and in 2020, NHTSA launched Automated Vehicle Transparency and Engagement for Safe Testing. (National Highway Traffic Safety, n.d., "NHTSA is Dedicated" section)

Currently, there are several levels of prototype testing underway. For example, Walmart has autonomous delivery between factory and distribution centers; it is essentially crewless box trucks traveling the same seven-mile route each day (Ramey, 2021). Uber began a trial under its partnership with Carnegie Mellon University National Robotics Center; unfortunately, a series of failed tests led to costly legal and financial hardship. They have since abandoned the project (Metz & Conger, 2020). Tesla autonomous driving systems have been relatively successful ("Autopilot and Full Self-Driving Capability," n.d.); however, their challenges lie within the intermediation of technology and human moral agency (Luppicini, 2009) because human drivers become distracted and disengaged with other amenities like cinema and lose control of the vehicle ("Patrol: Tesla," 2020).

TECHNOETHICAL DESIGN (TED)

Background

Technoethical design examines how newly proposed products or services are constructed to determine their overall efficacy in society, best use of resources, the probability for proper adoption versus alternate harmful use, and the degree of governance needed to sustain successful utilization. In an ideal setting, this assessment would occur before new products and services are introduced into society. However, since this is an open capitalistic economy, the ability to present many goods and services is unencumbered by regulation in some industries, so the evaluation becomes a retrospective in some instances. The goal is to decompose the product (or service) into design elements and evaluate each component's potential for success, safety, and sustainability. For example, the question arises as to whether or not integration of the new system will promote positive outcomes via intended usage, or if there will be high degrees of alternate use that is harmful. As noted earlier, using artificial intelligence (or robotic-assisted driving features) within the auto industry can result in adverse outcomes if misused, and the presence of a cinema feature in luxury cars ("Patrol: Tesla," 2020) created a distraction incentive for some drivers that is much too great to overcome and resulted in several accidents. Furthermore, the use of cloning or reproductive technologies (Ayala, 2015) has become an area of interest in dark society and underground markets; therefore, much work is needed to review and revise these design elements to reduce the potential for adverse outcomes while preserving their initial intended application and benefits.

Determining Overall Efficacy in Society

The tools and techniques for determining the overall efficacy of a newly proposed product (or service) in society are well known and documented as product market segmentation. They can be repurposed in the ethical arena to extend the analysis toward consumption rate, safety, and sustainability factors. For example, the first determination addresses the target population in a marketing and sales analysis. Then within that target population, the target user's characteristics (or distinguishing features) are defined.

After the first two questions are answered, a computation is used to calculate the percentage of potential consumers (or users) the company can expect to reach or adopt over the pilot introduction (usually 1 to 5 years). This analysis is followed by an examination of Michael Porter's Five Forces (2008) to determine potential threats, opportunities, substitutions, etc. The same framework can be applied to address an ethical assessment of the overall efficacy of products in society. Identify the total population of potential users and the target (most likely consumer) within that population, adding their personal and risk characteristics, and completing the study with a risk factor analysis that examines the probability and outcomes (impact) for adoption under the intended application, adoption using slightly alternate use, and adoption using harmful alternate use. Conclude by multiplying the probability times the impact from potential outcomes to determine those elements with the greatest level of volatility and requiring significant governance to deter misuse and harmful outcomes.

Best Use of Resources

Resource Management seeks to identify the sourcing, consumption, storage, and disposition of labor and materials used for product development throughout its lifecycle. The sourcing element addresses where the raw materials or capital expenditures are acquired, and storage looks at how and where they are temporarily held while awaiting processing. Consumption looks at how the harvesting of these resources affects society; it examines the environment in which they are removed to determine if this generates an undue burden and can they be easily restored to a condition of wholeness within a reasonable amount of time. Finally, disposition examines the degree of harm they present to the environment upon retirement. As noted earlier, Carlson (1962) discovered that the high residual value of DDT in the manufacturing settings presented an undue burden on natural and aquatic environments such that it was banned.

In today's society, deforestation is a significant problem in the war to combat climate change because removing the trees disturbs the natural order of climate mitigation through the absorption of carbon dioxide. This is particularly problematic in several areas, as noted in the United Nations "2020 State of the World's Forest, Biodiversity and People" report. They cite 1998 research by Ruf and Zadi that cocoa has been a historically significant product, however, it is also the reason for massive deforestation; "expansion into forests is often driven by low cocoa yields from established plantations since soils freshly cleared of natural vegetation are often more fertile" (Food and Agricultural Organization, 2020, p. 90). Efforts are currently underway to create a recovery vegetation system to restore losses. Palm oil is probably the most impactful contributor to deforestation because it produces some of society's most commonly used products (i.e., soap, detergent, lotions, margarine, biscuits, and pizza). "Palm oil has been and continues to be a major driver of deforestation of some of the world's most biodiverse forests, destroying the habitat of already endangered species like the Orangutan, pygmy elephant, and Sumatran rhino" ("8 Things," n.d., section 4). The United Nations reports that biodiversity conservation efforts will require as much as $200B in U.S. dollars annually to cultivate meat and vegetable alternative products and reduce deforestation activity (Food and Agricultural Organization, 2020).

Proper Adoption vs. Harmful Alternate

Companies are often confronted with the complicated dilemma of getting consumers (users) to conform or behave in the manner intended for new product releases. What one says does not always translate into what actions one takes, and conversely, the steps one takes are not always consistent with the stated inten-

Table 1. Risk factor analysis of automotive design elements

Risk Identification		Risk Factor Analysis				Risk Response
Risk	**Risk Category**	**Probability**	**Impact**	**Risk Score**	**Risk Ranking**	**Risk Response**
20211	**Cruise Control**	**0.5**	7.5	3.75	1	Accept
20212	-Intended Use	0.375	8	3	1b	Exploit
20213	-Alternate Use	0.125	1	0.125	1c	
20214	**Automatic Seatbelt**	**0.3**	10	3	2	Decline
20215	- Intended Use	0.225	8	1.8	2c	
20216	- Alternate Use	0.075	10	0.75	2b	Use Policing Policy
20217	**Autonomous Driving**	**0.2**	10	2	3	Defer
	Total	1				

tions of the object. Companies expect users to read the manual and perform the instructions as intended to realize the business and safety value. Unfortunately, the liberal desires of human agency to divest from these principles create a series of outcomes with small and large consequences. Returning to the earlier comments of product efficacy, after the market segmentation is complete, it is advised to produce a risk factor analysis of the probability of adoption and potential use to determine the next best steps.

A sample demonstration (see Table 1) illustrates the analysis for utilization of cruise control, automated seat belts, and autonomous vehicles in the automotive industry. The three elements are identified as positive (intended use) and negative (alternate use) risks. A probability of 100% or 1 point was distributed across the elements. A perceived 50% adoption rate is designated for cruise control, 30% for automatic seat belts, and 20% for autonomous driving. Impacts were selected at 7.5 for cruise control in general, with a score of eight for effectiveness if used as intended and one if used alternatively because it is a system-controlled feature with no user-specified parameters. Impacts for effectiveness for the Automatic Seat Belt were designated at 10 in general, the impact for the intended use (positive risk) was lowered to 8 (to account for occasional system deficiency), and the impact for alternate use (negative risk) was designated at 10 (because the potential for alternate use, modification of the object and total absence of any seat belt protection produces a significant hazard for the driver and passengers). Finally, the probability of adoption of autonomous driving was designated at 20% due to impending results from current prototyping and limited trials. An impact of 10 was assigned to the reflect the significantly transformative nature this element presents in the transportation industry. The computer risk score and ranking indicated the following outcomes. The acceptance and exploitation of positive risk implementing cruise control elements, declining and assigning alternate actions (use of policing policy) for automatic seatbelts design elements, and deferment of the autonomous driving element due to insufficient information at this time.

Degree of Governance for Sustainable Use

The behavioral traits and emotional intelligence demonstrated when people interact with any technological component are called human agency. The degree of accountability or ethics acknowledged during this interaction is known as human moral agency. Many of these perspectives are captured in Luppicini's (2008) *Law of Technoethics*, as the author asserts "ethical rights and responsibilities assigned to technology and its creators' increases as technological innovations increase their social impact ... [there is] the need for social and ethical responsibilities among those working with and affected by the paradoxical consequences of technological development" (Luppicini, 2009, p. 16).

Fortunately, human moral agency is a dynamic concept that will evolve and mature as people are better educated, informed, and attentive to the potential for harmful outcomes and the need for greater conformance to prevent issues with sustainability for their generation and future ones. For example, a series of consumer agencies were created to educate and inform the population of any previously known or recently reported issues with product launches. Additionally, several prominent manufacturers have proactively created user groups, think tanks, and sustainability initiatives to seek out better constructs for product design, development, and disposition of manufacturing or production waste. Finally, an Office of Sustainability was developed at the federal level (Office of the Federal Chief, n.d.) and chartered to coordinate policy to promote energy and environmental sustainability across Federal Government operations. The U.S. Department of State's Office of Management Strategy and Solutions develops and reports on environmental best practices and partnerships that work for environmental innovation and the development of eco-diplomacy (U.S. Department of State, n.d.).

ETHICS IMMERSION IN HIGHER EDUCATION

The preceding identified the framework for technoethics and the need for its application in government and industry to protect current society and promote social well-being for future generations. Some of the scenarios discussed present ethical dilemmas that were both known (and unknown) to the scientific and technology community. Also, some situations indicate varying levels of attentiveness to the ethical dilemmas demonstrating a range of behaviors from total indifference to limited active participation toward corporate social responsibility and greater degrees of sustainability and moral agency. This indicates that there might be different viewpoints regarding the need for demonstrating care amongst various entities or industries.

Background

Thornley et al. (2018) posits that "the current lack of a coherent ethical framework for Information and Communications Technology (ICT) is in contrast to other professions such as medicine or law, which have code of ethics and possible penalties in place for non-compliance" (p. 57). Most people are familiar with the widely-recognized component of the Hippocratic Oath, "do no harm" however, they may not be aware of the additional efforts towards continuous improvement through the use of forensic and clinical post-mortems. "The autopsy ... remains an important step towards the improvement and development of medicine, and remains the most certain way of finding an accurate and complete diagnosis for any known or suspected pathology" (Costache et al., 2014, p. 265). Some may attribute this ethical oversight in ICT due to time or costs; Gupta and Kumar (2018) argue that the cost of ethically-produced products creates conflict within society, because only the wealthy can afford to pay for these enhancements whereas middle class or poor people would prefer cheaper products regardless of their sourcing. Others recognize that even though implementing ethical frameworks within business settings can be costly, the civic-minded company can be rewarded from customer appreciation. Setiyaningrum and Aryanto (2016) clarify that "companies implementing business ethics practice and actively involved on corporate social responsibility (CSR) activities are commonly perceived as the good corporate citizens by consumers. This recognition allowed the companies to differentiate themselves from the competitors and attract customer loyalty" (p. 73).

Absence of Ethics in Learning

Regardless of the reason, there is a distinct absence of ethical framework in areas of ICT, which could explain why there is such a broad spectrum of attentiveness in areas of TEA, TED, and societal well-being. Deb Gearhart (2012) acknowledges this concern in *Ethics for eLearning: Two sides of the ethical coin* addressing the concern for intellectual depth between ground and online technology students, citing that young adults are not mentally predisposed to handle the complications of morality in their post-secondary educational experiences because they are too psychologically distant from the situation to ascertain the potential harm presented (Gearhart, 2012).

The "psychological distancing" that occurs in online classroom dynamics appears to be similar to the dysfunctional behavior that arises in ICT environments when there is a noted distance between the developer and consumer such that the former does not consciously reflect enough on the potential for alternate (misuse) outcomes or adverse reactions (series of related events that cause substantive detriment to a community or environment).

Additionally, the absence of ethical education in higher education creates a gap in the intellectual foundation of future scientists and technologists because moral character and agency are not fully developed before graduation. In 1979, Douglas Sloan published the "Teaching of Ethics in the American Undergraduate Curriculum, 1876-1976." This publication sought to explain the storied history of morality in higher education, highlighting the rise and fall of its popularity over the last hundred years from its original intent (moral philosophy), transformed goals (ethical reasoning), and points of fallacy. The curriculum moral philosophy was initially embraced by early American colleges like Harvard, William and Mary, and Yale but began to wane again during 1880. Although it seemed to hold a consistent position in smaller private schools, it eventually disappeared in 1905. Many students and instructors felt it took too much time away from their more formidable studies. Harvard professor and minister Francis Peabody was able to revitalize their internal focus on ethics in 1906 and maintained some presence in this arena until his retirement in 1913. By 1925, it became commonplace for many colleges and universities to offer at least one ethics course under the department of philosophy. About two years after Rachel Carson's 1962 publication, *Silent Spring*, a professor performed an audit using a sample of 100 colleges and universities, of various sizes and secular orientation, discovering that only about 27% required any ethics or philosophy for graduation; the professor also noted in their own institution of 12, 000 students, the average annual enrollment in ethics courses was around eleven students (Sloan, 1979).

Deni Elliott and Karlana June (2018) resumed the meta-analysis of teaching ethics in education. In this study, the findings from The Hasting Center Sloan Report were reaffirmed and expanded to include examples of how the importance of ethics continued to fluctuate and then explode between 1980 to 2015 in scholarly literature and lay discussions alike (Elliott & June, 2018). Scores of research studies, journals, scholarly debates, and literary discourse were developed to address the need for ethical education and moral guidance in universities. Figures 2 through 4 reflect the dispersion of journal articles categorized by topic and release date. Figures 5 through 8 reflect the ethics journals organized by the intended discipline of thought for the same time periods. Lastly, Figure 9 reflects the top ethics journal articles indexed by category and release date. A review of this meta-data reflects two key themes: student behavior and

Figure 1. Flagship Journal Articles by Category and Decade
Data Source: The Evolution of Ethics (2018)

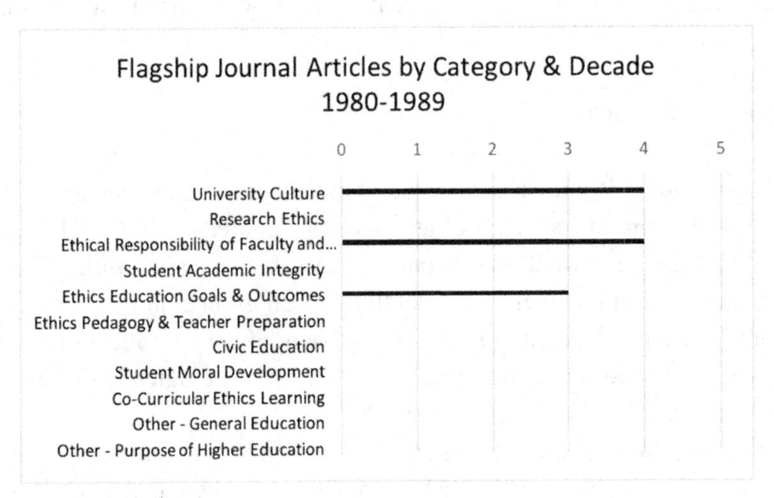

student preparedness. Student behavior reflects the actions, values, and moral conduct demonstrated within the university system that warrant correction and instructional deferent.

Figure 1 indicates that from 1980 to 1989, the primary issues of concern were a poor ethical culture within the university, the ethical responsibilities of faculty administrators, student moral development, and the need for comprehensive ethics education. Figure 2 demonstrates the introduction of publications towards student academic integrity and civic education.

Figure 2. Top Flagship Journals: Articles by Category and Decade
Data Source: The Evolution of Ethics (2018)

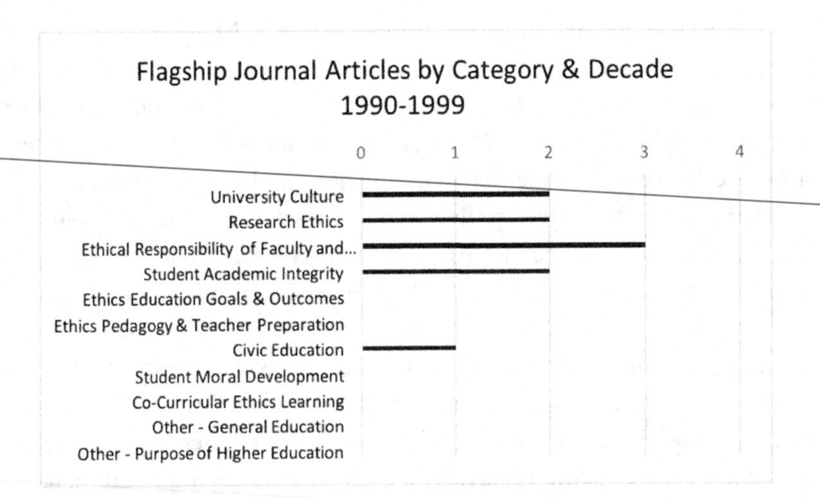

During the following decade (2000 – 2009), the issue of student academic integrity was significant, which adversely affected university culture and civic responsibility, as demonstrated in Figure 3. This focuses on research ethics, student moral development, and purpose of higher education increased during

Figure 3. Top Flagship Journals: Articles by Category and Decade
Data Source: The Evolution of Ethics Education (2018)

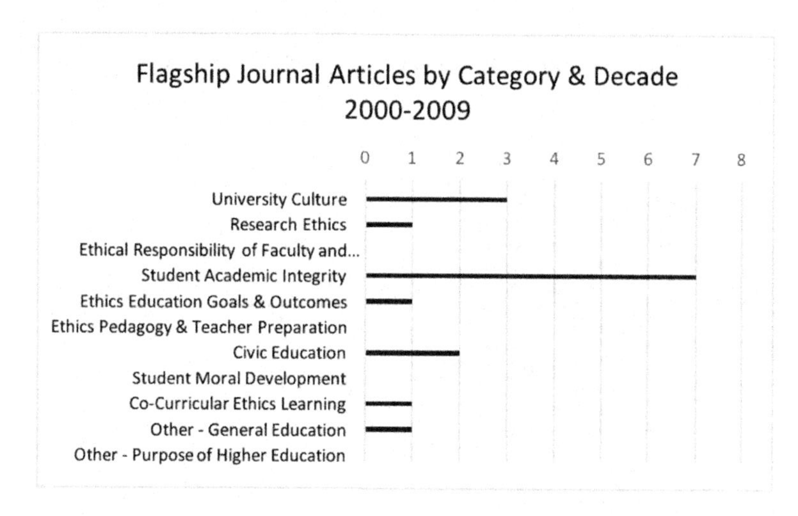

the following decade; however, university culture and student academic integrity waned as demonstrated in Figure 4.

Figure 4. Top Flagship Journals: Articles by Category and Decade
Data Source: The Evolution of Ethics Education (2018)

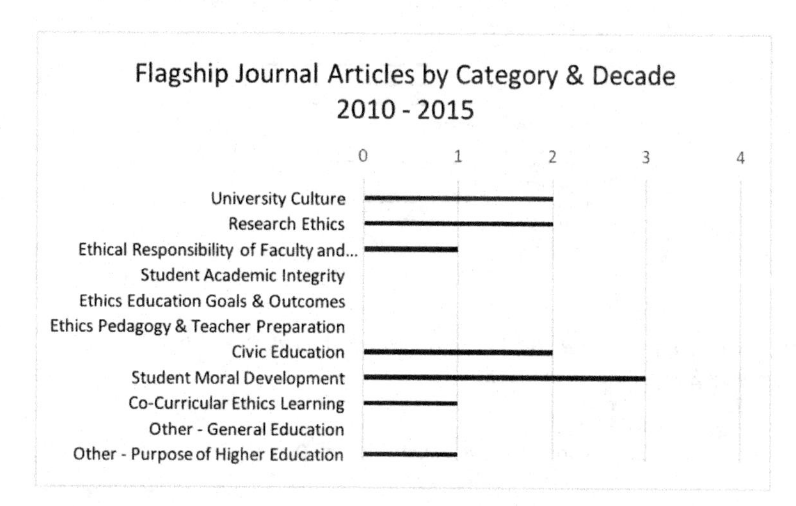

Figure 5 demonstrates that the spectrum of ethics journals was limited to student moral development from 1980-to 1989. However, the focus on ethics began to expand to university culture, ethics education initiatives, and ethics pedagogy in the next decade, as noted in Figure 6.

And, between 2000 and 2009, the depth of immersion in ethics pedagogy significantly increased, followed by a focus on university culture, student moral development, ethics educational goals, and research ethics as noted in Figure 7.

Figure 5. Top Ethics Journals: Articles by Category and Decade
Data Source: The Evolution of Ethics Education (2018)

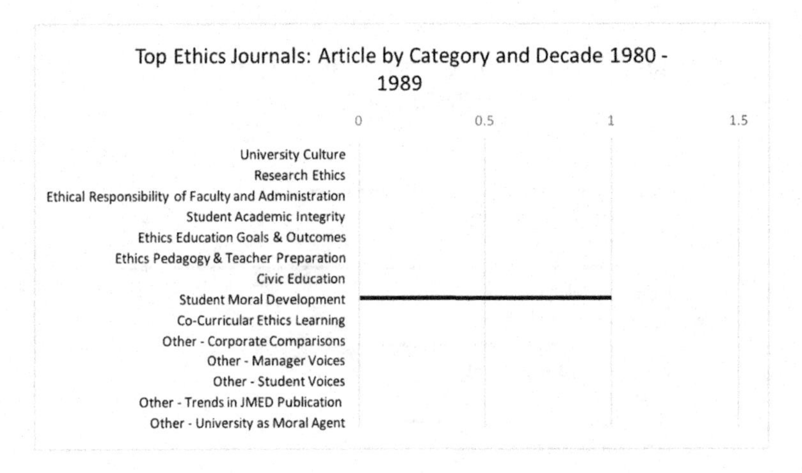

Figure 6. Top Ethics Journals: Articles by Category and Decade
Data Source: The Evolution of Ethics Education (2018)

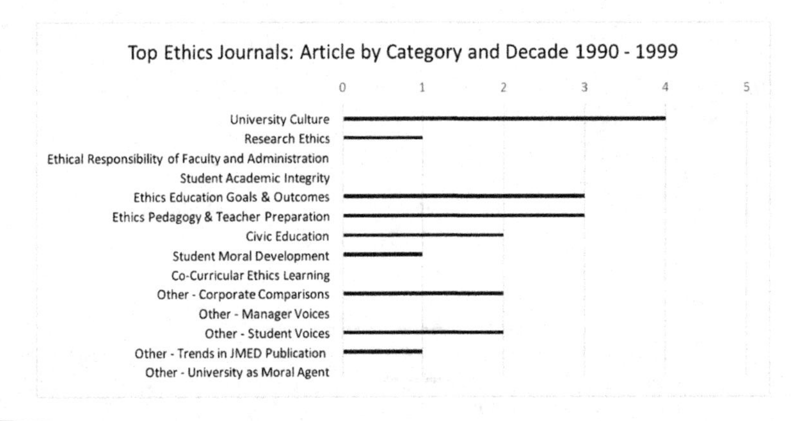

Figure 7. Top Ethics Journals: Articles by Category and Decade
Data Source: The Evolution of Ethics Education (2018)

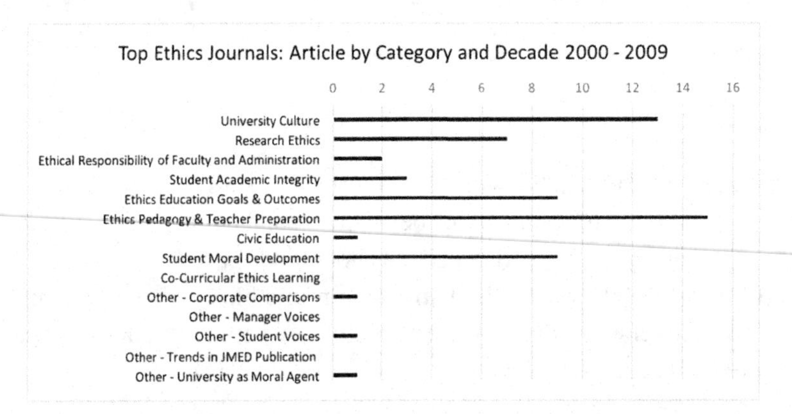

Figure 8. Top Ethics Journals: Articles by Category and Decade
Data Source: The Evolution of Ethics Education (2018)

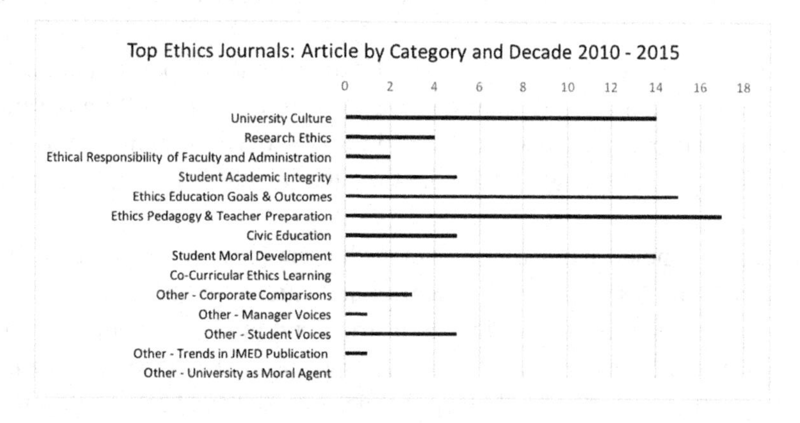

Figure 9. Ethics Journal by Discipline
Data Source: The Evolution of Ethics (2018)

Ethics Journals by Discipline	
Bio, Medical	33
Business & Economics	24
Moral Philosophy	20
Political Science	9
Higher Education & Teaching	7
Information Sciences & Technology	7
Social Sciences; Sociology	7
Religions & Theology	6
Environmental	6
Law & Criminal Justice	5
Psychology	5
Communications & Media	4
Military	3
Engineering	2
Public Health & Safety	2
Sports & Games	2
Public Administration	1
Animal Ethics	1
Social Services & Welfare	1
TOTAL	**145**

Focuses on research ethics fell almost 50% whereas ethics educations increased about 50% in the next decade; other notable increases include civic education, manager voices, student voices, and JMED as noted in Figure 8.

Finally, Figure 9 reaffirms the findings from the Hasting Center Sloan Report indicating that the majority of the literary focus in these ethics journals was on medical settings (biology and medicine – 33 journals), others include business and economics settings (24 journals) and general moral philosophy (20 journals). Information, Computing, and Technology had seven (7) journals, and Engineering had two (2) journals.

Elliott and June (2018) concluded that the focus of the literary production was more internally focused on improving or resolving issues of academic integrity and student morality rather than on integrating the discussion of ethics into specific disciplines to instruct and guide student populations on expectations and preparedness for future career endeavors. These perspectives seem to be consistent with others like the past Secretary of Education, who sharply criticized schools and universities for their failure to instruct students on moral values during that timeframe in his publication "The Teacher, the Curriculum, and Values Education" (Bennett, 1980) and as reported in the 1987 U.S. News and World Report, *A Nation of Liars* in which it was claimed "71% of those polled are dissatisfied with our current standards of honesty and conduct in the United States, and that 54% thought we are less honest today than ten years ago" McLouoghlin et al, 1987, pp. 54-60).

The last example in the Absence of Ethics in Learning section is an exploration of a study conducted by a high-profile research institution seeking perspective on the ethical curriculum provided to the student population. This literary review revealed another attempt at auditing the degree to which ethics was taught on a much smaller scale. A large metropolitan university that has significant government funding and high exposure to public health epidemiology conducted a study to ascertain the answer to two questions:

Where is ethics taught at our institution? Has where ethics is taught changed over time? We recognize the vast existing literature on whether ethics can be taught, and the extent to which ethics should be taught, but see less scholarly emphasis on how and where ethics is taught. (Beever et al., 2021, p. 217)

Although a valiant and formidable effort, many challenges were encountered that compromised the quality of the findings. For example, there were issues with data congruency (consistent definitions for ethics or moral-based courses) across curricula or disciplines and the inability to discern precise student demographics (individual vs. populace). Finally, the scope of the study was limited to just the undergraduate student population, so the relevance to the topic of ethics in the government-funded high-exposure research environment would not be revealed because this populace is too far removed from that level of engagement in scientific discovery.

Importance of Ethics for the Information Age

Richard Mason (1986) developed a publication titled the Four Ethical Issues of the Information Age designed to prepare individual citizens for responsible use as communities transitioned to a more digital state. Later, Crowell et al. (2007) developed a four-component model that described the posture of moral agency that would be necessary for ethical decision-making. Gearhart (2012) captures the key points as follows:

- **Ethical sensitivity:** Perceiving the relevant elements in the situation and constructing an interpretation of those elements. This component looks at what actions are possible, who and what might be affected by the action, and how the involved parties might be affected by the action. (p. 233)
- **Ethical judgment:** Relates to reasoning about the possible actions and deciding which is most moral or ethical. (p. 233)
- **Ethical motivation:** Involves prioritizing what is considered to be the most moral of ethical action over all others and being intent upon following that course. (p. 233)
- **Ethical action:** Combines the strength of will with the social and psychological skills necessary to carry out the intended action. (pp. 233-234)

The model stresses the importance of developing ethical standards for all citizens. The goal is to seek a greater understanding of the characteristics of human moral agency because society was transitioning into a new era, the Information Age. One where communication and access to information were seamless and timeless and where technological limitations of the past disappeared and capabilities of the future were yet to be discovered. However, with this power came significant risk and vulnerability to both those in positions of authority (and leadership) as well as those being separated and divided from the whole.

The Statista Research Department conducted three surveys on this topic ("Top Challenges," 2022); the first survey addressed the top challenges organizations face regarding ethics and compliance training (ECT); 44 percent of the respondents cited the limited hours available for training as one of their top three challenges. The second survey evaluated the methods used to provide ECT, 72 percent of the respondents stated they use person/live training to provide this training for their organization's board members as a preferred method ("Methods Used," 2022). The final survey focused on the most important objectives of the ECT; 72 percent of the respondents stated that creating a culture of ethics and respect was important, making this the most widely-held objective ("Most Important," 2022). This sentiment seems to be consistent with those of corporations and organizations that eventually onboard higher education graduates because challenges with ethics and compliance training exist globally.

STRATEGIES FOR BUILDING GREATER DEPTH IN ETHICS EDUCATION

The preceding identified the need for ethics training in collegial settings. The historical references revealed that attempts to integrate this guidance previously resulted in limited degrees of effectiveness: focusing solely on university culture and student academic integrity or narrow disciplines such as biology and medicine. This problem has persisted for some time, but there are some areas where greater depth has been successfully integrated and worthy of further explanation.

Case study situational analysis is needed as a preface to the examination of these different programs. This learning technique seeks to teach an individual decision-making skills by presenting a framework (short story or description of a condition within a company, university, research, etc.) and inserting an unethical or immoral incident within and then offers the student several potential solutions or allows the student to develop their solution and evaluates the efficacy of that outcome-based upon ethical or legal standards. Since the Information Age, the complexity and degree of challenge associated with ethical dilemmas have increased. However, several entities either capture or build databases of prior or potential ethical scenarios and use them as a teaching mechanism.

Case Study With Situational Analysis

National Institutes of Health

The National Institutes of Health (NIH) Office of Intramural Research has a database used for this purpose. There are currently 21 themes that explore facets of ethical dilemmas like scientific misconduct, authorship, data management, social responsibility, nepotism, research reproducibility, implicit and explicit biases, civility and harassment, lab management, and science under pressure. There is a recommended course of study which involves facilitation with group discussion. The NIH instructions (for the facilitator leading the case discussion) indicate the responsibility should include:

1. Reading the case aloud;
2. Defining and redefining, as needed, the questions to be answered;
3. Encouraging discussion that is "on-topic";
4. Discouraging discussion that is "off-topic";
5. Keeping the pace of discussion appropriate to the time available;
6. Eliciting contributions from all members of the discussion group;
7. Summarizing both majority and minority opinions at the end of the discussion (National Institutes of Health, 2021, "For Facilitators" section).

There is generally no right or wrong answer, rather an exploration of all possible solutions to mitigate the situation, reduce the exposure, and limit the legal and financial consequences of harm. The recommended set of questions each team should address includes who is affected, what interests are represented (such as financial or ethical), if actions taken are acceptable to those involved, if sanctions are appropriate for infractions and who would be in charge of determining this, what options are available along with their outcomes, and how could conflicts be avoided (National Institutes of Health, 2021).

American Psychological Association

The American Psychological Association (APA) offers ethics training for psychologists on the 89 Standards of their Ethics Code using commentary and case illustrations. The scenarios are for individuals in the following careers: consulting, forensics, education, training, research, organizational development, and public and private practice (Campbell et al., 2010).

National Institutes of Health PubMed Central

The NIH PubMedCentral (PMC) offers journal literature on a wide variety of biomedical and life sciences containing similar case studies (situational analyses) for training purposes. For example, a featured article, "Ethical Issues of AI " uses a case study and the Delph study to ascertain ethical dilemmas on artificial intelligence. Within, the study reveals the empirical accounts of issues related to the topic, such as:

- Cost to Innovation;
- Harm to physical integrity;
- Lack of trust;

- Lack of quality data;
- Disappearance of jobs;
- Lack of transparency;
- Potential for military use;
- Lack of informed consent;
- Bias and discrimination;
- Violation of fundamental human rights;
- Unintended, unforeseeable adverse impacts (Stahl, 2021, pp. 38-39).

Colleges and universities, large and small, religious or secular, can engage these materials to provide greater depth beyond the collegial dynamics of academic integrity to enhance a student's understanding of how to engage potential ethical dilemmas in their chosen field.

Additionally, some universities have developed their own database and training tools. The Illinois Institute of Technology has a Center for the Study of Ethics in the Professions, and Santa Clara University has a Markkula Center for Applied Ethics ("Technology ethics cases," n.d.). Using their database of cases, they offer case studies (using situational analysis) for a broad spectrum of professional fields like business, health care, biotechnology, government, and technology. They also offer faculty and students the opportunity to share new perspectives for case studies to keep the repository current with newly discovered ethical dilemmas (Center for the Study of Ethics, n.d.). Finally, the Association of Bioethics Program Directors offers a crowdsourced database that provides information about graduate programs with bioethics concentrations, organized by program type, institution, department, and location (Association of Bioethics Program Directors, n.d.).

There is no singular solution to such a complex and multi-dimensional problem. Every university, department, and faculty member has to determine what form of curriculum will address their needs for each given program, considering their available bandwidth and learning constraints. Because sensitive topics require strong facilitation and controlled settings, it would not be wise for distance learning unless the audience was a mature group of graduate-level study. However, it is ideal for ground or face-face education for undergraduates and students entering first-year residency (or internships).

CONCLUSION

The history of technoethics, beginning with an exploration of the denomination defining this phenomenon Mario Bunge to the perspectives of modern-day contemporaries Rocci Luppicini have thus been examined. TEA and TED were evaluated, which are recommended for assessing the introduction of technology into society by determining the impact of its introduction, the probability for intended use, the potential for misuse, and the anticipated inter-relations with biophysical, social, and environmental environments. Chronicling historical perspectives of critical implications associated with technoethics is an important next step, noting issues in environmental injustices, product and agency liability issues, instances of multi-national abuses like orbital space debris, and examples of conflict between technology design and human moral agency as seen in the introduction of assisted and autonomous driving.

The issues of product overall efficacy, best use of resources, proper adoption of new technology versus alternate use, and the degree of governance needed for sustainable use led one to consider the degree to which proper education and ethics guidance was provided for upcoming technologists. A meta-analysis of the immersion of ethics in education followed to determine to what extent this subject was administered in collegial settings to protect current society and promote social well-being for future generations. This revealed a fragmented history of application in areas of academia, with most concentrations toward students in areas of medicine or law. Fortunately, there is significant improvement in the current-day curriculum for a small set of colleges and universities. These institutions offer in-depth training on fundamental concepts and policy, case study libraries, and situational analysis to students in business, education, law, medical studies, and technology. These schools serve as a model for others and offer hope in establishing pedagogy toward better societal outcomes.

Unfortunately, this phenomenon's challenges increase exponentially with the introduction of new products, services, and technology that bypass traditional forms of creation and can be introduced into un-conforming (or non-mainstream markets that do not require governmental oversight or regulation). For example, the concept of cryptocurrency financial instruments (or Bitcoin) quickly emerged in the financial industry with mixed reception from consumers. In contrast, concept designs of autonomous travel are emerging everywhere with possible redundancy in efforts. Bitcoin investment appeared to experience a slow start and rapid ascent and is now experiencing a choppy and uncertain future (Kharpal, 2022).

Autonomous aircraft teams embarked on a series of TEA initiatives, then organized into an association, where independently-financed or wealthy entrepreneurs design products and services for various applications. It is uncertain whether Bitcoin will survive the challenges of the current inflation-ridden economy; however, the electronic vertical flight society (Evtol) is making significant strides ("Vertical Flight Society," n.d.).

The challenges all new and legacy industries face is ensuring the proposed design and targeted use for products (and services) is thoughtfully considered in the absence of formal policy or government regulation. Therefore, new college graduates and entrepreneurs need some form of TEA and TED training and ethics immersion to help formulate better designs, more robust logic towards averting the lack of human moral agency, and greater guidance (knowledge of precedence in areas of known harm). There are many venues to consider for this scenario; it could be that certifications are created in areas of academia and their alternates (trade schools) to address this gap. Another consideration is that a governing body is created that authorizes some form of licensure to introduce and market new products and services to consumers. Either method offers a remedy that could deter the creation of harmful products or services for others or reduce the opportunity for others to create harm to themselves.

A trendy quote says, "repeating behaviors and expecting a different result is a characteristic of insanity" and many people attribute its ownership to Albert Einstein. Historians discovered that Einstein never actually said that (Pruitt, 2018); however, society constantly reverberates this misconception year after year, so society is trapped in an unethical dilemma of misrepresentation by will. Conversely, the increased presence of harmful events, unfortunate circumstances, and morally unacceptable behaviors will continue if people fail to heed Bunge's advice (1977), take responsibility for technological change in society, and ensure it is performed responsibly.

Consider a collective effort between global leaders in space exploration towards participative technology assessment (TA) and TED for resolving orbital space debris and deforestation. In this scenario, there is a possibility that mutuality could lead to a transformative discovery in intergalactic exploration such that members of earth can cease excessive consumption of its resources and collectively search for distant planets for agricultural and household sourcing needs, so this planet can heal and continue to provide shelter for current and future generations.

REFERENCES

8 things to know about palm oil. (n.d.). World Wildlife Fund. https://www.wwf.org.uk/updates/8-things-know-about-palm-oil

Association of Bioethics Program Directors. (n.d.). *Graduate bioethics education programs results*. https://www.bioethicsdirectors.net/graduate-bioethics-education-programs-results

Autopilot and full self-driving capability. (n.d.). Tesla. https://www.tesla.com/en_AE/support/autopilot-and-full-self-driving-capability

Ayala, F. J. (2015, July 20). Cloning humans? Biological, ethical, and social considerations. *Proceedings of the National Academy of Sciences of the United States of America, 112*(29), 8879–8886. https://doi.org/10.1073/pnas.1501798112

Beever, J., Kuebler, S. M., & Collins, C. (2021, October). Where ethics is taught: An institutional epidemiology. *International Journal of Ethics Education, 6*, 215–238. https://doi.org/10.1007/s40889-021-00121-7

Bennett, W. J. (1980). The teacher, the curriculum, and values education. *New Directions for Higher Education, 1980*(31), 27–34. https://doi.org/10.1002/he.36919803106

Billinger, M. S. (2009). A technoethical approach to the race problem in anthropology. In R. Luppicini & R. Adell (Eds.), Handbook of Research on Technoethics. IGI-Global., https://dx.doi.org/10.4018/9781605660226.ch004

Bunge, M. (1977, January). Towards a technoethics. *The Monist, 60*(1), 96–107. https://doi.org/10.5840/monist197760134

Campbell, L., Vasquez, M., Behnke, S., & Kinscherff, R. (2010). *APA Ethics Code commentary and case illustrations*. American Psychological Association. https://psycnet.apa.org/record/2009-08922-000

Carson, R. (1962). *Silent spring*. Houghton Mifflin.

Center for the study of Ethics in the Professions. (n.d.). Illinois Institute of Technology. https://ethics.iit.edu/research

Costache, M., Lazariou, A. M., Contoleco, A., Costache, D., George, S., Sajin, M. & Patrascu, O. M. (2014, September). Clinical or post-mortem? The importance of the autopsy: A retrospective study. *Maedica (Bucur), 9*(3), 261–265. https://www.ncbi.nlm.nih.gov/pmc/articles/PMC4305994/pdf/maed-09-261.pdf

Crowell, C., Narvaez, D., & Gomberg, A. (2007, January). Moral psychology and information ethics. In L. A. Freeman & A. G. Peace (Eds.), *Information Ethics: Privacy and Intellectual Property* (pp. 19–37). Information Science Reference/IGI-Global. https://www.researchgate.net/publication/314456458_Moral_Psychology_and_Information_Ethics

Eid, M. (2014, January). Ethics, media, and reasoning: Systems and applications. In R. Luppicini (Ed.), Evolving Issues Surrounding Technoethics and Society in the Digital Age (pp. 188–197). IGI-Global. https://dx.doi.org/10.4018/978-1-4666-6122-6.ch012

Elliott, D., & June, K. (2018, May 9). The evolution of ethics education 1980–2015. In E. Englehardt & M. Pritchard (Eds.), Ethics Across the Curriculum—Pedagogical Perspectives. Springer., https://doi.org/10.1007/978-3-319-78939-2_2

Erwin, S. (2022, February 10). Space Force eager to invest in debris removal projects. *SpaceNews.* https://spacenews.com/space-force-eager-to-invest-in-debris-removal-projects

European Space Agency. (2019, September 12). *ESA commissions world's first space debris removal.* https://www.esa.int/Safety_Security/Clean_Space/ESA_commissions_world_s_first_space_debris_removal

Food and Agricultural Organization of the United Nations. (2020). *The state of the world's forests 2020.* https://www.fao.org/documents/card/en/c/ca8642en

Garcia, M. (Ed.). (2021, May 26). *Space debris and human spacecraft.* NASA. https://www.nasa.gov/mission_pages/station/news/orbital_debris.html

Gearhart, D. (2012, October). Lack of ethics for eLearning: Two sides of the ethical coin. *International Journal of Technoethics (IJT), 3*(4), 33-40. doi:10.4018/jte.2012100103

Gibson, B. (n.d.). *Systems theory.* Encyclopedia Britannica. https://www.britannica.com/topic/systems-theory

Gregersen, E. (n.d.). *History of technology timeline.* Encyclopedia Britannica. https://www.britannica.com/story/history-of-technology-timeline

Gupta, P., & Kumar, D. (2018). Ethical behavior and the development paradigm. In I. Oncioiu (Ed.), Ethics and Decision-Making for Sustainable Business Practices (pp. 258–267). IGI-Global. https://dx.doi.org/10.4018/978-1-5225-3773-1.ch015

Hutagalung, J. M., Tobing, C. I., Debastri, J., & Amanda, R. T. (2020, April). Space debris as environmental threat and the requirement of Indonesia's prevention regulation. *IOP Conference Series: Earth and Environmental Science, 465*(1), 1. doi:10.1088/1755-1315/456/1/012081

Kharpal, A. (2022, June 3). Crypto firms say thousands of digital currencies will collapse, compare market to early dotcom days. *CNBC.* https://www.cnbc.com/2022/06/03/crypto-firms-say-thousands-of-digital-currencies-will-collapse.html

Kim, S. E. (2021, August 25). Can the world's first space sweeper make a dent in orbiting debris? *Smithsonian Magazine.* https://www.smithsonianmag.com/science-nature/can-worlds-first-space-sweeper-make-dent-orbiting-debris-180978515

Luppicini, R. (2008). The emerging field of technoethics. In R. Luppicini & R. Adell (Eds.), Handbook of research on technoethics (pp. 1–18). Information Science Reference/IGI-Global. https://doi.org/10.4018/978-1-60566-022-6.ch001

Luppicini, R. (2009). Technoethical inquiry: From technological systems to society. Global Media Journal - Canadian Edition, 2(1), 5-21.

Luppicini, R. (2010). *Technoethics and the evolving knowledge society: Ethical issues in technological design, research, development, and innovation*. IGI-Global. doi:10.4018/978-1-60566-952-6

Mason, R. O. (1986, March). Four ethical issues of the information age. *Management Information Systems Quarterly*, *10*(1), 5–12. https://www.researchgate.net/publication/242705009_Four_Ethical_Issues_of_the_Information_Age

McLeod, S. (2021). *Pavlov's dogs study and Pavlovian conditioning explained*. Simply Psychology. https://www.simplypsychology.org/pavlov.html

McLoughlin, M., Sheler, J. L., & Witkin, G. (1987, February 23). A nation of liars? *U.S. News & World Report*, *103*(20), 54–60.

Methods used to provide E&C training to boards globally 2018. (2022, July 6). Statista. https://www.statista.com/statistics/896596/methods-used-to-provide-ethics-and-compliance-training-to-boards

Metz, C., & Conger, K. (2020, December 7). Uber is giving self-driving car project to a start-up. *The New York Times*. https://www.nytimes.com/2020/12/07/technology/uber-self-driving-car-project.html

Most important objectives of ethics and compliance training globally 2018. (2022, July 6). Statista. https://www.statista.com/statistics/896556/metrics-for-measuring-effectiveness-of-compliance-programs

Nader, R. (2004, May 1). Legislating corporate ethics. *Journal of Legislation*, *30*(2), 193–204. https://scholarship.law.nd.edu/jleg/vol30/iss2/1

National Highway Traffic Safety Administration. (n.d.). *Automated vehicles for safety*. https://www.nhtsa.gov/technology-innovation/automated-vehicles-safety

National Institutes of Health. (2021, June 23). *Annual review of ethics (case studies)*. https://oir.nih.gov/sourcebook/ethical-conduct/responsible-conduct-research-training/annual-review-ethics-case-studies

Office of the Federal Chief of Sustainability Officer. (n.d.). https://www.sustainability.gov

Patrol: Tesla autopilot driver was watching movie, crashed. (2020, August 28). *ABC News*. https://abcnews.go.com/Technology/wireStory/patrol-tesla-autopilot-driver-watching-movie-crashed-72685378

Pope, A. (1711) *An essay on criticism*. [Quote] https://www.quotes.net/quote/38071

Porter, M. E. (2008, January). The five competitive forces that shape strategy. *Harvard Business Review*, *86*(1), 78–93.

Pruitt, S. (2018, September 20). *Here are 6 things Albert Einstein never said*. History.com. https://www.history.com/news/here-are-6-things-albert-einstein-never-said

Prusak, L. (2011, January 28). 25 years after Challenger, Has NASA's judgment improved? *Harvard Business Review*. https://hbr.org/2011/01/25-years-after-challenger-has

Ramey, J. (2021, November 9). *Walmart is already using driverless trucks*. Autoweek. https://www.autoweek.com/news/technology/a38198243/walmart-autonomous-delivery-trucks-gatik/

Rip, A. (1986). Controversies as informal technology assessment. *Knowledge (Beverly Hills, Calif.)*, *8*(2), 349–371.

Sears, D. (2018, March 8). The sightless visionary who invented cruise control. *Smithsonian Magazine*. https://www.smithsonianmag.com/innovation/sightless-visionary-who-invented-cruise-control-180968418

Setiyaningrum, A., & Aryanto, V. D. W. (2016). Corporate ethics and corporate social responsibility in reinforcing consumers bonding: An empirical study in controversial industry. *International Journal of Technoethics*, *7*(1), 1–5. https://doi.org/10.4018/IJT.2016010101

Sloan, D. (1979). The teaching of ethics in the American undergraduate curriculum, 1876-1976. *The Hastings Center Report*, *9*(6), 21–41. https://doi.org/10.2307/3561673

Stahl, B. C. (2021, March 18). Ethical issues of AI. *Artificial Intelligence for a better future: An ecosystem perspective on the ethics of AI and emerging digital technologies*, 35-53. doi:10.1007/978-3-030-69978-9_4

Survivors of Laos' worst dam disaster still struggling two years later. (2020, July 22). *Radio Free Asia*. https://www.rfa.org/english/news/laos/xe-pian-xe-namnoi-two-year-07222020211103.html

Tannert, C. (2006, May). Thou shalt not clone. An ethical argument against the reproductive cloning of humans. *EMBO Reports*, *7*(3), 238–240. https://doi.org/10.1038/sj.embor.7400653

Technology ethics cases. (n.d.). Markkula Center for Applied Ethics, Santa Clara University. https://www.scu.edu/ethics/focus-areas/technology-ethics/resources/technology-ethics-cases

Thomson, A. (n.d.). *U.S. Naval Space Command space surveillance system*. FAS Space Policy Project. https://spp.fas.org/military/program/track/spasur_at.htm

Thornley, C., Murnane, S., McLoughlin, S., & Carcary, M. (2018, October). The role of ethics in developing professionalism within the global ICT community. *International Journal of Human Capital and Information Technology Professionals*, *9*(4), 56–71. https://dx.doi.org/10.4018/IJHCITP.2018100104

Top challenges with ethics and compliance training globally in 2018. (2022, July 6). Statista. https://www.statista.com/statistics/896563/metrics-for-measuring-effectiveness-of-compliance-programs

Tucker, S. (2021, August 2). *Tesla update adds Wi-Fi while driving, new streaming entertainment*. Kelley Blue Book. https://www.kbb.com/car-news/tesla-update-adds-wi-fi-while-driving-new-streaming-entertainment

U.S. consumers and cybercrime - statistics & facts. (2022, July 6). Statista. https://www.statista.com/topics/2588/us-consumers-and-cyber-crime

U.S. Department of State. (n.d.). *Sustainability at the U.S. Department of State*. https://www.state.gov/sustainability-at-the-u-s-department-of-state

van Est, R., & Brom, F. (2012, December). Technology assessment, analytic and democratic practice. In R. Chadwick (Ed.), Encyclopedia of Applied Ethics (2nd ed.) pp. 306–320. https://doi.org/10.1016/B978-0-12-373932-2.00010-7

Vertical Flight Society. (n.d.). VTOL. https://vtol.org

Wells, F., & Farthing, M. (Eds.). (2009). *Fraud and misconduct in biomedical research* (4th ed.). Taylor and Francis.

Chapter 9
What Do You Mean My Website Isn't Accessible?
Why Web Accessibility Matters in the Digital World

Florence Wolfe Sharp
Purdue University Global, USA

Paige R. Sharp
Pasco-Hernando State College, USA

ABSTRACT

Today's world is increasingly based on digital access to information. People conduct essential aspects of life online through their web browsers and mobile applications: education, healthcare, banking, shopping, entertainment, and even jobs are conducted through the internet. To be cut off from the digital world is to miss these essential connections; this is exactly what happens to people with disabilities when the websites and content they try to use have accessibility barriers. People and organizations creating web content need to understand the elements of accessibility, important laws and regulations that guide accessibility efforts, and ways to improve the accessibility of web content. Eliminating these barriers is an important step towards a more inclusive society.

INTRODUCTION

Today, roughly 15 percent of people across the globe have been diagnosed with a disability ("33 Accessibility Statistics," 2021). In a world that has become increasingly more connected by virtual means, it is essential to consider how accessible internet web pages, online school assignments, and other virtual information exchanges are for people with differing abilities. Presently, 90 percent of internet web pages are not fully accessible for use by those with disabilities, creating an ethical dilemma in the form of a grave disadvantage between the availability of online information compared to what is available to those who do not have a disability ("33 Accessibility Statistics," 2021). Inaccessible digital content is

DOI: 10.4018/978-1-6684-5892-1.ch009

the online equivalent of multilevel buildings without ramps or elevators. It excludes a significant portion of the population and causes unnecessary difficulties, sometimes entirely preventing them from accessing content.

Imagine these scenarios:

Rajiv has low vision and uses JAWS, a screen reader application that vocalizes the contents of a web page to the user. The website Rajiv is viewing has a lot of images; he knows this because JAWS announces each image it encounters. However, the website developers did not include alternative text content, so Rajiv has no idea what the images represent and whether they are important. Alternative text is an attribute included in the image tag in hypertext markup language (HTML) that provides details about the image to the screen reader.

Shani does not use a mouse due to limited mobility in her hands. Instead, she relies on keyboard commands, like the tab and arrow keys, to navigate. The form she is filling out jumps randomly to different fields because the tab order has not been set up correctly. Shani gets frustrated because some fields validate based on previous entries, and she must spend a lot of extra time tabbing to the correct fields.

Conner has a hearing impairment. The course they are taking has several videos, but most of them do not have closed captioning or a transcript. Course assignments are based on these videos, but Conner has no idea what the dialog in the video is saying.

These are only a few situations representing the barriers people with disabilities may encounter with inaccessible web content. Individuals without disabilities often do not notice these issues, which can present problems when they are responsible for creating and managing web content. Therefore, web content providers must understand web accessibility, their obligations, and how to recognize and address accessibility barriers. Anything less creates a barrier to full and equal access for people with disabilities.

OVERVIEW OF WEB ACCESSIBILITY

Digital accessibility is not new. In 1998, Congress passed Section 508 of the Rehabilitation Act of 1973, which requires federal agencies to make their digital and electronic content (U.S. Access Board, 2022). Since then, the rules have expanded to most organizations and agencies that serve or receive funding from the United States government. In addition, commercial websites have come under scrutiny based on the Americans with Disabilities Act (ADA), which prohibits discrimination based on an individual's disability (U. S. Department of Justice, 2016).

One of the earliest notable accessibility lawsuits, filed in 2006, alleged that Target Corporation's website was not accessible to users with blindness primarily due to the lack of alternative text for images and navigation links. Alternative text allows a screen reader program to describe non-text elements to a user. The lack of these elements excludes users with blindness or low vision from accessing the website content, putting them at an unfair disadvantage. In the Target case, the court determined that online users should have a shopping experience equal to that at the physical store location (Ozeran, 2018). This comparable access claim, or "nexus," became a test for future claims.

Later cases, most notably those against Five Guys restaurants and Blick® Art Supplies, an online retailer, in 2017, found that websites are places of public access subject to the ADA independent of the physical storefront. They also determined businesses that operate online only are still subject to accessibility requirements (Arenth, 2019; Ozeran, 2018). In both cases, the plaintiffs alleged that a visual interface that lacked commonly available accessibility features prevented users with blindness from using the websites to place orders (Andrews v. Blick Art Materials, LLC, 2017; Marett v. Five Guys Enterprises LLC, 2017). The Blick case also noted that the website provided features and promotions only available online, thereby denying users with blindness from enjoying the same benefits as sighted users. Thus, publicly available websites have become considered points of public access and, therefore, subject to the ADA's equal access provisions. The United States Department of Justice, in 2018, confirmed its position that websites constitute places of public access.

In all these cases and many others not mentioned here, the critical element is equity: full and equal access for people with disabilities. Lawsuits arise because individuals with disabilities are denied the same rights as those without disabilities. A lack of equitable access equates to discrimination. Aside from the legal aspects of discrimination, consider the ethical impact of telling 10 to 20 percent of the population it is not welcome. Their business is not wanted; organizations do not care about excluding them. That represents a significant loss to the affected individuals and the deficient organizations.

Amazingly, accessibility lawsuits continue to this day. Ozeran (2018) noted that by mid-2018, nearly 700 ADA lawsuits had been filed. Around 10,000 such suits were filed between 2017 and early 2021 (Horelick, 2021). The volume of cases is partly related to differences in interpretation of the meaning of "public access" across the various U. S. circuit court districts. Some districts, particularly the First, Second, and Seventh circuits, consider virtual spaces, such as websites or mobile apps, equally subject to public accommodations as physical spaces (Kassim & Lawless, 2021). Uncertainty and inconsistency among courts have led to reversals, challenges, and appeals in cases involving big companies like Netflix, Domino's Pizza, and Winn-Dixie supermarkets. Other types of organizations are not immune from lawsuits, either. Since 2010, several higher education institutions have been the subject of accessibility lawsuits: Arizona State University, Massachusetts Institute of Technology, Harvard University, and others. These cases involved non-compliant e-readers that blind students could not navigate to complete their assigned readings and digital media that lacked closed captioning (Taylor, 2019). When big-name organizations are at risk, smaller organizations and individuals should take note and take action to move toward accessibility. Every organization with a website or mobile application and producing web content should know what web accessibility is, who is affected, and how to identify and address inaccessible content.

Accessibility means that digital content–typically delivered through a web browser or mobile app–is designed so that users with disabilities can interact with the information without barriers. For example, images and graphics are marked so screen readers can read them; tables and text are organized in a logical and linear reading order; clear, intuitive navigation is provided; media, such as audio or video, includes captioning, transcripts, and user controls. In addition, web accessibility rules extend to linked documents, such as Microsoft® Word® or Portable Document Format (PDF) files, and embedded content, such as e-learning modules. Ethically, accessibility means exceeding a user's needs while delivering inclusive functionality (Troester, 2021).

Accessibility is an important aspect of an inclusive society. It is so important that the United Nations recognized internet access as a human right in 2011 (Bacsa Palmer & Palmer, 2018). By extension, then, digital accessibility is a human right. Creating accessible web content, therefore, is the right thing to do. Accessible digital content makes it easier for people with disabilities to have the same experience as people without disabilities. It brings equity and inclusion into the equation. Often, it has the added benefit of improving the online experience for all users.

THE AFFECTED POPULATION

According to data from the 2017 American Community Survey conducted by the U.S. Census Bureau, nearly 41 million Americans live with a disability. That equals approximately 12 percent of the population. Broken down by category, the impact of disabilities is clear:

- Hearing: 11.5 million
- Vision: 7.6 million
- Cognitive: 15.4 million
- Ambulatory: 21 million
- Self-Care: 8 million
- Independent living: 14.6 million

(Note that many individuals reported living with multiple disabilities.)

More than one-tenth of the U. S. population has at least one disability that affects how they access information, services, and entertainment online. On a global scale, that number climbs to about 15 percent ("33 Accessibility Statistics," 2021). Nearly 41 million Americans risk being cut off from the services essential to their daily lives because the web content they want and need is not always accessible due to their disability status. Creating barriers or allowing them to stand sends the message that individuals with disabilities are somehow less than or not as important as the rest of the population (Bacsa Palmer & Palmer, 2018). Worse, barriers can prevent folks from performing to the best of their abilities and exclude them from education, employment, promotion, and other opportunities. With significant aspects of daily life moving online, barriers prevent people with disabilities from meeting their basic needs. In addition to increased online sales, Bacsa Palmer and Palmer (2018) noted that 40 percent of major healthcare systems have moved their scheduling online, and more than 30 percent of undergraduate students were enrolled in online classes (pre-pandemic).

Imagine running a business and intentionally excluding 10 to 15 percent of the target audience. Or perhaps holding a public event and preventing a similar segment of the population from attending. That is not ethical or something most organizations would want to do. Therefore, excluding people with disabilities through inaccessible digital content is also inappropriate and undesirable. In some situations, it is also illegal.

LAWS AND STANDARDS

Accessibility laws and standards exist around the world. In the United States, various federal, state, and local laws apply to accessibility; many state and local laws are based on or complementary to federal laws. Some research is necessary to determine which laws apply to specific situations. Furthermore, laws may apply differently to public and private entities. For digital information, the two primary federal laws in the United States are the Americans with Disabilities Act (ADA) and Section 508 of the Rehabilitation Act of 1973. These are not the only laws to know, but together they form the basis for access, non-discrimination, and accessibility requirements in support of people with disabilities.

How those requirements are put into action varies. While the ADA and Section 508 specify who needs access and where that access applies, neither is specific about how to achieve it. The U. S. Access Board takes responsibility for leading the development of guidelines for accessible design in the United States. In 2017, the Access Board revised its accessibility guidance, aligning with other national and international standards (U. S. Access Board, 2022). Providing consistent direction across the international community is where the Web Wide Web Consortium (W3C) and Content Accessibility Guidelines (WCAG) enter. The W3C, an international collaboration that develops and maintains web standards, manages the WCAG. Each of these laws and standards contributes to the accessibility landscape. Remember, they are not the only resources for achieving accessibility but the foundational guidance.

Americans With Disabilities Act (ADA)

Before the enactment of the ADA, people with disabilities faced unmitigated discrimination. Aside from discrimination in healthcare, education, and employment opportunities, people with disabilities faced daily problems such as access to public places. At times, people with disabilities were not even treated with basic human dignity. This is the foundation of an ethical dilemma. People with disabilities, their parents or guardians, and others who advocated for them had to fight to make any gains.

On the heels of the civil rights movement in the 1960s, efforts increased to secure rights for the population of people with disabilities. Protests, investigations, and legislation followed. On March 12, 1990, in Washington, D. C., a massive demonstration took place. More than 1,000 protestors participated, with scores of them abandoning their mobility devices to climb the Capitol steps. One young protestor was just eight years old (Kaufman, 2015). The event, known as the "Capitol Crawl," highlighted the accessibility barriers in public places. A few months later, President George H. W. Bush signed the ADA into law in July 1990. The new law was intended to mitigate the widespread and deeply rooted discrimination people with disabilities faced. The ADA addresses employment, public services, public transportation, public accommodations, architectural and construction standards, and communication devices and services (U. S. Department of Justice, 2008).

Recognized mainly for providing physical accommodations in the past, the ADA has provided regulations for how businesses and organizations can make their buildings and services more accommodating to Americans living with disabilities. In addition, given the fusion of technology with commerce in today's world, the ADA has been amended to include provisions for making the online environment and the technology it uses more accessible.

Numerous commercial websites have come under scrutiny based on the ADA. Suits like the ones against Five Guys, Blick, Domino's Pizza, and Winn-Dixie are all part of the case law contributing to the concept that websites are, themselves, places of public access. Furthermore, websites that operate online only are also subject to accessibility requirements (Ozeran, 2018).

Section 508 of the Rehabilitation Act of 1973

The Rehabilitation Act of 1973 is a foundational piece of legislation prohibiting discrimination against people with disabilities. It addresses employment and programs offered by federal agencies and contractors receiving federal funding. Congress amended the Act in 1998 to apply to electronic and information technology (EIT). That amendment is Section 508 (U. S. Access Board, 2022). This revolutionary act established accessibility standards for internet content and accountability for compliance (WebAIM, 2022). Section 508 has been updated and revised over the years, and as recently as 2018, to keep up with the changing technology landscape.

As with the original Rehabilitation Act, Section 508 applies to federal agencies and organizations that receive federal funding. It addresses not only website access but other forms of EIT such as phones, mobile devices, televisions and related equipment, software, computers, office equipment, video games, digital training, and online services such as call centers. The legislation is quite far-reaching in its application.

Although Section 508 targets federal agencies and federally funded programs, given the extent of its coverage, it also affects the commercial marketplace. Consider computing devices: the same equipment sold to federal agencies and contractors is also offered to commercial organizations and consumers. For computer manufacturers selling to federal buyers, the logical manufacturing decision is to make all their equipment Section 508 compliant. This decision benefits commercial and individual consumers. Similarly with web content, when a website is compelled to comply with federal accessibility requirements, the accommodations also help private and public entities that use the same website.

U.S. Access Board

Section 502 of the Rehabilitation Act of 1973 established the Access Board. The board was initially tasked with addressing architectural and transportation barriers (U.S. Equal Employment Opportunity Commission, 1973). Over the years, the Access Board's scope has expanded to include all aspects of accessible design. In addition to architectural and transportation design, the Access Board provides guidance for EIT, technical assistance and training, and enforcement of accessibility requirements. Its members include representatives from most federal departments and presidential appointees, many of whom are people with disabilities (U. S. Access Board, n.d.b). The Access Board issues its accessibility standards and guideline under the ADA, Section 508, and other applicable laws and regulations.

World Wide Web Consortium (W3C)

As noted earlier, the W3C is an international collaboration that develops and maintains web standards and manages the WCAG. Tim Berners-Lee, credited as the world wide web inventor, founded the W3C in 1994. Berners-Lee created the HTML standard for creating hyperlinked documents and the hypertext transfer protocol (HTTP) for retrieving those documents (Raggett et al., 1997). Web users will recognize HTTP or HTTPS (secure HTTP) as the first component of a web address. Although the concept of interconnected computing was not new, the definition of HTML and HTTP meant that networked documents could link directly to each other. Over time and with a great deal of further development, including creating browsers capable of retrieving and rendering HTML documents, the world wide web became a reality. The W3C emerged in the interest of managing the HTML and HTTP standards and keeping them open. Berners-Lee continues to lead the W3C efforts (Facts about W3C, n.d.a).

According to the W3C, "The W3C mission is to lead the World Wide Web to its full potential by developing protocols and guidelines that ensure the long-term growth of the Web" (W3C Mission, n.d.b, para. 1). The W3C continues to focus on designing and rendering web content (HTML, HTTP, cascading style sheets [CSS], and other standards), but also manages standards related to devices, web architecture, browsers, web services, and more. A key initiative is web accessibility for people with disabilities. The WCAG is an international standard for creating accessible web content. The current version, as of this writing, is 2.1; WCAG 2.2 is expected to be published in September 2022 (Henry, 2022b). Organizations in the United States and worldwide rely on the WCAG as the foundation for their accessibility guidelines.

Web Content Accessibility Guidelines (WCAG)

The WCAG standards target content and tool developers. The standards are an international resource that aids in making web content more easily accessible. Federal agencies and contractors in the United States are required to meet the latest version of the WCAG standards. Other organizations and individuals are encouraged to follow the standards. The standards are backward compatible, meaning content that meets the latest version of the standards also meets earlier versions. The WCAG standards also apply to mobile content delivery. So, developers working across platforms and devices can rely on WCAG for guidance.

The standard consists of 12 to 13 guidelines organized under the principles of perceivable, operable, understandable, and robust. Each guideline has success criteria at levels A, A.A., and AAA. Conformance with the standards means that content meets the specified success criteria (Henry, 2022a). The success criteria level an organization chooses to meet may be contractually defined; for non-federally funded organizations, the level is optional and will be based on the organization's circumstances and goals. The W3C does not recommend requiring level AAA conformance, however, because achieving that level for all content across an entire site is not possible (Accessibility Guidelines Working Group, 2018). Therefore, a target accessibility level for many organizations as of this writing is WCAG 2.1 Level A.A. Developers and organizations that are just starting out, however, should not be discouraged if they are not at Level A.A. The WCAG provides explanations and techniques for meeting each level.

Outside of the United States, other countries have proposed laws or policies that align with WCAG to make their online landscape more navigable for those needing accessibility options (Akinyemi, 2021). The European Union (E.U.) Web Accessibility Directive includes the EN 301 549 Standard, which incorporates the need for at least the AA WCAG requirement to be met (Akinyemi, 2021). For further examples, Canadian web accessibility standards and the Australian Disability Discrimination Act have similar requirements (Akinyemi, 2021).

Remember conformance to the WCAG or any accessibility law or standard is voluntary in most situations. Even where compliance is contractually required, oversight is lacking. Furthermore, laws can be vague. While many accessibility lawsuits have identified websites as places of public access, others have ended in unclear conclusions. Finally, other than the court system, no enforcement exists. No consequences occur unless a user files a lawsuit if a website is inaccessible. Worldwide, guidelines for web accessibility are similar to the United States; most depend on the W3C and WCAG for guidance. But again, no enforcement mechanism is in place. So, the responsibility for implementing accessible design rests with individual organizations and designers. Couple that with the international reach of digital content, contributing to the ethical aspect of accessibility.

ADDRESSING ACCESIBILITY BARRIERS

The lack of strong laws or enforcement creates a need for businesses and individual designers to assume responsibility for creating accessible content (Bacsa Palmer & Palmer, 2018). To address accessibility barriers, developers must know what those barriers are and how to correct them. Peters and Bradbard (2010) identified the following as significant accessibility errors: the lack of alternative text, visually based navigation, misuse of structural elements (headings, tables, and so forth), lack of closed captioning or transcripts for audiovisual media, non-linear table formatting, and low color contrast. Those issues have not changed significantly over time. For example, in a survey of college websites in India, color contrast was the most frequent violation, accounting for 58 percent of the observed violations. The subsequent two most frequent violations were the lack of alternative text for images and visible (contextual) hyperlinks (Ismail & Kuppusamy, 2022). Addressing these three violations could fix nearly 80 percent of the accessibility barriers on those websites.

In another study of higher education websites in Texas, USA, four common errors caused more than 56 percent of the accessibility issues (Taylor, 2019).

1. The hyperlink title attribute duplicated text inside the link; this creates redundancy for screen readers.
2. The landing page I.D. attribute was duplicated, which can create problems for assistive technologies if the I.D. is being used for interface control.
3. Links were formatted with no text, providing no context for the assistive technology.
4. Invalid references in hyperlinks, proving the no information about the type of link (another web page, a file, video, etc.) for assistive technology.

In both studies, a few mistakes caused most of the errors. Those mistakes were not complicated and could have been easily addressed. Knowing some of the most common accessibility issues is important in an ethical context. Even if the person observing the issue cannot correct it, they will be able to seek help to address it, making the pages more accessible. The following topics look at several common accessibility errors and provide techniques to correct them.

Fonts and Colors

Not all vision issues are related to total blindness. Dyslexia is an example of a visual impairment that does not involve blindness, instead relating specifically to a person's ability to read. It can cause difficulty in a person's ability to process sounds and spell correctly, among other things (The Dyslexia Foundation, n.d.). Color blindness, eye diseases or injuries, and other visual impairments can also compromise a user's ability to see the content on a screen. Not all users with visual impairments employ a screen reader to aid them in web page navigation and content comprehension. Even for screen readers, some textual attributes can create confusion if improperly formatted.

Taking dyslexia as an example, users with dyslexia may find certain typefaces easier to read than others. These users also benefit from web page designers providing adequate space between characters and using simple fonts with distinct shapes for each letter. This practice can decrease the likelihood of users with dyslexia experiencing difficulties with reading comprehension of the written language. In addition, simple fonts can be dually helpful in aiding users with impaired vision, but not total absence of vision, to make distinctions between characters while reading. While no single font is hailed as being universally easy to read, some of the simple, distinct, and well-spaced fonts include Verdana and Arial (Canadian National Institute for the Blind [CNIB], n.d.).

Consistency is vital when it comes to typefaces. WebAIM (n.d.d), a non-profit organization based at Utah State University that provides web accessibility expertise, offers several recommendations for working with typefaces. The more typefaces introduced, the longer it may take a user to read the text. When multiple typefaces are used, the brain needs to put more cognitive effort into reading the text as it changes shape. This is true especially if the letters change from a simple, strong design to something thinner, decorative, and stylized.

For colorblindness or other similar visual impairments, it is important to avoid fonts or colors that increase halos, shadows, or much blending of the text with its background. For example, WebAIM (n.d.d) notes that italics, boldlines, all capital letters, and font sizes that are very large or very small can introduce shadows or halos around characters, thereby blending the text together or into the background. While this does not mean that italics, boldlines, or all caps are never to be used, developers should be mindful of the context and typeface in which they are applied.

Regarding text color, contrast is the main element to keep in mind. A contrast ratio can help determine whether the contrast of text color on a particular background is high enough and thus easily read by users. For example, a ratio of 1:1 would indicate white text on a white background while 21:1 would indicate black text on a white background. The minimum ratio one could hope to achieve for regularly sized text is 4.5:1, which is roughly equivalent to red text on a white background (WebAIM, n.d.b).

Alternative Text

Many accessibility errors are caused by missing alternative text, which is used to identify and describe non-text objects. A lack of alternative text was the primary complaint in the Target lawsuit. In Rajiv's scenario explained previously, the missing alternative text created confusion as the screen reader read the web page. On a web page, alternative text is provided with the alt attribute of the <img> tag. In HTML, which is used to code web pages, alternative text would be presented as follows:

```
<img src="image.gif" alt="alternative text">
```

Decorative images can have an empty alt attribute, signaling the screen reader to skip the image. An empty alt attribute is simple to designate:

```
<img src="image.gif" alt="">
```

When creating web pages using an editor, the program will prompt for alternative text when an image is added. The program then forms the HTML code accordingly. If the alt attribute is omitted, the screen reader continues to announce each image and the related file name. Users, like Rajiv in the example, are left frustrated and confused, not knowing what the images represent or whether they are important.

Alternative text also applies to programs other than web pages. As with web pages, screen readers look for alternative text when encountering images in documents, spreadsheets, presentations, and e-learning content. In these programs, like web page editors, alt text is entered when an image is added. For example, in Microsoft Office applications, the option to add alt text displays on the Picture Format tab on the ribbon after an image is inserted. The process for adding alternative text may vary for other applications. Developers can search for accessibility options to determine how to add alternative text in any application.

Alternative text should be concise and descriptive. It should not only indicate the subject but give context as well. Although screen readers do not limit how much alt text they read, some browsers truncate lengthy alternative text. Therefore, a reasonable recommendation is that alternative text should be about as long as a tweet: 280 characters. For complex graphics, designers can provide a link to an external file with a detailed description using a caption or an in-text reference. The W3C provides a tutorial on how to describe complex images (Eggert & Abou-Zahra, 2022).

Hyperlinks

The issue with hyperlinks lies with how they are read. When using a screen reader, a uniform resource locator (URL), or web address, posted as-is to a site is read as written, including the odd slash, pound sign, or underscore. These URLs can be lengthy, tedious, and unhelpful in describing where the link will take the user (WebAIM, n.d.c). For this reason, URLs should be linked to a word or words on a page that a user can click to access the hyperlinked page instead. These words or phrases should be contextual and inform users where the link leads them.

For example, if an article discussing how to turn on automatic captions while using Zoom needed to provide readers with a link to download Zoom, the developer should avoid hyperlinking entire paragraphs or sentences to do so (WebAIM, n.d.c). Likewise, the developer should ensure the linked text is helpful to the user. To create an effective hyperlink, the developer needs to understand how users, both without and with visual impairment, skim a page for links. A user without visual impairments can scan a web page, identify links and key phrases (usually by color, underlining, or other visual cues), and understand where the link will take them. A user requiring a screen reader may skim through the page by tabbing or other means provided by the assistive technology. When the screen reader hits a link, it reads the link in its entirety or all the text associated with it. If the linked text is too long (like an entire sentence or paragraph), the user experiences a delay while the screen reader reads the linked text. Consider the following sentence:

"Before reading about Zoom's caption function, if you have not already downloaded Zoom, click here."

If hyperlinking the whole sentence is unhelpful, developers may think that hyperlinking just the phrase "click here" in this sentence would be better because it is short. This phase is essential to ensure that the screen reader does not have to read unnecessary phrases and that the hyperlinked text can be descriptive enough of where it is taking the user (WebAIM, n.d.c). While linking "click here" clearly indicates *where* the link is, this sentence may be more concisely written and linked to describe better where the link takes a user if read on its own as with a screen reader. Say the sentence is written this way instead:

"Download Zoom, if you have not already, before navigating the auto-caption function."

When the simple instruction "Download Zoom" is the only portion of the sentence hyperlinked, a user skimming with a screen reader to find a link to download the application can find it quickly.

Keyboard Compatibility

Many users with differing mobility capabilities do not rely on a mouse or trackpad to navigate websites or applications. They may instead use keyboard commands, some as simple as the tab key (W3C, 2005). Keyboard commands can help make navigation easier, even for programs or devices that do not originally have a keyboard. Devices and applications have the option to integrate a keyboard interface if they do not originally have a keyboard (W3C, 2008).

The next hurdle for keyboard input users is when items in an application or web page are out of order. This is what happened to the user, Shani, in one of the opening scenarios. Often, the tab order for a page is automatically generated, but this order is not necessarily the ideal order and, if left uncorrected, can create confusion (W3C, 2016). For example, suppose a user attempts to navigate using the tab key on their keyboard. In that case, their command of the page may be thrown from one side of the page to another and back in a way that can create confusion, forcing the user to go searching for where their keyboard command has taken them.

Lack of compatibility with keyboard devices or improper sequencing for tab controls can lead to confusion or greater time consumption navigating a site for users like Shani, who rely on these features to quickly and effectively navigate the internet. Lack of keyboard compatibility can create an experience that is significantly more difficult at worst and tedious and annoying at best.

Keyboard compatibility can be tested by attempting to navigate a web page using keyboard controls. Web page editors also let developers review and edit keyboard compatibility issues. For documents created in an auxiliary program, like a Microsoft application, the accessibility checker tool can assist. The tool is accessed from either the Review tab on the ribbon or the status bar at the bottom of the window.

Keyboard compatibility is often not a concern in basic web pages that use common HTML features like buttons and hyperlinks. More advanced features, like forms and tables or interactive content, sometimes need additional coding to ensure compatibility. For example, in HTML, the tabindex attribute with a numeric value can be added to an element to set the tab order. The tabindex is a more advanced feature of HTML and should be used carefully.

Audio and Video Transcripts

Audio and video have extensive uses on the internet and are commonly included in all types of web pages. From marketing content to entertainment and education, audio and video improve the user experience. Audio recordings require transcripts or closed-captioning to aid people with hearing disabilities. Videos can require accommodations for both the audio component of their content and the visual aspects, as applicable.

Where pre-recorded videos are concerned, whether a YouTube® video created by a marketing department or a lecture produced by a university professor, videos need to be fitted with closed captions. While applications such as YouTube, Kaltura®, and others automatically generate captions on request, those automatically generated captions do not always catch and interpret the speaker's exact words and will need to be reviewed and edited (W3C, 2019a). Captions are the preferred approach to accommodating people with hearing impairments and are necessary for WCAG 2.1 Perceivable Level A.A. compliance. Even live audio or video captions are necessary. Alternatives like lipreading or sign language may not be reliable or available, requiring a written companion to a speaker's words in the absence of captioning.

Asking students like Conner to rely on lipreading to follow the content of a lecture video can be problematic at best and an ethical issue at worst for several reasons. The first is that the visible portion of the video may not always focus on a physical speaker, thus leaving no lips to read. Even if a video were to maintain a clear view of the speaker and their mouth for the duration of the media, lipreading, while helpful to users who have a hearing impairment, may not be relied upon heavily as several sounds can appear similar or differ from speaker to speaker with changes in pronunciation ("What is Lipreading?," 2018). Sign language is also not a reliable alternative to captioning as it is not widely used in the hearing world ("What is Lipreading?," 2018).

Longer videos with extensive audio may also benefit from a basic transcript, for which services exist that can quickly and affordably provide a full transcript of audio content in several languages ("Transcription Services," n.d.). Including a descriptive transcript may provide descriptions of visual content in a lecture video–content that may be missed by a user with a visual impairment (W3C, 2019b). Transcripts can also aid users with cognitive disabilities and even individuals who prefer to read rather than watch a video or listen to an audio recording.

Tables

Users without visual disabilities can quickly and easily skim a table and make associations concerning the data and the logical organization with minimal formatting. Users who employ a screen reader, however, may benefit from additional formatting (WebAIM, n.d.a). Screen readers read table text linearly: left to right, top to bottom. If the table is not marked up using headings and row and column headers, the screen reader will move through the table like a paragraph.

When building a table, it is vital to provide identifiers like row and column headers and captions (WebAIM, n.d.d). Providing headers can help visually impaired users quickly identify where data starts and what the data is concerning. The appropriate data must be listed under the proper header. Captions can provide a short summary of the table's contents (WebAIM, n.d.a). Web page editor and other applications can guide developers in marking up tables correctly. Additional concerns are merged cells (row or column headers that span multiple rows) or nested tables (a table inside a table). Developers should consider whether multiple tables are a better way to present the data.

SPECIAL CONSIDERATIONS

Some organizations may think, "This is not a federal agency. It is a small business. What does all this mean?" Educators and other contributors may even feel they are exempt. Even if an organization is not legally bound to comply with accessibility laws, accessibility efforts benefit everybody. Purposely leaving web pages and online information sources inaccessible to various disabled populations is an ethical dilemma that can be corrected easily in most cases to provide a more equitable online experience for all. Before closing discussion, there are two circumstances that often lead to questions and confusion about accessibility and complicate the ethical dilemma some organizations and individuals face.

Curriculum Development and Online Learning

Section 504 of the Rehabilitation Act, as amended, expands the definition of disability and extends the protections of the Rehabilitation Act to students (U. S. Department of Education, 2020). Public colleges and universities, public K-12 schools, and many private institutions have long accepted that the ADA and Section 508 apply to them due to receiving federal funding. Decisions regarding EIT, such as Learning Management Systems and computing equipment, must be made in favor of ADA compliance. The risk for non-compliance is legal action, which no school wants. More importantly, schools recognize the importance of providing an inclusive learning environment.

Buying compliant systems has limitations, however. During the 2020 pandemic-driven "online pivot," schools quickly discovered how inclusive their learning systems were. From non-compliant content to inadequate software, instructors were challenged with overcoming accessibility obstacles–and quickly. These challenges arose from several circumstances:

- Existing content or systems thought to be compliant had limitations that only became apparent under pressure.
- New online content simply was not compliant.

- Publisher materials that worked in the face-to-face classroom were not accessible for online distribution.
- Teachers were not fluent in accessibility requirements.
- The school lacked a knowledgeable technology coordinator.
- Schools and teachers had to implement online solutions quickly and did not have time to research and confirm accessibility.
- Programs that were never intended for online delivery had to be converted quickly.

These problems only compounded the technology and scheduling issues many teachers and families felt: lack of adequate internet connections, outdated computers, NO computers, and limited parental supervision and support. Add to that the disciplines that were less likely to offer ADA-compliant materials, like mathematics or physical education. As a result, instructors were on their own in many cases. Schools and independent instructors, however, can take steps to improve access for students with disabilities.

Awareness is the first step (Huss, 2019; Ismail & Kuppusamy, 2022). Knowing the accessibility requirements under the ADA and Section 508 is a starting point. Another essential step is understanding key accessibility issues, like those described here. Seeking out compliant content from publishers and other content providers is also supportive. When speaking with publishers or content providers, instructors can ask about their compliance with the ADA, Section 508, or WCAG. Instructors can also request a publisher's Voluntary Product Accessibility Template (VPAT), a standardized document that identifies how the content meets Section 508 requirements. This information can help instructors determine whether the content they want to use meets web accessibility standards.

Undue Burden

Discussions about whether and how to create accessible web content sometimes come back to the ethical concept of "undue burden." Both the ADA and Section 508 use this nebulous term when referring to providing adequate accommodations without imposing exceptional hardship on the agency. Therefore, understanding the undue burden criteria is essential.

Organizations sometimes assume that either their size or the anticipated cost of the accommodation may qualify as an undue burden. Neither assumption is entirely accurate. While Section 508 applies to Federal agencies or contractors regardless of size, the ADA Title I (Employment) applies to employers with 15 or more employees, and Title II (state and local government requirements) applies regardless of size (U. S. Department of Justice, 2020). Organizational size, therefore, is difficult to justify as an undue burden. Excessive expense may be grounds for an undue burden claim, but qualified organizations must document why and to what extent compliance would create an undue burden (U. S. Access Board, n.d.a). Relief is not guaranteed, and an organization claiming undue burden should consider the overall costs of avoiding compliance.

Lawfully speaking, many organizations have an obligation to make their web content accessible to the public. Even those organizations that are not legally bound to comply with accessibility requirements may be at risk of a lawsuit and should seriously consider the related costs: changes to websites, possible damages paid to plaintiffs, attorney's fees, court costs, and the public quandary of the situation. Logistically, it makes sense to provide accommodations and produce accessible web content. Doing so creates a favorable impression and expands an organization's base.

But logistics is not the only consideration. Ethically speaking, anyone requiring any accommodation is entitled to be part of the same world and access the same information and experiences as those who do not have a disability. So the question should not be whether it is *advantageous* to be accessible, but if it is *right* to be accessible.

CONCLUSION

A Win for All

Accessibility measures are often metaphorized as curb ramps, the physical accommodations used to provide access to a sidewalk. The ramp allows someone using a wheelchair or walker to navigate onto a sidewalk without having to breach a curb. However, the curb ramp is also helpful to others: parents pushing a stroller; cyclists, skaters, and runners; delivery people pushing or pulling a cart; and anyone with a temporary or minor mobility issue. So while some people might view those with disabilities as a small part of the population, the accommodations provided improve access for a much larger segment of the community.

Even the accommodations provided to address one type of disability sometimes help address other types of disabilities. For example, ensuring that content is screen-reader accessible helps address low vision and blindness, but it can also address cognitive conditions such as dyslexia. Some users with dyslexia find it helpful to use a screen reader and therefore benefit when web content is screen reader compatible.

The benefits of accessible design are not limited to individuals. Organizations that adopt and promote accessible design benefit as well. One of the key benefits is reputation. However, the improved functionality of an accessible design can mean an improved digital experience for all users. That, in turn, can translate into improved sales and market share (Bacsa Palmer & Palmer, 2018; Troester, 2021). In addition, organizations can move from addressing accessibility as an afterthought to bringing it to the forefront of their design efforts. Doing so gets them out of the mindset of retrofitting, which is an ineffective approach.

Creating accessible content and providing accommodations wherever possible creates a more organized and functional space for everyone. Web accessibility makes the online world easier to navigate for everyone, including the unimpaired, those born with a lifelong impairment, the temporarily impaired, or those who become impaired over the course of their lives. It breaks down barriers and eases hardships in communication. Even if an organization is not legally obligated to create this kind of a world, it should strive to seek out and participate in accessibility efforts. It is not fair nor equitable to deprive somebody of the ability to go to school, participate in local government, or purchase a product online simply because they require accommodation. Ethically speaking, it is problematic to leave digital content inaccessible when formatting it correctly will make it more navigable for those with disabilities and the general public. Moreover, it is not difficult or taxing to adjust to make these things possible or easier where it is currently difficult.

REFERENCES

Accessibility Guidelines Working Group (A.G.W.G.). (2018, June 5). *Web Content Accessibility Guidelines (WCAG) 2.1*. World Wide Web Consortium (W3C). https://www.w3.org/TR/WCAG21

33 Accessibility statistics you need to know in 2021. (2021, April 14). Monsido. https://monsido.com/blog/accessibility-statistics

Akinyemi, A. (2021, December 10). *International web accessibility laws and policies*. Who Is Accessible. https://www.whoisaccessible.com/guidelines/international-web-accessibility-laws-and-policies

Andrews v. Blick Art Materials, LLC, 1:17-cv-00767 (EDNY, 2017). https://www.classaction.org/media/andrews-v-blick-art-materials.pdf

Arenth, T. L. (2019). ADA web site accessibility claims on the rise: Practical strategies for defense. *Journal of Internet Law*, *23*(4), 12–14.

Bacsa Palmer, Z., & Palmer, R. H. (2018). Legal and ethical implications of website accessibility. *Business and Professional Communication Quarterly*, *81*(4), 399–420. https://doi.org/10.1177%2F2329490618802418

Canadian National Institute for the Blind (CNIB). (n.d.). *Accessibility at CNIB*. https://cnib.ca/en/accessibility-cnib?region=gta

Eggert, E., & Abou-Zahra, S. (Eds.). (2022, January 17). *Complex images*. W3C Web Accessibility Initiative (WAI). https://www.w3.org/WAI/tutorials/images/complex/#approach-3-structurally-associating-the-image-and-its-adjacent-long-description-html5

Henry, S. L. (Ed.). (2022a, March 18). *WCAG 2 overview*. https://www.w3.org/WAI/standards-guidelines/wcag

Henry, S. L. (Ed.). (2022b, March 18). *What's new in WCAG 2.2 working draft*. https://www.w3.org/WAI/standards-guidelines/wcag/new-in-22/#introduction-timeline-comments

Horelick, J. (2021). Four web accessibility developments that shaped the first half of 2021. *Labor & Employment Law*, *49*(2), 4–5.

Huss, J. A. (2019). Middle level education aims for equity and inclusion, but do our school websites meet ADA compliance? *Middle Grades Review, 5*(1). https://scholarworks.uvm.edu/mgreview/vol5/iss1/4

Ismail, A., & Kuppusamy, K. S. (2022). Web accessibility investigation and identification of major issues of higher education websites with statistical measures: A case study of college websites. *Journal of King Saud University - Computer and Information Sciences, 34*(3), 901-911. doi:10.1016/j.jksuci.2019.03.011

Kassim, A. J., & Lawless, L. (2021). The ADA and website accessibility post-Domino's: Detangling employers' and business owners' web and mobile accessibility obligations. *Tort Trial & Insurance Practice Law Journal*, *56*(1), 53–66.

Kaufman, S. (2015, March 12). *They abandoned their wheelchairs and crawled up the Capitol steps*. ShareAmerica. https://share.america.gov/crawling-up-steps-demand-their-rights

Marett v. Five Guys Enterprises LLC, 1:17-cv-00788 (SDNY, 2017). https://www.classaction.org/media/marett-v-five-guys.pdf

Ozeran, N. (2018, August 16). Insight: A mid-year review of the current state of ADA website accessibility lawsuits. *Bloomberg Law.* https://news.bloomberglaw.com/daily-labor-report/insight-a-mid-year-review-of-the-current-state-of-ada-website-accessibility-lawsuits

Peters, C., & Bradbard, D. (2010). Web accessibility: An introduction and ethical implications. *Journal of Information Communication and Ethics in Society*, *8*(2), 206–232. https://doi.org/10.1108/14779961011041757

Raggett, D., Lam, J., Alexander, I. F., & Kmiec, M. (1997). *Raggett on HTML 4* (2nd ed.). Addison Wesley Longman. https://www.w3.org/People/Raggett/book4/ch02.html

Taylor, Z. W. (2019). Web (in)accessible: Supporting access to Texas higher education for students with disabilities. *Texas Education Review, 7*(2), 60-75. doi:10.26153/tsw/2285

The Dyslexia Foundation. (n.d.). *The Dyslexia Foundation.* https://dyslexiafoundation.org

Transcription services. (n.d.). *uiAccess.* http://www.uiaccess.com/transcripts/transcript_services.html

Troester, M. (2021, May 20). *Accessibility in application design: Ethical, inclusive and good for the business.* Progress. https://www.progress.com/blogs/accessibility-in-application-design-ethical-inclusive-and-good-for-the-business

U.S. Access Board. (2022, March). *I.T. accessibility laws and policies.* General Services Administration. https://www.section508.gov/manage/laws-and-policies

U.S. Access Board. (n.d.a). *About the ICT accessibility 508 standards and 255 guidelines.* General Services Administration. https://www.access-board.gov/ict/#

U.S. Access Board. (n.d.b). *About the U.S. Access Board.* General Services Administration. https://www.access-board.gov/about

U.S. Census Bureau. (2017). *American Community Survey: Disability characteristics.* https://data.census.gov/cedsci/table?q=disability

U.S. Department of Education. (2020). *Protecting students with disabilities: Frequently asked questions about Section 504 and the education of children with disabilities.* https://www2.ed.gov/about/offices/list/ocr/504faq.html

U.S. Department of Justice. (2008). *Americans with Disabilities Act of 1990, as amended.* https://www.ada.gov/pubs/adastatute08.htm

U.S. Department of Justice. (2016, October 11). *Americans with Disabilities Act Title II regulations.* https://www.ada.gov/regs2010/titleII_2010/titleII_2010_regulations.htm#a35103

U.S. Department of Justice. (2020). *A guide to disability rights laws.* https://www.ada.gov/cguide.htm

U.S. Equal Employment Opportunity Commission (EEOC). (1973, September 26). *Rehabilitation Act of 1973 (original text).* https://www.eeoc.gov/rehabilitation-act-1973-original-text#

W3C. (2005, February). *Introduction to web accessibility*. https://www.w3.org/WAI/fundamentals/accessibility-intro

W3C. (2008, December 11). *Web Content Accessibility Guidelines 2.0*. https://www.w3.org/TR/2008/REC-WCAG20-20081211/#keybrd-interfacedef

W3C. (2016). *H4: Creating a logical tab order through links, form controls, and objects*. https://www.w3.org/TR/WCAG20-TECHS/H4.html

W3C. (2019a, September). *Captions/subtitles*. https://www.w3.org/WAI/media/av/captions

W3C. (2019b, September). *Transcripts*. https://www.w3.org/WAI/media/av/transcripts

W3C. (n.d.a). *Facts about W3C*. https://www.w3.org/Consortium/facts

W3C. (n.d.b). *W3C mission*. https://www.w3.org/Consortium/mission.html

WebAIM. (n.d.a). *Creating accessible tables*. https://webaim.org/techniques/tables/data

WebAIM. (n.d.b). *Contrast and color accessibility*. https://webaim.org/articles/contrast

WebAIM. (n.d.c). *Links and hypertext*. https://webaim.org/techniques/hypertext/link_text

WebAIM. (n.d.d). *Typefaces and fonts*. https://webaim.org/techniques/fonts

WebAIM. (n.d.). *United States laws: The Rehabilitation Act of 1973 (Sections 504 and 508)*. https://webaim.org/articles/laws/usa/rehab

What is lipreading? (2018). Hearing Link Services. https://www.hearinglink.org/living/lipreading-communicating/what-is-lipreading

KEY TERMS AND DEFINITIONS

Accessibility Barrier: Any website error that renders the site unusable to individuals with disabilities.

ADA: The Americans With Disabilities Act, which was signed into law in July 1990 and intended to mitigate the widespread and deeply rooted discrimination people with disabilities faced. The ADA addresses topics including employment, public services, public transportation, public accommodations, architectural and construction standards, and communication devices and services.

Alternative Text: A short text description provided for a non-text element, such as an image, on a web page.

HTML: Hypertext markup language, the markup system used to code web pages.

HTTP: Hypertext transfer protocol, the internet protocol for retrieving documents on the web.

Hyperlink: A link on a web page that leads to another web document or other content.

Section 508: An amendment to the Rehabilitation Act of 1973 that applies to electronic and information technology and established accessibility for internet content and accountability for compliance.

Web Content Accessibility Guidelines (WCAG): A set of standards developed to provide guidance on creating accessible web content; maintained by the World Wide Web Consortium.

World Wide Web Consortium (W3C): An international organization dedicated to developing and maintaining internet protocols and standards.

Section 4
Public Policy Development

Chapter 10
Analysis of Ethical Development for Public Policies in the Acquisition of AI-Based Systems

Reinel Tabares-Soto
Universidad Autónoma de Manizales, Colombia

Joshua Bernal-Salcedo
Universidad Autónoma de Manizales, Colombia

Zergio Nicolás García-Arias
https://orcid.org/0000-0003-2835-6167
Universidad Autónoma de Manizales, Colombia

Ricardo Ortega-Bolaños
https://orcid.org/0000-0002-4598-4133
Universidad Autónoma de Manizales, Colombia

María Paz Hermosilla
https://orcid.org/0000-0003-3570-7914
Universidad Adolfo Ibáñez, Chile

Gonzalo A. Ruz
https://orcid.org/0000-0001-7740-9865
Universidad Adolfo Ibáñez, Chile

Harold Brayan Arteaga-Arteaga
https://orcid.org/0000-0002-2341-5079
Universidad Autónoma de Manizales, Colombia

ABSTRACT

The exponential growth of AI and its applications in different areas of society, such as the financial, agricultural, telecommunications, or health sectors, poses new challenges for the government's public sector, mainly in regulating these systems. Governments and entities in general address these challenges by formulating soft laws such as manuals or guidelines. They seek full transparency, privacy, and bias reduction when implementing an AI-based system, including its life cycle and respective data management or governance. These tools and documents aim to develop an ethical AI that addresses or solves the aforementioned ethical implications. The revision of 22 documents within frameworks, guides, articles, toolkits, and manuals proposed by different governments and entities are examined in detail. Analyses include a general summary, the main objective, characteristics to be highlighted, advantages and disadvantages if any, and possible improvements.

DOI: 10.4018/978-1-6684-5892-1.ch010

INTRODUCTION

Artificial Intelligence (AI) is heard ever more in different fields of society. It is no longer an indifferent word in humans' daily lives. AI systems are present in financial, automotive, health, science, education, industry, and telecommunications, among other sectors. Its development continues to increase (Berryhill et al., 2019). AI permeates society, and today it is possible to find its uses almost everywhere: translation apps, recommendation systems, e.g., when people perform a search on Google or YouTube, voice assistants such as Siri or Alexa, traceability of optimized traffic routes, and the list goes on, because of this, the demand for AI systems is increasing as AI becomes a technology that specializes in specific tasks. AI promises to generate productivity gains, improve well-being and, in turn, solve complex challenges. It is also precise for their predictions, recommendations, or decisions. Also, it does not require a high economic cost ("Artificial Intelligence in Society," 2019).

The uses of AI can generate significant advances in different sectors of society. More than 60 countries are developing national AI strategies to maximize their potential. According to the "Government AI Readiness Index 2021", published by Stanford University, nearly 40% of the 160 countries have published or are drafting national AI strategies (Fuentes et al., 2022). It shows that AI is fast becoming a top concern for leaders globally. The economic sector of the United States, Japan, Germany, Finland, and eight other countries with developed economies would increase annual economic growth rates by approximately two percentage points. At the same time, improve labor productivity by around 40% by 2035 with AI systems implementations (Purdy & Daugherty, 2016; Wirtz et al., 2019). Governments play a crucial role in determining national strategic priorities, public investments, and regulations (Fuentes et al., 2022; "The Strategic and Responsible Use," 2022). However, the regulatory and social limits for the use of AI are becoming increasingly visible, and governments have two ways to address this problem. The first is the hard law, which are laws or regulations created by governments and are mandatory. However, this process entails high costs of time and economic resources; also, in the end, it does not present quick answers to emerging problems. The second way is soft law, which exists in programs that set high expectations but the governments cannot directly enforce. In other words, they are non-binding and can coexist without jurisdiction and be developed, modified, and adapted by any entity (Gutierrez & Marchant, 2021).

For governments and interested people, using AI systems also implies having a clear notion of the challenges. The development or acquisition of an AI technology poses in all phases of the life cycle the following challenges ("Artificial Intelligence in Society," 2019):

- Planning and design.
- Data collection and processing.
- Model construction and interpretation.
- Verification and validation.
- Deployment.
- Operation and monitoring.

(Wirtz et al., 2019) proposed four main dimensions that model the challenges: AI technology implementation, AI law and regulation, AI ethics, and finally, AI society, identifying 15 sub-challenges (Wirtz et al., 2019); this reveals the public sector's challenges like the charge of establishing priorities, investments, and national regulations (Berryhill et al., 2019).

In particular, the compendium of associated ethical and regulatory risks or challenges mentioned by different organizations and companies addresses privacy and security, algorithmic discrimination or bias, and transparency or opacity (Buenadicha Sánchez et al., 2019; "Artificial Intelligence in Society," 2019). All these issues must have a responsible approach, reliability, and explainability. Above all, focus on the target population to design and implement the AI system. Therefore, it must be an ongoing process that identifies trade-offs, mitigates risk and bias, and ensures open and accountable processes or actions ("The Strategic and Responsible Use," 2022).

The complexity of this type of technology innovation in public policy areas requires specific audit actions. The need to develop reference frameworks or guides focused on the audit process at the internal and external levels of the projects is latent (Castillo, 2021), where a professional needs to do documentary processes, carry out an analysis of codes, data, and results obtained by the machine learning (ML) algorithms, the documentary part of frameworks development and toolkits, the information related to the context and relevance of the AI projects. Furthermore, breaking down and understanding the processes associated with obtaining, manipulating, and pre-processing the data (Castillo, 2021). Also, the nature of these data protection laws contemplation in each item named above, the biases, and possible mitigation processes part of the audits.

Developing frameworks and applications surrounding AI's ethical implications has also increased. These have focused on specific areas that seek to improve the impact of AI on populations: to evaluate data management, analyze risks that arise from biases and the compensatory measures or mitigating actions that allow these risks to be avoided, to open and understand the black boxes, e.g., in Deep Neural Networks (DNN) and Convolutional Neural Network (CNN) projects (Wexler et al., 2020). The frameworks and toolkits are an active part of developing more ethical algorithms, which respond to healthy frameworks and objectives with the population, making them increasingly necessary today. For this reason, it is pertinent to understand them and find development patterns the same that allows for optimizing and regularizing a tool that responds to the ethical needs of AI.

Raising questions about ethics is an important first step. Manuals, frameworks and guide, toolkits, and more are included below. Within the variety of documents, ethical frameworks proposed by the governments of Canada, the United Kingdom, and Colombia, manuals from the Inter-American Development Bank, guides, articles, tools, and white papers are shared.

RAISING ETHICAL ISSUES

The use of AI has brought a wide range of solutions to current problems, and there are likely to be many other improvements or benefits in the future. However, the generation of new knowledge and tools implies recognizing weaknesses or issues. In particular, AI and its applications carry a variety of ethical implications that can affect the quality of life of the target population if not used correctly or ethically. An example is what happened in Amazon, which implemented an AI to support the personnel selection processes. This AI had a high precision that, in principle, guaranteed its viability. However, a gender bias was discovered based on the database that favored men in the selection process (González et al., 2020). Another example is what happened in China, where the government implemented a facial recognition system in sunglasses worn by police officers. This system tracks citizens as part of the government's effort to implement a social behavior rating system, which has ethical implications, such as violating the right to privacy (Buenadicha Sánchez et al., 2019).

Table 1. Ethical implications addressed by each section.

Implications	Section
Opacity or Transparency	• Ethical Toolkits • Articles and White papers
Traceability or Accountability	• Manuals • Frameworks and Guides • Articles and White papers
Explainability	• Ethical Toolkits • Articles and White papers
Bias	• Manuals • Ethical Toolkits
Privacy	• Manuals • Frameworks and Guides
Equity	• Frameworks and Guides • Ethical Toolkits
Responsibility	• Manuals • Frameworks and Guides • Articles and White papers • Ethical Toolkits

Table 1 contains the most recurrent ethical implications in the AI life cycle (described below) and the following sections in which they are addressed, or a solution is proposed. It is essential to mention that the discussion of ethical implications that follow are not exhaustive but aim to provide some important resources that mention, propose, or provide solutions for these implications.

The ethical implications mentioned in Table 1 also play a crucial role as fundamental principles of ethical AI. The development of an AI-based system must always address these implications/principles (Eitel-Porter, 2020). The documents presented in the following sections serve as tools to provide knowledge and awareness of the importance of ethical governance in using AI (Cath, 2018) while providing some practical solutions.

MANUALS

The manuals analyzed to address the most common aspects related to the use of AI-based automated decision-making systems. Some manuals' target audience or orientation are AI project managers and technical teams developing and implementing algorithms. The orientation is towards solving the most common public policy problems (e.g., project feasibility, bias detection, mitigation, and transparency).

The manuals addressed have in common the analysis of the AI life cycle, a study of the possible problems that may arise, and a series of recommendations to solve them. They also show the current panorama on the handling of ethical standards in terms of development and their implementation. It explores the analysis of the possible challenges faced by various types of populations. It also emphasizes vital issues such as privacy, data protection, and communication to the target population about the use of their data and the expected results. It also mentions a series of instructions for the correct detection of errors and biases that will allow the development of a more robust ethical tool.

Responsible Use of AI for Public Policy: A Manual for Project Formulation

This manual establishes a guide to properly formulating an AI-based project in developing public policies. The manual also describes the phases necessary for the proper implementation of a project from the perspective of the person in charge of making public policy and management decisions. A detailed description of the process is provided from the problem definition stage to identifying the type of analysis required. Subsequently, the methods of data collection and processing, the measurement of the impact of the AI system, and observation to identify possible risks in the future are studied (Denis et al., 2021).

Components of an AI System for Public Policy

The creation and use of an automated decision support system represent a significant advantage during the life cycle of public policies (see Figure 1). This tool provides recommendations, predictions, and rankings, among other tasks, during social action intervention (Denis et al., 2021).

The public policy life cycle in Figure 1 is a tool that seeks to represent how policies will be developed and serve for planning and analysis during each of the project phases (Denis et al., 2021).

The first part of the cycle is identifying the problem, then formulating the different actions for its solution. In this phase, it is determined whether an AI-based system intervention is necessary.

Stages of the AI Life Cycle

A project has two phases and different stages (see Figure 2). The first phase is the conceptualization and design of the project. Its correct execution will ensure the sustainability and viability of the project and even help minimize risks during the operation of the tools (Denis et al., 2021). The manual provides a detailed description of each process involved in this first phase, from defining the problem, feasibility, and viability to identifying the type of analysis required.

This is followed by implementing the project (second phase), where data collection, processing, and analysis come into play to identify potential risks in the future. In addition, there is the importance of measuring the impact of the IA system (Denis et al., 2021).

The manual provides a series of documents, such as design sheets and profiles of example models, which allow a clear definition of the problem and a pre-feasibility analysis. They also allow to define the objectives, describe the actions, map the data and tools chosen, and ethical and legal considerations.

Responsible Use of AI for Public Policy: A Primer on Data Science

This handbook focuses primarily on the challenges related to technical processes throughout the lifecycle of AI-based systems, and that it intends to implement them in public policy domains. Its main objective is to help the technical teams in charge of work and develop ML algorithms oriented to public policies and provide a framework for the considerations to be taken into account in designing an AI-based automated decision-making tool (González et al., 2020). This manual analyzes some of the challenges in applying ML technologies to public policy decision-making. It proposes alternatives to avoid or mitigate errors and biases that may arise in the implementation of the tool.

Figure 1. Public Policy Life Cycle
Source: Adapted from Denis et al. (2021).

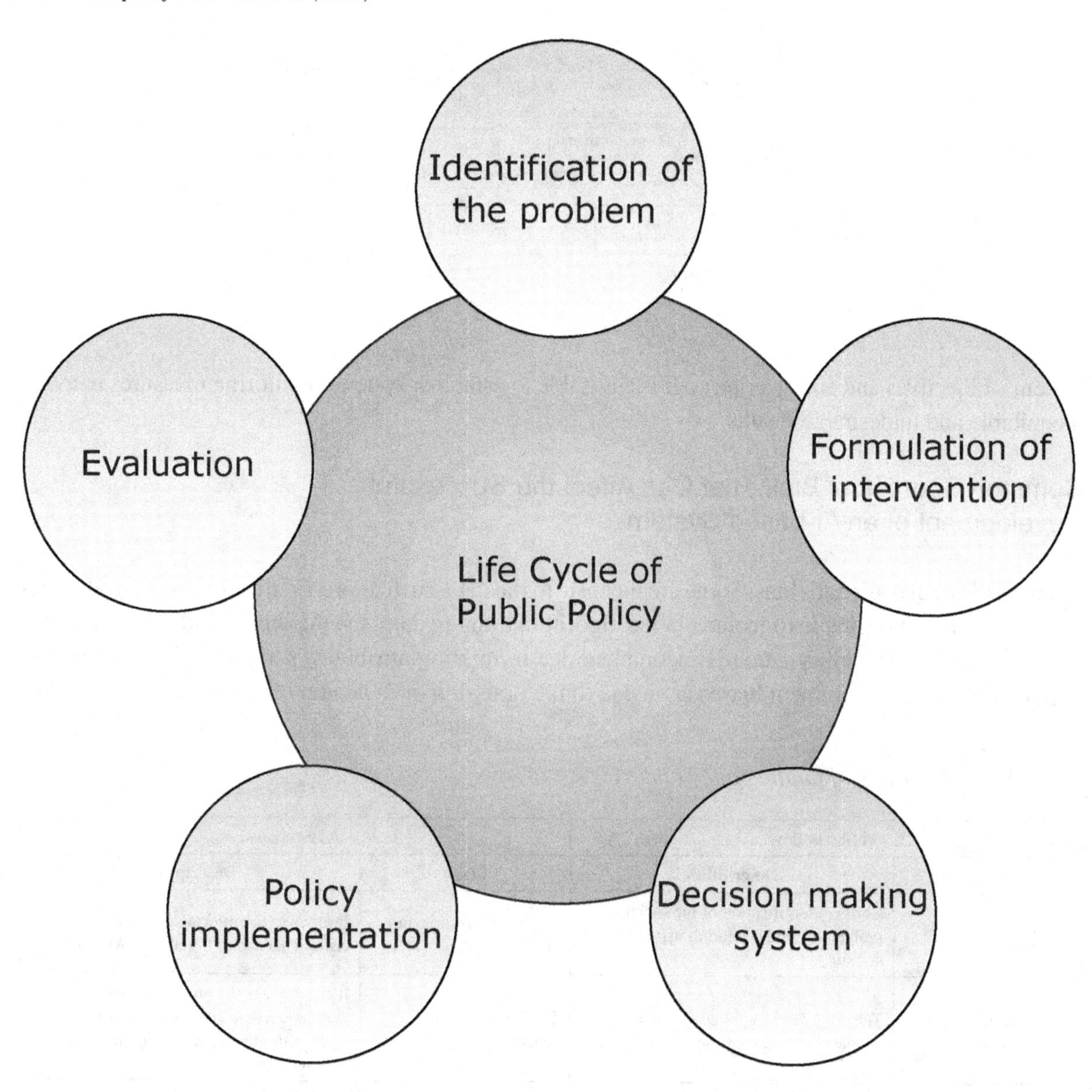

Components of an AI System for Public Policy

Table 2 shows the life cycle of public policy and AI proposed by the manual. Figures 1 and 2 show this information at a more general level.

One of the most important concepts regarding the challenges of AI-based systems is bias. It is recommended to conduct reviews during each stage of the AI life cycle to identify this problem early. In each of these reviews, it is essential to involve experts and end-users of the programs to verify assumptions throughout the process (González et al., 2020). In addition, the manual suggests clearly defining the

Figure 2. AI life cycle
Source: Adapted from Denis et al. (2021).

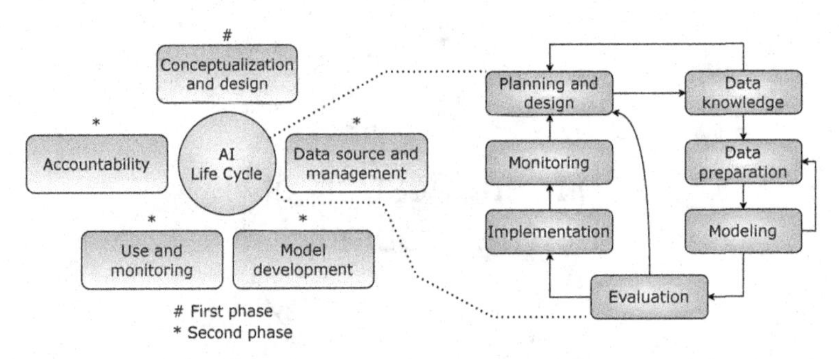

system's objectives and equity criteria. It is advisable to establish system monitoring measures to avoid inequitable and undesirable results.

Common Sources of Bias That Can Affect the Successful Development of an AI-based System

There are several sources of bias. Some are intrinsic to the data, such as pre-existing historical and social biases that are not desirable to include in the algorithm training data. On the other hand, representation bias occurs when the information is incomplete due to missing attributes, sample design, or missing population data. Measurement biases arise due to the omission or inclusion of variables that should or

Table 2. The life cycle of public policy and AI

Public Policy		AI Systems	
Stage	**Description**	**Stage**	**Description**
Identification of the problem	Correct identification of the social problem to find a solution using the system.	Conceptualization and design	It refers to the information and decision criteria needed to initiate the AI project.
Intervention formulation	The policy to be applied to specific people or processes is determined.	Data collection and processing	It emphasizes the process of data generation, selection, and control of sources to identify errors and mitigate deficiencies and biases.
Decision making or support system	The AI life cycle begins with the design of the decision-making system.	Model development and validation	These methods and principles are essential for constructing solid, robust, and adequately validated models.
Policy implementation	Implementation of the public policy.	Use and monitoring	The evaluation is to be made to the model in production and the performance monitoring to avoid errors and unexpected system degradation.
Policy evaluation	Evaluate public policy measures' efficiency, cost, feasibility, and consequences.	Accountability	Refers to measures of transparency and explainability to promote understanding of the mechanisms. Through which the AI system produces the results, the reproducibility of the outcome, and the ability of users to identify and challenge unexpected results.

Table 3. Recommended tools in the development of an AI system

Tool	Description
Robust and accountable AI checklist	It accentuates the main concerns for the risk dimension of the AI life cycle. The technical team should continuously review the checklist, accompanied by the decision-maker (González et al., 2020).
Data Profile	It is an exploratory analysis during the data collection and processing phase. It provides information to evaluate the quality and integrity of the data and possible biases or implications in its use (González et al., 2020).
Model Card	It is a tracking card that summarizes the main features of an AI-based decision-making system and highlights the mitigation measures implemented (González et al., 2020).

should not be in a model. Biases can also arise from errors during training, failures in the validation processes, metrics definition, and the evaluation of results. In addition, initial assumptions about the target population do not correspond to reality (González et al., 2020). Other biases arise when there is no proper handling in the augmentation of synthetic data.

AI System Development Utility Tools

The manual proposes three tools to accompany the development of the AI-based system (see Table 3).

Ethical Data Management

This handbook provides instructions, recommendations, and frameworks for ethical management and best practices to address potential challenges in data use. These include privacy, algorithmic discrimination, and opacity.

The manual is a guide from the Organization for Economic Cooperation and Development (OECD). In it are the guiding principles for safeguarding the privacy and the flow of personal data along with the comprehensive ethical framework of New Zealand. The most relevant issues are raised around three main axes: value, protection, and choice, where there are recommendations and limits are demarcated for cases in which the group in charge of developing the tool wishes to sell the data that have been used (Buenadicha Sánchez et al., 2019). Making recommendations and demarcating boundaries are important for cases where the group developing the tool wishes to sell the data used (Buenadicha Sánchez et al., 2019).

Data Privacy

The first risk that teams face when handling data is privacy. The following highlights the six basic principles for the processing of personal data from the General Data Protection Regulation (GDPR):

- Processing in a lawful, fair, and transparent manner.
- Collection for specified, explicit, and legitimate purposes.
- Adequate, relevant, and limited to what is necessary depending on use.
- Accurate and up-to-date.
- Time-limited conservation.

- Treatment with guaranteed safety.

Proposed Criteria for Ethical Data Management in the Public Sector

According to lessons learned from less successful cases due to erroneous ethical risk assessment, for example, when the French government implemented a university admission system with many biases and the case of the UK when it implemented a facial recognition system that had a high false-positive rate (Buenadicha Sánchez et al., 2019). This manual proposes a framework for ethical data management in the public sector that should permeate the entire data lifecycle. It is aimed primarily at public agencies in Latin America and the Caribbean (Buenadicha Sánchez et al., 2019).

The following are the main criteria for ethical data management in the public sector presented in the manual:

Main Criteria

- Creation of public value.
- Identification of the people benefited or affected and assessment of impact.
- Data diagnostics.
- Privacy by default/design.
- Transparency and accountability.
- Identification of international best practices.
- Design of institutional governance and definition of necessary capabilities.
- Pilot design and small-scale testing before system deployment.

The three manuals reviewed mention projects, regulations, frameworks, pilot studies, and guidelines from different countries. These provide a context or pathway for organizations or individuals interested in developing AI systems. Two resources mentioned are frameworks and guidelines, which are soft laws proposed by governments or organizations to regulate the use of AI in public policy. For this reason, the following section is addressed.

FRAMEWORKS AND GUIDES

The frameworks proposed by the governments of the UK, Canada, and Colombia are documents that aim to provide key aspects of safe, transparent, responsible, and ethical use of AI-based systems. Both the guidelines and the frameworks (except Canada's) are soft laws. Therefore, governments cannot enforce them. However, their use should be a priority. Unlike the frameworks, the guidelines provide a more detailed procedure on key considerations or aspects to ensure that an AI technology is reliable and responsible.

Ethics, Transparency, and Accountability Framework for Automated Decision-Making

The UK government proposes this framework for the ethical use of AI in the public sector as a mitigation measure to the apparent distrust in the regulation of advanced technology in society. The framework contains seven points with practical steps that aim to support government bodies in using automated systems and not leave the decision-making entirely to AI. Especially on complex problems (GOV.UK, 2021).

Its primary objective is to assist government departments with the safe, sustainable, and ethical use of automated or algorithmic decision-making systems. However, different groups or individuals planning to acquire an AI system may use it.

The seven guidelines or aspects proposed by the framework for automated decision making in service are:

1. Test to avoid undesired results or consequences.
2. Provide fair services for all citizens.
3. Be clear about who is responsible.
4. Handle data securely and protect the interests of citizens.
5. Help citizens understand how the system affects them.
6. Ensure compliance with the law.
7. Build something ready for the future.

Directive on Automated Decision-Making

This directive or framework on automated decision-making systems proposed by the Canadian government takes the fundamental principles of administrative law as its bases, such as transparency, accountability, legality, and procedural fairness. These directives aim to reduce the risks associated with implementing an AI system and ensure that the decisions made are effective, accurate, consistent, and interpretable under Canadian law (Government of Canada, 2021).

It is essential to mention that this directive is mandatory throughout Canada. For most government institutions, it set out five requirements that institutions must meet to ensure the responsible and ethical use of automated systems. Including AI or ML systems and digital systems that make or recommend decisions regardless of the technology used (Deshaies & Hall, 2021). The five requirements proposed by the directive are:

1. Algorithmic impact assessment.
2. Transparency.
    a. Notification before decisions.
    b. Provide explanations after decisions.
    c. Access to components.
    d. Source code release.
    e. Decision documentation.
3. Quality insurance.
    a. Test results and follow-up.
    b. Data quality.

   c. Peer review.
   d. Employee training.
   e. Contingency.
   f. Security.
   g. Legality.
   h. Guarantee human intervention.
4. Resources.
5. Reports.

Ethical Framework for Artificial Intelligence in Colombia

The Colombian Ministry of Education and national and international entities such as the Inter-American Development Bank (IDB), the Development Bank of Latin America (CAF), Co-Lab, the Berkman Klein Center for Internet & Society, and Colombian economic sectors created this ethical framework (soft law). It incorporates principles with a human rights approach. One of its main objectives is to recognize the need to protect and reinforce the rights of people in the development, use, and governance of AI systems (Guio Español et al., 2021).

The framework exposes and highlights the great importance of ethical considerations for designing, developing, and implementing AI in Colombia. Its main objective is to provide recommendations and suggestions to public entities when formulating and managing projects that include AI technologies.

The aspects proposed by the ethical framework are:

- Transparency and Explanation.
- Privacy.
- Human Control of the Decisions Proper to an AI System.
- Security.
- Responsibility.
- Non-Discrimination.
- Inclusion.
- Prevalence of Children and Adolescents' Rights.
- Social Benefit.

Each aspect mentioned above has three ethical considerations: data ethics, algorithm ethics, and practice ethics. The framework has a tool to monitor the implementation of the factors described. This tool is a Dashboard (GOV.CO, 2021) with a record of various information on the Digital Transformation and AI projects of public entities in Colombia. Additionally, the framework suggests that the entities carry out eight activities for which there are various resources, in addition to three recommendations as a final point for the public sector.

Comparison of the Frameworks Analyzed

In summary, Table 4 presents the main characteristics, advantages, and disadvantages of the three frameworks analyzed to help the reader determine which one best suit their needs. and the values presented by UNESCO (OECD.AI, 2019; UNESCO, 2021)

Table 4. Main features, advantages, and disadvantages of the analyzed frameworks

Country	Main Features	Advantage (A)/Disadvantage (D)
United Kingdom	• It contextualizes the reader on what is meant by automated decision-making and exemplifies it. • The framework applies to two forms of automated decision making: automated decisions only and assisted automated decision making. • It highlights the importance of considering whether an automated system is necessary for the context of the project and, if used, keeping in mind that risks also depend on policy domains. • Use the framework in conjunction with the guidelines of the Data Ethics Framework (GOV.UK, 2020a). • The framework emphasizes the importance of constant performance review and updating the algorithm and data.	• **A**. Contains an extensive list of resources for each of the seven guidelines, including links, documents, and guides. • **A**. It addresses different aspects at the moment of implementing automated decision-making systems, and the prioritization of the user or target population stands out. • **A**. Provides 4 case studies where it is possible to implement AI-enabled systems. • **D**. The framework overlooks that the automated system may fail and incur biases (even if the probability is minimal). It does not go into detail on actions to mitigate the impact on the user.
Canada	• Divides automated decision-making systems into four levels depending on their impact assessment. • The directive works in conjunction with an Algorithmic Impact Assessment Tool. • Depending on the level of impact, the directive determines the specific requirements: peer review, notice, and training, among others. • Specifies its scope, application, consequences of non-compliance, and contact lines for further information or clarification of doubts regarding the directive.	• **A**. Clearly and concisely explains each requirement, making it an easy-to-interpret directive. • **A**. It is a directive that is constantly updated (every six months), responding to the growing advance of emerging technologies such as AI. • **D**. It does not provide a case study or practical example of the application of the directive.
Colombia	• The aspects proposed in the ethical framework are closely related to the OECD's AI principles and the values presented by UNESCO (OECD.AI, 2019; UNESCO, 2021). • Its creation had two central axes: the prevalence of the rights of children and adolescents and the best practices to implement based on international experience. • Each of the nine aspects of the framework has three ethical contexts. • The ethical framework takes different policies, tools, and initiatives already applied in other countries to justify it.	• **A**. In addition to establishing the principles to be followed, the framework also provides tools for monitoring their implementation. • **A**. It offers an interactive Dashboard that keeps track of AI projects implemented in Colombia. • **A**. Eight activities to be carried out by project managers and how to carry them out with some proposed tools are specified. • **D**. Due to its recent creation and limited dissemination, the AI tracking tool has very few projects. Some do not offer complete information.

Two types of classes frame the following four guidelines. The first class focuses on the guidelines to consider when acquiring an AI system. The second class is oriented toward auditing or evaluating AI-based systems.

Guidelines for AI Procurement

In collaboration with the World Economic Forum (WEF), the UK government created this guide, introducing the definition of AI to contextualize the reader. It also shows how different governments use AI and the potential risks if unregulated or misused. It sets out ten primary considerations to follow when implementing an artificial intelligence system and four specific concerns when procuring AI.

Its main objectives are to provide a set of baseline principles for the procurement of AI-based technology and provide information on how to address challenges during procurement and commissioning. The guide clarifies the rigorous compliance with the principle of transparency in tools, data, and algorithms under any procurement (GOV.UK, 2020b).

Table 5. Main and specific considerations in the AI procurement process

<table>
<tr><th colspan="3">Considerations</th></tr>
<tr><th>Principal</th><th>Specifics</th><th>Sub-Specifics</th></tr>
<tr><td rowspan="4">• Include procurement within a strategy for AI adoption.
• Making decisions in a diverse multidisciplinary team.
• Conduct a data assessment before starting the acquisition process.
• Evaluate the benefits and risks of AI implementation.
• Engaging effectively with the market from the outset.
• Establishing the correct route to market and focusing on the challenge rather than a specific solution.
• Develop an information governance and assurance plan.
• Avoiding black-box algorithms and vendor lock-in.
• Focus on the need to address AN implementation's technical and ethical limitations during its evaluation.
• Consider AI system lifecycle management.</td><td>Preparation and planning</td><td>• Multidisciplinary teams.
• Data evaluation and governance.
• AI impact assessment.
• Preliminary market share.
• Procurement approach and vehicle.</td></tr>
<tr><td>Publication</td><td>• Drafting of the request.</td></tr>
<tr><td>Selection, evaluation, and award</td><td>• Strong practices as a requirement to suppliers.</td></tr>
<tr><td>Contract execution and ongoing management</td><td>• Process-based governance and auditability.
• Model testing.
• Knowledge transfer and training.
• End of the life cycle.</td></tr>
</table>

Table 5 shows the top ten considerations, which provide a basic set of principles for acquiring AI technology. It also shows the four specific considerations that teams should address during the acquisition process, considering the potential benefits of AI technology and the situations conducive to its use (GOV.UK, 2020b).

Internal and External Audit Guidelines

"Auditing machine learning algorithms" is a document created by the Supreme Audit Institutions (SAIs) of Finland, Germany, the Netherlands, Norway, and the UK. This manual summarizes the risks related to the use of ML in public services. It suggests an audit catalog that includes methodological approaches for audits of AI applications ("Auditing machine learning algorithms," 2020). Its main objective is to help SAIs and independent audit staff perform a proper ML algorithms audit. The need for this guidance also lies in the high use of AI in the public sector and the implication of new risks as an emerging technology ("Auditing machine learning algorithms," 2020).

The document and its methodology take the six phases of the Cross-Industry Standard Process for Data Mining (CRISP-DM) as a basis. From these, five sections or audit topics (see Table 6) are derived and assembled into a Data Mining Audit Catalog. Each section provides a set of guiding questions for the auditor, the most common risks of not complying with that section, and a table of responsibility charges.

The second document, "Algorithm Audit Guide", provides guidelines and methodologies for conducting audits of AI algorithms (including the use and collection of personal and sensitive data). It targets algorithm developers, data administrators, and the general public, not just those responsible for audits (Castillo, 2021).

The proposed methodology focuses on detecting, preventing, assisting, and correcting possible undesirable consequences during the life cycle of the algorithm and the processed data. It also ensures that the algorithm does not cause harm (mainly to vulnerable social groups). It should prioritize that it is more controllable, desirable, sustainable, socially fair, and responsible and does not violate data protection regulations (Castillo, 2021). Table 6 shows the structure of the guiding methodology, which considers both a technical analysis and a qualitative analysis.

The third and last document regarding internal audit guidelines is "AI Ethical Self-Assessment for Entrepreneurial Ecosystem Actors," created by the IDB and IDB Lab within the fAIr LAC initiative. It is a guide aimed at those who design, develop, deploy or use AI-based technologies and includes companies, investors, and accelerators. Its objective is to improve product development and, at the same time, identify the main areas of attention to prevent errors, biases, or discrimination. It also seeks to reference the main aspects of conducting an internal audit with an ethical approach (Rosales Torres et al., 2021).

Two innovative aspects of the guide are worth of inspection: self-regulation is not only for entrepreneurs but also for investors and accelerators, and the questions provided exist at three different levels of company development (early stage; Series A and B startups; Series C startups and corporations). The paper offers a progressive internal ethics audit matrix with guiding questions, three stages depending on company development, technology development level, and related stakeholders (Rosales Torres et al., 2021).

Table 6 shows the structure of the internal and external audit guides analyzed. The steps do not follow an order because an algorithmic audit is a dynamic process that adapts according to its context and the type of AI system or algorithm.

Once the guidelines and ethical considerations established by the frameworks and guidelines are understood, it is notable that some of them include or suggest the use of ethical tools. These allow for minimizing unwanted biases and ensuring fairness in data management and project development.

ETHICAL TOOLKITS

Currently, the processes for ethical AI development have been under the spotlight since, as mentioned, it is increasingly common to find AI algorithms and systems in the context of public policies. Tools for technical analysis of data and fairness of algorithms. These tools have marked a before and after using ethical toolkits from questionnaires that allow evidence of the project's relevance and possible failures in data acquisition (Government of Canada, 2022; Saleiro et al., 2018; Wexler et al., 2020).

The following ethical toolkits seek to improve, optimize and guarantee a level of quality in different steps of the life cycle of an AI. To obtain a final product with a high added value that complies with current ethical standards.

Surveys

The first ethical toolkits are questionnaires that establish a set of questions according to their main objective. It can be helpful in internal audit processes (e.g., to measure the impact that the AI can achieve).

Depending on their objective, the questionnaires show a pattern in questions that establish groupings or sections in the stages of the AI life cycle. It is possible find context or governance, data acquisition and processing, autonomy and level of human intervention, impact, and risks and mitigation measures in these partitions.

Table 6. Stages or sections of the internal and external auditing guides

Guide	Stages / Sections	Considerations	Resources of Interests
Auditing Machine Learning Algorithms	Project Management and Governance	• Misalignment/deviation from project objectives. • Lack of business readiness/inability to support the project. • Legal and ethical issues. • Inappropriate use of ML. • Transparency, explainability, and fairness. • Privacy. • Autonomy and Responsibility. • Risk assessment.	• Auditability checklist. • Auxiliary Audit Tool (MS Excel).
	Data	• Personal data and GDPR in the context of ML and AI. • Risk assessment. • Possible audit tests.	
	Model Development	• Development and performance process. • Cost-benefit analysis. • Reliability. • Quality assurance. • Risk assessment. • Possible audit tests.	
	Production Model	• Risk assessment. • Possible audit tests.	
	Evaluation	• Transparency and explainability. • Equal treatment and equity. • Security. • Risk assessment. • Possible audit tests.	
Algorithm Audit Guide	Preliminary Study	• Parties involved. • New or known problem. • Exchange of information. • Audit journal.	• Sample algorithmic audit report template. • Example of a risk assessment table. • Examples of recommendations for system improvement.
	Mapping	• Level of system development. • List of minimum requirements. • Expectations and main issues to be analyzed.	
	Analysis Plan	• Definition of audit terms and deadlines. • Choice of methodology and audit team. • Consensus of the analysis plan.	
	Analysis	• Research. • Execution, follow-up, and readjustment of the analysis plan. • Obtaining and analyzing results.	
	Audit Report	• Interpretation of results. • Conclusions and final assessment of the system. • Recommendations for improvement.	
AI Ethical Self-Assessment for Entrepreneurial Ecosystem Stakeholders	Conceptualization and design	• Objective of AI in business operations. • Evaluation of the digital ecosystem. • Social and environmental impact.	• Ethical internal audit matrices for each consideration. • Description of the stages of development of the companies.
	Governance and Security	• Corporate structure for AI governance. Risk management and internal controls.	
	Human Involvement in AI Systems	• Human-AI interaction: determining the level of human supervision.	
	AI life cycle	• Data source and management. • Algorithm development.	
	Relevant Actors	• Transparency. • Consumers and users of the system. • Access to financing.	
	Communications	• Communications with stakeholders. • Expectations about the technological scope.	

Figure 3. Outline of risk definition questions and areas of mitigation

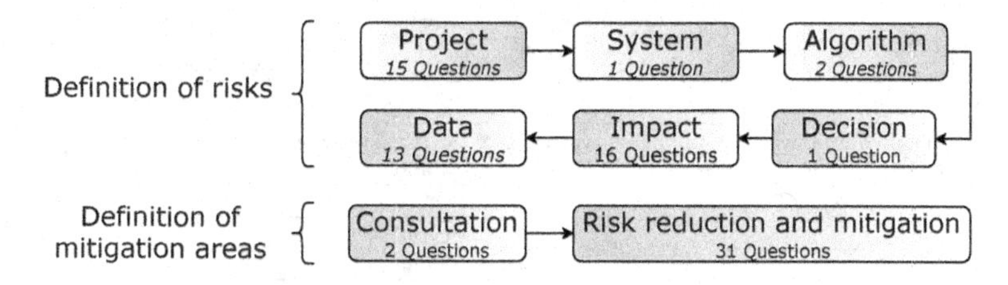

Algorithmic Impact Assessment Tool

These are tools developed by the Canadian government whose objective is to measure an algorithm's impact on a population when used in public policies. It is a questionnaire divided into two parts. The first is the definition of risks, and the second is the definition of mitigation areas (Government of Canada, 2022).

Risk Definition

This area reviews the project's relevance to the population's needs, emphasizing the parts that make up the AI, whether autonomous or supporting, how it obtains the data and their treatment. Figure 3 shows the factors of the whole process. Three areas are of great importance: project, impact, and data.

Definition of Mitigation Areas

This part deals with the consultation/audit process carried out internally or externally, the quality of the data, and how the users' right to privacy complies (see Figure 3)

Scoring

The questionnaire has a scoring form given by the relevance of the questions. The highest concentration of scores is in the first section, with one hundred and seven points. The second section dealing with the mitigation area has forty-five points for one hundred and fifty-two points. Subsequently, the points obtained allow placing the project in an impact ranging from very low to very high.

AI Risk Assessment Tool

This tool is part of a WEF project that aims to provide comprehensive support in an AI technology procurement process. The project is called AI Procurement in a Box and contains four texts. However, it has chosen the questionnaire type tool, whose focus is risk assessment. The objective of this tool is to allow the user to assess the risks faced by an entity in developing and acquiring technology in tandem with the development of a Request for Proposals (RFP) (Gerdon et al., 2020).

Initial Questionnaire

This part of the tool allows eight questions to establish the first step of relevance in the risk assessment and then map it and establish the themes that guide the tool.

Figure 4. Key issues when evaluating AI systems during the procurement process

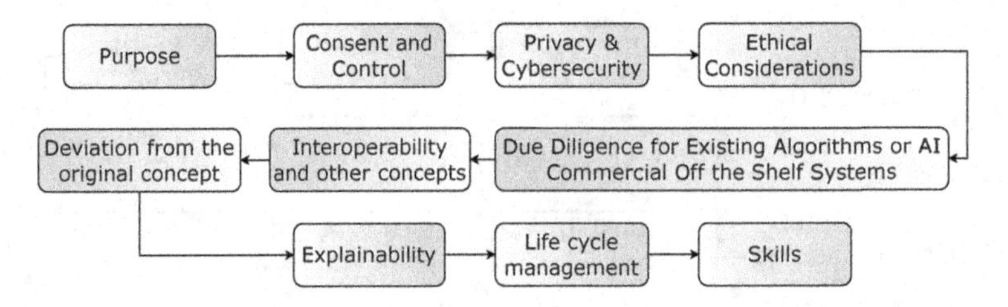

Proportional Requirements

After answering the questions in the initial questionnaire and stating the importance of the tool's focus guidelines, the requirements to be evaluated through the set of questions developed are clear. For each question, there is one essential requirement. Additional requirements depend on the answer (Yes or No).

Risk Matrix

Depending on the purpose of using the tool for control or visibility, suggested or proposed approaches have been established that range from more minor to more complex to perform. It is an excellent resource to consider when assessing certain risks.

Key factors to consider when starting the process of acquiring an AI-based solution:

1. Use procurement processes focusing on precise identification of the problem.
2. Define the public benefit of using AI and assess the risks.
3. Align the procurement process with relevant government strategies and help improve them.
4. Incorporate relevant legislation and codes of practice into your RFP.
5. Coordinate the technical feasibility and governance factors of access to relevant data.
6. Address the technical and ethical limitations of the intended uses of the data to avoid bias.
7. Work with a diverse, multidisciplinary team.
8. Focus on mechanisms that ensure algorithmic results and transparency standards.
9. Establish ongoing collaboration between the purchasing entity and the AI solution provider to address different issues.
10. Create the conditions for a level and fair playing field among AI solution providers.

Evaluation and Specifications

Ten segments group the evaluation with descriptions of examples and formulable questions. Figure 4 shows the general segments.

Best Practices for Government Procurement of Data-Driven Technologies

This document is a tool that seeks to facilitate or implement good practices by governments to acquire data-driven technology such as AI. It emphasizes the need to establish a protocol or guide for this process. One of the main problems of governments lies in the inability to identify whether AI technology-based systems help provide solutions to their needs. Therefore, a set of crucial points framed in a questionnaire divided into the following sections is established (Richardson, 2021).

Figure 5. Application approaches

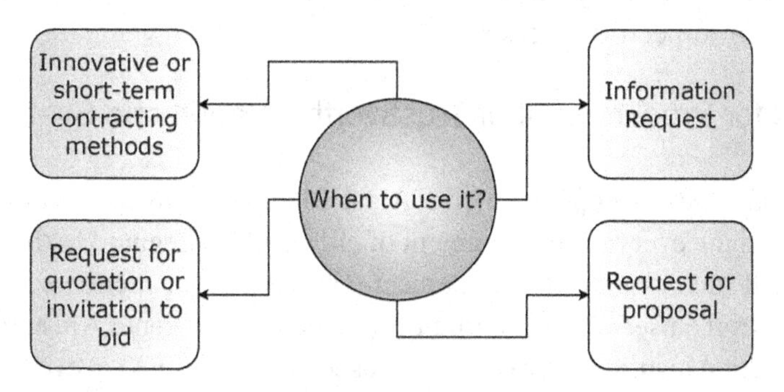

Document your Pre-solicitation Technology Assessment

The purpose of this stage is to prepare and contextualize the process. It argues the relevance of using AI-based technology and systems to address questions related to the government agency's needs in the procurement process. The goal is to determine the opportunity to use an AI-based system, clarify which agency department needs such a solution and whether it matches its mission or vision. For example, it is neither consistent nor ethical for a department studying gender-based violence to have an AI system and only use physical injury data without considering the psychological spectrum. This section underpins the following steps and is critical to the proper development of the acquisition process.

Assess all Solicitation Approaches

There are various requests that the government entity or agency must consider if it wishes to acquire technology. These vary depending on the level of certainty needed or the company's stage of a technology exploration. Figure 5 shows the types of requests.

Each type of request is different in terms of the term (short, medium, or long time) and the buyer's commitment to pay for the technology. However, it also depends on how much information the buyer has about the technology and whether it is for the agency's own internal or external use.

Proposal Evaluation and Contract Negotiations

This section deals with intervention in data acquisition, handling, storage, capturing, and participation by third parties. It also addresses issues such as streamlining the movement of data externally and internally, approaches when formalizing the contract with the client, and data ownership confidentiality. Figure 6 shows the three key points with their respective questions and scenarios.

Figure 6. Outline of key considerations and questions

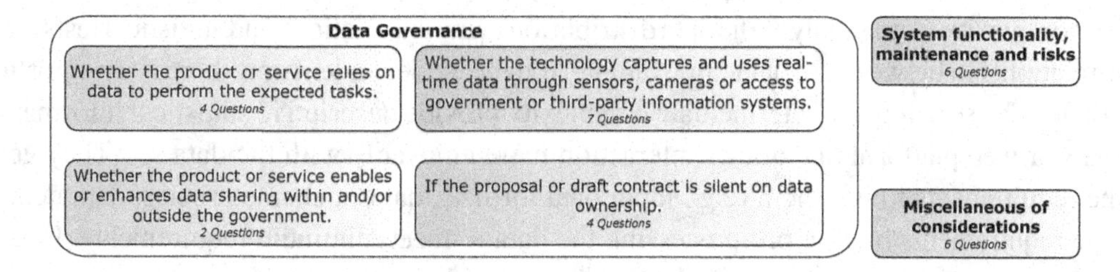

Finally, the document lists existing resources for the proper acquisition or development of AI systems, eight resources with their respective descriptions.

Standard Clauses for Procurement of Trustworthy Algorithmic Systems

These types of tools are clauses proposed by the Amsterdam government. These are born to observe the constant use of AI in the everyday environment of citizens, to safeguard both rights of citizens, and companies have created such clauses. The document consists of four sections. Three of them contain information that involves the user with the entire development (Gemeente Amsterdam, 2019).

The introductory section justifies and explains the user's need to use AI in everyday life. Why clauses are necessary, when to use them, and explain what a memorandum is to help the reader understand the context in which it is correct to use clauses and why.

Explanation and Definition

This section works through each part of the project, from its algorithm and decision making to transparency and explainability. Each unit aims to help the reader understand the formation of the project, the model of the chosen system, the purpose of its creation, the intended use, and the target population. All this allows people to improve the acceptance of the project, managing more effectively the population permission. Depending on the wording of the project, the clauses can be adapted to meet the objective of improving transparency and accessibility to project information.

Article-by-Article Comments

It consists of nine articles on applicability and data quality explaining the following topics:

- Population data processing and management.
- Rights that users have over their data.
- Quality, governance, and transparency of the algorithmic system.
- Risks and management strategies in the development process.
- Audit process or another process that has involved the system.
- Project costs.

Open-source Code

These tools allow analyzing the model under development or finalized, defragment or understanding of the ML or Deep Learning (DL) black box, and auditing the data and the algorithms used in the AI lifecycle. They also allow proposal processes to mitigate risks and biases that may have occurred.

The tools addressed were mainly implemented in Python or R code. Their design varies according to the level of complexity. Some analyze the data distribution using calculations and statistical tests, allowing to see the relationship between the data, for example, if they are dependent or if there is a proportionality between them. The structures of intermediate complexity provide descriptive statistical information and have an API or web platform that allows interaction more enjoyably with the data set. They generally have some requirements to use them (e.g., input data formats, dataset characteristics, data debugging, etc.). These requirements involve pre-processing the data to meet minimum requirements. Finally, the

Figure 7. Input data format

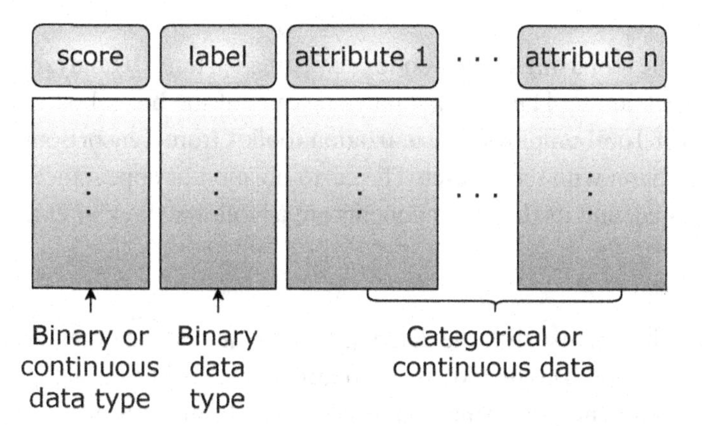

higher-level tools have more complex features and complete support options. Some of their functions allow corrections from the data structure to the configuration of hyperparameters in the original model, thus giving a wide range of options to obtain a quality AI.

Aequitas

Aequitas is a tool aimed at detecting biases in the dataset. The tool requires a dataset, which can be uploaded directly on the web page or loaded through Python code (library and other dependencies are needed). Figure 7 shows the dataset format (Saleiro et al., 2018).

The attribute "score" refers to the prediction of the model used above, and "label" is the expected value to be predicted. Next, the attributes to be analyzed are listed, so there can be "n" attributes. All data must be in a character-type format (including numeric values). The tool does not accept NaN values, so a pre-processing is required.

When the tool processes the dataset, the tool allows you to select the protected and reference attributes. It can also do this with the most representative or the least biased group. It is then possible to choose the following metrics that audit the dataset,.

- Prevalence.
- Proportional parity.
- False Positive Rate Parity.
- False Discovery Rate Parity.
- False Negative Rate Parity.
- False Omission Rate Parity.

Finally, Aequitas proposes a (modifiable) tolerance percentage depending on the margin of inequity that the user allows or tolerates in the project. However, it is recommended to use 80%. Finally, the tool creates a report with the results of each metric and its respective explanation. The report shows tables with the mentioned metrics, sorted according to the selected attributes and reference groups.

What-If Tool

It is possible to deploy the tool using Python code or in Google Cloud. It analyzes the datasets of trained models. It is a post-execution tool and allows understanding of the black box behavior in DNN, CNN, or ML models. 'What-if Tool' employs a visualization toolkit from TensorBoard. It allows immediate visualization and interaction with the dataset. The customization or appearance of TensorBoard in this tool is different than usual, and its three components are as follows (Wexler et al., 2020).

Datapoint Editor

In this section, it is possible to visualize the data used by the model and organize them according to the task (classification or regression). The tool also supports models working with images or words.

The tool allows to make changes in the graphs according to the axis attribute and observe the data simultaneously as their behavior according to the automatic training process. It is possible to make adjustments to the labels and the dispersion of the data. The results vary if the model is an ML model or a neural network model. In the latter case, the tool displays optional functions such as Dropout or two types of regularization (L1 and L2).

Performance and Fairness

Here it is possible to observe the performance, the fairness, adjust the established thresholds, and how they have interacted. The tool displays a Receiver Operating Characteristic (ROC) curve and a Precision-Recall (PR) curve, true positive rate, and F1 score, among other metrics. The dashboard also displays one or more confusion matrices depending on the number of models. Finally, the section allows adjusting bounds, optimization strategies, and other settings to analyze changes in the described metrics.

Features

This last section or component makes it possible to observe the data distribution. The tool displays statistical variables such as mean or standard deviation if the data are numerical. If they are categorical data, the tool displays counts or top values.

AI Fairness 360

This tool was developed to provide support to AI projects. Its central focus is to reinforce the idea of reliable and bias-free systems or, at least, to reduce them to levels where the systemic disparity is socially acceptable (Bellamy et al., 2018).

The tool has a repository on GitHub (Trusted-AI, 2022), which contains Jupyter notebooks and relevant documentation to facilitate its use. The notebooks show data preparation to pre-processing performed by the library functions (AIF360). The tool analyzes sensitive data or features to address systemic discrimination in real-life or sampling processes. AIF360 has several fairness metrics and bias mitigation algorithms such as reweighting, optimized pre-processing, or adversarial debiasing (Kenna, 2021; Tanton et al., 2011). It is essential to highlight that the creators of AIF360 mention that the tool's intended use is in risk allocation or risk assessment problems with well-defined protected attributes. All of the above to obtain equality's statistical or mathematical idea ("AI Fairness 360," 2022).

Figure 8. Bias mitigation algorithms and their recommended use

Pre-processing Algorithms
Disparate Impact Remover
Learning Fair Representations
Optimized Preprocessing
Reweighing
If the user can modify the training data

In-processing Algorithms
Adversarial Debiasing
ART Classifier
Gerry Fair Classifier
Meta Fair Classifier
Prejudice Remover
Exponentiated Gradient Reduction
Grid Search Reduction
If the user can change the learning algorithm

Post-processing Algorithms
Calibrated Equalized Odds
Equalized Odds
Reject Option Classification
If the user can only treat the trained model as a black box

Algorithms

The tool has 14 bias mitigation algorithms framed in three categories (see Figure 8): pre-processing, internal processing, and post-processing. The recommendation is to use the earliest category if the user has access to this stage since biases can be corrected flexibly and promptly ("AI Fairness 360," 2022).

Fairness Metrics

AIF360 has four general types of metrics framed around the following concepts: Individual vs. Group Fairness (or both), group fairness (Data vs. Model), group fairness (We're All Equal vs. What You See Is What You Get), and group fairness (Ratios vs. Differences). The library contains more than 80 metrics specific to datasets or algorithms that may be appropriate for a given application. However, the context of the problem to be solved and the vulnerable population must be clear ("AI Fairness 360," 2022). Figure 9 shows the general groups of equity metrics and their respective number of specific metrics.

Figure 9. Diagram of equity metrics

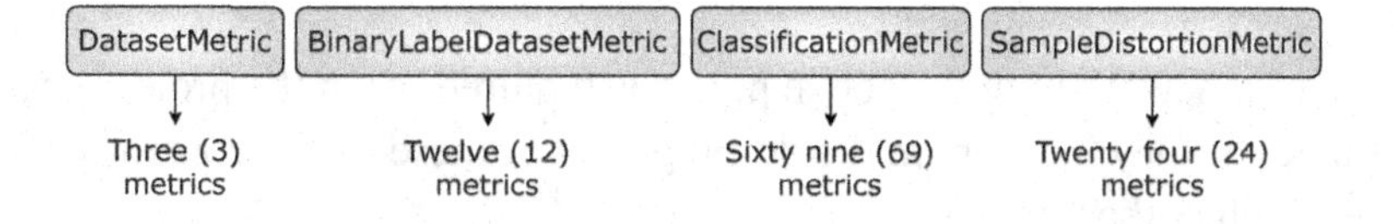

In addition, the tool provides explanations of metrics in text or JSON files. It features five different types of databases: Adult Census Income Dataset, Bank Marketing Dataset, ProPublica COMPAS Dataset, German Credit Dataset, and Law School GPA Dataset, as well as reference databases that depend on whether the data is your own or you opt for exemplification to train the model you want to create.

Many considerations mentioned in the three previous sections (Manuals, Frameworks, Guides, Ethical Toolkits) have intellectual foundations in research articles. These establish the importance of following an ethical line in projects related to public policies involving AI, as shown in the following section.

ARTICLES AND WHITE PAPERS

The articles and white papers explained below influence the importance of the ethical development or acquisition of AI-based algorithms and systems. They also clarify the need for standards and guidelines in the use of data, which are increasingly subject to hard laws. Table 7 shows the objectives, relevant characteristics, and proposals for improvement arising from the interaction with these documents. The documents gathered in Table 7 also aim to improve the existing image of AI in society. In addition to clarifying that the backbone of any AI project used in public policy is the target population.

DISCUSSION

From the previously shared information, it infers a notorious relationship between manuals, guides, frameworks, ethical toolkits, papers, and white papers. This relationship determines the intrinsic need for guiding frameworks that guarantee quality standards, mitigate biases and eliminate algorithmic discrimination when implementing AI-based tools or systems. The legislative requirement for procurement and traceability processes is essential to ensure the rights of vulnerable groups.

Despite a great variety of points already explored, the four predominant lines that mark a trend among them can be distinguished. These are:

1. Raise the need for guidelines in the development, auditing, and contracting processes.
2. The objective and central axis must benefit the population above the companies or particular interests.
3. The documents agree on the need for ethical data management at all stages to avoid bias or algorithmic discrimination and guarantee digital rights.
4. Warn the importance of structuring projects to be auditable. To address traceability and accountability processes.

These lines form the basis for justifying good practices when developing or acquiring AI-based systems. Therefore, is suggest the following considerations:

- A standardized framework with the above points is required when the project's purpose is related to public policies or may interfere in the development of these—also contemplating private companies that may affect the population.
- Forming a multidisciplinary work team and its respective process of distribution of responsibilities can improve the auditability of the project and raise its quality standards.
- Fully understand the context and the problem to be solved to identify whether or not it is necessary to implement an AI-based system.
- The application of ethical toolkits in the development of AI-based systems is of utmost importance as they generate documentation conducive to auditing processes. They are improving the aspects of traceability, opacity, and accountability.

It is necessary to mention certain limitations on the proposed considerations, such as the regions of origin of the texts since most of them are of European or American origin.

Table 7. The main objective and relevant characteristics of Articles and White Papers

Article/ White Paper	Main Objective	Relevant Features	Proposals for Improvement
A Global Perspective of Soft Law Programs for the Governance of Artificial Intelligence (Gutierrez & Marchant, 2021).	Compile and analyze global trends on how governments, non-profits, and the private sector use soft law to manage AI methods and applications.	• Six hundred thirty-four soft laws were identified, screened, and classified according to their type of nature. • A three-step methodology was followed: identification, selection, and classification. • Two data types have been differentiated regarding the information extracted from soft laws: variables and topics. • The compendium of soft laws provides a typology of the structural or procedural dimensions.	• Update the database to include soft laws for 2020 and 2021. • Include soft laws written not only in English.
"The Human Body is a Black Box": Supporting Clinical Decision-Making with Deep Learning (Sendak et al., 2020).	To raise awareness of the ethical challenges involved in building an AI model and present a clinical support tool's development, implementation, and evaluation.	• Four fundamental values and practices when developing AI systems have been identified: defining the problem in context, establishing relationships with stakeholders, respecting professional discretion, and creating feedback loops. • System development is approached by involving the interconnected social and technical dimensions of technology. • Four frameworks are highlighted, which provide essential information on how to adopt innovations in healthcare.	Extend the model datasheet, including the complete model architecture, API for training, and data cleansing.
Towards Accountability for Machine Learning Datasets: Practices from Software Engineering and Infrastructure (Hutchinson et al., 2021).	Propose a rigorous framework for responsible and transparent database management in ML-based on Software and Infrastructure Engineering practices.	• Propose a similarity between data sets and technical infrastructures, the processes that give rise to databases, and fundamentally goal-oriented engineering. • Five types of "accounts" are recommended. Each related to a stage of the dataset development cycle: requirements, design, implementation, testing, and maintenance accounts. • The importance of establishing a non-linear dataset development cycle, with five phases: requirements analysis, design, implementation, test/test, and maintenance, is emphasized.	Include an example of using the three templates provided, and see the type of record or account used for implementation and maintenance.
The Global Landscape of AI Ethics Guidelines (Jobin et al., 2019).	Conduct analysis and mapping of the current corpus of principles and guidelines on AI ethics.	The findings of this study highlight the importance of guidelines with substantive ethical analysis and strategies for proper implementation.	Include in the review documents produced by Latin American governmental entities, not only from the United States, Europe, and Asia.
The Ethics of Algorithms: Key Problems and Solutions (Tsamados et al., 2022).	Contribute to the debate on identifying and addressing ethical implications in algorithms and provide an updated analysis of current concerns and regulations for the proper development and implementation of AI algorithms.	• Work on the ethics of algorithms experienced significant growth in 2016 as national governments and non-governmental organizations. Public and private companies began to play an essential role in the conversation about the ethics of AI algorithms. • This article focuses on the ethical issues raised by algorithms as mathematical constructs, their implementations, and applications.	Update specific references and detailed information about the literature reviewed, and add updated literature on the ethical considerations that governmental bodies have made in recent years.
The ethics of algorithms: Mapping the debate (Mittelstadt et al., 2016).	Expose the ethical issues that arise from algorithms and how they apply to themselves.	The main feature is the "ethics map," which develops all the content at epistemological and normative levels.	Construct an annex with the map format that allows for dynamic changes, considering the relevant points applicable to the epistemic and normative sections.
Public Procurement and Innovation for Human-Centered Artificial Intelligence (Naudé & Dimitri, 2021).	Justify public procurement and public policy innovation to enable the development of HCAIs.	• The construction that makes around the need for public procurement improves the perception and development of HCAI. • Innovation processes began to establish the need for governments to carry out a public procurement of projects, in this case, focused on AI. • It advocates that it is necessary to generate an investment of responsibility, which produces public procurement focused on innovation and the development of HCAI.	• Establish a guiding framework for the development of the HCAI to ensure compliance with its objectives. • Generate a framework focusing on AI acquisition and a breakdown of HCAI characteristics in the development sections.
White Paper on Data Ethics in Public Procurement of AI-based Services and Solutions (Hasselbalch et al., 2020).	Establish a reference framework for public procurement in the EU that meets quality and accountability standards.	• Generates legal relevance by addressing the importance of dignity, privacy, data protection, fairness, and strategic procurement. • Relevance in data acquisition processes by prioritizing the human being and being able to sustain traceability that responds to other ethical qualities. • Risk detection as the root of mitigation processes. • Potential service providers and the awarding of AI development projects in public policies.	Generate a web add-on based on this document that would generate a report to evaluate the procurement process and create a document with recommendations.

CONCLUSION

According to the texts discussed, the public policies that predominate in the regulation of AI-based systems are mostly soft laws. Includes manuals, guides, frameworks or guidelines, and recommendations that do not constitute a mandatory framework (excluding the framework proposed by the Canadian government). The soft laws addressed here are characterized by the following commonalities when developing or acquiring an AI system: privacy, transparency, bias, traceability, and security, applicable from the data collection phase to the deployment and continuous monitoring of the system. Most of the texts highlight the importance of incorporating the target population in the development of the system. Making them aware of how the algorithm works in broad terms, how their data are processed, and the benefits of implementing the model.

When reviewing the texts, it was found that most of them lack case studies or practical examples that better understand the application or use of the strategies. Some of them contain little information on the actions to be taken in case of possible system failures after deployment. These shortcomings become points for improvement.

The development of an ethical tool, divided into two parts, will be considered for future work. The first stage will generate a context or profiling report to measure the potential impact of AI and build mitigation processes for possible biases in the project. The second stage will be an open-source code that will allow evaluating the quality of the data used with metrics that ensure fairness in the data.

ACKNOWLEDGMENT

The authors acknowledge to Universidad Autónoma de Manizales (UAM) from Colombia under project No. 589-089, and Universidad Adolfo Ibáñez from Chile, for their contributions and funding through the project "Oportunidades de Mercado para las Empresas de Tecnología - Compras Públicas de Algoritmos Responsables, Éticos y Transparentes" with code CH-T1246 from Banco Interamericano de Desarrollo (BID) Representación Chile. Also is acknowledge to ANID PIA/BASAL FB0002 project.

REFERENCES

AI Fairness 360 – Resources: Guidance on choosing metrics and mitigation. (n.d.). IBM Research Trusted AI. http://aif360.mybluemix.net/resources#guidance

Artificial Intelligence in society. (2019, June 11). OECD. doi:10.1787/eedfee77-en

Auditing machine learning algorithms. (2020, November 24). Supreme Audit Institutions of Finland, Germany, the Netherlands, Norway, & the UK at Auditing Algorithms. https://www.auditingalgorithms.net

Bellamy, R. K. E., Dey, K., Hind, M., Hoffman, S. C., Houde, S., Kannan, K., Lohia, P., Martino, J., Mehta, S., Mojsilovic, A., Nagar, S., Ramamurthy, K. N., Richards, J., Saha, D., Sattigeri, P., Singh, M., Varshney, K. R., & Zhang, Y. (2018). AI fairness 360: An extensible toolkit for detecting, understanding, and mitigating unwanted algorithmic bias. *IBM Journal of Research and Development*, *63*(4/5), 1–15. doi:10.48550/arxiv.1810.01943

Berryhill, J., Heang, K. K., Clogher, R., & McBride, K. (2019). Hello, World: Artificial Intelligence and its use in the public sector. *OECD Observatory of Public Sector Innovation*, *36*, 1–148. doi:10.1787/19934351

Buenadicha Sánchez, C., Galdon Clavell, G., Hermosilla, M. P., Loewe, D., & Pombo, C. (2019, March). *La gestión ética de los datos.* Inter-American Development Bank. doi:10.18235/0001623

Castillo, C. (2021, January). *Guía de Auditoría Algorítmica.* Eticas. https://www.eticasconsulting.com/eticas-consulting-guia-de-auditoria-algoritmica-para-desarrollar-algoritmos-justos-y-eficaces/

Cath, C. (2018, October 15). Governing Artificial Intelligence: Ethical, legal and technical opportunities and challenges. *Philosophical Transactions - Royal Society. Mathematical, Physical, and Engineering Sciences*, *376*(2133), 20180080. https://doi.org/ doi:10.1098/rsta.2018.0080 PMID:30322996

Denis, G., Hermosilla, M. P., Claudio, A., Ávalos, R. S., Alarcón, N. G., & Pombo, C. (2021, August). *Uso responsable de IA para política pública: Manual de formulación de proyectos.* Inter-American Development Bank. doi:10.18235/0003631

Deshaies, B., & Hall, D. (2021, December 1). *Responsible use of automated decision systems in the federal government.* Statistics Canada. https://www.statcan.gc.ca/en/data-science/network/automated-systems

Eitel-Porter, R. (2020, September). Beyond the promise: Implementing ethical AI. *AI Ethics*, *1*(1), 73–80. doi:10.100743681-020-00011-6

Fuentes, P., Rogerson, A., Westgarth, T., Lida, K., Mbayo, H., Finotto, A., Rahim, S., & Petheram, A. (2022). *Government AI readiness index 2021.* Oxford Insights. https://static1.squarespace.com/static/58b2e92c1e5b6c828058484e/t/61ead0752e7529590e98d35f/1642778757117/Government_AI_Readiness_21.pdf

Gemeente Amsterdam. (2019). *Standard clauses for procurement of trustworthy algorithmic systems.* https://www.amsterdam.nl/innovatie/digitalisering-technologie/algoritmen-ai/contractual-terms-for-algorithms

Gerdon, S., Katz, E., LeGrand, E., Morrison, G., & Torres Santeli, J. (2020, June). *AI procurement in a box: Workbook.* World Economic Forum. https://www3.weforum.org/docs/WEF_AI_Procurement_in_a_Box_Workbook_2020.pdf

González, F., Ortiz, T., & Sánchez Avalos, R. (2020, October). *Uso responsable de la IA para las políticas públicas: Manual de ciencia de datos.* Inter-American Development Bank. doi:10.18235/0002876

GOV.CO. (2021). *Dashboard - Seguimiento Marco Ético de IA.* https://inteligenciaartificial.gov.co/dashboard-IA/

Government of Canada. (2021, April 1). *Directive on automated decision-making.* https://www.tbs-sct.canada.ca/pol/doc-eng.aspx?id=32592

Government of Canada. (2022, April 19). *Algorithmic impact assessment tool.* https://www.canada.ca/en/government/system/digital-government/digital-government-innovations/responsible-use-ai/algorithmic-impact-assessment.html

GOV.UK. (2020a, September 16). *Data Ethics Framework.* https://www.gov.uk/government/publications/data-ethics-framework/data-ethics-framework-2020

GOV.UK. (2020b, June 8). *Guidelines for AI procurement.* https://www.gov.uk/government/publications/guidelines-for-ai-procurement/guidelines-for-ai-procurement

GOV.UK. (2021, May 13). *Ethics, transparency and accountability framework for automated decision-making.* https://www.gov.uk/government/publications/ethics-transparency-and-accountability-framework-for-automated-decision-making/ethics-transparency-and-accountability-framework-for-automated-decision-making

Guio Español, A., Tamayo Uribe, E., Gómez Ayerbe, P., & Mujica, M. P. (2021). *Marco Ético para la Inteligencia Artificial en Colombia.* https://dapre.presidencia.gov.co/TD/MARCO-ETICO-PARA-LA-INTELIGENCIA-ARTIFICIAL-EN-COLOMBIA-2021.pdf

Gutierrez, C. I., & Marchant, G. E. (2021, May). A global perspective of soft law programs for the governance of Artificial Intelligence. SSRN *Electronic Journal.* doi:10.2139/ssrn.3897486

Hasselbalch, G., Olsen, B. K., & Tranberg, P. (2020). *White paper on data ethics in public procurement of AI-based services and solutions.* Data Ethics EU. https://dataethics.eu/wp-content/uploads/dataethics-whitepaper-april-2020.pdf

Hutchinson, B., Smart, A., Hanna, A., Denton, E., Greer, C., Kjartansson, O., Barnes, P., & Mitchell, M. (2021, March). Towards accountability for Machine Learning datasets: Practices from software engineering and infrastructure. *Proceedings of the 2021 ACM Conference on Fairness, Accountability, and Transparency*, 560–575. 10.1145/3442188.3445918

Jobin, A., Ienca, M., & Vayena, E. (2019, September 2). The global landscape of AI ethics guidelines. *Nature Machine Intelligence, 1*(9), 389–399. doi:10.103842256-019-0088-2

Kenna, D. (2021, July). *Using adversarial debiasing to remove bias from Word Embeddings.* doi:10.48550/arXiv.2107.10251

Mittelstadt, B. D., Allo, P., Taddeo, M., Wachter, S., & Floridi, L. (2016, December 1). The ethics of algorithms: Mapping the debate. *Big Data & Society, 3*(2). doi:10.1177/2053951716679679

Naudé, W., & Dimitri, N. (2021). Public procurement and innovation for Human-Centered Artificial Intelligence. SSRN *Electronic Journal.* doi:10.2139/ssrn.3762891

OECD. AI. (2019). *OECD AI Principles overview.* https://oecd.ai/en/ai-principles

Purdy, M., & Daugherty, P. (2016). *Why Artificial Intelligence is the future of growth.* Accenture. https://dl.icdst.org/pdfs/files2/2aea5d87070f0116f8aaa9f545530e47.pdf

Richardson, R. (2021). Best practices for government procurement of data-driven technologies: A short guidance for key stages of government technology procurement. SSRN *Electronic Journal.* doi:10.2139/ssrn.3855637

Rosales Torres, C. S., Buenadicha Sánchez, C., & Tetsuro, N. (2021, May 1). *Autoevaluación Ética de IA para actores del Ecosistema Emprendedor.* doi:10.18235/0003269

Saleiro, P., Kuester, B., Hinkson, L., London, J., Stevens, A., Anisfeld, A., Rodolfa, K. T., & Ghani, R. (2018). *Aequitas: A bias and fairness audit toolkit.* doi:10.48550/arxiv.1811.05577

Sendak, M., Elish, M. C., Gao, M., Futoma, J., Ratliff, W., Nichols, M., Bedoya, A., Balu, S., & O'Brien, C. (2020, January). "The human body is a black box": Supporting clinical decision-making with deep learning. *Proceedings of the 2020 Conference on Fairness, Accountability, and Transparency*, 99–109. 10.1145/3351095.3372827

Tanton, R., Vidyattama, Y., Nepal, B., & McNamara, J. (2011, April 8). Small area estimation using a reweighting algorithm. *Journal of the Royal Statistical Society. Series A, (Statistics in Society)*, *174*(4), 931–951. doi:10.1111/j.1467-985X.2011.00690.x

The strategic and responsible use of Artificial Intelligence in the public sector of Latin America and the Caribbean. (2022, March 22). OECD & CAF. doi:10.1787/1f334543-en

Trusted-AI. (2022). *AI Fairness 360 (AIF360) - Examples.* Git Hub. https://github.com/Trusted-AI/AIF360/tree/master/examples

Tsamados, A., Aggarwal, N., Cowls, J., Morley, J., Roberts, H., Taddeo, M., & Floridi, L. (2022). The ethics of algorithms: Key problems and solutions. *AI & Society*, *37*(1), 215–230. doi:10.100700146-021-01154-8

UNESCO. (2021, June). Draft text of the Recommendation on the Ethics of Artificial Intelligence. *Intergovernmental Meeting of Experts (Category II) related to a Draft Recommendation on the Ethics of Artificial Intelligence*. https://unesdoc.unesco.org/ark:/48223/pf0000377897

Wexler, J., Pushkarna, M., Bolukbasi, T., Wattenberg, M., Viégas, F., & Wilson, J. (2020, January). The What-If Tool: Interactive probing of Machine Learning models. *IEEE Transactions on Visualization and Computer Graphics*, *26*(1), 56–65. doi:10.1109/TVCG.2019.2934619 PMID:31442996

Wirtz, B. W., Weyerer, J. C., & Geyer, C. (2019). Artificial Intelligence and the public sector—Applications and challenges. *International Journal of Public Administration*, *42*(7), 596–615. doi:10.1080/01900692.2018.1498103

KEY TERMS AND DEFINITIONS

Artificial Intelligence (AI): Refers to the set of algorithms or computational methods that aim to give computers the characteristics or abilities of human intelligence.

Convolutional Neural Networks (CNN): It is a type of deep learning model commonly used for images-related tasks. It uses the mathematical operation of convolution to extract features from images.

Data Governance: Is the set of processes, responsibilities, policies, standards, or metrics integrating the availability, usability and security of these in business systems.

Deep Learning (DL): Corresponds to a subset of artificial intelligence techniques that compromises models based on artificial neural networks.

Deep Neural Networks (DNN): Is a type of deep learning model commonly used for classification tasks. It uses mathematical operations of linear and non-linear functions.

Machine Learning (ML): AI subfield and evolving branch of computational algorithms that are designed to emulate human intelligence by learning from the surrounding environment.

Parity Metrics: Set of metrics that indicate the state of equality and the quality of the database used.

Soft Laws: Government programs that establishes substantive expectations but are not mandatory such as recommendations, guides, directives, manuals, frameworks, among others.

Section 5
When Technology Gets Personal

Chapter 11
Facial Recognition Technology: Ethical and Legal Implications

Ellen Marie Raineri
Pennsylvania State University, USA

Erin A. Brennan
Pennsylvania State University, USA

Audrey E. B. Ryder
Pennsylvania State University, USA

ABSTRACT

Facial recognition technology (FRT) is a type of biometric technology that uses a digital image of one's face and uses algorithms to match that image in a database. This type of technology has been adopted by individuals when using their mobile devices and automobiles and to access restricted areas or events. FRT has also been adopted by law enforcement and the government to support crime detection and prevention. In addition to the benefits, the associated cyber security problems, legal issues, and ethical challenges of privacy and discrimination are explored. Understanding ethical theories permits the public and decision-makers to make informed choices to influence changes in law to support the changing environment for FRT. Recommended solutions are included along with future research that addresses face connect in automobiles, state law, and ethical comparisons.

INTRODUCTION

Facial Recognition Technology (FRT) is used by social media providers, law enforcement, and private businesses. It impacts the daily existence of American citizens in ways that are both obvious and obscure. Digital cameras are omnipresent in today's society. They are widely available, easy to use, and reasonably priced. Databases that include facial identification information are growing, and the consequences of that stored information can be problematic (Roussi, 2020). A human's face is a unique and reliable indicator of identity (Chen, 2020). FRT leverages that unique data in a variety of ways, ranging from uses that

DOI: 10.4018/978-1-6684-5892-1.ch011

provide superficial convenience to those that provide profound security to individuals and communities (Chilson & Barkley, 2021). Each and all of these uses raises ethical implications and considerations. In turn, these ethical considerations influence the development of legal policies. It is important for consumers, business owners, employees, and all members of society to understand the fundamental uses of FRT, the ways in which FRT use presents ethical challenges, and how society's collective response to those ethical challenges is shaping the law.

FRT BACKGROUND INFORMATION

Defining FRT

Facial recognition is a biometric technology that combines machine learning and artificial intelligence to "identify, recognize, and interpret images of faces" (Chilson & Barkley, 2021, p. 87; Ivanova & Borzunov, 2020, Kostka et al., 2021). The first phase is facial detection in which a face is detected solo or in a crowd. Next, the facial image is analyzed, examining areas like the distance between the eyes or the distance from the chin to the forehead. Then, the image is converted into a unique numerical code or faceprint. Last, the faceprint is compared to other images that are stored in a database to see if a match exists (Chilson & Barkley, 2021; Kaspersky, n.d.).

History of FRT

FRT is in wide global use today, but the concepts beyond the technology did not emerge until the 1960s. Over the course of the last two decades, the algorithms that drive the technology have developed. During that same period, the cost of cameras has decreased and access to high quality broadband has increased (Keener, 2022). The result is the extensive and daily use of FRT in and across global communities.

Identifying Uses of FRT

FRT is used in both private and public settings to create convenient consumer experiences and safer societal experiences (Chilson & Barkley, 2021). For example, private businesses and individuals use FRT daily to access smartphones, work areas, and automobiles. Businesses regularly use FRT to monitor worksite admission, provide security, and create targeted marketing. Police and government actors also use FRT to assist in crime prevention and law enforcement (Summa Linguae Technologies, 2021). The general public has grown reliant on the convenience, accuracy, and safety that many of these uses provide. At the heart of each of these measures is the uniquely identifiable human face, and its image that is captured in a moment.

CURRENT SOCIETAL ISSUES RESULTING FROM FRT USE

Having identified what FRT is and some primary ways of its use, consideration of the impact of that use in the context of security, privacy, and discrimination is necessary. Such use is constantly evolving as is its resultant impact. A discussion of some current issues follows.

FRT Use Raises Security Concerns

FRT can ensure security by authenticating the identity of an individual, but it can also implicate security concerns. Specifically, cyber security is an area of concern that results from FRT use. Breaches of information can have a wide-reaching impact. For example, a breach of BioStar 2 (a security platform for attendance and access) occurred in 2019 and impacted over 1,000,000 individuals (Symanovich, 2019). In 2020, the data of FRT software firm, Clearview AI was breached. The perpetrators obtained stored facial data along with other stored information (Kumar, 2021). In 2021, the surveillance camera provider, Verkada suffered a breach. Perpetrators accessed up to 4,530 of Verkada's security cameras and were able to view areas like hospital rooms, prison cells, police interrogation rooms, private homes, and gyms for up to 90 minutes. Additionally, 87 video archives were accessed (Ikeda, 2021; Verkada, 2022). Although the use of stolen facial data is unknown, perpetrators could have used such data to access the true user's secure mobile phones or ATMs and secure entrances such as buildings, conference rooms, or labs.

FRT Use Raises Privacy Implications

FRT can protect privacy interests, but it can also impede upon one's privacy. Certainly, FRT provides many privacy protections that benefit society daily. For example, one's home, car, smartphone, and computer contain personal information that the average person would like to keep secure, and FRT provides an easy way to do that. However, these convenient safety measures are offset by the privacy concern of FRT being used by private businesses without consent, without knowledge, or without freedom of choice. The ubiquitous nature of digital cameras means that images of the human face are frequently being captured, often without knowledge or consent (Van Noorden, 2020). The ease and accuracy with which FRT connects these captured images to individual identities create privacy concerns. Those privacy concerns implicate fundamental autonomy concerns (Smith & Miller, 2021). For example, if smartphone access universally requires facial recognition, little choice remains but to acquiesce, even if the consumer does not realize the full extent of a decision that may add the consumer's unique face to an unregulated database.

Another privacy concern that results from FRT use is artificially-created and/or altered facial images, otherwise known as "deep fakes." These can cause inaccuracy when FRT algorithms apply. Deep fakes and CGI or AI-altered faces have successfully been utilized in popular movies such as the Star Wars franchise, to recreate characters when their portrayers have died. However, deep fakes have also been utilized to accomplish troubling goals (Walsh, 2016). Advanced AI technology and FRT allow faces to be superimposed onto other bodies almost seamlessly. This practice may implicate people in a crime, destroy their public image, or influence the public in its perception of what someone has done or said. Commonly, deep fakes have been utilized to target celebrities by, for example, creating pornographic material that appears to involve the celebrity. More dangerously it has been used to create misleading or inaccurate footage of political candidates. This can have the collateral impact of threatening democracy and impacting elections, both in the present and in the future (Galston, 2022; Lee, 2018). Deep fakes are seemingly imperceptible from the real face, and as a result, they are able to fool current FRT algorithms. This opens the door for breaches in privacy, and the potential implication of people in crimes or ethical violations that they did not commit (Wiggers, 2021).

Popular social media apps that use FRT also implicate privacy concerns. For example, FaceApp is one of hundreds of apps that allows consumers to digitally alter their faces or add filters. While this app's popularity soared in 2019, with thousands of users posting copies of their digitally-aged faces to social media, most consumers were unaware that their photos and data were also uploaded to clouds and consumers were unable to delete them. While FaceApp denied any wrongdoing, the inability to have control over what happens to your information and facial image is certainly a privacy concern (Baraniuk, 2021; Fowler, 2020; O'Flaherty, 2020). Other companies have more openly stated their assumed ownership of what data and photos have been posted to social media, though it appears rare that consumers actually read through the terms of service. An example of such terms follows:

Twitter uses the following language in its Terms of Service:

By submitting, posting or displaying Content on or through the Services, you grant us a worldwide, non-exclusive, royalty-free license (with the right to sublicense) to use, copy, reproduce, process, adapt, modify, publish, transmit, display and distribute such Content in any and all media or distribution methods now known or later developed (for clarity, these rights include, for example, curating, transforming, and translating). This license authorizes us to make your Content available to the rest of the world and to let others do the same. (Twitter, 2022, "Your Rights and Grant of Rights in the Content" section)

Social media use may not be necessary, but its popularity in business, hiring processes, and as a means of communication make avoidance of all social media platforms difficult. That means that many people consent to their images, likenesses, and faces being utilized simply because they have no other choice. Even if an individual voluntarily consents to his/her likeness and data being used within a singular social media app, they may not consent to information being utilized by other companies or law enforcement, who then are able to obtain their personal information, and this is a current risk.

FRT Use Can Result in Discrimination

The benefit of FRT use is offset by the frequency with which its use results in discriminatory outcomes (Haber, 2021). Numerous vendors have created FRT algorithms, but the accuracy of matching faces within a database varies. When a match does not occur when it should have, the result is a false negative. When that happens, an individual may be excluded and denied access to an authorized area, such as the work place (Lwuozor, 2022). Conversely, when a match occurs that should not have occurred, a false positive results. Grother et al. (2019) found that false positives vary with a factor of 10 to 100+ times. Data has also shown that the misidentification is reflective of biases. For example, the misidentification is most likely to occur with women and within the Asian and African American populations (Grother et al; Lunter, 2020). Because of false positives, numerous individuals have been falsely accused and convicted of crimes. For this reason, rather than relying upon FRT for criminal accusations or convictions, Hoan Ton-That, founder of Clearview AI FRT software that is extensively used by law enforcement officers, recommends that FRT be used not as an investigative tool but as evidence ("Clearview AI's founder," 2020).

Not only is there the potential for discrimination based on flawed algorithms and/or error rates, FRT can be selectively used to target certain races. For example, the NYPD maintains a facial database of 99% Black and Latinx "gang affiliates" without cause for presuming gang affiliation (Najibi, 2020). The "Green Light Project" in Detroit employs cameras, almost exclusively in predominantly Black neighborhoods, to track, monitor, and scan the identity of residents (Hill, 2020; Najibi, 2020). Data sets that lack demographic diversity reinforce stereotypes and historical biases (Georgopoulos et al., 2021).

Some cities (i.e. Baltimore and Portland) have prioritized individual liberty concerns and banned the current use of FRT (Opilo, 2021; Wray, 2020). Because FRT is currently the least accurate biometric technology, governments such as these have chosen to rely on other forms of biometric technology, rather than risk the potential that someone else could be unintentionally implicated for a crime. Meanwhile, some states, Virginia among them, have instead opted to prioritize law enforcement by maintaining a statewide database for all residents and driver's license holders. This data is used alongside FRT to track criminal activity (Chen, 2020; Van Noorden, 2020). Either of those extreme examples presents its own unique privacy concerns, but the ethical concerns reach beyond privacy. Broad academic research supports the discriminatory impact of FRT on minorities. This discrimination enforces historical biases and inequalities (Zalnieriute, 2021). Although FRT has generally high overall accuracy, the 2018 Gender Shades Project indicated higher error rates in accurately identifying women (20.6%) than occurred when identifying men (8.6%). Additionally, significantly higher error rates exist when FRT is used to identify people of color (up to 19.2% error rate differentiation). This can produce error rates of 34.4% in comparing white male demographics with darker female demographics (Buolamwini & Gebrui, 2021).

ETHICAL THEORIES

Ethical theories aid individuals in making ethical choices and in understanding the decision-making processes that others use in the design or deployment of technologies. Seven ethical theories are important to examine. Specifically, considering FRT use through various ethical lenses is important. In order to do that, one must first have a basic understanding of various ethical theories. Having reached that understanding, a critical thinker should then consider particular FRT uses and their consequences through one or more of the various ethical lenses to determine ethical action.

Utilitarianism

The Principle of Utility or the Happiness Principle evaluates choices or actions based upon the consequence that produces happiness (Scarre, 2020). Utilitarianism is further expanded into Act Utilitarianism and Rule Utilitarianism. Act Utilitarianism simply considers the action that would produce the greatest good or happiness for the greatest number of people. As an example of Utilitarianism, Carzo (2010) discussed the value of FRT benefiting the greater good in the area of public safety as he includes the scenario of FRT being used to identify that a teenager wants to purchase alcohol. Consequently, the sales clerk does not sell alcohol which may impact many others who are now not negatively impacted by the teenager's intoxicated behavior. Additionally, when examining the overall convenience of room/building access as well as accuracy applications such as verifying patient medical records, FRT benefits the greater good (Katsanis et al., 2020). Rule Utilitarianism presumes that selecting an action that produces the greatest good for the greatest number of people must be aligned with moral principles or rules that also consider long-term, not just short-term, consequences (Mulgan, 2020).

Ethical Egoism

When faced with an ethical dilemma, an individual using Ethical Egoism would select the choice that most aligned with his/her own self-interest. Ethical egoism could be aligned with the use of FRT because an individual's personal good can be supported. Consider that with the use of FRT, an individual's self-interest is served by having the convenience of easy and quick access to entry points like rooms or offices (Katsanis et al., 2020). Additionally, an individual's good is promoted since personal safety is enhanced through the use of FRT monitoring for criminal activities (Chilson & Barkley, 2021).

Kantianism

Kantianism is based upon the beliefs of Immanuel Kant. According to his Principle of Universalizability, if an individual makes an ethical choice, then that choice can be made into a universal law (Kahn, 2014). For example, if an individual determines that it is acceptable to engage in lying, then lying can be a universal law. Kant's Imperative (CI) instructs individuals to avoid using others as a mere means to an end. As an example, Kantianism would regard FRT as a means to an end when peoples' faces are used to eventually group individuals into categories or to identify suspicious activities (Wang, 2018).

Rawls Theory of Rights

This theory is inspired by John Rawls' beliefs. He shares that first, an ethical decision is based upon the belief that all individuals have access to basic rights such as freedom of speech, freedom to vote, and freedom to think. Second, the ethical decision must be supportive of an outcome that is fair and equitable to others. For example, all individuals should be paid fairly and have the same opportunity to access high-paying jobs (McCartney & Parent, n.d.). Third, according to Rawl's Difference Principle, inequalities can exist internationally, as long as the outcome of such inequalities benefits the least advantaged individuals (Zha, 2012). Laws, such as the California Consumer Privacy Act, that require employers to disclose information collected using FRT and allow employees to access and or delete that information find support in this ethical theory (California Consumer Privacy Act, 2018). An employee's right to privacy is protected and a fair and equitable outcome is ensured. Conversely, consider the examples of FRT discrimination when individuals are denied access or wrongfully accused of a crime. In such instances, all individuals are not treated fairly and as a result of such inequality, a person experiencing discrimination is not at an advantage (McCarthy & Parent, n.d.; Zha, 2012).

Virtue Ethics

Under this theory, an individual making an ethical decision would contemplate what a virtuous person would do using virtues such as honesty, compassion, integrity, courage, fairness, loyalty, and self-control (Velasquez et al.,1988). The individual is then less focused on making a choice based on rules. As an individual practices using different virtues in decision-making, the individual inherently becomes more ethical. Additionally, the decision-maker is influenced by an active community (i.e. the employer or a professional organization) and supports the virtues. Last, as the decision-maker utilizes virtues, they disseminate such virtues back into the active community (Dobson, 2007). The degree of a particular virtue varies by individual such that the use of FRT may or may not be ethical. For example, consider an individual who is responsible for FRT monitoring. If this individual has biases against different races and the individual notices loitering among those different races, the monitoring individual may act in a manner that is not fair by staying focused on viewing that group of people (Carzo, 2010).

Divine Command Theory

This theory explains that an individual making an ethical decision would consider which option is aligned with being obedient to God as well as which option is most aligned with the principles in the holy book of God. As an example, individuals utilizing Divine Command Theory would not steal from their employer. This theory is subjective in its interpretation of God's rule and the associated holy book. Additionally, even with a selected holy book like the Bible, multiple editions exist, such as the Catholic Bible, the King James Version, the New International Version, or the English Version, each of which may contain different interpretations of actions aligned with God. Also, making a decision based upon Divine Command Theory may be difficult to justify in some workplaces that do not have a belief in God. Last, individuals may reject the law to favor making a decision aligned with the Divine Command theory as is the case of police offers who do not follow a court order to remove protestors in front of an abortion clinic (McCartney & Parent, n.d.).

The Bible has several verses that frown upon the threat of individual privacy or discrimination associated with FRT. In Matthew 7:12 individuals are instructed "So in everything, do to others what you would have them do to you" (Bible Gateway, n.d.a, para. 1). In Matthew 22:39, individuals are instructed to "Love your neighbor as yourself" (Bible Gateway, n.d.b, para. 1). Using the Divine Command Theory, if an individual would not want their privacy violated or to be discriminated against due to the use of FRT, then FRT should not be used to violate the privacy of others. Additionally, FRT would not be used if discrimination occurred or privacy were compromised because that would detract from loving neighbors (Carzo, 2010).

Cultural Relativism

An individual acting in accord with cultural relativism makes an ethical decision based upon his/her culture's beliefs. This decision may be viewed as morally acceptable in some societies, while viewed as morally wrong in other societies. Some examples in which the morality of actions varies based on cultures include bribery, political repression, child labor, or gender inequality (Velasquez et al., 1992). This ethical theory does not encourage diversity of thinking to explore options and can lead to a stalemate in decision making if individuals from different cultures must make a cohesive decision. Furthermore, ethical decision-making can be complicated when a company has international locations that have different moral beliefs. Last, cultural relativism has considered several decisions to be morally correct while much of society believed the decision to be extremely wrong or inhumane, as was the case with the Holocaust, slavery, and war crimes (Bostian, 2005).

The perceived ethical use of FRT varies by country. The expanded government's use of FRT in other countries such as China would not be ethically acceptable in the U.S. For example, China has utilized FRT to enforce its view of civilized behavior. It has captured individuals who wear pajamas in public and then displayed the pictures of the names and IDs of those individuals to shame the individuals to adopt more civilized behavior ("Pyjamas in public," 2020). In another example, China has utilized FRT along with AI to reveal an individual's loyalty to the government (Eliot, 2022).

Table 1. Foundational differences between ethics and law

Ethics	Law
Highly individualized, personal, and variable	More static in nature and representative of the whole as opposed to the individual
Varying in application	Uniform in application
Unenforceable	Enforceable
Lacks uniform and meaningful consequences	Characterized by uniform and meaningful consequences

ANALYZING THE ETHICAL AND LEGAL IMPACT OF FRT USE

FRT provides societal benefits while simultaneously testing individual freedoms and societal ethics. In the United States, no current comprehensive federal regulations govern the various uses of FRT, and state laws vary in substance and effect (Haddad, 2021). This same pattern is reflected globally (Kouroupis, 2021). As the use of FRT by both private companies and law enforcement evolves and increases, society is able to watch ethical issues play out in real-time. Societal ethics ultimately impact the direction and scope of legislation and the formulation of law. For example, United States citizens, as members of a democratized society, exert their ethical beliefs by voting to elect lawmakers with similar priorities and petitioning representatives for reforms that reflect their ethical priorities. Differentiating between ethics and law is necessary in order to contemplate ways in which each broad concept can be used to drive fair and equitable use of FRT in a manner that supports a sustainable society. The solution to resolving the various types of ethical issues associated with FRT that have been reviewed is found in informed discussion and development of robust and uniform laws that govern the field.

When considering law and ethics in the context of FRT, it is important to reflect on the foundational differences between law and ethics. Some key differences between ethics and law are set forth in the following table.

Since societal ethics drive the development of law, ethical action is also legal action in many instances. However, remembering that ethics and law are not always in accord is important. Drawing on an earlier example, a liquor store owner who prioritizes utilitarianism might determine that using FRT is the best way to promote public safety since it will allow him to identify minors and refuse to sell alcohol to them. In turn, this will protect the public from the intoxicated behavior of a minor. The store owner's personal ethics may direct him to sacrifice employee and customer privacy in order to serve the greater good of eliminating the negative impact of drunken behavior on society. The law, on the other hand, might prioritize privacy and prohibit the use of FRT in such a setting. In that instance, the business owner might be acting ethically while simultaneously acting illegally.

Having established this foundational groundwork, a comprehensive review of current ethical and legal considerations is necessary. This review and analysis focus on the ethical and legal approaches to FRT in the United States but also considers some examples of global impact. The analysis that follows is instructive to those who will frame the future use of FRT.

Ethical Issues Framing the Development of Laws Applicable to FRT Use

In many respects, FRT makes daily lives easier and more secure. Those clear ethical benefits of FRT are offset by false negatives, privacy, and bias concerns (Chilson & Barkley, 2021). Informed discussion of benefits and detriments needs to drive legal policies and user practices. When FRT issues occur, they can produce anything from minor inconveniences, like the inability to log into work computers, to major civil liberty concerns, like being falsely arrested as a result of FRT misidentification. Individuals and businesses who use (or are subjected to) FRT should be aware of the ethical concerns associated with FRT so their usage can be guided by ethics and ensure the most likely chance of compliance with developing laws. Further, awareness of ethical concerns is imperative so legal regulations are effectively created and enforced in a relevant manner. The various specific issues that have been examined can be grouped into two (2) broad categories: (1) privacy concerns; and (2) bias/discrimination concerns.

Having reviewed various pros and cons of FRT use and explored specific ethical constructs, the next step is to consider an overview of the broad application of ethics to FRT and how this application reflects societal norms.

Privacy Concerns

Individual privacy and autonomy are highly valued ethical principles in the United States and other democracies (Smith & Miller, 2021). National privacy rights are fundamental and grounded in the United States Constitution. This is contrasted by the ethical priorities of countries such as Russia or China, where FRT is legal and its use is widely accepted as a part of daily life (Kouroupis, 2021). Societies that value individual privacy rights may also prioritize security, community safety, and convenience (Smith & Miller, 2021). These ethical priorities often find themselves in direct opposition when it comes to the use of FRT. Society as a whole is determining if it is comfortable with the use and potential compilation of highly individualized biometric data in return for safety and convenience. At present, societal response to that question is that people are largely comfortable with it in some instances but remain very divided in other instances. This is the current collective stance, both regarding the use of FRT by private companies and government entities. The ethical and legal tasks that lay ahead of us involve striking a balance between privacy and security considerations in both the public and private sector (Smith & Miller, 2021).

When considering how to reach that balance, it is important to consider real-life scenarios, like those already studied above. It is also helpful to recognize some basic realities of the legal and ethical history of various societal communities. While privacy is a right that is highly valued in the United States, it has never been an absolute right. For example, an individual most certainly has a right to privacy in their individual home. If a woman wishes to keep her home and its contents secreted away from the public, then she may do so. However, when probable cause to support a search warrant of that home for illegal drug production exists, then that individual right to privacy yields. As a society, people accept and welcome that limitation on individual privacy because it contributes to greater community safety. That safety comes at a cost to individualized privacy, but legal safeguards are in place that limit that intrusion and society seems comfortable with that balance. To that end and in that context, United States jurisprudence has found the balancing point between privacy and safety.

Finding a comparable balancing point when it comes to the use of FRT will evolve gradually as informed discussion evolves and laws develop and are applied. Some broad concepts to consider will involve consent, usage, and storage of the biometric data gathered through FRT. The European Union provides an example of how such a balance might be accomplished. Its legislation conditions the use of FRT upon the existence of a documented, imminent, and substantial risk (Kouroupis, 2021). That precondition has the result of eliminating the use of FRT by private companies in broad and uncontrolled settings (Kouroupis, 2021).

Bias and Discrimination Concerns

In addition to privacy concerns, FRT use also implicates significant bias and discrimination concerns. Broad academic research establishes the discriminatory impact of FRT on minorities. This discrimination enforces historical biases and inequalities (Zalnieriute, 2021). When FRT produces inaccurate identifying information, it can result in wrongful criminal accusations (Brey, 2004). Society values the right to be free from unwarranted criminal charges. The concept that biometric data gathered by FRT could support a criminal charge against an innocent person raises significant ethical concerns. Imagine the difficulty in refuting a criminal charge that was supported by biometric data that was wrongfully identified. This basic ethical concern is compounded by the reality that FRT exhibits higher error rates in accurately identifying women and people of color (Buolamwin & Gebrui, 2021). In the United States, the national ethical stance regarding the equality of women and people of color has been slow to evolve, but that equality is now prioritized in society. The disproportionate error rate when FRT is applied to these groups results in a step backward from the work that societal ethics have driven.

Legal Developments That Have Resulted From Societal Ethics

A discussion about the ethical implications of FRT cannot be free-standing. By its nature, it must also include consideration of law. Ethics are highly personal in nature, fluid in their evolution, and unenforceable by society. One individual's ethics may vary greatly from the next. Ethics certainly differ between and among nations. Even when an ethical majority exists, ethics do not allow for a procedure for enforcement and consequences. The law, however, provides this vehicle for enforcement and consequences. Ethics inform the creation of the law and rely on that law to compel individuals into action. This analysis focuses on the development of laws in the United State while also acknowledging legal trends in other nations.

In the United States, the law in the area of privacy and discrimination is well-developed and comprehensive. However, deficiencies in addressing issues stemming from the use of technology are present. This does not suggest that the law is incapable of addressing these issues. Rather, it is an acknowledgment that the development of law is largely reactive and therefore requires time to address the privacy and discrimination issues that have resulted from the use of FRT. As the general public has become more aware of the ethical concerns associated with FRT, the legal process has accelerated. This is particularly the case when it comes to government use of FRT. While the evolution of the law in regard to private use of FRT may be slower in progression, there is progression, nonetheless.

The European Union similarly values privacy and its legal framework has been guided accordingly. FRT is used in the European Union, but its use is regulated. For example, France uses cameras and video surveillance to monitor compliance with COVID 19 regulations, but the images captured are neither transmitted nor stored (Kouroupis, 2021). Juxtapose this with practices in Russia, which likewise uses FRT for COVID 19 enforcement but without restriction on transmission or storage of images (Kouroupis, 2021. Outcomes and societal ethics may differ between and among the United States, the countries in the European Union, and Russia but the process that ultimately achieves legal policies in each instance is similar in that it is framed by the ethics of the particular country.

Returning the focus to the United States, a plurality of Americans thinks that widespread use of FRT by police is a good idea (Rainie et al., 2022). This indicates a societal ethical stance that supports FRT use and will color future laws. However, generally speaking, individuals who have heard a lot or read a lot about FRT tend to conclude that the use of FRT by police is a bad idea (Rainie et al., 2022). This tends to indicate that society's ethical stance may shift as the public knowledge base grows. Additionally, 57% of Americans opine that crime in the United States would stay about the same even if police used FRT (Rainie et al., 2022). Opinions vary based on political affiliation and race and up to 64% of Americans are in favor of police use of FRT as long as that use is accompanied by relevant training (Rainie et al., 2022). In sum, the ethical positions of at least a plurality of Americans support the use of FRT by law enforcement, although that support is tempered by substantial support for limitations on that use of FRT. Interestingly, this position might result in a legal scenario similar to China's Red-Light Record System, which uses traffic cameras to identify and fine e-bike users who violate traffic laws. This system eliminates the need for an actual interface with traffic police and has been highly successful in reducing traffic violation rates (Yang et al., 2022). While outcomes are unclear, it is certain that societal ethics will certainly drive future laws. Further, the variances between and among races, cultures, and political parties indicate that it may be a long road to achieve comprehensive federal regulation over the use of FRT by police.

While many Americans are willing to sacrifice their privacy rights in return for what they perceive to be a safer society and better policing, they are not as willing to sacrifice their privacy in the interest of advancement of private industry. At this time, 57% of Americans are opposed to the use of FRT by social media outlets to automatically identify people in photos. However, approximately 50% of Americans support the use of FRT by retailers to confirm the identity of credit card users and by building owners to monitor security issues (Rainie et al., 2022).

In light of these varying ethical priorities, it should be no surprise that the current laws in the United States that apply to FRT are underdeveloped and inconsistent. In the absence of comprehensive federal regulation, cities and states have been enacting their own laws pertaining to the use of FRT. In 2021, Virginia passed one of the most restrictive bans in the country on the use of FRT. The ban barred local police and campus security from even purchasing FRT unless such purchase was explicitly authorized by the Virginia Legislature. Notably, this ban did not apply to the State Police, which continued to compare unknown images of people to an existing database of mugshots of individuals who were previously arrested. To that end, even Virginia's restrictive ban had many loopholes. Eight months after the ban, Virginia changed its stance and approved a bill to lift the ban. The new legislation will allow law enforcement to use FRT to identify a person upon reasonable suspicion that the person committed a crime. It also allows FRT to be used to identify crime victims and witnesses, as well as victims of sex trafficking (Lavoie, 2022). The developing law in Virginia provides a good example for the dynamic and changing direction that future laws may take. In late 2021, Maine enacted what is considered to be one of the strictest state laws regulating the use of FRT by the government. In effect, the law prohibits government use of FRT unless a specific exception applies. Notably, one of these exceptions is the existence of probable cause that an unidentified person has committed a serious crime (Act to Increase Privacy and Security by Regulating the Use of Facial Surveillance Systems by Departments, Public Employees and Public Officials, 2021). This carve-out follows the historical jurisprudence approach that requires warrants based on probable cause to conduct a search of one's home and is consistent with Virginia's revised approach.

California, Illinois, and Texas are among the few states that have also taken action to regulate the use of facial recognition by private actors (Keener, 2022). The California Consumer Privacy Act requires employers to disclose the type of information, including biometric information, that is collected from employees and why this data is collected. Employees also have rights to request access to and/or deletion of this data. Further, employers who allow this data to be compromised can be subject to penalties. Similar provisions apply to the collection of biometric information from consumers (California Consumer Privacy Act, 2018). Illinois has likewise adopted laws requiring notice requirements when biometric data is gathered by private entities (Biometric Information Privacy Act, 2018). Texas' "Capture or Use of Biometric Identifier Act" limits the use of biometric data and imposes financial penalties for violations (Keener, 2022). In February 2022, the Texas Attorney General used this law as the basis for a legal action against Facebook, which allegedly collected biometric data of more than 20 million Texans without first obtaining consent (Keener, 2022).

Local laws also must be considered. In the absence of clear federal and/or state guidance, cities and municipalities may take their own legal action. By way of example, Portland, Oregon prohibits private entities from using FRT in public places unless an exception applies (Kouroupis, 2021). This patchwork framing of laws is not unique to the United States. In the absence of a unified legal framework for FRT use in traffic safety, cities across China have enacted and enforced their own standards (Yang et al., 2022).

Just as state and local laws are constantly adapting in reaction to the use of FRT, so too are federal laws. Comprehensive federal regulation of the use and storage of biometric information does not yet exist, but bills addressing the use and storage of such information in specific instances are being pursued. For example, the No Biometric Barriers to Housing Act was introduced in 2019 in an effort to prohibit the use of biometric data in certain federally assisted housing units (No Biometric Barriers to Housing Act, 2019). This bill particularly addressed the bias, discrimination, and civil rights concerns of members of vulnerable communities. The United States Government Accountability Office (2021) has recommended policies and regulations to be adopted by thirteen (13) government agencies that use FRT (Facial recognition technology: Federal law enforcement agencies should better assess privacy and other risks). Federal legislators have also introduced a bill specifically addressing consumer accessibility to notice and consent when tech companies use and/or store biometric data. If passed, this bill would require online companies to make their terms-of-service more accessible and understandable to consumers (Terms-of-Service, Labeling, Design, and Readability Act, 2022).

FUTURE RESEARCH CONSIDERATIONS RELATED TO FRT

FRT and its uses are constantly evolving and developing. Likewise, society's ethical priorities do not remain constant. Finally, the law develops in a reactive manner that responds to the developments in technology and fluctuations in societal ethics. To that end, there remains much work to be accomplished in considering and surveying the ethical and legal impact and responses to FRT.

Future research could survey the scope and breadth of relevant laws at the state level and explore connections between those state laws and the ethical tendencies of each state. A similar study at the federal level and across federal agencies would also be valuable. That data could also be tracked to political priorities. Further study examining the reasons behind societal ethical responses to FRT use would also be valuable. Specifically, a case study based on a singular instance of use/misuse of FRT resulting in a denial of civil liberties, would be particularly enlightening.

Companies currently responding to the results the Gender Shades study of 2018, such as IBM, Microsoft and Face+, acknowledged that additional inclusive testing to fix the current "biased" algorithmic errors would be necessary in order to truly roll out a product that could be considered reliable in both private and public settings (Buolamwini & Gebru, 2018). Research chronicling the progress of such testing would provide guidance as to the state of reliability of existing technology.

In addition to future research related to sociology and/or law, one might explore FRT developments and catalog technology changes/advancements that result in a more reliable end product. If more reliable technology is created, then a study of expanding and future usages of the technology might be explored.

FRT and its regulation is a dynamic area of study. As more data relative to FRT use is gathered, the areas of applicable research will expand. Some particularly impactful and interesting areas of growing technology include the automotive industry. As an advancement to Smart Keys, the Genesis Face Connect Systems uses a Near-Infrared camera to identify the driver's face. Such technology permits the driver to open doors and load personalized adjustments (i.e. side mirrors, seating and Infotainment settings). Additionally, a fingerprint authentication system then permits the driver to start the car. Owners are permitted to set up FRT for two drivers (Choksey, 2021). FRT not only offers convenience to drivers, it also offers protection. For example, in 2020, the cost of automobile theft was $6 billion. If FRT were used, the cost of automobile theft would be reduced (Hye-ran & Tae-gyu, 2021).

Based on the reduction in theft, future research could involve insurance companies to discover if they will adjust policy prices based on drivers having Face Connect Systems. Additionally, insurers could be queried to explore their likelihood and rationale for adopting FRT to determine if policyholders submitting automobile claims are truthful. Similar work is currently being undertaken by automobile insurers in China ("Facial Recognition," 2020).

Additionally, future research with consumers could determine their interest and/or willingness to pay extra for FRT, and whether theft-reduction or convenience most influence their decision. Also, focus groups can explore customers' concerns around Face Connect Systems such as possible insurance discounts, the ability to change those who registered, if the car can open in an alternative way (i.e. owner loans out the car).

As these future areas of research are developed, findings can be considered in light of societal ethics to help shape and support future laws. For example, as use of FRT to regulate compliance with COVID restrictions expands, the societal ethics of individual communities will inform the development of laws that define and potentially limit that use.

CONCLUSION

FRT use and its regulation are constantly expanding and developing. FRT impacts endless aspects of daily life. For that reason, it is important to both understand what FRT is and the types of ethical issues that commonly arise when FRT is used, both in the private and public sectors. A fundamental understanding of basic ethical theories is needed to properly assess, analyze, and potentially resolve these ethical issues. Consideration of societal ethics is also crucial in tracking and predicting the future regulation and use of FRT which will continue to be a part of people's lives, workplaces, and jurisprudence. Informed discussion of the foundational ethical issues associated with FRT is crucial to ensuring responsible and ethical use of FRT in the future.

REFERENCES

Act to Increase Privacy and Security by Regulating the Use of Facial Surveillance Systems by Departments, Public Employees and Public Officials, Publ. L. No. 2021, c. 394, §1, 25 MRSA Pt. 14 (2021). https://legislature.maine.gov/legis/statutes/25/title25sec6001.html

AirSlate Legal Forms, Inc. (n.d.). *State law and legal definition.* U.S. Legal. https://definitions.uslegal.com/s/state-law

Baraniuk, C. (2019, July 17). Can you trust FaceApp with your face? *BBC News.* https://www.bbc.com/news/technology-49018103

Bible Gateway. (n.d.a). *Passage lookup: Matthew 7:12.* https://www.biblegateway.com/passage/?search=Matthew+7%3A12%2C&version=NIV

Bible Gateway. (n.d.b). *Passage lookup: Matthew 22:39.* https://www.biblegateway.com/passage/?search=Matthew+22%3A39&version=NIV/

Biometric Information Privacy Act, Ill Comp. Stat. Ann., ch. 740, § 14/1 *et seq.* (2018). https://www.akingump.com/a/web/101105/Biometric-Information-Privacy-Act-740-ILCS-14-1-et-seq.pdf

Bostian, I. L. (2005). Cultural relativism in international war crimes prosecutions: The international criminal tribunal for Rwanda. *ILSA Journal of International & Comparative Law*, *12*(1), 1.

Brey, P. A. E. (2004). Ethical aspects of facial recognition systems in public spaces. *Information. Communication & Ethics in Society*, *2*(2), 97–109. doi:10.1108/14779960480000246

Buolamwini, J., & Gebru, T. (2018, February 9). *Gender shades: Intersectional accuracy disparities in commercial gender classification.* Gender Shades Project. http://proceedings.mlr.press/v81/buolamwini18a/buolamwini18a.pdf

California Consumer Privacy Act, Cal. Civ. Code, Title 1.81.5, ch. 55, § 3 (2018). https://leginfo.legislature.ca.gov/faces/codes_displayText.xhtml?division=3.&part=4.&lawCode=CIV&title=1.81.5

Cambridge University Press. (n.d.a). Algorithm. In *Cambridge English Dictionary online.* https://dictionary.cambridge.org/us/dictionary/english/algorithm

Cambridge University Press. (n.d.b). Privacy. In *Cambridge English Dictionary online.* https://dictionary.cambridge.org/us/dictionary/english/privacy

Carzo, R. (2010). Under the watchful eye: The highly intrusive nature of facial recognition technology. *The Review: A Journal of Undergraduate Student Research, 12*(2010), 1-5.

Chen, Y. (2020). Your face is commodity, fiercely contract accordingly: Regulating the capitalization of facial recognition technology through contract law. *Notre Dame Journal of Law, Ethics & Public Policy*, *34*(2), 501–528.

Chilson, N., & Barkley, T. (2021, December). The two faces of facial recognition technology. *IEEE Technology and Society Magazine*, *40*(4), 87–99. doi:10.1109/MTS.2021.3123752

Choksey, J. S. (2021, October 4). *What is Genesis Face Connect and how does it work?* J. D. Power. https://www.jdpower.com/cars/shopping-guides/what-is-genesis-face-connect-and-how-does-it-work

Clearview AI's founder Hoan Ton-That speaks out [*Extended interview*]. (2020, March 6). [Video file]. YouTube. https://www.youtube.com/watch?v=q-1bR3P9RAw

Dobson, J. (2007, November 12). Applying virtue ethics to business. The agent-based approach. *Electronic Journal of Business Ethics and Organizational Studies, 12*(2). http://ejbo.jyu.fi/articles/0901_3.html

Eliot, L. (2022, July 5). AI ethics perturbed by latest China devised AI party-loyalty mind-reading facial recognition attestation that might foreshadow oppressive autonomous systems. *Forbes*. https://www.forbes.com/sites/lanceeliot/2022/07/05/ai-ethics-perturbed-by-latest-china-devised-ai-party-loyalty-mind-reading-facial-recognition-attestation-that-might-foreshadow-oppressive-autonomous-systems

Facial recognition in insurance. (2020, December 16). HDI. https://www.hdi.global/en-za/infocenter/insights/2020/facial-recognition/

Fowler, G. A. (2020, June 18). You downloaded FaceApp. Here's what you've just done to your privacy. *The Washington Pos*t. https://www.washingtonpost.com/technology/2019/07/17/you-downloaded-face-app-heres-what-youve-just-done-your-privacy

Galston, W. A. (2022, March 9). *Is seeing still believing? The deepfake challenge to truth in Politics*. Brookings. https://www.brookings.edu/research/is-seeing-still-believing-the-deepfake-challenge-to-truth-in-politics

Georgopoulos, M., Oldfield, J., Nicolaou, M. A., Yannis, P., & Pantic, M. (2021, May 15). Mitigating demographic bias in facial datasets with style-based multi-attribute transfer. *International Journal of Computer Vision*, *129*(7), 2288–2307. doi:10.100711263-021-01448-w

Grother, P., Ngan, M., & Hanaoka, K. (2019, December). *Face recognition vendor test (FRVT) Part 3: Demographic effects*. NIST. https://nvlpubs.nist.gov/nistpubs/ir/2019/NIST.IR.8280.pdf

Haber, E. (2021, October). Racial recognition. *Cardozo Law Review*, *43*(1), 71–134. http://cardozolaw-review.com/wp-content/uploads/2021/12/2_Haber.43.1.1.pdf

Haddad, G. M. (2021). Confronting the biased algorithm: The danger of admitting facial technology results in the courtroom. *Vanderbilt Journal of Entertainment and Technology Law*, *23*(4), 891–918.

Hye-ran, K., & Tae-gyu, K. (2021, September 17). *Hyundai to feature facial recognition in electric vehicles*. UPI. https://www.upi.com/Top_News/World-News/2021/09/17/Hyundai-facial-recognition-Genesis/5491631899253

Ikeda, S. (2021, March 15). *Verkada data breach exposes feeds of 150,000 security cameras; targets include health care facilities, schools, police stations and a Tesla plant*. CPO. https://www.cpomagazine.com/cyber-security/verkada-data-breach-exposes-feeds-of-150000-security-cameras-targets-include-health-care-facilities-schools-police-stations-and-a-tesla-plant

Ivanova, E., & Borzunov, G. (2020). Optimization of machine learning algorithm of emotion recognition in terms of human facial expressions. *Procedia Computer Science*, *169*, 244–248. doi:10.1016/j.procs.2020.02.143

Johansen, A. G. (2019, February 8). *Biometrics and biometric data: What is it and is it secure?* Norton. https://us.norton.com/internetsecurity-iot-biometrics-how-do-they-work-are-they-safe.html

Kahn, S. (2014, January 31). Can positive duties be derived from Kant's formula of universal law? *Kantian Review*, *19*(1), 93–108. doi:10.1017/S1369415413000319

Kaspersky. (n.d.). *What is facial recognition: definition and explanation.* https://www.kaspersky.com/resource-center/definitions/what-is-facial-recognition

Katsanis, S. H., Claes, P., Doerr, M., Cook-Deegan, R., Tenenbaum, J. D., Evans, B. J., Lee, M. K., Anderton, J., Weinberg, S. M., & Wagner, J. K. (2021, October 14). A survey of U.S. public perspectives on facial recognition technology and facial imaging data practices in health and research contexts. *PLoS One*, *16*(10), e0257923. doi:10.1371/journal.pone.0257923 PMID:34648520

Keener, E. B. (2022, April 25). Facial recognition: A new trend in state regulation. *Business Law Today.* https://businesslawtoday.org/2022/04/facial-recognition-new-trend-state-regulation

Kostka, G., Steinacker, L., & Meckel, M. (2021, March). Between security and convenience: Facial recognition technology in the eyes of citizens in China, Germany, the United Kingdom, and the United States. *Public Understanding of Science (Bristol, England)*, *30*(6), 671–690. doi:10.1177/09636625211001555 PMID:33769157

Kouroupis, K. (2021). Facial recognition: A challenge for Europe or a threat to human rights? *European Journal of Privacy Law & Technologies*, *2021*(1), 142–156. https://universitypress.unisob.na.it/ojs/index.php/ejplt/article/view/1265/667

Kumar, V. (2021, March 11). *What will happen when a facial recognition firm is hacked?* Analytics Insight. https://www.analyticsinsight.net/what-will-happen-when-a-facial-recognition-firm-is-hacked

Lavoie, D. (2022, March 10). Virginia lawmakers ok lifting ban on facial technology use. *Associated Press*. https://www.msn.com/en-us/news/us/virginia-lawmakers-ok-lifting-ban-on-facial-technology-use/ar-AAUU3rK

Law Insider, Inc. (n.d.). *Federal or state law definition.* https://www.lawinsider.com/dictionary/federal-or-state-law

Lee, D. (2018, February 3). Deepfakes porn has serious consequences. *BBC News.* https://www.bbc.com/news/technology-42912529

Lunter, J. (2020, October). Beating the bias in facial recognition technology. *Biometric Technology Today*, *2020*(9), 5–7. doi:10.1016/S0969-4765(20)30122-3

Lwuozor, J. (2022, February 12). *Best facial recognition software for enterprises in 2022.* eSecurity Planet. https://www.esecurityplanet.com/products/facial-recognition-soft

Martinez, A. M. (2009). Face recognition, overview. In S. Z. Li & A. Jain (Eds.), *Encyclopedia of Biometrics*. Springer. doi:10.1007/978-0-387-73003-5_84

McCartney, S., & Parent, R. (n.d.) *Ethics in law enforcement*. Pressbooks. https://ecampusontario.pressbooks.pub/ethicslawenforcement

Merriam-Webster. (n.d.a). Bias. In *Merriam-Webster.com dictionary*. https://www.merriam-webster.com/dictionary/bias

Merriam-Webster. (n.d.b). Deepfake. In *Merriam-Webster.com dictionary*. https://www.merriam-webster.com/dictionary/deepfake

Merriam-Webster. (n.d.c). Discrimination. In *Merriam-Webster.com dictionary*. https://www.merriam-webster.com/dictionary/discrimination

Mulgan, T. (2020). *Utilitarianism*. Cambridge University Press. doi:10.1017/9781108582643

Najibi, A. (2020, October 26). *Racial discrimination in face recognition technology*. Science in the News. https://sitn.hms.harvard.edu/flash/2020/racial-discrimination-in-face-recognition-technology

No Biometric Barriers to Housing Act H.R. 4008, 116th Cong. (2019). https://www.congress.gov/bill/116th-congress/house-bill/4008

O'Flaherty, K. (2020, June 19). FaceApp privacy: What you need to know about the viral Russian app. *Forbes*. https://www.forbes.com/sites/kateoflahertyuk/2020/06/19/faceapp-privacy-what-you-need-to-know-about-the-viral-russian-app

Opilo, E. (2021, June 14). Baltimore city council approves moratorium on facial recognition technology; City police exempt from ban. *The Baltimore Sun*. https://www.baltimoresun.com/politics/bs-md-ci-baltimore-council-facial-recognition-20210614-xbooqalr6be7zhzljcnpeb3cqm-story.html

Pyjamas in public. Chinese city apologises for 'shaming' residents. (2020, January 21). *British Broadcasting Corporation (BBC)*. https://www.bbc.com/news/world-asia-china-51188669

Rainie, L., Funk, C., Anderson, M., & Tyson, A. (2022, March 17). *AI and human enhancements: Americans' openness is tempered by a range of concerns*. Pew Research Center. https://www.pewresearch.org/internet/2022/03/17/public-more-likely-to-see-facial-recognition-use-by-police-as-good-rather-than-bad-for-society

Roussi, A. (2020, November 18). Resisting the rise of facial recognition. *Nature*, *587*(7834), 350–353. doi:10.1038/d41586-020-03188-2 PMID:33208966

Scarre, G. (2020). *Utilitarianism*. Routledge. doi:10.4324/9781003070962

Smith, M., & Miller, S. (2021, April 13). The ethical application of biometric facial recognition technology. *AI & Society*, *37*(1), 167–175. doi:10.100700146-021-01199-9 PMID:33867693

Summa Linguae Technologies. (2021, October 13). *Facing the future: Innovative uses of facial recognition*. https://summalinguae.com/language-technology/facial-recognition-uses/

Symanovich, S. (2019, August 18). *Biometric data breach: Database exposes fingerprints, facial recognition data of 1 million people.* Norton. https://us.norton.com/internetsecurity-emerging-threats-biometric-data-breach-database-exposes-fingerprints-and-facial-recognition-data.html

Terms-of-Service Labeling, Design, and Readability Act, S.B. 3501, 117th Cong. (2022). https://trahan.house.gov/uploadedfiles/tldr_act.pdf

Twitter. (2022, June 10). *Twitter Terms of Service.* https://twitter.com/en/tos

United States Government Accountability Office. (2021, June). *Facial recognition technology: Federal law enforcement agencies should better assess privacy and other risks, document GAO-21-518.* https://www.gao.gov/assets/gao-21-518.pdf

Van Noorden, R. (2020, November 18). The ethical questions that haunt facial-recognition research. *Nature, 587*(7834), 354–358. doi:10.1038/d41586-020-03187-3 PMID:33208967

Velasquez, M., Andre, C., Shanks, T. J. S., & Meyer, M. J. (1988, January 1). *Ethics and Virtue.* Markulla Center for Applied Ethics. https://www.scu.edu/ethics/ethics-resources/ethical-decision-making/ethics-and-virtue

Velasquez, M., Andre, C., Shanks, T. J. S., & Meyer, M. J. (1992, August 1). *Ethical relativism.* Markkula Center for Applied Ethics. https://www.scu.edu/ethics/ethics-resources/ethical-decision-making/ethical-relativism

Verkada. (2022). *Summary: March 9, 2021 security incident report.* https://www.verkada.com/security-update/report

Walsh, J. (2016, December 16). Rogue One: The CGI resurrection of Peter Cushing is thrilling – but is it right? *The Guardian.* https://www.theguardian.com/film/filmblog/2016/dec/16/rogue-one-star-wars-cgi-resurrection-peter-cushing

Wang, J. (2018). *What's in your face? Discrimination in facial recognition technology* [Master's thesis, Georgetown University]. https://repository.library.georgetown.edu/bitstream/handle/10822/1050752/Wang_georgetown_0076M_14043.pdf

Wiggers, K. (2021, March 5). *Study warns deepfakes can fool facial recognition.* VentureBeat. https://venturebeat.com/2021/03/05/study-warns-deepfakes-can-fool-facial-recognition

Wray, S. (2020, September 17). *Portland bans private companies from using facial recognition technology.* Cities Today. https://cities-today.com/portland-bans-private-companies-from-using-facial-recognition-technology

Yang, Y., Yin, D., Easa, S. M., & Liu, J. (2022). Attitudes toward applying facial recognition technology for red-light running by e-bikers: A case study in Fuzhou, China. *Applied Sciences (Basel, Switzerland), 12*(211), 211. doi:10.3390/app12010211

Zalnieriute, M. (2021, September 1). Burning bridges: The automated facial recognition technology and public space surveillance in the modern state. *The Columbia Science and Technology Law Review, 22*(2), 284–307. doi:10.52214tlr.v22i2.8666

Zha, W. (2012, September 23). *The difference principle: Inconsistent in Rawlsian theory?* E-International Relations. https://www.e-ir.info/2012/09/23/the-difference-principle-inconsistency-in-rawlsian-theory/

ADDITIONAL READING

Castelvecchi, D. (2020, November 18). Is facial recognition too biased to be let loose? *Nature, 587*(7834), 347–349. doi:10.1038/d41586-020-03186-4 PMID:33208976

Chochia, A., & Nässi, T. (2021, December). Ethics and emerging technologies – facial recognition. *IDP: Revista De Internet. Derecho y Política*, (34), 1–12. doi:10.7238/idp.v0i34.387466

Facial-recognition research needs an ethical reckoning. (2020, November 18). *Nature (London), 587*(7834), 330. doi:10.1038/d41586-020-03256-7

Hirose, M. (2017). Privacy in public spaces: The reasonable expectation of privacy against the dragnet use of facial recognition technology. *Connecticut Law Review*, *49*(5), 1591.

Kukielski, K. (2022, February 17). The first amendment and facial recognition technology. *Loyola of Los Angeles Law Review*, *55*(1), 231.

Mann, M., & Smith, M. (2017). Automated facial recognition technology: Recent developments and approaches to oversight. *The University of New South Wales Law Journal*, *40*(1), 121–145. doi:10.53637/KAVV4291

Sarabdeen, J. (2022, March). Protection of the rights of the individual when using facial recognition technology. *Heliyon*, *8*(3), e09086. doi:10.1016/j.heliyon.2022.e09086 PMID:35309394

Savage, C. W. (2020, July/August). Washington enacts first-in-nation state law regulating governmental use of facial recognition technology. *Intellectual Property & Technology Law Journal*, *32*(7), 21–23.

Schuetz, P. N. K. (2021). Fly in the face of bias: Algorithmic bias in law enforcement's facial recognition technology and the need for an adaptive legal framework. *Law & Inequality*, *39*(1), 221–254. doi:10.24926/25730037.626

Shore, A. (2022, May). Talking about facial recognition technology: How framing and context influence privacy concerns and support for prohibitive policy. *Telematics and Informatics*, *70*, 10185. doi:10.1016/j.tele.2022.101815

Tariq, S., Jeon, S., & Woo, S. S. (2021, March 2). *Am I a real or fake celebrity? Measuring commercial face recognition web APIs under deepfake impersonation attack.* Cornell University ArXiv. https://arxiv.org/abs/2103.00847

Tsoukas, H. (2018). Strategy and virtue: Developing strategy-as-practice through virtue ethics. *Strategic Organization*, *16*(3), 323–351. doi:10.1177/1476127017733142

KEY TERMS AND DEFINITIONS

Algorithm: Rules and processes often created by computers as a set of mathematical commands and directives that calculate answers to problems (Cambridge University Press, 2022a).

Bias: In this particular paper, a bias is a personal inclination towards making a prejudiced or unfair judgment against something, someone, or a group (Merriam-Webster, n.d.a).

Biometric Data: Measurements that are taken to capture a person's physical traits in order to match an identity. Examples of biometric data can include fingerprints, retinal scans, face capturing, or even voice recognition (Johansen, 2019).

Deepfake: A convincing image or video of someone or something that has been altered to distort or misrepresent someone's actions or words (Merriam-Webster, n.d.b).

Discrimination: Within this paper, discrimination is acting upon a personal bias or prejudice attitude (Merriam-Webster, n.d.c).

Face Detection (aka Face Capture, Face Match, Face Identification, Face Verification, Face Clustering, Face Tracking): The technological means to replicate the biological process of recognizing faces by the means of artificial intelligence. This employs two forms of technology: 1) A scanner (i.e., a camera) and 2) A computer algorithm that can match the scan of a person's face to their identity (Martinez, 2009).

Federal Law: The body of law that governs an entire country, regardless of city, state, or province (Law Insider, Inc., n.d.).

Privacy: A personal right to keep one's personal matters, information, and relationships secret (Cambridge University Press, 2022b).

State Law: The statutes, common law, and regulations that comprise the body of law in a particular state. State Law runs parallel to Federal Law, though in cases of conflict between State and Federal Law, Federal law prevails under Article IV, Section 2 of the United States Constitution (AirSlate Legal Forms, Inc, n.d.).

Chapter 12
Technology, Ethics, and Elements of Pervasive Digital Footprints

Lynne Williams
Purdue University Global, USA

Andrew J. Campbell
https://orcid.org/0000-0002-2217-2589
The University of Sydney, Australia

ABSTRACT

Philosophically speaking, technology has evolved primarily as an ethically agnostic construct with the emphasis being placed on how well the technology works rather than how the technology affects its users. This "practical rather than ethical" focus presents special concerns when considering data that are intimately attached to an individual. Law enforcement increasingly uses investigative genetic genealogy (IGG) cross-matched with commercial DNA databases to definitively identify perpetrators. This overlap between judicially overseen data and commercially obtained data creates ethical issues surrounding an individual's right to privacy and informed consent. While the ethical use of GPS data has been debated since the emergence of location tracking, devices that are easily carried on an individual's body add an intimate understanding of not only where that individual has been but inferences about their motivation for going there. DNA databases, biometrics, and GPS tracking data are explored here as they pertain to ethical concerns related to personal autonomy.

INTRODUCTION

To properly examine the ethical issues surrounding technology that can identify an individual and their movements, some definition and distinction between ethics and rights must be made. Any ethical construct must establish whether a given action violates fundamental ethical principles in some manner. Questions to be asked might include whether the action causes harm to an individual or whether the

DOI: 10.4018/978-1-6684-5892-1.ch012

action somehow exploits the individual to the benefit of others. A deeper examination of such questions exposes the potential consequences of the action to determine whether beneficial consequences to the individual are outweighed by harmful consequences ("DNA Technology in Forensic Science," 1992).

For example, Smith and Miller (2021) note that DNA is inherited and thus carries with it the risk that other individuals, besides the subject, may be implicated in any examination of the subject's DNA data. Further, DNA can extract health information that might impact the subject's ability to obtain insurance or even employment in some fields.

Ethics are not quite the same thing as rights, although the concepts are certainly related. An individual's rights in relation to a given technology are usually determined judicially. Examples of judicially defined rights to personal data, such as DNA, include the General Data Protection Regulation (GDPR) and the Health Insurance Portability and Accountability Act (HIPAA). Involuntary DNA collection has been ruled by the European Court of Human Rights to be an invasion of privacy, constituting the violation of an individual's reasonable expectation of privacy (Tuazon, 2021). While the voluntary collection of DNA material would not necessarily fall under this protective legal umbrella, once the DNA is stored in a commercial database, such as that maintained by AncestryDNA®, usually some sort of use or principles statement is provided ("Ancestry Guide," n.d.).

Given the practitioner emphasis of technology, it makes sense to examine the ethics of technology in applied form. Taking the framework of the judicial rights of a user of technology, most vendors of various technologies either provide some sort of statement of ethical principles, such as Google's "Don't Be Evil" code of conduct or attempt to code ethical principles directly into the algorithms used by software. While the former ethical framework has its weakness in depending on the user to follow the principles, the latter concept also has some serious weaknesses. As Green (2021) notes, "In 2016, ProPublica revealed that an algorithm used in criminal courts was biased against Black defendants, mislabeling them as future criminals at twice the rates of white defendants" (p. 210).

Facebook (also known as Meta©) has frequently been taken to task for deliberately allowing bias to be used by advertisers. Hao (2019) describes Facebook job postings for teachers being shown to higher percentages of women job seekers while job postings for janitors are more often shown to minorities. Although it is beyond the scope here to examine all the various means by which bias creeps into algorithms, the most obvious culprit is the manner in which machine learning (on which many algorithms are based) seeks patterns found in massive amounts of data then leverages those patterns for decision making. For example, when an advertiser sets up a Facebook ad, they are given a choice of objectives such as the number of views given the ad, the amount of engagement with the ad, and the number of sales generated by the ad. These choices are disconnected from any humanly selected form of bias, but bias is still introduced because if the algorithm determines it will generate more engagement by showing the ad to a particular demographic, then discrimination against other demographic groups will naturally occur. (Ali et al., 2019)

Tsamados et al. (2021) explicitly state that "algorithms are not ethically neutral" (p. 215). They warn that despite growing attempts to address these ethical issues through design, "the number of algorithmic systems exhibiting ethical problems continues to grow" (Tsamados et al., 2021, p. 215)

THE ETHICS OF BUSINESS DATA

Biometrics use the physical and behavioral characteristics of an individual to identify that person in a manner that cannot be easily repudiated. Fingerprints were one of the earliest uses of biometrics, but many other characteristics can now be included, such as facial recognition imaging, voice recognition, iris and retina recognition, DNA, body odor, walking gait, ear recognition, and keystroke biometrics (Dey et al., 2020). The emphasis here is on DNA biometrics.

In the United States, there are two main categories of DNA databases; recreational DNA databases maintained by commercial entities such as 23andme©, AncestryDNA®, and law enforcement DNA databases. Law enforcement was previously confined to using government-run DNA databases simply because in the early days of DNA technology, the processing of DNA material was expensive and time-consuming, thus beyond the means of commercial entities. U.S. courts, in reference to government-run DNA databases, tended to rule that incarcerated individuals had less of a right to privacy than the general public. Even with a suspicion that a person of interest had committed a crime, a warrant was and is required to collect DNA material for storage and cross-reference in a government-run DNA database. In the case of a person arrested or jailed for a felony, jail personnel may be authorized "to collect DNA samples ... and then profile them" (Slobogin & Hazel, 2021, p. 753), without being seen to violate the offender's Fourth Amendment right prohibiting unreasonable searches.

When it comes to public DNA databases such as GEDmatch or AncestryDNA, because the DNA samples are given voluntarily by individuals, ethical and legal governance become considerably murkier. As Slobogin and Hazel (2021) note, "DNA samples surrendered to a public database... are truly shared, because the services provided are not indispensable to participation in modern society, rather the consumer can forego sharing the information without significant hardship" (p. 755).

There are several broad concerns entangled within the ethics of DNA databases. DNA databases have become central to various nation-state governance policies that administer processes such as immigration, societal welfare schemes, humanitarian relief, policing, and even elections (Boy et al., 2018). Some nation-states, such as Kuwait, have made biometric registration compulsory by law while others, such as India, have made biometric registration a necessity for accessing crucial national services. Whether legally or de facto compulsory, pressure to comply with biometric collection exposes possible violation of an individual's basic rights to privacy, autonomy, confidentiality, and legal due process.

DNA analysis provides both a broad, phenotypical view of an individual's general traits and a narrow view specific to that particular individual. These foci are used for different purposes. Most non-geneticist laypeople tend to think of the latter view when considering the use of DNA material, but the former view is more typical of use by law enforcement. Wienroth et al. (2021) provide an example:

To date, in the USA, hundreds of investigations have drawn on intelligence from Forensic DNA Phenotyping (FDP) and Biogeographic Ancestry Testing (BGA) analyses since 2015. In Europe, these techniques (especially BGA), have found application in cases since at least 1999, including in the Marianne Vaatstra case; the investigation of the Madrid train bombings of 2004; the Phantom of Heilbronn investigation in 2008; and the detection of the murder of Eva Blanco Puig in Spain in 2015. In the Vaatstra case in 1999, an early form of BGA was used to the effect of taking public pressure off asylum seekers who had been considered prime suspects. (para. 31)

The Vaatstra case clearly demonstrates the ability of genetic phenotyping to eliminate a population that does not share characteristics in common with the perpetrator. While identifying an entire population of similar characteristics has its own ethical considerations, being able to identify a specific individual is of more pressing concern.

Cyranoski (2020) describes another example of the ethical concerns inherent to mass DNA profiling related to China's ongoing project collecting DNA from broad sections of the male population. Law enforcement is collecting DNA from men and boys across China with the intention of storing the genetic profiles of approximately 10% of all male citizens including those who have no criminal history (Cyranoski, 2020). Although the publicized reason is to help capture criminals, the indiscriminate collection from those without a criminal history implies potential intent to tighten the government's grip on social control. The type of DNA collection being used in China allows law enforcement to single out individuals who may not have provided their own DNA samples but who are related to someone in the DNA database. While the database has been used for its purported goal of investigating criminal activity, the manner and type of DNA collected for the database would also allow the government to identify and punish family members of individuals who have criticized the government (de Groot et al., 2021).

The intersection of Big Data and data held in DNA databases creates a unique ethical vulnerability. While there is no set definition of Big Data, it is often characterized by volume, velocity, and variety. Data sets incorporated within Big Data tend to cover entire populations and can identify specific individuals within a population. These data sets are also relational so data gleaned from a variety of sources can be used to refine results. Finally, Big Data can incorporate fresh data continually thus further refining results (Machado & Granja, 2020).

Being able to quickly and accurately obtain results from the analysis of Big Data is not just of interest to law enforcement but has significant financial value to commercial DNA entities such as 23andme. Stoeklé et al. (2016) note that 23andme has created a kind of interface between private individuals and researchers seeking access to large quantities of DNA data establishing a "two-sided platform, with two kinds of consumers: people who want information about their own genes, and researchers and others who want access to genetic, web behavior, and self-reported information for a large number of people" (p. 3).

In this two-sided model, it should be noted that the individual is considered by the commercial DNA database business to be a "consumer," not a "patient." The ethical construct of consent has different definitions pertaining to each role. If the individual is giving consent as a patient, in most countries, they will be advised on the benefits and risks of genetic testing by a qualified health professional. Moreover, a patient can request to have their personal health data removed at a later point in time if not deidentified on collection. However, given that the user is viewed as a "consumer," the DNA database business is under no such constraints and is only required to provide the individual with such information as is deemed appropriate to the business environment. This has profound implications to ethical considerations such as the right to privacy and an individual's autonomy. Most ordinary people are not sufficiently scientifically literate to interpret their results correctly, much less discern what benefits and risks may be involved. Further, the individual may have no knowledge or control over how the commercial DNA database stores, anonymizes or distributes their genetic data.

A patchwork of regulatory oversight has been implemented to govern the use, both judicial and commercial, of DNA information in recent years. In the US, the Genetic Information Nondiscrimination Act of 2008 (GINA) was enacted to address this need, but GINA has since proven inadequate to protect consumers. Other U.S. agencies, such as the Food and Drug Administration (FDA) and Federal Trade Commission (FDA) also regulate limited aspects of DTC genetics testing. May (2018) describes the current environment as the "Wild West," with direct-to-consumer (DTC) genetics testing companies allowing a third party to send in a sample and receive the results without confirming that the third party is conclusively identified as the owner of the sample.

From a personal point of view, whether the individual is seeking enlightenment about one's own genetically exposed health risk factors or researching ancestry, determining how a particular DTC vendor may protect the consumer's privacy and autonomy largely depend on the vendor's stated policies. As an example, Majumder et al. (2021) found that company practice was variable with some vendors refusing to disclose information without a warrant while other DTC companies leave the decision up to the consumer to opt-in or opt out of providing law enforcement with their data. In all cases within the current environment, it is up to the consumer to research the policies of a particular DTC vendor and determine if the benefits outweigh the risks within their circumstances, truly a case of caveat emptor.

THE ETHICS OF LOCATION TRACKING DATA

Location tracking physically locates people and objects such as cars or shipping containers. While commonly associated with GPS, location tracking can be done with a variety of techniques. The Global Positioning System uses radio signals to stream data from satellites orbiting the earth to tracking devices on the planet's surface. The tracking devices then calculate their location based on the satellite data. Location tracking can also be done by triangulating cellular signals between cell towers in the vicinity of a tracking device. Radio frequency tracking (RF) also works by triangulation, while infrared tracking (IR) uses light frequencies to calculate a tracking device's location (Wang et al., 2022).

Most people are familiar with using GPS to guide them to a location which is a typically benign use. However, GPS can also be used for surveillance, allowing a third party to track another person. In the case of wandering children or dementia patients, this use is considered benign, but at what point does location tracking become actively invasive or harmful? An increasing number of people have been found guilty of criminal activity because of their location data; this may be a legally justified use of location data, but there is the potential for the use of location data for stalking activity or surveillance of personnel by an employer. Michael et al. (2006) argue that "the innovative nature of the technology should not be cause to excuse it from the same judicial or procedural constraints which limit the extent to which traditional surveillance technologies are permitted to infringe privacy" (p. 7).

Part of the difficulty in defining an acceptable set of ethical constructs for location tracking is the ubiquitous nature of location tracking devices. Global Positioning System (GPS) devices are just one subset of Geospatial Technologies (GST) which also includes remote sensing devices as well as Internet of Things (IoT) sensors. This broad range of GST usage can include cell phones, wearable devices, vehicles, motion sensors, and webcams. As Apte et al. (2019) note, the main ethical dilemma concerning GST (and thus GPS) is the sheer quantity of information that can be derived from tracking an individual's movements. A variety of inferred information may also be taken from this geospatial data, particularly if triangulated by one or more other data sets from other individuals.

In 2015, Richard Dabate claimed that a masked intruder invaded his house, restrained him, and tortured him. Dabate insisted the intruder shot and killed his wife when she came home and interrupted the intruder. His account did not hold up when investigators dug deeper. Police charged Dabate on evidence gathered from the fitness tracker his wife was wearing at the time of the attack. Although the area was thoroughly searched, the police could not find a suspect. There was also no evidence of forced entry and no sign of a burglary. Search warrants for Ms. Dabate's Fitbit were obtained along with both of their cell phones, computers, and logs from the home security system. When the logs were examined, a timeline of what actually happened emerged (Hauser, 2017).

At 9:01 a.m. Dabate checked his email using an IP address assigned to the router at the house. At 9:04 a.m., he told his supervisor via email that a security alarm had gone off at his house. Movement was tracked on Ms. Dabate's Fitbit at 9:23 a.m., this would have been the time that she arrived at the house and entered from the garage door into the home (Hauser, 2017).

Between 9:40 and 9:46 a.m., she posted several videos to her Facebook page using her iPhone. The iPhone used the IP address assigned to the Dabates' home router. Ms. Dabate's Fitbit tracked an overall distance of 1,217 feet in movement between 9:18 a.m. and 10:05 a.m. If Dabate's claims concerning the intruder are genuine, the total distance it would take the victim to walk from her vehicle into the house, then down to the basement where she died, would be no more than 125 feet. "The Fitbit showed her last living movement was at 10:05 a.m." (Hauser, 2017, para. 7).

Cross-examining the router logs and the data from the Fitbit, the incident could not have occurred as Dabate claims. As Hauser (2017) comments in the article, personal tracking devices give researchers and investigators a uniquely granular look at an individual's movements and location as well as data concerning the type of movement such as sleeping, walking, running, and even struggling. In the Dabate case, lack of heartbeat was also recorded.

While the ability to synchronize two sources such as the router logs and the Fitbit tracking data is clearly efficacious to law enforcement, the overall process of accessing the data, determining what types of data to analyze, and how the data are acquired are rigidly proscribed by a legally specified collection of evidence rules. In cases not covered by those rules, such as data mining, only regional statutes such as the General Data Protection Regulation (GDPR) serve as a protection for an individual's right to privacy and autonomy over their personal data. Anonymization offers a slight remedy to privacy and autonomy concerns but, as discussed by Apte et al. (2019), even this may not be possible. If an investigator, whether human or algorithm, can access the location history of a device, the individual's location and movements can be observed and associated with that specific person thus nullifying anonymization.

Using contact tracing apps presents an ethical puzzle within the larger location tracking sphere. Contact tracing has been around for decades in various forms and is used for a variety of infectious diseases such as tuberculosis or measles. The advent of the ubiquitous use of smartphones has made contact tracing via an app on the phone much more efficient. All types of disease-related contact tracing are typically governed by whatever regulations must be followed by local health authorities. Whether done manually or technically, contract tracing has to balance the benefit accruing to society by identifying a vector of infection against the restrictions caused by quarantining infected individuals. Bernard and Sim (2020) state that "contact tracing presents particularly vexing challenges of balancing societal versus individual interests" (p. 395), specifically when considering COVID-19.

One of the most disturbing challenges involves the sharing of data gathered by a contact app with inappropriate third parties.

Serious privacy breaches have been identified in many COVID-19 apps, all contrary to the apps' stated privacy policies. The North Dakota app Care19 shared information with a digital advertising firm, including the unique advertising identifier that allows targeted advertisements in other apps. (Bernard & Sim, 2020, p. 397)

Klar and Lanzerath (2020) comment on the differences between traditional, manual methods of contract tracing and the type of digital contract tracing used for tracking COVID-19 cases. Because COVID-19 is exceedingly infectious and can be spread by asymptomatic individuals, manual tracking methods are not effective. Digital COVID-19 tracking apps (CTAs) have been rapidly developed to overcome these problems. A key difference between traditional tracking and CTAs is the combination of proximity tracing with contact tracing. Proximity tracing typically uses Bluetooth which is a wireless, short-range form of connectivity. If two individuals have Bluetooth enabled on their devices and both individuals also have a CTA on their devices, the CTAs will automatically connect via Bluetooth to record the proximity of the device owners as well as the length of time spent in proximity. Should one of the individuals report that they have received a positive result for a COVID-19 test, all other individuals will receive some sort of alert advising them of health risks, testing requirements, and contact information for the local health authorities.

While there are concerns about the overall effectiveness of CTAs, including the need for widespread adoption and a variety of technical difficulties surrounding their adoption, several ethical issues relating to privacy have also arisen. Given the rapid development of CTAs and the lack of controlled trials, CTAs tend to be vulnerable to hacking. Because CTAs enable large-scale data collection, this data is particularly vulnerable to misuse by government authorities who seek to retain tracking data on people for surveillance. The data could also allow various entities to monitor population health thus causing a "health dictatorship" that would seek to apply draconian measures such as detaining people because of their viral status or entering their homes without a warrant.

A more subtle ethical concern is that of the role of machine learning incorporated into CTAs. As mentioned earlier, artificial intelligence can introduce algorithmic bias and machine learning is a subset of AI. Within the COVID-19 context, machine learning can be deployed to infer the asymptomatic spread of the virus, make models of the effect of quarantine policies, and can even be used with facial recognition when performing emergency room triage. Effectively, machine learning within the COVID-19 context is being used to manage risk and to determine how to respond to that risk. As Moss and Metcalf (2020) note, these machine learning examples have been deployed with the best of intentions, but some effects create unequal outcomes. Moss and Metcalf (2020) state that:

> *Any contract tracing is a sociotechnical system that depends on how different parts of social life fit together — telling people they should stay home does not mean that they will be able to stay home. For contact tracing to work at all, its designers must be attuned to the context of social life in which such systems can produce harmful, difficult-to-foresee effects that replicate or amplify inequalities already present in society. (p. 3)*

In the US, an example of this type of disparity would be Black and Latin American communities, whose members tend to work in jobs that cannot be done from home and who also have inadequate sick leave or healthcare options. As a result, those in Black or Latin American communities would not significantly benefit from CTAs because even if they receive an alert, they cannot afford to self-quarantine and lack societal resources to support them in doing so. Imposing a "health dictatorship" on these communities, without providing adequate support resources, creates an unfair burden on this demographic while doing little to reduce individual risk. This also raises the ethical issue of the use of rudimentary digital health solutions in order to achieve broad-vs-focused public health efficacy. A double standard of public health is perpetuated given current machine learning recommendations from CTAs. In one sense, the technology aims to protect the greater community. In reality, it protects a disproportionate amount of the community (those resourced) at the detriment of those less resourced. Paradoxically, while the employment of the technology is viewed as a safeguard for the wellbeing of society, it ultimately cannot achieve public health impact at a high success rate, unless it factors solutions for differing levels of socioeconomic status of individuals. Notably, this is a problem that is not solved by technology advancements alone. Notwithstanding this, current CTA effectiveness in developed nations such as Australia indicates it has been a low uptake <22%, and a low positive predictability factor of <39% (Vogt et al, 2022).

Subbian et al. (2021) point out that CTAs create privacy and autonomy concerns. Devices running CTAs contain both carrot and stick measures; not only do they alert users to risk but they are also used to check whether users are self-quarantining in compliance with local COVID-19 policy. CTAs generate large amounts of Big Data which can be sold for commercial purposes (as seen in the North Dakota example) which causes autonomy concerns in that the individual loses control of their own data. CTAs can also provide coordinates that allow identification of a particular person further raising concerns about privacy.

This ability of location tracking to generate Big Data almost effortlessly brings up an ethical concern not seen before the advent of ubiquitous mobile devices. While individual privacy issues have long been apparent, the idea of "group privacy," where violations of privacy can occur for an entire demographic, is quite new. Now that the movement of large groups of people can be tracked, Big Data can reveal not just location but behavioral patterns within populations. This has far-ranging implications from benign to malign.

For all the instances of humanitarian intention, including responding to environmental or conflict crises, there is equal leeway for more harmful uses. Being able to track the movements and behavior of groups functioning as dissident networks would allow an unfriendly government to stifle opposition to authority. Taylor (2016) comments that other concerns such as ethnic violence and discrimination against a population may occur. Besides privacy violations, since group privacy poses a risk to groups, new issues surrounding autonomy and informed consent come into play. If one member of a group is identified, the analysis methods of Big Data make discovering other individuals within the group relatively trivial.

Compared to the macroscopic view of group tracking, the microscopic view of intense individual tracking via the millions of sensors embedded within the Internet of Things (IoT) also plays a growing part in ethical issues attached to location tracking. Karale (2021) lists use cases of IoT devices to include AI assistants, smart home appliances, IoT-enabled door locks, CTAs, emotion detection sensors, fitness trackers, facial detection devices, CCTV, and smart meters related to energy consumption.

The movement of individuals is being increasingly tracked and surveilled by an almost invisible mesh of sensors that not only record where a person is at any given time, but their behavior and activity at that time and in that place as well. Involuntary location data can be taken from a number of sources. For example, speed cameras at road intersections can record a vehicle's speed while taking a photograph of the vehicle's license plate. This may take place several times in the course of a driver's journey, leaving a data trail of the vehicle's travels.

This ability to track such granular activity specific to an individual blurs the distinction between a person's public and private life and makes it difficult-to-impossible to define, at what point, the right to privacy is violated. The devices and their sensors collect data indiscriminately without concern for boundaries. Once the data have been transmitted, not necessarily back to the owner of the device but to the vendor, the individual loses control of the data and no longer has autonomy over it.

The triangulation of location data spread across a mesh of several IoT sensors can create a highly accurate data trail of which the person is completely unaware. The voluntary nature of many IoT activities, such as the use of fitness trackers, makes building an ethical construct for this type of location tracking difficult. People purchase IoT devices to make their lives more convenient so, on one level, the blurring of boundaries is deliberate. In addition, the user may assume that the history of their usage is erased by the object when, in fact, the device stores the data across its entire life cycle.

The key question is whether users truly understand just how much of their privacy and autonomy are being sacrificed to convenience. Boutet and Gambs (2019) demonstrated how easy it is to profile an individual based on their location history from a smart device. The authors refer to exposing a location profile without the user's consent as an inference attack. In their demonstration of an inference attack, they accurately predicted personality traits by deriving those traits from mobility data. Boutet and Gambs (2019) note that "by crossing information from reports of [a] national statistics institute and both the home and the working places, we are able to predict the gender, the age, and the salary" (p. 2863).

The June 24, 2022, decision by the United States Supreme Court to strike down the Roe v. Wade decision of 1973 abruptly provided another use case of the ethical dilemmas posed by location data. The court's decision immediately triggered a number of state level laws banning abortion. In many of these same states, there are additional laws being proposed that would forbid female residents from traveling across state borders to seek procedures that remain legal elsewhere. In the meantime, many women use a variety of mobile phone apps that help them track menstrual cycles and may allow the user to seek information about a variety of reproductive health issues. This reproductive health related digital activity has now become of special interest to law enforcement.

Hill (2022) explains:

Investigators could also potentially use smartphone location data if states pass laws forbidding women to travel to areas where abortion is legal. Information about people's movements, collected via apps on their phones, is regularly sold by data brokers. (Hill, 2022, para. 14).

When The New York Times investigated the supposedly anonymized data on the market in 2018, it was able to identify a woman who had spent an hour at a Planned Parenthood in Newark. In May, a journalist at Vice was able to buy information from a data broker about phones that had been carried to Planned Parenthoods over the course of a week for just $160. (After Vice's report, the data broker said it planned to cease selling data about visits to the health provider.) (Hill, 2022, para. 15).

In this case, the user may not only be sacrificing her privacy but may also be unknowingly putting herself in legal jeopardy.

COMBINING BIOMETRIC AND LOCATION TRACKING DATA

Biometrics and location tracking, when examined separately, can each pose significant risk to an individual's privacy and autonomy. When the two technologies are combined to create a highly detailed digital footprint that cannot be repudiated, the result can so significantly breach the boundary between public and private life that the individual has no privacy in a practical sense. This elimination of the delineation between public and private life can be done in several ways.

Digital forensics, data analytics, and linkage attacks share many of the same techniques for aggregating data to form a profile of a user, varying mostly by the intent of use. Digital forensics extends traditional physical evidence gathering and is associated with the investigation of data found on a variety of devices and networks. Data analytics is not necessarily concerned with the identification of an individual or that person's activities but is rather a means of deriving patterns and relationships from large data sets i.e., Big Data. Linkage attacks can be defined as the "de-anonymization" of data provided by or about an individual, derived by analyzing data relationships in two or more unrelated databases. Merener (2012) comments that anonymization typically removes "the variables that uniquely associate records to the corresponding individuals, such as name, email address, social security number" (p. 378). The author then notes that this step does not prevent algorithmic linkage of other variables in the data that can still serve as a type of digital footprint.

A growing sector for the convergence of location tracking and biometrics is attendance systems. Zelenák (2020) comments that attendance systems are almost ubiquitous in the workplace and that 38% of employers using manual systems encounter a high number of timesheet errors. Manual attendance systems also tend to be complicated and time-consuming. Moving to a digital attendance system that combines location tracking and biometrics is much more efficient and can be beneficial for a business. The ethical question, in this case, is whether using biometrics and location tracking is intrusive to the employee. A digital attendance system could be considered invasive of the employee's privacy.

The most troublesome concerns surrounding digital attendance systems are a feeling of pervasive surveillance on the part of the employee as well as employees not being clearly informed of the extent of data collection, or the uses to which the data may be put. Non-visual biometric data, such as DNA or fingerprints, can reveal genetic disorders of which even the data's owner may not be aware. Collecting such data risks providing certain types of data to the employer without the explicit consent of the employee. The inadvertently exposed information exposes the employee to the possibility of discrimination in matters of promotion or employer-provided healthcare coverage. Holland and Tham (2020) note "this also puts into question whether certain personal boundaries are being crossed without the knowledge and consent of the employee" (p. 503).

There is an inherently uneven power balance between employer and employee when it comes to personal data collection. Employees typically need the work and wish to be recognized as good employees, so when faced with a requirement to provide personal data as a condition of employment they effectively give up their right to privacy to keep their jobs. This lack of power and a lack of knowledge concerning the extent of the data collected means that some sort of outside regulatory agency is required to prevent boundary creep regarding privacy.

Holland and Tham (2020) describe the case of Jeremy Lee v. Superior Wood Pty Ltd as an example of the clash between an employee's rights connected to their personal data and the employer's desire to streamline its attendance system. The employee (Lee) objected to Superior Wood's implementation of a biometric scanner as a requirement for recording attendance. Although Lee lost the initial case, he won on appeal when the Full Bench found that the Australian Privacy Act 1988 had been breached by the company. The breach was caused when the company failed to give employees full disclosure concerning how this data might be used independently of the attendance system. Superior Wood also neglected to provide information regarding the storage and control of the data by a third-party vendor. In other countries, such as Austria, implementing this type of biometrically enhanced attendance management system requires that any business deploying such a system must pass a judicial review to ensure that employees are well informed of the implications.

As noted above, the most obvious combination of biometrics with location tracking is embodied by various COVID-19 tracking applications. Governmental and healthcare entities around the globe have implemented a range of these applications in an attempt to enforce quarantine mandates and to alert users when they have been in proximity to someone who tested positive for the virus. Usually, this type of tracking application that combines biometric data with location tracking is provided by a tech company such as Google or Apple.

The ethical concern appears when the legitimate need to track infections is handled by a business entity with a model based on targeted advertising that depends on intensively collecting a variety of consumer data for commercial gain. Early in the COVID-19 pandemic, Google launched a website allowing consumers to access pandemic information based on state data, safety and prevention resources, and other virus-related data. This information was not provided for free but required a Google account to view the content. This linking of a Google account to the data meant that user searches on the website might be monetized with ads. Although Google insisted that the website's data were stored separately from data intended for commercial use and not associated with other Google products, the privacy policy associated with the website revealed that using a Google account to access the data still allowed the data to be shared with third-party entities. This raises the additional concern that not only are there privacy issues but the possibility of the user losing all control over data that is likely to be sensitive.

Biometrics are also associated with the use of virus tracking apps. Some apps use facial recognition to identify the user which is then combined with the user's location. Thermal imaging can be used with some tracking apps for remote temperature monitoring. Pandit et al. (2020) refer to this combination of biometrics and location as "biometric surveillance" and state that the technology's serious invasiveness poses a major threat to the user's privacy and autonomy over highly sensitive personal data.

Ironically, the combination of biometrics and location tracking is being extensively researched to improve cell phone security (if not privacy or autonomy). Most users find robust mobile security inconvenient and will avoid taking appropriate precautions in favor of ease of use. In the case of cell phone authentication, the application developers deliberately profile the user as accurately as possible to ensure that only the owner of the phone will be authenticated on the device.

Location tracking on a smartphone can be done with a variety of wireless technologies. As the user carries the handset, the device may use Wi-Fi, Bluetooth, or cellular signal to connect with a telecom network. Particularly with a cellular signal, as the user moves around, the phone pings nearby cellular radio masts to maintain a connection. The location data are periodically transmitted to a Global Positioning System (GPS) satellite to narrow down the phone's location more accurately.

The connection history can give a highly accurate picture of the user's behavior, allowing an application to build a profile specific to that particular user. This profile is then combined with the user's touch gestures and other biometric data to create a distinct proof of identity that is almost impossible to counterfeit. Acien et al. (2020) refer to this combination of biometric data and location tracking as Active Authentication (AA). The researchers discovered that continuous usage of a cell phone with AA effectively trained the phone to recognize only the primary user for authentication.

While having a mobile phone that only recognizes the genuine owner is more secure and certainly more convenient, the idea begs the question of who is in control of the owner's profile data. Users should know where this data is stored, if third parties can access it and for what purpose, and what, if any, autonomy they have over their own profiles.

CONCLUSION

Individual rights issues over personal data use will continue to accelerate, especially during major world events such as the COVID-19 pandemic, terrorist attacks, war, natural and financial crises. Given this, a growing gap between individual knowledge of rights to privacy and data control, as well as an inability for users to pre-empt and attempt to control how technology and other commercial sectors may attempt to use individual data will remain. Organizations such as the World Health Organization have long stated that individual health data, for example, is the individual's property. However, given the rise of biometrics, GPS and other data capture technology used particularly in consumer technology, the divide between health ethics and consumer ethics has emerged, giving technology creators and other corporate entities a loophole to exploit data laws simply due to the terminology attributed to the user. This raises the issue of whether "human data ethics" should be the ubiquitous legal framework to provide agency to users around their data rights regardless of their user status. With this, classifications such as stating that data was collected by a "patient" or a "consumer" would be made redundant. By no means is this a simple way forward to extend greater human rights to global digital users, as changing a terminology must be agreed to, and enforceable, by national and global legal entities. However, tackling rapidly changing technology data collection methods is unlikely to resolve the key issue raised here – that being, ethics and data control need to be more closely aligned to human rights rather than "categories" of users.

REFERENCES

Acien, A., Morales, A., Vera-Rodriguez, R., & Fierrez, J. (2020). Mobile active authentication based on multiple biometric and behavioral patterns. In T. Bourlai, P. Karampelas, & V. M. Patel (Eds.), *Securing Social Identity in Mobile Platforms. Advanced Sciences and Technologies for Security Applications* (1st ed.) pp. 161–177. Springer. doi:10.1007/978-3-030-39489-9_9

Ali, M., Sapiezynski, P., Bogan, M., Korolova, A., Mislove, A., & Rieke, A. (2019, November). Discrimination through optimization: How Facebook's ad delivery can lead to skewed outcomes. *Proceedings of the ACM on Human-Computer Interaction, 2019*(3), 1–17. doi:10.1145/3359301

Ancestry Guide for Law Enforcement. (n.d.). Ancestory.com. https://www.ancestry.com/c/legal/lawenforcement

Apte, A., Ingole, V., Lele, P., Marsh, A., Bhattacharjee, T., Hirve, S., Campbell, H., Nair, H., Chan, S., & Juvekar, S. (2019, June). Ethical considerations in the use of GPS-based movement tracking in health research – lessons from a care-seeking study in rural west India. *Journal of Global Health, 9*(1), 1–7. doi:10.7189/jogh.09.010323 PMID:31275566

Bernard, L., & Sim, I. (2020). Ethical framework for assessing manual and digital contact tracing for COVID-19. *Medicine and Public Issues, 174*(3), 395–400. doi:10.7326/M20-5834 PMID:33076694

Boutet, A., & Gambs, S. (2019, November 3). Inspect what your location history reveals about you: Raising user awareness on privacy threats associated with disclosing his location data. *CIKM '19: Proceedings of the 28th ACM International Conference on Information and Knowledge Management*, Beijing, China. https://dl.acm.org/doi/abs/10.1145/3357384.3357837

Boy, N., Jacobsen, E. K. U., & Lidén, K. (2018). *Societal ethics of biometric technologies* (PRIO Paper 2018). Peace Research Institute Oslo. https://www.prio.org/publications/11199

Cyranoski, D. (2020, July 7). *China's massive effort to collect its people's DNA concerns scientists.* Nature. https://www.nature.com/articles/d41586-020-01984-4

de Groot, N. F., van Beers, B. C., & Meynen, G. (2021). Commercial DNA tests and police investigations: A broad bioethical perspective. *Journal of Medical Ethics, 47*(12), 788–795. doi:10.1136/medethics-2021-107568 PMID:34509983

Dey, A., Tharmavaram, M., Pandey, G., Rawtani, D., & Hussain, C. M. (2020). Conventional and emerging biometrics techniques in forensic investigations. In D. Rawtani & C. M. Hussain (Eds.), *Technology in forensic science: Sampling, analysis, data and regulations* (1st ed.) pp. 177–389. Wiley Online Library. doi:10.1002/9783527827688.ch9

DNA Technology in Forensic Science. (1992). *National Academies Press.* https://www.ncbi.nlm.nih.gov/books/NBK234542

Green, B. (2021). The contestation of tech ethics: A sociotechnical approach to technology ethics in practice. *Journal of Scientific Computing, 2*(3), 209–225. doi:10.48550/arXiv.2106.01784

Hao, K. (2019, April 9). *Facebook's ad-serving algorithm discriminates by gender and race.* MIT Technology Review. https://www.technologyreview.com/2019/04/05/1175/facebook-algorithm-discriminates-ai-bias

Hauser, C. (2017, April 27). In Connecticut murder case, a Fitbit is a silent witness. *The New York Times.* https://www.nytimes.com/2017/04/27/nyregion/in-connecticut-murder-case-a-fitbit-is-a-silent-witness.html

Hill, K. (2022, June 30). Deleting your period tracker won't protect you. *The New York Times.* https://www.nytimes.com/2022/06/30/technology/period-tracker-privacy-abortion.html

Holland, P., & Tham, T. L. (2020, April 27). Workplace biometrics: Protecting employee privacy one fingerprint at a time. *Economic and Industrial Democracy, 43*(2), 501–515. doi:10.1177/0143831X20917453

Karale, A. (2021, September). The challenges of IoT addressing security, ethics, privacy, and laws. *Internet of Things, 15*, 1–20. doi:10.1016/j.iot.2021.100420

Klar, R., & Lanzerath, D. (2020). The ethics of COVID-19 tracking apps – challenges and voluntariness. *Research Ethics Review, 16*(3), 1–9. doi:10.1177/1747016120943622

Machado, H., & Granja, R. (2020). DNA databases and Big Data. In Forensic Genetics in the Governance of Crime (pp. 57–70). Palgrave Pivot. doi:10.1007/978-981-15-2429-5_5

Majumder, M. A., Guerrini, C. J., & McGuire, A. L. (2021, January). Direct-to-Consumer genetic testing: Value and risk. *Annual Review of Medicine, 72*(1), 151–166. doi:10.1146/annurev-med-070119-114727 PMID:32735764

May, T. (2018, August 2). Sociogenetic risks–ancestry DNA testing, third-party identity, and protection of privacy. *The New England Journal of Medicine*, *379*(5), 410–412. doi:10.1056/NEJMp1805870 PMID:29924688

Merener, M. M. (2012, August). Theoretical results on de-anonymization via linkage attacks. *Transactions on Data Privacy*, *5*(2), 377–402.

Michael, K., McNamee, A., & Michael, M. G. (2006, June 26). The emerging ethics of humancentric GPS tracking and monitoring. *2006 International Conference on Mobile Business*. Copenhagen, Denmark. https://ieeexplore.ieee.org/xpl/conhome/4124088/proceeding

Moss, E., & Metcalf, J. (2020, October 9). High tech, high risk: Tech ethics lessons for the COVID-19 pandemic response. *Patterns*, *1*(7), 1–8. doi:10.1016/j.patter.2020.100102 PMID:33073256

Pandit, C., Kothari, H., & Neuman, C. (2020, November 22). Privacy in time of a pandemic [Paper presentation]. *2020 13th CMI Conference on Cybersecurity and Privacy (CMI) - Digital Transformation - Potentials and Challenges*, Copenhagen, Denmark. https://ieeexplore.ieee.org/abstract/document/9322737

Slobogin, C., & Hazel, J. W. (2021). "A world of difference"? Law enforcement, genetic data, and the Fourth Amendment. *Duke Law Journal*, *70*(4), 705–774.

Smith, M., & Miller, S. (2021, December 11). The rise of biometric identification: Fingerprints and applied ethics. In *Biometric Identification, Law and Ethics* (pp. 1–19). Springer. https://link.springer.com/chapter/10.1007/978-3-030-90256-8_1

Stoeklé, H.-C., Mamzer-Bruneel, M.-F., Vogt, G., & Hervé, C. (2016, March 31). 23andMe: A new two-sided data-banking market model. *BMC Medical Ethics*, *17*(1), 1–11. doi:10.118612910-016-0101-9 PMID:27059184

Subbian, V., Solomonides, A., Clarkson, M., Rahimzadeh, V. N., Petersen, C., Schreiber, R., DeMuro, P. R., Dua, P., Goodman, K. W., Kaplan, B., Koppel, R., Lehmann, C. U., Pan, E., & Senathirajah, Y. (2021, January 15). Ethics and informatics in the age of COVID-19: Challenges and recommendations for public health organization and public policy. *Journal of the American Medical Informatics Association: JAMIA*, *28*(1), 184–189. doi:10.1093/jamia/ocaa188 PMID:32722749

Taylor, L. (2016, April 1). No place to hide? The ethics and analytics of tracking mobility using mobile phone data. *Environment and Planning. D, Society & Space*, *34*(2), 319–336. doi:10.1177/0263775815608851

Tsamados, A., Aggarwal, N., Cowls, J., Morley, J., Roberts, H., Taddeo, M., & Floridi, L. (2021, February 20). The ethics of algorithms: Key problems and solutions. *AI & Society*, *37*(1), 215–230. doi:10.100700146-021-01154-8

Tuazon, O. M. (2021, January-June). Universal forensic DNA databases: Acceptable or illegal under the European Court of Human Rights regime? *Journal of Law and the Biosciences*, *8*(1), 1–24. doi:10.1093/jlb/lsab022 PMID:34188945

Vogt, F., Haire, B., Selvey, L., Katelaris, A. L., & Kaldor, J. (2022, February 4). Effectiveness evaluation of digital contact tracing for COVID-19 in New South Wales, Australia. *The Lancet. Public Health*, *7*(3), e250–e258. doi:10.1016/S2468-2667(22)00010-X PMID:35131045

Wang, J., Ranganathan, V., Lester, J., & Kumar, S. (2022, March). Ultra low-latency backscatter for fast-moving location tracking. *Proceedings of the ACM on Interactive, Mobile, Wearable and Ubiquitous Technologies*, *6*(1), 1–22. doi:10.1145/3517255

Wienroth, M., Granja, R., Lipphardt, V., Amoako, E. N., & McCartney, C. (2021, November 24). Ethics as ived practice. Anticipatory capacity and ethical decision-making in forensic genetics. *Genes*, *12*(12), 1–17. doi:10.3390/genes12121868 PMID:34946816

Zelenák, M. (2020). *Mobile application for attendance monitoring* [Master's thesis, Masaryk University]. ProQuest Dissertations and Theses Global. https://is.muni.cz/th/tpg1p/Mobile_Application_for_Attendance_Monitoring_Archive.pdf

Chapter 13
Embedding RFID Chips in Human Beings:
Various Uses With Benefits and Ethical Concerns

Tamara Phillips Fudge
https://orcid.org/0000-0002-8682-9711
Purdue University Global, USA

Linnea Hall
American Journal Experts, USA

Kathleen McCain
Mississippi State University, USA

ABSTRACT

RFID (Radio Frequency Identification) chips can contain a variety of information and are placed in debit and credit cards, embedded in products in the supply chain, planted in our pets as "microchips," and enable badge access to workspaces. They are being used in hospitals to ensure proper medications are given, help libraries keep track of holdings, and are used in many other ways. The fact that data can be transported easily and wirelessly presents many opportunities. Use in humans themselves, however, is a relatively new concept, and along with some benefits come several serious ethical questions that need to be addressed.

INTRODUCTION

Radio frequency identification (RFID) technology is hardly new considering its development can be traced back to the mid-1800s when a mathematician and scientist named James Clerk Maxwell proposed the idea of electromagnetic waves. In 1888 Heinrich Hertz, building on Maxwell's theories, invented a device capable of producing and detecting microwaves (Landt, 2001). These discoveries led to the

DOI: 10.4018/978-1-6684-5892-1.ch013

identify friend or foe system (IFF), which used transmitters on British aircraft to transmit signals that were picked up by radar to identify aircraft as "friends," thus forming the essential basis for RFID technology (Rieback et al., 2006).

RFID uses radio frequency transmission in combination with tags, readers, and software to identify specific items. Today, RFID technology is used in many applications such as logging tolls at no-contact toll booths (Kalantri et al., 2014), tracking products in warehousing and logistics (Chang et al., 2008), and ensuring a beloved pet is returned home safely (Saeed & Green, 2019). Combining RFID with other technologies can also enable real-time monitoring in supply chains providing greater visibility of package movement for customers (Kumar et al., 2021). RFID chips are also usable in a variety of environmental conditions (Kaur et al., 2011) allowing their use in both indoor and outdoor applications.

Over the last quarter of a century, the idea of human RFID implants has been gathering increasing interest. There are many potential benefits to RFID implants from being able to pay for something without carrying an additional device, to conveying important information such medical or victim identification. However, ethical considerations such as privacy, security, and health concerns continue to prevent its widespread adoption and use. RFID technology, its uses, especially in regards to living beings, the benefits, and ethical implementations for human use thus require additional exploration and understanding. This paper will review RFID technology and address some of the ethical, legal, and moral concerns related to its use in humans.

THE TECHNOLOGY

An examination of how the technology works includes an understanding of components, necessary devices, network connections, radio frequencies, and various common usage.

Components and Devices

In brief, data are pulled by an *antenna* through wireless means, and a *transceiver* is required to read the data; together these are called an *interrogator*. Readers work to convey data via a third main component called a *transponder* (Amsler, 2021). This may be hand-held or affixed (Rouse & Shea, 2017). There are also integrated RFID readers, which include a port to enable an extra antenna ("The Beginner's Guide," 2019).

Tags are a kind of label affixed to an item to be scanned. As tags do not hold much information on their own (usually less than 2,000 KB each), they contain mostly unique identifying data (Amsler, 2021). EPCs (electronic product codes) store unique identification in each tag and can specify millions of products, manufacturers, and other data (Taghaboni-Dutta & Velthouse, 2006). Notably, tags are often hidden from plain view but can still be read even through materials such as plastic and cardboard that may encase a product (Kaur et al., 2011).

The three types of tags are *active* (possessing its own power source), *passive* (power coming from the antenna's electromagnetic waves), and *semipassive*, in which the circuitry is battery-operated and the reader takes care of communication power (Amsler, 2021). The technology can read and write data to and from a database without direct contact ("Technical Explanation," 2018). Some other facts may be of interest:

Table 1. Frequencies

	Low Frequency	High Frequency	Ultra-high Frequency (UHF)
Frequency range	30-500 KHz	3-30 MHz	300-960 Mhz
Typical frequency	125 KHz	13.56 MHz	433 MHz
Range	A few inches to almost 6 feet	A few inches to several feet	Up to 25 or more feet
Transmission	Electromagnetic induction	Electromagnetic induction	Radio waves
Some uses	Access control, animal tracking, car key fob chips	Library holdings, contactless payment, ID cards	Inventory
General cost (2019)	$.75 to $5	$.20 to $10	$.05 to $.15
Notes	Resistance to moisture and metals	Most expensive, better memory	Lower costs, reads quickly

Sources: Amsler (2021), "The Beginner's Guide," (2019), "Technical Explanation" (2018), and "Comparing Different Types" (2018).

- Because active tags have a more robust power source, they typically allow for a longer reading range (Rouse & Shea, 2017); they also tend to be more expensive ("Comparing Different Types," 2018) and larger.
- It has been said that due to the power methods, passive tags may be more environmentally-friendly (Taghaboni-Dutta & Velthouse, 2006).
- The semipassive tags' batteries alert the tag when a signal is acknowledged but otherwise remains dormant ("Comparing Different Types," 2018).

Cost may well be a factor in determining which form of the technology to employ. In 2019, active tags cost as much as $25-50 each; these might be used for tracking vehicles, construction, and other manufacturing needs. In comparison, passive tags are far less expensive, starting at 9 cents a tag, and have many uses including inventory, toll road payments, and tracking pets. Obviously, return on investment and feasibility analyses must be made before making any RFID decisions ("The Beginner's Guide," 2019).

The frequency of radio waves is also an important business and technology consideration: low, high, UHF waves, and even microwave systems are designed to capture a chip's data from various ranges:

There is a medium-frequency band available that uses a frequency range of 300 KHz to 3 MHz and electromagnetic coupling instead of induction ("Technical Explanation," 2018), and microwave systems that can use a frequency of 2.45 GHz, with a range of 30 or more feet (Amsler, 2021).

RFID tags can additionally be classified by function (read-write or only read data) and are also identified by the many potential shapes known as the "form factor," including inlet, label, card, square, round, cylindrical, spherical, and box ("Technical Explanation," 2018; "The Beginner's Guide," 2019). Form factor terminology varies depending on the manufacturer.

Several examples of RFID tags are shown below. In the first image, the size in centimeters is shown (which is approx. 1.5" in U.S. measurement). The microchip is in the innermost ring, and the other lines are the integrated antenna.

The next image shows a special and simple RFID tag called a "smart label" (Amsler, 2021). The components are affixed to an adhesive label that can be placed on a plastic product bag. The other side of the label has a barcode and some printed wording for the consumer.

Figure 1. Product inventory control showing relative size in centimeters; bar code on reverse
Source: Maschinenjunge (2008)

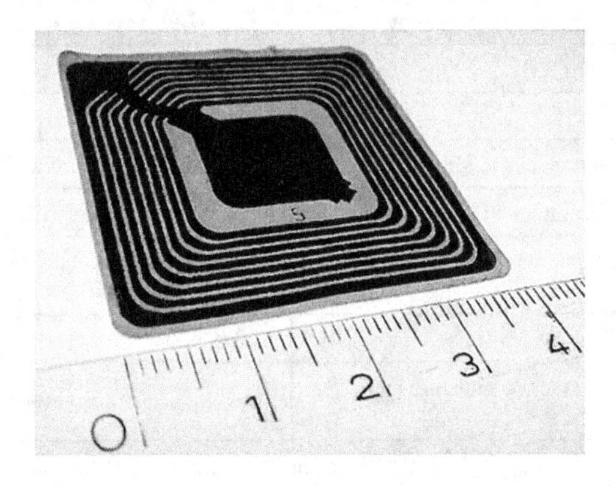

Figure 2. Commercial packaging "smart label" RFID; bar code on reverse
Source: Fudge (2022b)

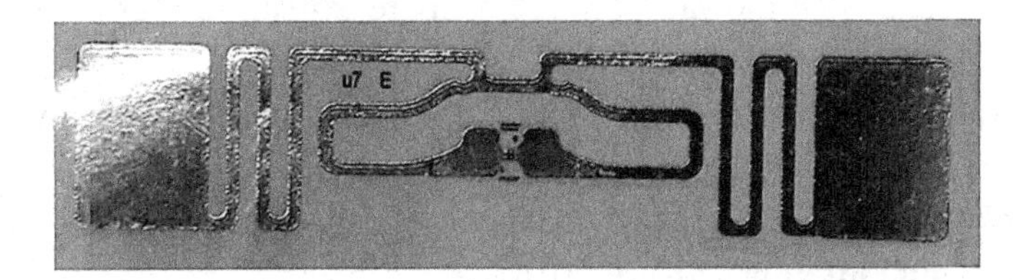

In this image, it is clear the integrated circuit (the "chip") is the dark spot just above the + sign. The wavy lines are the antenna. This tag might have been used for inventory.

Figure 3. Amazon Dot Echo RFID
Source: Spekking (2021)
Required attribution: © Raimond Spekking / CC BY-SA 4.0 (via Wikimedia Commons)

RFID antennas are considered "dumb" in that they are not self-powered. The transmitted waves either follow a horizontal or vertical plane – this polarity can make a difference in how far-reaching the antenna is in attempting to read a tag. There are also circular-polarity antennas, which use both horizontal and vertical planes ("The Beginner's Guide," 2019).

More About the Dataflow Process

Crookes (2018) explains that "typically, the data are stored in the form of a unique serial number, which is in turn used to unlock information stored elsewhere" (p. 38). This means any gathered data are not stored in the chip; a computer network with a separate database is needed to connect to the information.

The data collected through the transponder moves within a wireless local area network (WLAN). LANs have been around since the 1970s and by the 1980s have been subject to IEEE standards. The wireless versions have been in use since the early 1990s ("What is a Local Area Network," n.d.). This brief history helps to put perspective on data transmission: there are standards, and changes in the future are always possible.

The network sends the data to servers. A LAN server is a computer which maintains files and the database that holds the data collected by the transmission from the tag or chip. The server also holds applications allowing for users to see the data ("What is a Local Area Network," n.d.); this can be done via a website, mobile application, or some other interface.

Comparisons to Other Technologies

Near-field communication (NFC band) is a subset of RFID technology. It uses a high frequency, only works bidirectionally, and the range is very small (Amsler, 2021).

Barcodes have been in use for inventory for quite some time, but compared to RFID tagging, barcodes need to be closer for the scanning process and in direct line-of-sight, reading is slower, and since it is printed and placed on the outside of an item, the code is susceptible to damage (Amsler, 2021). Additionally, due to the line-of-sight requirement, the barcode must be placed such that the packaging does not occlude any portion of the product label. Barcodes are scanned for reading in one direction ("QR Code Security," n.d.). Reprogramming a chip when necessary is a quicker process with RFID than changing bar codes (Kaur et al., 2011). Moreover, RFID reads data faster, can read multiple tags at the same time, and works within a larger physical range (Álvarez López et al., 2018).

QR codes, also known as 2-dimensional code ("Technical Explanation," 2018) or Quick Response, are composed of a block of pixels and can be read without a special device; many smart phones can read these block-shaped codes. Invented in 1994 by a Japanese car manufacturer to manage parts, QR codes can be read top-to-bottom and left-to-right, this means more data can be stored in a QR code than in a bar code ("QR Code Security," n.d.). They still need to be printed or otherwise visually displayed.

VARIOUS BUSINESS USES

RFID is widely used and accepted today to manage inventory, complete transactions, and allow access to buildings, hotel rooms, and workspaces; residents of gated neighborhoods can easily enter their communities, wristbands can identify babies in hospitals, and many other identification and access needs are met through this technology. A few other common applications are illustrated below in an effort to show the benefits of this technology, and some history is included for perspective.

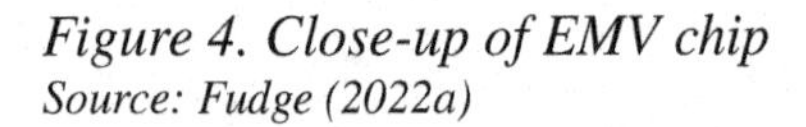

Figure 4. Close-up of EMV chip
Source: Fudge (2022a)

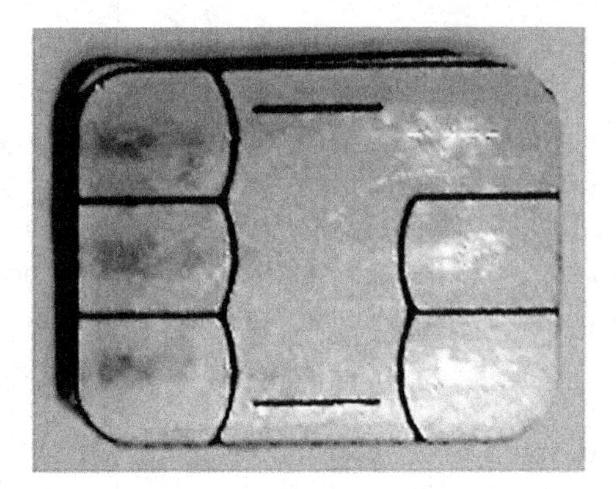

Banking

A very common usage is in the banking industry. The plastic credit card did not appear until 1959; in the following decade, a magnetic strip was added to contain data. The EMV chip shown below was added in the 1980s but not standardized until 1996 when Europay, Mastercard, and Visa together developed a standard; these can use encryption (Steele, 2022).

While EMV chips are not involved in RFID technology, they are still widely in use and like magnetic strips, serve as backups for when contactless payment is not available. Contactless payment has been possible since approximately 2010 and is identified by the wireless waves symbol, shown in the image below to the right of the EMV chip (Steele, 2022).

Identity

For quite some time now, RFID tags have been included in student ID cards, and in some countries, in national identity cards. Tags have even been added to schoolbags and children's clothes in some schools in Asia as a safety precaution (Kamal, 2013). Passports also include RFID technology, and tagging helps travelers move more quickly through checkpoints and decreases instances of counterfeiting (Taghaboni-Dutta & Velthouse, 2006).

It should be noted the U.S. states of Michigan, Minnesota, New York, Vermont, and Washington offer an "enhanced" driver's license (EDL) that includes RFID biometric data. This card can be used as an alternative to obtaining a passport for travel to and from Canada, and Canadian citizens in British Columbia Manitoba, Ontario, and Quebec likewise can apply for an EDL (U.S. Department of Homeland Security, 2022).

Libraries

Library books have in the past had magnetic strips inserted for reading checkout information; this does not work with some items such as video tapes, however, and only one item at a time can be checked out

Figure 5. Debit/credit card showing chip
Source: Mattes (2018)

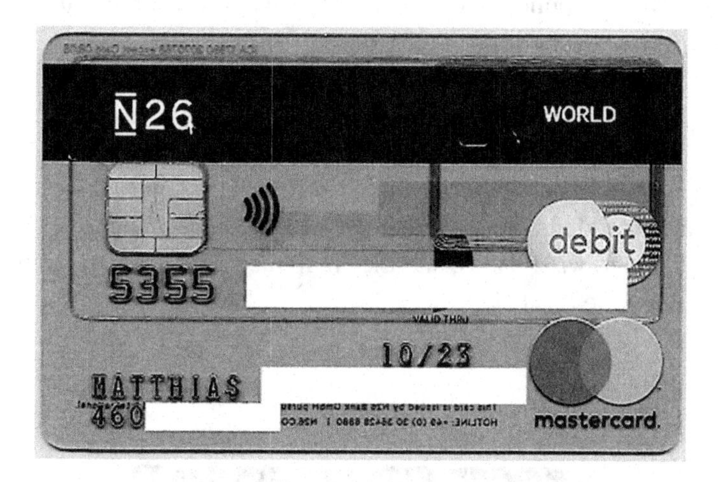

if only magnetic strips are used (Molnar & Wagner, 2004). Today, RFID tags are more widely employed for checkout purposes. A variety of shapes are shown below.

Figure 6. RFID used in libraries
Source: Grika (2006)

Medicine

Hospital wristbands and medicine containers can be fitted with RFID tags or chips; tracking patients with this technology can reduce the risk of grievous errors caused by issues such as medicines having similar packaging and patients who may have allergic reactions to particular medications; it helps to verify patients are actually receiving the care that has been prescribed (Álvarez López et al., 2018). This also helps with inventory and protection against theft (Kamal, 2013). Paaske et al. (2017) also noted the ability to watch over location and temperature data. Wearables such as bracelets may also be helpful for ensuring the location and safety of patients with dementia or Alzheimer's disease (Kamal, 2013).

Additionally, employee badges may contain RFID technology to help identify the closest staff to assist in patient emergencies. Mobile applications can be connected to the monitoring system for ease of use and caretaker mobility (Álvarez López et al., 2018).

Sports

The Boston Marathon has been using RFID since 1999 (Taghaboni-Dutta & Velthouse, 2006) and most races today use it for tracking individual participants. The tags can be pinned to runner numbers, attached to runners' shoes, or affixed to wrist bands, and they help race organizers to monitor each participant's location and timing in real time.

Figure 7. Race timing RFID bracelet
Source: Tomkarlo (2009)

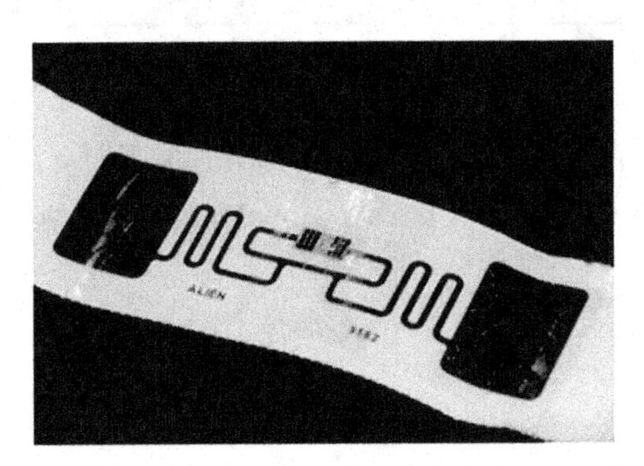

Transportation

Toll roads use RFID technologies to collect tolls; rather than force every vehicle to stop at every booth, systems such as FasTrack and E-Z Pass use hard tags to help save travel time (Taghaboni-Dutta & Velthouse, 2006). Other positive effects include alleviating traffic congestion in tool booth areas and reducing fuel consumption (Bhaskar Rao et al., 2021), thereby providing an ecological benefit.

TRACKING LIVING BEINGS

An interesting use of RFID is the tracking of living beings, both animal and human. This approach differs from the typical business use of item tracking. Unlike merchandise, people and animals can move between locations under their own power, and often of their own free will. In certain situations, it makes sense to track missing beings. For instance, a person may be interested in ensuring the safety of a beloved pet, a farmer may have an interest in monitoring livestock health, zookeepers might need to collect information to better care for their animals, or scientists may be interested in real-time observations of wildlife behaviors.

Figure 8. Apparatus for inserting chips into pets and capsule-shaped pet microchip
Source: Reinraum (2003)

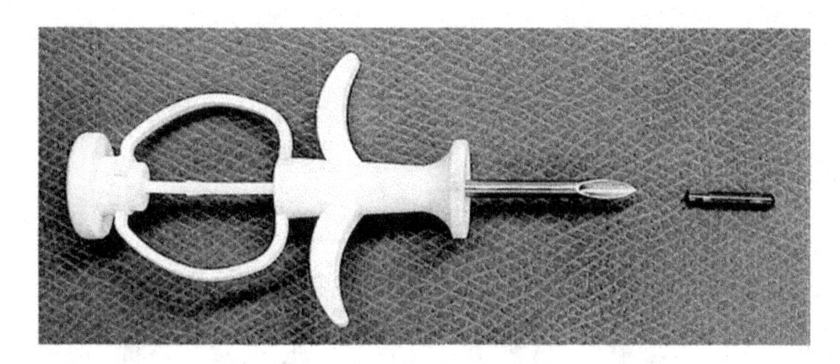

Animals

RFID chips are used in each of these situations, and can be either carried or implanted. When applied to animals, a carried chip is "carried" or attached in some way to the animal. This can be something that is worn, such as a collar or an ankle bracelet, or it can be more permanent, such as an ear tag worn by piercing the animal's ear. This is distinguished from an implanted chip (Saeed & Green, 2019), which is obligatory for pets in some countries and localities (Walsom, 2014).

Microchips for this purpose are small (usually about the size of a grain of rice), capsule-shaped devices implanted using a special hypodermic syringe near the animal's shoulder blades, within the subcutaneous tissue (Legallet et al., 2017). Lost pets found by a good Samaritan can thus be returned to the owners, although owners need to ensure personal data are kept up-to-date with the company that holds the data. Other than owner transfer or lack of up-to-date contact information, common problems include failure of a reading device to read the chip, swelling or infection at the injection site, and migration of the chip within the body (Legallet et al., 2017).

Additional benefits to this identification for companion animals are the cost savings for shelters and rescues, as they can more quickly return pets to their rightful homes, and the potential reduction in euthanasia for animals not connected to their owners (Walsom, 2014).

Folger (2016) notes it is not always the legal owner who registers the chip; it could be a shelter or other agency, and transfer to a true owner does not always occur. This can create ethical and legal dilemmas for veterinarians who find a chip that is not registered to the person who brought in a pet for care.

Pets are not the only animals that may experience this technology. Chips have been suggested for dairy cow temperature monitoring to assist farmers in identifying the early stages of illnesses. Rising body temperatures in bovines may be due to an immune response to infection and disease, and automating the collection of temperature data can increase herd management efficiency as well as be less invasive and frightening for the cow than having to endure rectal temperature measurement (Woodrum Setser et al., 2020).

Scott et al. (2016) promotes the use of RFID tagging in zoos to supply real-time data, continual work despite weather conditions, and provide data that can be analyzed for better understanding of captive animal social behaviors and lead to better management strategizing.

The World Wildlife Fund (WWF) tracked white-lipped peccaries in the Amazon using RFID ear tags in an effort to understand how the large-herd animals work within the rainforest ecosystem. To be able to read the tagged animals, the scientists placed readers next to strategically placed salt licks (Spivey, 2016).

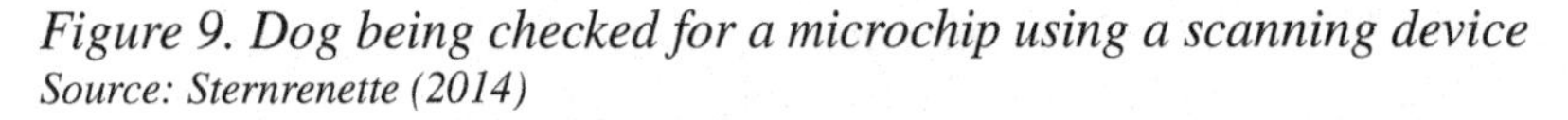

Figure 9. Dog being checked for a microchip using a scanning device
Source: Sternrenette (2014)

Testud et al. (2019) were dissatisfied with camera traps in evaluating a wildlife corridor's usefulness, so employed passive tagging to successfully study the movement of toads and two species of ground beetles. They also mentioned previous studies involving salamanders, fish, lizards, and voles.

RFID Implanted in Humans

The invasion of a chip into the human body presents ethical, legal, and moral questions. Microchipping human beings has been under consideration for some time, and performed since at least 2015 (Tangermann, 2018). Like chips in pets, the device is capsule-shaped. It is inserted in the fleshy area between the thumb and forefinger via a special syringe. The images below show the location and relative size of the chip.

Figure 10. Dog being checked for a microchip using a device with different design
Source: Nickel (2015)

Figure 11. Human hand immediately after insertion of chip (skin coloration due to antiseptic application)
Source: Graafstra (2005)

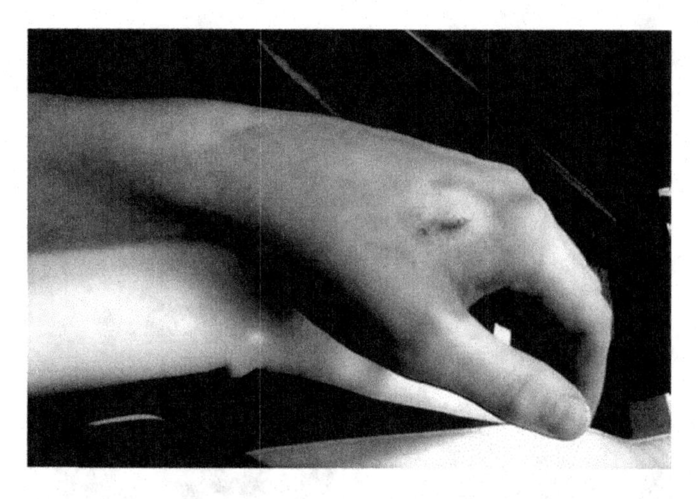

Benefits

There can be some benefits to human RFID implants. Individuals in Sweden voluntarily underwent microchip implantation so they could have quick and easy methods of unlocking office doors, getting into their cars without a key fob, purchasing train tickets, or attempting access to any number of other things (Tangermann, 2018). Once they learned of what the Swedes had been doing, a company in River Falls, Wisconsin microchipped those of its employees who agreed to the procedure in 2017. Now their employees can gain access to their building, work computers, and even vending machines by simply waving a hand (Metz, 2018) – without the need for keys and passwords (Turner, 2020).

Microchipping humans also facilitates easy availability of medical data, which can be especially useful in case of emergencies (Merrill, 2007; Nicola, 2018). As long as the data are kept up-to-date, identification of medical needs, patient tracking, and safety might be improved (Paaske et al., 2017), even if a patient is not conscious (Žnidaršič et al., 2021) or cannot communicate due to Alzheimer's disease – although obtaining permission from an Alzheimer's patient to implant a chip in the first place may be difficult and raises ethical, if not legal, concerns (Merrill, 2007). If the patient has been declared incompetent but adamantly asserts that they do not wish to have a chip implanted, one questions if the wishes of the patient should be followed, or those of the caregiver who is only thinking of the patient's best interests and safety. Also to be considered is whether the patient has executed a healthcare power of attorney thus clearly defining the legality, but this could be argued that it does not mitigate the ethical dilemma.

Humans already endure some invasive technology such as pacemakers and insulin pumps (Žnidaršič et al., 2021), and these are widely accepted. It has also been suggested that RFID microchips can be used to provide "COVID vaccine passports" (Theisen, 2021), although this concept has not gained public acceptance at this time.

In Japan, a child monitoring system has been in place since 2004. Even the most caring and diligent caregivers can lose track of a child while performing other necessary duties. According to Zedner (2003), such surveillance systems offer caregivers a sense of wellbeing and security knowing the monitoring system is in place. An embedded chip is also not something that can be lost, like an ID card, bracelet, or child's backpack. From a moral standpoint, this seems to be appropriate, especially for small children;

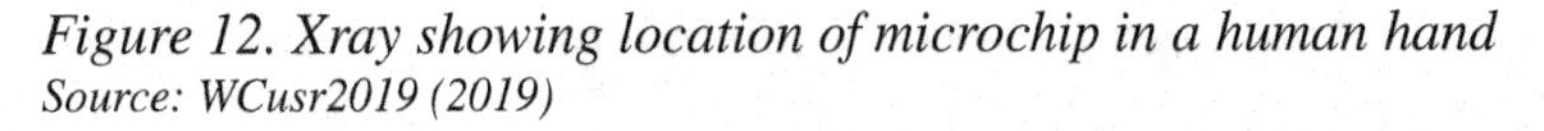

Figure 12. Xray showing location of microchip in a human hand
Source: WCusr2019 (2019)

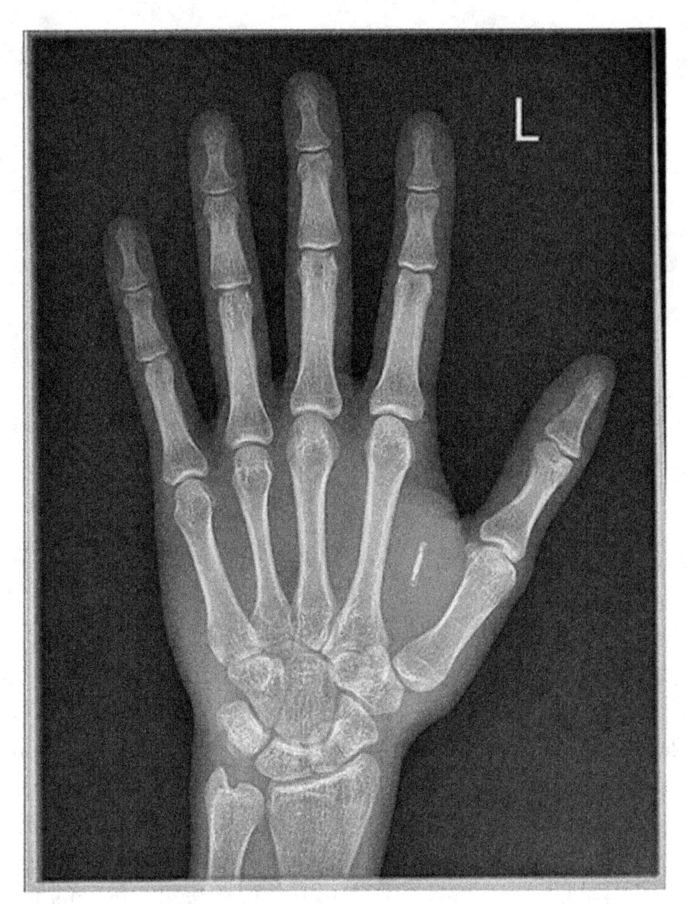

however, as the child ages, the ethicality of tracking a child, even a minor, begins to be called into question. At age 5, the idea of tracking a child may lack any apparent ethical concern, but questions arise such as determining at what age the child should have a say in whether (and how) they are tracked, if it is ethical to track them until they are a legal adult, and when the child can exercise the right to have the chip removed.

Concerns and Analysis

While there is very little question about the ethics of using RFID to monitor merchandise or animals and microchipping people does have some benefits, such tracking raises both ethical and legal issues. Unlike pets, livestock, zoo animals, and wildlife, people have the capacity to choose whether they will be tracked and to understand the ramifications of those actions. Thus, it is important to consider the motivation for being tracked as well as the benefits and challenges for both the tracker and the person being tracked.

When a person with an embedded microchip is fully aware and has accepted a situation – for example, waving their hand to gain access to their workplace – there is no ethical alarm. However, there are some situations in which ethics must be questioned, such as

1. An involuntary RFID reading at the behest of a third party, wherein a benefit to the person being tracked may not be readily or immediately apparent.
2. An involuntary reading wherein the person does not have a choice in whether they are tracked.
3. An undeclared reading, wherein a person is either unaware they are being tracked, or aware but choose to ignore the consequences as the action is a part of the person's habits or due to social norms that make tracking acceptable.

In the first case, a person is tracked for their benefit at the behest of a third party. This includes children and medical patients or people with a diminished mental faculty. According to the FBI, nearly 350,000 people under the age of 21 and nearly 30,000 individuals with a proven mental or physical disability went missing in 2020 (U.S. Federal Bureau of Investigation, 2020). Although RFID is different from global positioning system (GPS) technology, which requires a satellite, the growth in ubiquitous computing makes RFID tracking a realistic option. RFID requires a reader to obtain a signal. Such readers can easily be placed in strategic locations such as entrances to schools, shopping malls and stores, gas stations, amusement parks and other public locations children or individuals with mental impairments may frequent. Similar readers can be placed at the entrances to hospitals, assisted living and senior care facilities, and in retirement communities. The benefit to such a system is clear. By monitoring children's movements, a parent (or the police) has the ability to determine a child's location should the child wander off (or be taken), or to enable a level of independence as the child matures. Likewise, those with cognitive impairment or diminished mental faculties can be tracked, allowing them to live alone, or with some level of independence while ensuring their safety.

Involuntary tracking occurs when a person is not given a choice with regards to wearing the RFID tag. An example of this is prisoner tracking, both within and outside a prison. Within smart prisons, both the guards and the inmates are required to wear RFID tags. This provides information including inmate and guard movement within the prison, interactions between guards and inmates, and interactions between inmates and objects within the prison (Sun, 2022). Involuntary tracking may also be used in conjunction with a house arrest sentence. Using RFID equipped ankle bracelets allows an offender to complete their sentence at home rather than in a correctional facility (Nellis, 2021).

The third situation is unique in that it involves a passive acceptance to surveillance. One example is in hospital use of RFID tagging; patients should be told what is being surveilled (Parks et al., 2014), although patients who arrive in a state of unconsciousness would not be able to sign any documentation to prove their acceptance of any risks. Another example of this would be the use of a shopping card or a cell phone. This is generally the result of something that is either a habit, or a social norm generally accepted in society. In this case, the individual may not be aware that they are being tracked, or they may be aware, but choose to ignore the consequences either out of habit or because they are receiving some benefit. For example, people choose to use "loyalty" cards; these are linked to a specific store and generally offer discounts or targeted advertisements based on the user's shopping behavior. The information collected as a result of card use is supposed to be used for the shopper's benefit, but also for the benefit of the store. Parks et al. (2014) note that there have been "protest cases" in Germany related to hidden RFID in loyalty cards. A third example is use of a cellphone. Most people keep their cellphones with them at all times and many have applications therein that allow them to be tracked, either explicitly (Find My™) or implicitly (Google™). Such applications were advantageous for COVID contact tracing during the pandemic (Lewis, 2021). However, due to the COVID contact tracing effort, the public has become more aware of ways in which personal information is collected using various technologies, and similarly more averse to tracking technology. A 2017 Deloitte survey found that 91% of consumers fail to read the terms of service (97% for consumers age 18-34) so they are unaware their information is being collected (Guynn, 2020) or they choose to ignore the consequences because of the perceived benefits of using their phone apps.

While there have always been questions as to the ethical, legal, and social implications of tracking people, these have become more relevant in recent years due to the pandemic. Once medical professionals determined the coronavirus was primarily spread through close contact with other people, contact tracing became a key tool in limiting the spread of the virus. In areas where numbers were low, manual tracing served well enough, but with surges and larger numbers in many areas, a more automated system using RFID tagging and cell phones could be implemented (Rajasekar, 2021). By identifying infected individuals automatically, those who had been in contacted with infected individuals could be quickly located and quarantined to slow the spread. Even though the coronavirus presented a clear public health danger and because the technology was not in place, accurate contact tracing proved difficult. Some countries took strict measures to determine where a person had been and with whom they might have been in contact. For instance, China accessed mobile and bank data to identify individuals who may have been exposed to an infected individual (Sternlich, 2020). Other countries (such as the United States and Australia), however, implemented a more voluntary approach by providing apps that users could download onto their phones to help improve contact tracing and notification (Kretzschmar et al., 2020). Despite the severity of the virus, people were generally reluctant to download and use these apps (Barbaschow, 2020). When that was not effective, Apple and Google included Bluetooth-based contact tracing in their operating systems, which were automatically installed on users' phones with updates. These required the user to take action to turn them off, rather than requiring the user to choose to install the app. However, this created a considerable pushback from users who felt that it was unethical, and possibly illegal, to add tracking features to their phones without their explicit consent. This reluctance to be tracked can be traced back to concerns related to privacy, and lack of trust (Chan & Saqib, 2021). As explained by Banafa (2021) this mistrust is related the human's desire for autonomy and, when possible, anonymity. Surveillance by governments or hackers is surely possible using RFID chip technology, and identify theft is possible. The U.S. Constitution's Fourth Amendment protects against government searches, but not private-sector searches (Turner, 2020). However, ethical and legal questions arise when the government requests (or requires) private-sector companies to turn over tracking information.

Cyberthreats – here called *biohacking* since it also includes genetic material of living beings – are also of great concern. Banafa (2021) explains both health and data safety as potential issues. Depending on the transmission of data from a person's chip, a hacker might be able to steal someone's identity. Nicola (2018) reminds that all technology has the potential for being hacked, and implanted chips could still be hacked to send malicious code. As mentioned regarding pets, microchips are bodily invasive, and pose a threat via movement or through the technological sensors that "[touch] upon your hand, your heart, your brain and the rest of your body – literally" (Banafa, 2021, para. 6). Tangermann (2018) states this more bluntly as a "digital security nightmare" (para. 9).

In addition to viruses, eavesdropping, spoofing, denial-of-service, and other typical network attacks, power consumption can be watched to set up an attack (Smiley, 2016). Waiting at a bus stop can give a hacker an opportunity to "sniff" at a microchipped person's data (Metz, 2018). As noted in Loss Prevention Magazine, with the introduction of RFID enabled credit cards, there was a considerable concern about people surreptitiously stealing credit card information simply by passing by. In contrast to the concerns, however, there was no rash of stolen credit card data and new technologies in place have made it much more difficult for thieves to "sniff" RFID data (Seivold, 2022). Reverse engineering is also possible, although Smiley (2016) notes this would be a time-consuming process for the hacker. Korinchak (2021) adds that devices might be surreptitiously replaced with false readers, tag location can be tracked, and organizations using the technology must ensure both encryption as well as a strong patching process to update software whenever needed. Data expert Friedemann Ebelt notes "you can switch off and put away an infected smartphone, but you can't do that with an implant" (Nicola, 2018, p. 23).

There is the concern of "reader collision" and "tag collision" wherein there is overlap of signals due to simultaneous readings resulting in confusion (Arora et al., 2019). While there are ways to try to prevent collisions from occurring, they are still possible, and it is an extra cost for companies to add anticollision algorithms into the coding; Kaur et al. (2011) noted these are typically patented. Signal reflection (where the signal bounces back to the origination) is possible, and cellphones, security systems, and other devices have been known to interfere (Kaur et al., 2011).

Another potential technical issue is that not all countries use the same frequencies for the same purposes, although there is work towards uniformity (Kaur et al., 2011). Whether a chip would work as expected on one continent then might not work at all on another, making some travel expectations dubious.

Additionally, not all private companies take heed of laws. An incident in 2003 made headlines when Walmart and Proctor & Gamble used RFID tags and cameras to surreptitiously watch customers surveying a lipstick display (Taghaboni-Dutta & Velthouse, 2006).

ID cards and passports with RFID chips could potentially collect information while traveling, the technology "offers almost limitless surveillance capacities" (Kamal, 2013, p. 92) if security is not included in development and implementation. Indeed, encryption is necessary to ensure those who do not have sufficient clearance cannot obtain access to data (Kaur et al., 2011). Likewise, concerns of RFID sniffing will continue to increase as the Internet of Things becomes more ubiquitous.

A few scenarios illustrate what can happen when seemingly minor information is stolen:

- A shopping trip can end up with personal habits being collected and sold to third parties with the intent of bombarding the shopper with unwanted advertising through text messages, mail, and email (Turri et al., 2017).
- Workplace conversations can be tracked for tone and vocal volume even from a nonimplanted badge (Sheng, 2019). If a worker had an RFID implant, the line between workplace surveillance and parking-lot listening or other locations may be blurred.
- A person expecting a promotion could potentially be passed over due to gathered data about physical activities and sleep patterns the employer has scrutinized as unhealthy; this can already be collected via Fitbits but RFID can furtively collect the data (Turner, 2020).
- A trip to a drug store for prescription medication can result in the information about the medical condition being shared with potential employers (Turri et al., 2017).

Sheng (2019) reported prepandemic that workplace surveillance has been on the rise; computer usage, employee movement within the workplace, and email monitoring is common. The data may have changed due to the pandemic move to virtual work for many employees, but the fact remains employees have a right to be concerned about privacy. With more employers allowing employees to work from home, implanted RFID to unlock employer-provided technology could be used to surreptitiously monitor other employee activities.

Indeed, there are already many electronic devices and technologies in use for surveillance in the workplace. Turner (2020) cited the United Parcel Service (UPS)'s "telematics" system to record parking, speed (even when backing up the truck), how much time is spent in buckling seatbelts, and generally everything a delivery person does. Walmart and Amazon also are mentioned for micromanaging employees (Turner, 2020). While some of these are managed through GPS and other sensors, the idea that all of these actions and activities could possibly be monitored so closely even when away from work is alarming and creates ethical concerns.

Similar to physical risks with animals, humans may find infection at the point of insertion; there is potential for components to corrode (Banafa, 2021). According to Eliaz (2019) the body creates a harsh environment and it is difficult to identify metals that will not corrode when implanted. Corrosion of implants can release toxins into the body which can lead to serious health issues. When a medical device is implanted, the following reactions occur: "injury, blood-material interactions, provisional matrix formation, acute inflammation, chronic inflammation, granulation tissue development, foreign-body reaction (FBR) and fibrosis (fibrous capsule development)" (Anderson et al., 2008, p. 87). These reactions fundamentally change the area of implantation. Additionally, electrical current can change cellular behavior, corrosion can affect the chemical makeup of the surrounding tissue, and cellular metabolism can be negatively affected. There is also concern of the chip migrating away from the point of insertion (Žnidaršič et al., 2021). Le Calvez et al. (2006) documented chip migration in animals and found tumors sometimes formed near or encased the implant.

As mentioned above, insulin pumps may also be embedded into the human body – and they can be subject to attacks by hackers, as Medtronic customers found out in 2019 when their MiniMed devices were recalled (Picchi, 2019). Likewise, poorly-designed RFID chips may need to be recalled, causing additional concern and potential physical pain. If a chip is specific to a particular workplace, a person could be subject to chip removal upon leaving employment, or even multiple chip implants for access or tracking for different reasons. For instance, a person may have an implant for work access, one for access to their credit and payment information, and yet another for access to their housing complex. As previously mentioned, a lack of standardizing, and required access to an external server and/or database for information may make multiple chips necessary.

Implanted chips may also need to be removed due to the need for necessary technology updates (Metz, 2018), and those who refuse a chip in the future could possibly be refused a job or a crucial service if the social norm is to be chipped (Banafa, 2021). Suggested alternatives for many situations include devices that are wearable instead of implanted, although they would be more susceptible to damage, theft, or loss (Žnidaršič et al., 2021).

Tangermann (2018) reported "implant parties," during which attendees all receive a chip. This kind of peer pressure, whether familial or workplace-related, is ethically questionable.

REGULATING THE TECHNOLOGY

Arora et al. (2019) blames a previous lack of standards with the slow introduction of RFID technology, and Landt (2001) claimed over two decades ago that there were already more than 350 RFID patents.

Today there are clear standards set by the International Organization for Standardization (ISO). A search of the organization's database currently reveals 353 results, with published standards focusing on everything from general RFID tags to interrogator requirements, data protocols, data capture methods, and many specific implementations of the technology (International Organization for Standardization, n.d.). More can be expected to be written as new uses are proposed, processes refined, and other changes occur.

There are some state laws regulating the use of RFID technology in ID cards and tracking of students, and the European Union, Japan, and Canada have laws that apply to the use of this technology. In terms of human microchipping, however, there are no federal U.S. laws, although several states have passed laws to prohibit workplaces from *requiring* employees to be microchipped, including North Dakota, California, Oklahoma, and Wisconsin (Turri et al., 2017).

The U.S. Federal Trade Commission (FTC) is also known to be watching for privacy concerns (Turri et al., 2017). The Electronic Communications Privacy Act and Communications Act helps to regulate companies' official use of data; states also have wiretapping laws to follow (Sheng, 2019). However, the wording of such laws has not always kept up with current technologies and is vague at times. In addition, bring your own device (BYOD) policies may reduce the expectation of privacy even further (Turner, 2020). The person who has an embedded chip may be connected to far more devices than they know.

This means the RFID industry runs mostly on self-regulation at this time; while it provides flexibility in development and implementation of new ideas, it also means legally there are no demands to conform (Turri et al., 2017). Important guidelines have been suggested by the International Chamber of Commerce: consumers must be educated about their rights and potential issues of RFID, anonymity must be preserved where possible, data collection must be limited to only that which is necessary, tracking must be performed only with consent in writing, and an expiration date should be set so the data does not "live" forever (Turri et al., 2017).

While there are some well-laid foundations, it is clear the law needs to catch up with the newest versions and uses for the technology and there are likely issues that have yet to be addressed. More research into the legal and ethical concerns is necessary.

CONCLUSION

While there are clear benefits in both health and convenience to implanted RFID use, to gain widespread acceptance, the health and privacy related issues must be adequately addressed. Implanting a foreign object into the body creates serious concerns. There are numerous examples of medically approved implants being recalled due to health issues (see e.g., breast implant illness, surgical mesh, Lap-band, ASR metal hip implant, vena-cava filters, implantable defibrillators.) RFID chip implants can cause health issues in pets from bruising and swelling at the injection site to "chip failure, migration, abnormal mass and tumor formation, infection, rejection and death at levels higher than originally believed" ("Protect Pets," n.d., More Risks section). Other potential health issues include allergies and possible effects on emotional behavior and the nervous system (Fram et al., 2020).

Additionally, legal issues need to be adequately addressed, not just locally, but worldwide. Because an implanted chip cannot be easily removed and left at home, individuals with implants will be subject to the RFID tracking laws of whatever country they visit. This is an issue that must be considered before widespread chip implants are implemented.

As with any "new" technologies, consumer acceptance is crucial; this is often based on the perception of feeling safe from harm and an expectation of privacy (Turri et al., 2017). Adequately addressing the health, legal, and privacy concerns may facilitate acceptance if people do not feel their rights are being compromised.

There is unease about the future of using embedded RFID chips in humans including the many questions about undue dissemination of data, loss of personal autonomy, and a need for legislation to control the potential complications. Whether the benefits outweigh the concerns requires further investigation.

REFERENCES

Álvarez López, Y., Franssen, J., Álvarez Narciandi, G., Pagnozzi, J., González-Pinto Arrillaga, I., & Las-Heras Andrés, F. (2018, August 13). RFID technology for management and tracking: e-Health applications. *Sensors (Basel), 18*(8), 2663. doi:10.339018082663 PMID:30104557

Amsler, S. (2021, March). *RFID (radio frequency identification).* TechTarget: IoT agenda. https://internetofthingsagenda.techtarget.com/definition/RFID-radio-frequency-identification

Anderson, J. M., Rodrguez, A., & Chang, D. T. (2008, April). Foreign body reaction to biomaterials. *Seminars in Immunology, 20*(2), 86–100. doi:10.1016/j.smim.2007.11.004 PMID:18162407

Arora, A., Gupta, S., & Sehgal, U. (2019, April). Reader collision and tag payoff matrix resulting in a Wal-Mart RFID. *International Journal of Computer Science and Network, 8*(2), 154–156.

Banafa, A. (2021, April 5). *Technology under your skin: 3 challenges of microchip implants.* BBVA Open Mind. https://www.bbvaopenmind.com/en/technology/innovation/technology-under-your-skin

Barbaschow, A. (2020, December 9). *Australian committee calls for independent review of CovidSafe app.* ZDNet. https://www.zdnet.com/article/australian-committee-calls-for-independent-review-of-covidsafe-app

Bhaskar Rao, J., Hrithik, P. S. S., Sushma, I., Karthikeya, M. S. S. G., Meghana, T., & Anurag, K. (2021, October). Toll collection system using image processing. *IUP Journal of Computer Sciences, 15*(4), 34–44.

Chan, E. Y., & Saqib, N. U. (2021, June). Privacy concerns can explain unwillingness to download and use contact tracing apps when COVID-19 concerns are high. *Computers in Human Behavior, 119*, 106718. doi:10.1016/j.chb.2021.106718 PMID:33526957

Chang, S.-I., Huang, S.-Y., Yen, D. C., & Chen, Y.-J. (2008, September). The determinants of RFID adoption in the logistics industry - A supply chain. *Communications of the Association for Information Systems, 23*, 197–218. https://aisel.aisnet.org/cais/vol23/iss1/1 doi:10.17705/1CAIS.02312

Comparing different types of RFID tags. (2018, November 12). Resource Label Group. https://www.resourcelabel.com/comparing-different-types-of-rfid-tags

Crookes, D. (2018, September 19-October 2). Microchip implants. *Web User,* (458), 38-39.

Eliaz, N. (2019). Corrosion of metallic biomaterials: A review. *Materials (Basel, Switzerland), 12*(3), 407. doi:10.3390/ma12030407 PMID:30696087

Folger, B. (2016, September). Microchips, ownership, & ethics. *Veterinary Team Brief, 4*(8), 38–42.

Fram, B. R., Rivlin, M., & Beredjiklian, P. K. (2020). On emerging technology: What to know when your patient has a microchip in his hand. *The Journal of Hand Surgery*, *45*(7), 645–649. doi:10.1016/j.jhsa.2020.01.008 PMID:32164995

Fudge, T. P. (2022a, March 15). *Chip removed from debit/credit card casing* [Image]. Academic Press.

Fudge, T. P. (2022b, April 28). *RFID from commercial packaging* [Image]. Academic Press.

Graafstra, A. (2005, March 23). *RFID hand 2* [Image]. Licensed by Creative Commons Share-Alike 2.0 Generic. No changes made to the image. https://commons.wikimedia.org/wiki/File:RFID_hand_2.jpg

Grika. (2006, March 30). *RFID tags* [Image]. Licensed by Creative Commons 2.5 Generic. Image was rotated but otherwise unchanged. https://commons.wikimedia.org/wiki/File:RFID_Tags.jpg

Guynn, J. (2020, January 29). What you need to know before clicking 'I agree' on that terms of service agreement or privacy policy. *USA Today*. https://www.usatoday.com/story/tech/2020/01/28/not-reading-the-small-print-is-privacy-policy-fail/4565274002

International Organization for Standardization. (n.d.). *Online browsing platform [search for RFID]*. Retrieved April 30, 2022 from https://www.iso.org/obp/ui/#search

Kalantri, R., Parekar, A., Mohite, A., & Kankapurkar, R. (2014). RFID based toll collection system. *International Journal of Computer Science and Information Technologies*, *5*(2), 2582–2585.

Kamal, A. H. M. (2013, July-December). Legal aspects of the security concerns surrounding Radio Frequency Identification (RFID) technology. *ASA University Review*, *7*(2), 91–110.

Kaur, M., Sandhu, M., Mohan, N., & Sandhu, P. S. (2011, February). RFID technology principles, advantages, limitations & its applications. *International Journal of Computer and Electrical Engineering*, *3*(1), 151–157. doi:10.7763/IJCEE.2011.V3.306

Korinchak, D. (2021, October 22). *RFID security vulnerabilities.* Cyber Experts. https://cyberexperts.com/rfid-security-vulnerabilites

Kretzschmar, M. E., Rozhnova, G., Bootsma, M. C. J., van Boven, M., van de Wijgert, J. H. H. M., & Bonten, M. J. M. (2020, July 16). Impact of delays on effectiveness of contact tracing strategies for COVID-19: A modelling study. *The Lancet. Public Health*, *5*(8), e452–e459. doi:10.1016/S2468-2667(20)30157-2 PMID:32682487

Kumar, S. G., Prince, S., & Shankar, B. M. (2021, May). Smart tracking and monitoring in supply chain systems using RFID and BLE. *2021 3rd International Conference on Signal Processing and Communication (ICPSC)*, pp. 757-760. https://ieeexplore.ieee.org/document/9451700

Landt, J. (2001, October 1). *Shrouds of time: The history of RFID*. The Association for Automatic Identification and Data Capture Technologies (AIM). https://transcore.com/wp-content/uploads/2017/01/History-of-RFID-White-Paper.pdf

Le Calvez, S., Perron-Lepage, M. F., & Burnett, R. (2006, March 6). Subcutaneous microchip-associated tumours in B6C3F1 mice: A retrospective study to attempt to determine their histogenesis. *Experimental and Toxicologic Pathology*, *57*(4), 255–265. doi:10.1016/j.etp.2005.10.007 PMID:16427258

Legallet, C., Mankin, K. T., Spaulding, K., & Mansell, J. (2017, July/August). Granulomatous inflammatory response to a microchip implanted in a dog for eight years. *Journal of the American Animal Hospital Association*, *53*(4), 227–229. doi:10.5326/JAAHA-MS-6418 PMID:28535132

Lewis, D. (2021, February 26). Contact-tracing apps help reduce COVID infections, data suggest. *Nature*, *591*(7848), 18–19. doi:10.1038/d41586-021-00451-y PMID:33623147

Maschinenjunge. (2008, March 11). *RFID Chip 003* [Image]. Licensed by CC Share-Alike 3.0 Unported. No changes made to the image. https://commons.wikimedia.org/wiki/File:RFID_Chip_003.JPG

Mattes. (2018, November 24). *MasterCard (transparent) 2018, Bank- N26* [Image]. Licensed under Creative Commons 2.0 Germany. No changes made to the image. https://commons.wikimedia.org/wiki/File:MasterCard_(transparent)_2018,_Bank-_N26.jpg

Merrill, M. (2007, July 30). *Human-implantable RFID chips: Some ethical and privacy concerns.* Healthcare IT News. https://www.healthcareitnews.com/news/human-implantable-rfid-chips-some-ethical-and-privacy-concerns

Metz, R. (2018, August 17). *This company embeds microchips in its employees, and they love it.* Technology Review. https://www.technologyreview.com/s/611884/this-company-embeds-microchips-in-its-employees-and-they-love-it

Molnar, D., & Wagner, D. (2004, October 25). Privacy and security in library RFID: Issues, practices, and architectures. [Conference presentation] *Proceedings of the 11th ACM Conference on Computer and Communications Security*. Washington DC. 10.1145/1030083.1030112

Nellis, M. (2021, July 30). Towards predictivity? Immediacy and imminence in the electronic monitoring of offenders. In B. Arrigo & B. Sellers (Eds.), *The pre-crime society: Crime, culture and control in the ultramodern age* (pp. 341–364). Bristol University Press. doi:10.1332/policypress/9781529205251.003.0016

Nickel, S. (2015, March 2). *Patriot puppies, 731st AMS port dogs open doors to new passengers 150302-F-FT438-003* [Image]. In Public Domain. https://commons.wikimedia.org/wiki/File:Patriot_puppies,_731st_AMS_port_dogs_open_doors_to_new_passengers_150302-F-FT438-003.jpg

Nicola, S. (2018, October 22). Invasion of the body hackers. *Bloomberg Business Week,* (4589), 22-23.

Paaske, S., Bauer, A., Moser, T., & Seckman, C. (2017, June). The benefits and barriers to RFID technology in healthcare. *On-Line Journal of Nursing Informatics*, *21*(2), 10–11.

Parks, R., Hsu, C.-H., & Xu, H. (2014, September 10). RFID privacy issues in healthcare: Exploring the roles of technologies and regulations. *Journal of Information Privacy and Security*, *6*(3), 3–28. doi:10.1080/15536548.2010.10855891

Picchi, A. (2019, June 28). Medtronic recalls insulin pumps because hackers could hijack device. *CBS News*. https://www.cbsnews.com/news/medtronic-insulin-pump-recall-fda-says-hackers-could-hijack-device

Protect pets. (n.d.) ChipMeNot. https://chipmenot.info

QR code security: What are QR codes and are they safe to use? (n.d.). Kaspersky. https://usa.kaspersky.com/resource-center/definitions/what-is-a-qr-code-how-to-scan

Rajasekar, S. J. S. (2021, January 18). An enhanced IoT based tracing and tracking module for COVID-19 cases. *SN Computer Science, 2*(1), 42. doi:10.1007/s42979-020-00400-y

Reinraum. (2003, January 1). *134 2khz rfid animal tag* [Image]. In Public Domain. https://commons.wikimedia.org/wiki/File:134_2khz_rfid_animal_tag.jpg

Rieback, M. R., Crispo, B., & Tanenbaum, A. S. (2006, January-March). The evolution of RFID security. *IEEE Pervasive Computing*, *5*(1), 62–69. https://ieeexplore.ieee.org/document/1593573

Rouse, M., & Shea, S. (2017, December 12). *RFID (radio frequency identification)*. Internet of Things (Tech Target). https://internetofthingsagenda.techtarget.com/definition/RFID-radio-frequency-identification

Saeed, N., & Green, A. R. L. (2019, January 29). *RFID pet monitoring & identification system with RFID*. University of West London. https://core.ac.uk/download/pdf/195384891.pdf

Scott, N. L., Hansen, B., LaDue, C. A., Lam, C., Lai, A., & Chan, L. (2016, August). Using an active Radio Frequency Identification Real-Time Location System to remotely monitor animal movement in zoos. *Animal Biotelemetry*, *4*(16). https://doi.org/10.1186/s40317-016-0108-5

Seivold, G. (2022, March 13). Are RFID-blocking wallets necessary to prevent credit card theft? *Loss Prevention Magazine*. https://losspreventionmedia.com/are-rfid-blocking-wallets-necessary-to-prevent-credit-card-theft

Sheng, E. (2019, April 15). Employee privacy in the US is at stake as corporate surveillance technology monitors workers' every move. *CNBC*. https://www.cnbc.com/2019/04/15/employee-privacy-is-at-stake-as-surveillance-tech-monitors-workers.html

Smiley, S. (2016, June 14). *7 types of security attacks on RFID systems*. Atlas RFID Store. https://www.atlasrfidstore.com/rfid-insider/7-types-security-attacks-rfid-systems

Spekking, R. (2021, May 16). *Amazon Echo Dot (RS03QR) - case - RFID-0589* [Image]. Licensed by Creative Commons Share-Alike 4.0 International. No changes made to the image. https://commons.wikimedia.org/wiki/File:Amazon_Echo_Dot_(RS03QR)_-_case_-_RFID-0589.jpg

Spivey, M. (2016, September 27). *How RFID tags can change wildlife conservation*. Gateway RFID Store. https://gatewayrfidstore.com/rfid-tags-can-change-wildlife-conservation

Steele, J. (2022, January 27). *The history of credit cards*. Credit Cards. https://www.creditcards.com/statistics/history-of-credit-cards

Sternlich, A. (2020, April 30). South Korea's widespread testing and contact tracing lead to first day with no new cases. *Forbes*. https://www.forbes.com/sites/alexandrasternlicht/2020/04/30/south-koreas-widespread-testing-and-contact-tracing-lead-to-first-day-with-no-new-cases

Sternrenette. (2014, January 7). *Ablesen eines Mikrotransponders bei einem Hund* [Image]. Licensed by Creative Commons Share-Alike 3.0 Unported. No changes made to the image. https://commons.wikimedia.org/wiki/File:Ablesen_eines_Mikrotransponders_bei_einem_Hund.JPG

Sun, P. (2022). Prison IOT. In *Smart Prisons.* Springer Singapore. doi:10.1007/978-981-16-9657-2_3

Taghaboni-Dutta, F., & Velthouse, B. (2006, November). RFID technology is revolutionary: Who should be involved in this game of tag? *The Academy of Management Perspectives*, *20*(4), 65–78.

Tangermann, V. (2018, May 14). *All the rage in Sweden: Embedding microchips under your skin.* Futurism. https://futurism.com/sweden-microchip-trend

Technical explanation for RFID systems. (2018, May 8). *Omron.* https://www.ia.omron.com/data_pdf/guide/47/rfid_tg_e_2_1.pdf

Testud, G., Vergnes, A., Cordier, P., Labarraque, D., & Miaud, C. (2019, October 22). Automatic detection of small PIT-tagged animals using wildlife crossings. *Animal Biotelemetry*, *7*(1). https://doi.org/10.1186/s40317-019-0183-5

The beginner's guide to RFID systems. (2019). Atlas RFID Store. https://rfid.atlasrfidstore.com/hs-fs/hub/300870/file-252314647-pdf/Content/basics-of-an-rfid-system-atlasrfidstore.pdf

Theisen, T. (2021, December 26). Microchip implanted under skin could be your COVID vaccine passport. *Orlando Sentinel.* https://www.nny360.com/news/publicservicenews/microchip-implanted-under-skin-could-be-your-covid-vaccine-passport/article_21a09e0a-c63c-5f23-85b7-76f65a4a8932.html

Tomkarlo. (2009, August 9). *Back side of disposable RFID tag used for race timing* [Image]. In Public Domain. https://commons.wikimedia.org/wiki/File:Back_side_of_disposable_RFID_tag_used_for_race_timing.png

Turner, W. (2020, March 31). Chipping away at workplace privacy: The implementation of RFID microchips and erosion of employee privacy. *Washington University Journal of Law and Policy*, *61*(1), 275–297. https://openscholarship.wustl.edu/law_journal_law_policy/vol61/iss1/18

Turri, A. M., Smith, R. J., & Kopp, S. W. (2017, January 13). Privacy and RFID technology: A review of regulatory efforts. *The Journal of Consumer Affairs*, *51*(2), 329–354. https://doi.org/10.1111/joca.12133

U.S. Department of Homeland Security. (2022, April 7). *Enhanced drivers licenses: What are they?* https://www.dhs.gov/enhanced-drivers-licenses-what-are-they

U.S. Federal Bureau of Investigation National Crime Information Center. (2020). *2020 NCIC missing person and unidentified person statistics.* https://www.fbi.gov/file-repository/2020-ncic-missing-person-and-unidentified-person-statistics.pdf

Walsom, C. (2014, June 5). Microchipping: a brief history. *Companion Animal, 19*(6), 288–290. doi:10.12968/coan.2014.19.6.288

WCusr. (2019, March 28). *X-Ray of RFID implant* [Image]. Licensed under the Creative Commons CC0 1.0 Universal Public Domain Dedication. https://commons.wikimedia.org/wiki/File:X-Ray_of_RFID_Implant.jpg

What is a Local Area Network? LAN definition, history and examples. (n.d.). *CompTIA.* https://www.comptia.org/content/guides/what-is-a-local-area-network

Woodrum Setser, M. M., Cantor, M. C., & Costa, J. H. C. (2020, August). A comprehensive evaluation of microchips to measure temperature in dairy calves. *Journal of Dairy Science*, *103*(10), 9290–9300. https://doi.org/10.3168/jds.2019-17999

Zedner, L. (2003, March). The concept of security: An agenda for comparative analysis. *Legal Studies*, *23*(1), 153–176. https://doi.org/10.1111/j.1748-121X.2003.tb00209.x

Žnidaršič, A., Baggia, A., Pavliеček, A., Fischer, J., Rostański, M., & Weber, B. (2021, December). Are we ready to use microchip implants? An international cross-sectional study. *Organizacija*, *54*(4), 275–292.

Chapter 14
A Study on Ethical Issues and Related Solutions for Smart Home Technologies

Roneeta Purkayastha
Adamas University, India

ABSTRACT

In today's ever-changing modern world, everyone yearns for a life that scores high on the quality of living (QoL) index. Smart home technologies aim to improve the quality of our lives by leveraging advanced technologies such as the internet of things (IoT) and sensors. The smart home is an automation system that mainly addresses ambient assisted living, energy management, and home security services. It is seen that smart home technologies are not easily accepted especially by the older adults of society. Ethics plays a crucial role in enhancing its wide adoption across the globe. Ethical problems surrounding these technologies have to be minimized to increase its adoption. The solutions proposed towards these ethical issues encompass secured data transfer and general awareness towards smart home technologies. The primary ethical issues surrounding smart home technologies are highlighted, and certain solutions for its wider adoption across all age groups of society are thus recommended.

INTRODUCTION

With the proliferation of Information and Communication Technology (ICT), newer emerging technologies such as the Internet of Things (IoT), Cloud Computing platforms, etc. have permeated into the lives of people and contributed to changing their lifestyles completely (Yigitcanlar & Kamruzzaman, 2019; Yigitcanlar et al., 2019a; Yigitcanlar et al., 2019b). Smart Home (SH) is a buzzword in the major media and technical forums nowadays. Older adults account for a good percentage of the total world population. They need special care with respect to their health problems as most of them suffer from chronic illness. If these health issues can be taken care of at home itself, then it will be a lifesaving situation for older adults. The Smart Home enables smart healthcare through ambient assisted living technologies. These technologies help to monitor the activities of the elderly related to physical impairments

DOI: 10.4018/978-1-6684-5892-1.ch014

and cognitive impairments and provide tools for enhanced social involvement (Pirzada et al., 2022). Smart Home provides energy management techniques for smart monitoring, billing, and conservation of energy consumption. Renewable energy sources have a great role to play in the area of smart energy conservation techniques. Smart Home aims to provide end-users with smart security services such as alert systems, and burglar alarms to name a few. With the advent of the Internet of Things (IoT), the delivery of Smart Home security services to the end-users has been possible through constant monitoring and communication among the inter-connections. Figure 1 illustrates the different functionalities of Smart Home such as alert system, health monitoring, etc.

Figure 1. Functionalities of the Smart Home

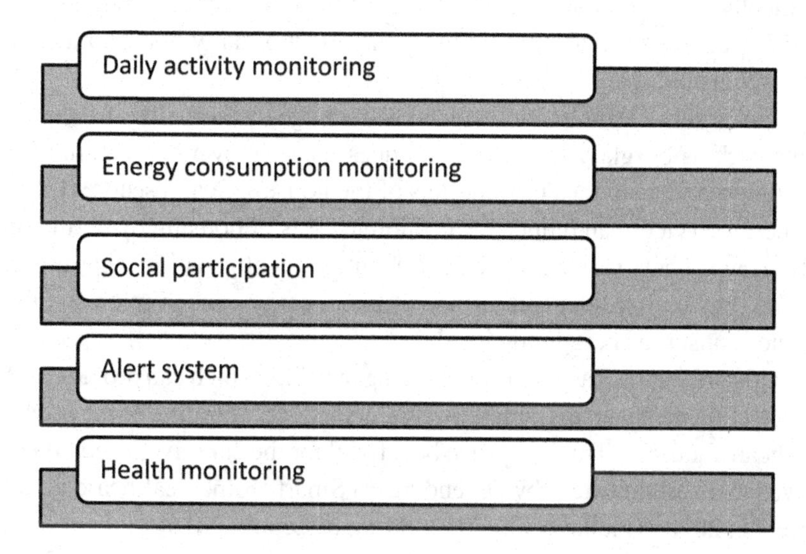

Smart Home technologies are subject to acceptance and adoption issues by the different age groups in society, especially the older adults. Ethical considerations play a major role in this area, which is discussed within this work. Ethical issues like privacy concerns, higher costs, the prior notion of relevant benefits, etc. have been highlighted. The ethical issue of privacy includes some of the common security threats that affect Smart Home technology-based products. Reliability and trust factors help in enhancing the acceptance rate. Here, the unobtrusiveness criterion is the most expected by end-users. Support for schemes to provide Smart Home services to the older population can be given by the government. Social awareness for SH products among the older population has to be developed to prevent hesitation, impatience, and discouragement. These ethical issues have to be minimized as far as possible by following best practices, responsible use by manufacturers and service providers, providing customized solutions to the end-users, and general awareness among end-users.

LITERATURE REVIEW

Advantages of the Smart Home

The overall Quality of Life (QoL) of residents can be improved by features of Smart Home technology which assists in their daily lives. Smart Home services can be divided into four types: convenience, safety, security, energy, and healthcare (Balta-Ozkan et al., 2013; Baudier et al., 2020).

Smart Home convenience services enhance the comfort of the residents by supporting their daily lifestyle. The most appropriate functionality representative of this feature is environmental control. Under environmental control, the different components of the house can be either remotely controlled or scheduled automatically. These components include kitchen items, lighting equipment, ventilators, etc., (Wilson et al., 2017; Hubert et al., 2018). It reduces the effort of daily household chores and provides comfort for the house inhabitants.

Smart Home security services help the residents to manage the security of their house and prevent untoward incidents such as burglary etc. These services detect movements within the house to warn about potential intruders and keep the surroundings of the house safe and secure (Coskun et al., 2018).

Smart Home energy services can reduce the economic costs of housing by optimizing or decreasing energy consumption rates. These services provide residents with information regarding energy consumption in the house and thus keep residents informed of their energy consumption so that they can keep a check on it and hence consume energy judiciously.

Smart Home healthcare services aid in the monitoring of health which can provide useful for especially older adults. Fall detection mechanisms, smart thermometers, etc. are some of the Smart Home products in the area of healthcare facilities. This can prove beneficial for the caregivers and other family members who can get relieved to a certain extent by depending on Smart Home healthcare facilities (Allameh et al., 2012; Reeder et al., 2020; Thielke et al., 2012; Wong & Leung, 2016).

Limitations of the Smart Home

Smart Home technologies suffer from certain limitations which need to be addressed (Edu et al., 2020). The authentication process is performed using wake-up words chosen by the user from a set of pre-defined options. There is no additional way of authenticating the user. Any command following the wake-up word will be accepted by the device, thus making it easier for anyone in the vicinity to issue commands to the device. To record the wake-up word, the voice command interpreter listens to the user's conversations constantly, which affects the user's privacy. Confidential conversations can get leaked accidentally and adversaries may get access to sensitive information.

In a multi-user environment, it is difficult to define correctly what and how resources are to be accessed. Additionally, it is difficult to specify how access should be granted. Anyone in the household can put the smart home device into recording mode and enable it to perform actions on commands. Without the primary user's permission, it is possible for others in the household to modify the access to the controls of the Smart home device.

User profiling derives relevant personal information from data gathered from users. The profiled data is related to the behaviors, interests, and preferences of the targeted users. Although the network traffic may be encrypted, en-route profiling attacks can be made and access the user network traffic.

The Smart home device at first has to understand what the user uttered, what the user wants and then fulfill the user's request. This is called a speech recognition system that utilizes Artificial Intelligence techniques such as Natural Language Processing (NLP) and Machine Learning (ML). These techniques can further introduce the issues of denial of service attacks (Vaidya et al., 2015) or reduce the quality and performance of the Machine Learning model.

Ameliorative Measures for Smart Home Problems

Some of the ameliorative measures for the Smart Home problems are outlined below (Touqeer et al., 2021).

- **Countermeasure against phishing attacks**: Gupta et al. (2018) proposed network protection based on a blacklist scheme in which erroneous measures are blocked either on the client-side or server-side. Awareness among users for differentiating between a phishing website and a normal website has to be developed.
- **Detection of malicious code**: As per Wei et al. (2018), the normal running time versus abnormal behavior is analyzed when malicious code is deployed.
- **Countermeasure against eavesdropping**: A system has been proposed in which activities of eavesdroppers are monitored as stated in Li et al (2016). Different effects like path loss effect etc. are considered to propose a formal analytical model (Classen et al., 2015).
- **Countermeasure against sniffing**: Sniffing can be avoided by connecting only to trustworthy networks. End-users must not connect to public places network. Encryption helps protect the network track that encrypts all the data that leaves the IoT system.
- **Network monitoring**: The network protocol running on the IoT-based smart homes is mostly targeted by Denial of Service (DoS) attacks. The intrusion Detection System (IDS) helps in detecting, classifying, and monitoring these attacks.

Deciding Criteria for Acceptance of Smart Home Technology

Researchers and technical developers opine that Quality of Life (QoL) can be improved considerably by applying knowledge through technologies for convenience, safety, healthcare, etc. as discussed earlier. Despite these potential benefits of Smart Home technology services, if the concentration is only on the technical side of product development, then they may disappear even before they become a part of daily life (Cook, 2012). For the wide adoption of technology, it is highly essential to have a systematic understanding of users' needs and preferences in daily life. In the case of healthcare for older adults, home telehealth services aim to enhance the Quality of Life (QoL) in the home and facilitate independent living (Onor et al., 2008; Choi et al., 2018). Using these services requires an understanding of modern IT-based solutions and techniques to a certain extent. Elderly people face difficulties and challenges in using these techniques (Cimperman et al., 2013).

The security-based solutions under Smart Home technologies have access to the personal data of Smart Home users for their health and physiological monitoring. Research shows that many users decline to use these technologies (Lund & Nygård, 2003). This may be considered the failing point of Smart Home technologies. Hence, for the successful accomplishment of Smart Home technology, it is required to understand the important requirements of the potential users, instead of focusing on the technological side of products only.

SMART HOME TECHNOLOGY SERVICES

Ambient Assisted Living (AAL)

An individual can age successfully if the physical and cognitive function is higher, less probability of chronic diseases, and highly active participation in social activities. However, the hindrance is posed by the high probability of chronic diseases which is unavoidable in the aging process. Elderly people face limitations in their daily activities like standing in queue for long periods, stooping, kneeling, walking for long hours, etc (Holmes et al., 2009; Tkatch et al., 2016)..Other daily activities that the older face problem with is dressing, eating, shopping, etc. These are caused due to the risk of falling, which may happen as a result of muscle weakness, weak vision and gait, etc (Lipardo & Tsang, 2018; Doi et al., 2015; Muir et al., 2010; Moreland et al., 2004). The risk of falls is higher in dementia patients than in normal healthy people.

In aged people, the cognitive functioning activities such as speed processing, attention, etc. start to decline with time. Serious medical conditions like dementia, Alzheimer's disease, stroke, etc. cause the decline in cognitive functioning in the elderly population. Millán-Calenti et al. (2011) pointed out that there is a correlation between depression and cognitive impairment, and both these conditions are related to reduced daily activities in older people.

Active social participation is one of the key factors for healthy aging in older people. Janoski and Wilson (1995) stated that social participation includes communication with friends and family, volunteering, etc. The problem is that older people will face difficulties in maintaining social networks and social relationships, however (Hao et al., 2017). There are numerous factors that contribute to low social participation such as poor socioeconomic status, health issues, etc. These pose a hindrance in performing daily routine activities of life.

Ambient Assisted Living for the Older Population

The equipment and services that are used to improve the physical and cognitive functions and participation in the social life of the aging and differently-abled population are termed assistive technology (World Health Organization, 2019). The Ambient Assistive Living Technology (AALT) is an emerging assistive technology that encompasses artificial intelligence, information and communication technology, and networking. It helps the aging population to live comfortably in various environments and participate in social activities without any hassles.

Ambient Assisted Living for Physical Impairments

With the proliferation of Information and Communication Technologies (ICT), bio-medical engineering, and wireless communications, human activities can be monitored with the help of wearable and non-wearable sensors. These wearable sensors are portable, light-weight, low cost and suitable for assisting the elderly in performing self-care activities (Liu et al., 2017). Demir et al. (2017) pointed out that the sensors of a kitchen, toilet, bathroom, bedroom, etc. can be used to confirm the position of older people while eating or sleeping by using a force resistive sensor. Fall detection is another important area of concern that can assist the elderly in performing daily activities in the home and social environment. A home-based Ambient Assisted Living for fall detection, multi-modality fall detection, and tele-care system, called "Home Healthcare Sentinel System (HONEY)", was introduced by Zhang et al. (2013). The developed system consisted of multimodality signal sources, on-demand video techniques, audio, and image.

Ambient Assisted Living for Cognitive Impairments

Cognitive impairments in elderly people contribute to their limited functioning. These kinds of people are hospitalized three times more than some other medical conditions (Alzheimer's Association, 2019). They face difficulties in recognizing familiar people and places, will repeat their words frequently and will make poor decisions and judgments. Apart from the problem associated with these daily activities, depression can create a considerable impact on the basic life activities of elderly people. Both wearable and non-wearable sensors have been suggested as a solution for monitoring basic daily life activities. Among the wearable sensors are accelerometers, wrist-worn unit sensors, and smartphones with multiple sensors (Eisa & Moreira, 2017). The sleeping conditions in dementia patients can be traced by using non-wearable sensors. Memory impairment in the old is another factor related to cognitive function. The memory issues can be tackled by leveraging Ambient Assisted Living technologies and tools. These tools can issue reminders to the elderly through speaking watches, digital calendar, digital event organizers, Global Positioning System (GPS), etc. (Dupuy et al., 2017; Holthe et al., 2018). Thus Smart Home technologies help enable a comfortable environment for the elderly and assess their physical health and cognitive functioning at home. With the use of low-cost sensors, actuators, mobile phones, computers, and appliances, these technologies are becoming a viable part of the aged population.

Ambient Assisted Living Tools for Enhanced Social Involvement

Social participation is essential for elderly people to keep loneliness and mental health issues at bay. Poor social involvement is an indication of deterioration of quality of life. With the physical and cognitive impairments, it is difficult for aged people to participate in socializing through various social media. Ambient Assisted Living tools of video tele-planning and simplified e-mailing systems can be delivered to the aging population which will also reduce the family member or caregiver burden.

GENERAL ETHICS PRINCIPLES

The following are major principles of general ethics which are to be followed by the computing professionals and manufacturers of Smart Home technology-based products (Gotterbarn et al., 2018).

- **Contribution towards society and well-being of people**: It is an obligation of computing professionals to utilize their skills for the good of society, its members, and its surrounding environment. Various threats to privacy and personal security have to be minimized by computing professionals. It is the moral obligation of the developers to promote fundamental human rights. The work of computing professionals should be targeted toward the societal benefit and should be widely accessible and usable across different segments of society. Sustainability of the environment is another important perspective that is to be promoted across society.

- **Social engineering-related attack**: In this type of attack, the humans are attacked instead of network devices. Here, the attacker interacts with the victim directly and compels the victim to act as per instructions from the intruder. This can lead the victim to visit some fake website and leak personal confidential information to the intruder, thus causing huge personal loss (Ghafir et al., 2016; He et al., 2018; Shamsoshoara et al., 2020).

The Privacy issues in Smart Home technologies result from the following few reasons:

- **Close observation**: Some Internet of Things (IoT) devices like smart doorbells allow close observation of the activities of the residents. The smart doorbell is one of the best starter devices for a smart home enthusiast. The smart doorbell provides an event history with a video and audio footage about the activities outside the door. This kind of smart device allowed the residents to open the door if they forgot the door keys. On the contrary, some of the residents may feel that this kind of smart device is obtrusive. It may give a feeling that someone is always watching them. The smart doorbell keeps the resident informed of events happening right outside the door without the need for asking or enquiring of neighbors or others. It offers direct control rather than automated control. Since it is under direct control, it permits the owner to observe closely and monitor the environment, which may lead to other partners of the house feeling being tracked by obtrusive technology.
- **Prudent observation**: Although prudent observations may sound similar to the previous type of observation i.e. close observation, it is different as it is unobtrusive and does not take into account observing other residents. Here, the primary aim is to improve the systems within the house. The residents of the house are always able to keep a check on the house even when they are away from the house. The monitoring is done discreetly here while skipping any interactions with the other residents, which includes observing the status of the house and other residents.
- **Requirement of being adaptive to technology use**: Some of the technical models require the users to adopt them such as Voice User Interfaces (VUI). Smart speakers are a medium of interaction with Smart Home. Some Smart Home users may find it to be a convenient medium of interaction, whereas others may feel that it hampers their privacy and increases their dependence on technologies. New technology may attract technical enthusiasts, but at the same time, it may fail to offer satisfaction and a feeling of convenience to other segments of society. This segment does not prefer keeping their smartphones always with themselves. Per the vision of Smart Home technologies, however, they are not free to keep away from the phone as the notifications for keeping a check on the home will continue getting delivered on the phone.
- **Tuning commodity usage**: Commodities of regular use such as hot water are normally easily accessible to the inhabitants of the home. Fortunately, with the usage of Smart Home products, these commodities become regulated with the aid of technology. For example: If an inhabitant wishes to take a warm shower outside normal hours, then he can do so only if he has notified the system well in advance. This system of preventing users from having a warm shower at any given time is designed to save energy and manage energy consumption at home efficiently.

Smart Home Energy Management Technology Service

During the last few decades, the standard of living has raised highly due to the accelerated growth in human inventions. This has resulted in the increased consumption of electricity and increased demand for energy in the utilities. A good percentage of consumption is generated from the residential load around 13-37% of the total load. The residential load comes from homes. There is generally a pattern for power consumption which depends on factors like humidity, temperature and holiday, etc. It is possible to forecast the consumption of a particular home by analysing these factors. The scheduling of appliances and implementation of demand response techniques can be done with the help of load forecasting. Therefore, forecasting can play a major role in the optimized working of the Smart Home Energy Management System (SHEMS) (Liu et al., 2016). The decline in the cost of renewable energy sources has enabled consumers to produce their electricity. While working out implementation solutions of Smart Home, renewable energy sources have to be considered too.

Smart Home Security Service

Smart Home Technologies (SHT) aims to enhance the quality of humans' everyday lifestyle and household activities and to provide people with home security. By using Smart Home Technologies (SHT), everyday household objects such as smart bulbs, speakers, security cameras, consumer electronics such as TVs, computer systems, sensors, etc. can be connected to a hub that aids the communication among them and with the end-users via their smartphones. These interconnections can be monitored, accessed, and controlled to assist the end-users of Smart Home Technologies (SHT) (Gerber et al., 2018; Zeng et al., 2017; Balta-Ozkan et al., 2013; Davis, 1989).

Perceived Benefits

If the notion of perceived benefits is higher for users, then they show more acceptance towards Smart Home technology. They think that these technologies not only increase their independence but also reduce the load on their family and caregivers. It is very important that the benefits provided by the technology match with the user's expectations. Such Smart Home technologies should deliver the functionality as expected by the users and should avoid unnecessary or redundant features. Unobtrusiveness is one of the Smart Home features expected by most users. Hence, the technology should be able to enhance the Quality of Life (QoL) of users and monitor and assist in their daily activities in an unobtrusive manner.

Reliability is one of the major criteria that persuade older adults towards accepting assistive technologies. The users should first be able to trust the devices before they use them.

High Cost

Smart Home technologies involve high costs as they contain several types of sensors and related technologies. Cost can be a limiting acceptance factor for older adults as they live on a pension. Smart Home services should be designed keeping in mind the inclusiveness of the user community. These should be accessible for maximum users irrespective of their having a high income. The government can come forward with some schemes to provide support to the older population in using these technologies. The existing services have to be utilized in constructing newer Smart Home solutions. In this way, the expense of producing new Smart Home services can be reduced.

Social Taboo and Other Issues

The low adoption rate for assistive technologies can have multiple reasons, one of them being the social taboo associated with using Smart Home products. According to some Smart Home users, acceptance of such products by the public shows the obsession of an individual towards health. The family members might spread negative emotions regarding Smart Home use leading to impatience, discouragement, and hesitation among older adults. On the contrary, if the family members encourage and support Smart Home use, then the older adults can become comfortable with using Smart Home products for daily use.

Awareness is another criterion that affects the adoption of Smart Home products (Guo et al., 2016; Shirani et al., 2020). According to Guo et al. (2016), males were more familiar with the term "IoT" and "Smart Home" than females.

RELEVANT SOLUTIONS AND RECOMMENDATIONS:

The ethical issues surrounding the Smart Home technologies range from privacy concerns, the higher cost involved and the prior notion of related benefits, etc. The minimization of these ethical issues is highly essential for the wider adoption and acceptance of Smart Home technologies. The recommended solutions for these ethical issues are discussed below:

- **Best practices**: The Smart Home potential users become skeptical of using these technologies as their personal data may be subject to misuse by intruders. These privacy issues have to be handled by the makers of Smart Home technologies carefully so that the end-users can place trust in them. Trust plays a major role in the acceptance factor of Smart Home technologies. Some of the Smart Home technologies like smart surveillance at home are thought to be obtrusive technology by the older adults especially. They find it uncomfortable to be monitored constantly at home, which hinders their privacy. The makers of Smart Home technologies have to come out with policies and best practices that can address these privacy concerns (Pirzada et al., 2022). One of them is the use of personal data gathered from Smart Home devices in a legitimate and transparent way.
- **Responsible use by manufacturers and service providers**: Great responsibilities rest on the shoulders of the Smart Home manufacturers and service providers. They have to be honest in their part of the work, which means that the regulations for using Smart Home technologies should be included in the rule book clearly and transparently. They have to be clear of their rights and responsibilities. They have to address issues such as the gathering of personal information by smart devices without user consent. Proper cryptographic security protocols have to be designed to prevent data abuse by attackers.
- **Providing customized solutions**: Smart Home users should be able to select the services that they want from the plethora of services offered by Smart Home technologies. The service offering has to be flexible if Smart Home technologies have to be made popular and widely adopted among potential users. In some cases, it is seen that the Smart Home users are unable to switch to other Smart Home service providers due to the strict agreement between users and makers of Smart

Home technologies. These issues form a barrier in the path of adoption and acceptance of Smart Home technologies. Products available under Smart Home technologies should be focused on improving the Quality of Life (QoL) of older adults. If these products do not cater to serving the specific needs of the users such as smart energy consumption or smart health monitoring through the Internet of Things (IoT), then the adoption rate will become automatically low. While designing smart solutions, the targeted user group has to be kept in mind. If a particular user group of older adults has problems with reading small text on the screen, then larger text with visually understandable illustrations should be provided. In this way, the system becomes more personalized and specific to the targeted user group. Customized Smart Home solutions as per need should be encouraged among Smart Home manufacturers to tackle these issues.

- **General awareness among end-users**: In a large number of homes, the older adults are left alone to deal with their health issues as their wards are outside due to their work. Older adults have to be made aware of the benefits of technology and its implications in the long run. This will enable them to accept Smart Home technologies and convert from potential users to active users. For example, consider the case of the fall detection mechanism of Smart Home technologies. This is of great use as preventing falls in older adults can prevent fatal injuries and mishaps at home. The caregivers for the elderly will be relieved to some extent if such technologies are embraced by the elderly and accepted widely.

The older adults feel incapable of learning new technology and thus are hesitant to operate modern Internet-of-Things-based tools. They feel intimidated by complex user interfaces or system configurations. It is the responsibility of the developers and manufacturers of Smart Home products to take care of the user-friendliness factor in system design and build easy-to-use and understandable interfaces for frequent use.

- **Transparent communication**: All the communication regarding the Smart Home products should go to the stakeholders in a transparent manner. Here, stakeholders mean anyone related to the process directly or indirectly such as employees, employers, Smart Home users, clients, etc. Before using any Smart Home product, Smart Home users should be aware of the policies and principles relating to the product. The manufacturers should be alert about any negative consequence resulting from the product usage due to poor quality work and should not neglect their responsibility.
- **Professional competence**: Professional competence arises from individuals and teams who take extraordinary care in accomplishing high-quality work through their knowledge, skills, and technical competence. Ethical issues are an area of primary concern for such professionals. They should recognize these types of serious issues affecting Smart Home products and should know how to resolve them according to existing laws governing them. The computing professionals should upgrade their skills and expertise continuously which includes self-study, attending workshops or seminars, and practice of such acquired skills. Professional organizations should give serious thought to the upgrade of skills of their employees and facilitate such activities from time to time.
- **Review of work done**: The quality of computing work is dependent on professional review, both peer and stakeholder review during all phases of work. It is the moral duty of the reviewers to provide constructive and critical reviews of work done by others.
- **Design of secure systems**: Implementing system design with built-in security is of utmost importance for improving the Smart Home adoption rate. Smart Home builders and manufactur-

ers should take extensive steps to secure resources against deliberate misuse or modification. Appropriate remedial measures and techniques have to be adopted such as monitoring, patching, or timely reporting. Security features of the system should be easily understandable, explainable, and easy to use. These should not be confusing or hard to follow by the users. Security of systems plays a major role in enhancing the trust factor of users in the overall system. This has to be handled and implemented judiciously and with care, keeping in mind the legitimate use of the Smart Home products.

CONCLUSION

The advent of Information and Communication Technologies (ICT) has facilitated the growth of Smart Home technologies. Smart Home (SH) provides the benefits of healthcare monitoring, alert system, fall detection mechanism, etc. for older adults especially. The acceptance rate of Smart Home products is comparatively low among older adults. One of the major reasons for being apprehensive of welcoming Smart Home products into daily lives is less awareness about IT solutions such as the Internet of Things (IoT), 5G, and Embedded Systems among the users. Another reason is the negative emotion spread by family members and friends about embracing Smart Home products, which according to them are for the vulnerable and physically weak people only. These issues have to be handled wisely if Smart Home is to be made popular among users. Also, the trust factor has to be prioritized. Some of the ethical issues affecting Smart Home products and their relevant solutions have been discussed here. This list of recommended solutions may be enhanced further by performing a broader analysis of the ethical issues.

REFERENCES

Allameh, E., Heidari Jozam, M., de Vries, B., Timmermans, H., Beetz, J., & Mozaffar, F. (2012, August). The role of Smart Home in smart real estate. *Journal of European Real Estate Research*, *5*(2), 156–170. doi:10.1108/17539261211250726

Alzheimer's Association. (2019). *Alzheimer's Disease Facts and Figures*. https://www.alz.org/media/Documents/alzheimers-facts-and-figures-2019-r.pdf

Balta-Ozkan, N., Davidson, R., Bicket, M., & Whitmarsh, L. (2013, December). Social barriers to the adoption of smart homes. *Energy Policy*, *63*, 363–374. doi:10.1016/j.enpol.2013.08.043

Baudier, P., Ammi, C., & Deboeuf-Rouchon, M. (2020, April). Smart home: Highly-educated students' acceptance. *Technological Forecasting and Social Change*, *153*, 119355. doi:10.1016/j.techfore.2018.06.043

Choi, Y., Kwon, Y. H., & Kim, J. (2018, April). The effect of the social networks of the elderly on housing choice in Korea. *Habitat International*, *74*, 1–8. doi:10.1016/j.habitatint.2018.02.003

Cimperman, M., Brenčič, M. M., Trkman, P., & Stanonik, M. de L. (2013, September 30). Older adults' perceptions of home telehealth services. *Telemedicine Journal and e-Health*, *19*(10), 786–790. doi:10.1089/tmj.2012.0272 PMID:23931702

Classen, J., Chen, J., Steinmetzer, D., Hollick, M., & Knightly, E. (2015, September 11). The spy next door: Eavesdropping on high throughput visible light communications. *Proceedings of the 2nd International Workshop on Visible Light Communications Systems*, 9-14. 10.1145/2801073.2801075

Cook, D. J. (2012, March 30). How smart is your home? *Science*, *335*(6076), 1579–1581. doi:10.1126cience.1217640 PMID:22461596

Coskun, A., Kaner, G., & Bostan, İ. (2018, April 30). Is smart home a necessity or a fantasy for the mainstream user? A study on users' expectations of smart household appliances. *International Journal of Design*, *12*(1), 7–20.

Davis, F. D. (1989, September). Perceived usefulness, perceived ease of use, and user acceptance of information technology. *Management Information Systems Quarterly*, *13*(3), 319–340. doi:10.2307/249008

Demir, E., Köseoğlu, E., Sokullu, R., & Şeker, B. (2017). Smart home assistant for ambient assisted living of elderly people with dementia. *Procedia Computer Science*, *113*, 609–614. doi:10.1016/j.procs.2017.08.302

Doi, T., Shimada, H., Park, H., Makizako, H., Tsutsumimoto, K., Uemura, K., Nakakubo, S., Hotta, R., & Suzuki, T. (2015, August). Cognitive function and falling among older adults with mild cognitive impairment and slow gait. *Geriatrics & Gerontology International*, *15*(8), 1073–1078. doi:10.1111/ggi.12407 PMID:25363419

Dupuy, L., Froger, C., Consel, C., & Sauzéon, H. (2017, September 28). Everyday functioning benefits from an assisted living platform amongst frail older adults and their caregivers. *Frontiers in Aging Neuroscience*, *9*, 302. doi:10.3389/fnagi.2017.00302 PMID:29033826

Edu, J. S., Such, J. M., & Suarez-Tangil, G. (2020, August). Smart home personal assistants: A security and privacy review. *ACM Computing Surveys*, *53*(6), 1–36. doi:10.1145/3412383

Eisa, S., & Moreira, A. (2017). A behaviour monitoring system (BMS) for ambient assisted living. *Sensors (Basel)*, *17*(9), 1946. doi:10.339017091946 PMID:28837105

Gerber, N., Reinheimer, B., & Volkamer, M. (2018, August). Home sweet home? Investigating users' awareness of smart home privacy threats. *Proceedings of An Interactive Workshop on the Human aspects of Smarthome Security and Privacy (WSSP).*

Ghafir, I., Prenosil, V., Alhejailan, A., & Hammoudeh, M. (2016, August). Social engineering attack strategies and defence approaches. *2016 IEEE 4th International Conference on Future Internet of Things and Cloud (FiCloud)*, 145-149.

Gotterbarn, D. W., Brinkman, B., Flick, C., Kirkpatrick, M. S., Miller, K., Vazansky, K., & Wolf, M. J. (2018). *ACM code of ethics and professional conduct.* ACM. https://www.acm.org/code-of-ethics

Guo, X., Zhang, X., & Sun, Y. (2016). The privacy–personalization paradox in mHealth services acceptance of different age groups. *Electronic Commerce Research and Applications*, *16*, 55–65. doi:10.1016/j.elerap.2015.11.001

Gupta, B. B., Arachchilage, N. A. G., & Psannis, K. E. (2018). Defending against phishing attacks: Taxonomy of methods, current issues and future directions. *Telecommunication Systems*, *67*(2), 247–267. doi:10.100711235-017-0334-z

Hao, G., Bishwajit, G., Tang, S., Nie, C., Ji, L., & Huang, R. (2017, June 23). Social participation and perceived depression among elderly population in South Africa. *Clinical Interventions in Aging*, *12*, 971–976. doi:10.2147/CIA.S137993 PMID:28694690

He, D., Ye, R., Chan, S., Guizani, M., & Xu, Y. (2018, April 13). Privacy in the internet of things for smart healthcare. *IEEE Communications Magazine*, *56*(4), 38–44. doi:10.1109/MCOM.2018.1700809

Holmes, J., Powell-Griner, E., Lethbridge-Cejku, M., & Heyman, K. (2009, July). *Aging differently: Physical limitations among adults aged 50 years and over: United States, 2001-2007* (NCHS Data Brief No. 20). U.S. Department of Health and Human Services, Centers for Disease Control and Prevention, National Center for Health Statistics. https://www.cdc.gov/nchs/data/databriefs/db20.pdf

Holthe, T., Halvorsrud, L., Karterud, D., Hoel, K. A., & Lund, A. (2018, May). Usability and acceptability of technology for community-dwelling older adults with mild cognitive impairment and dementia: A systematic literature review. *Clinical Interventions in Aging*, *13*, 863–886. doi:10.2147/CIA.S154717 PMID:29765211

Hubert, M., Blut, M., Brock, C., Zhang, R. W., Koch, V., & Riedl, R. (2018). The influence of acceptance and adoption drivers on smart home usage. *European Journal of Marketing*, *53*(6), 1073–1098. doi:10.1108/EJM-12-2016-0794

Janoski, T., & Wilson, J. (1995, September). Pathways to voluntarism: Family socialization and status transmission models. *Social Forces*, *74*(1), 271–292. doi:10.2307/2580632

Li, X., Wang, H., Dai, H.-N., Wang, Y., & Zhao, Q. (2016). An analytical study on eavesdropping attacks in wireless nets of things. *Mobile Information Systems*, *431475*, 1–10. Advance online publication. doi:10.1155/2016/4313475

Lipardo, D. S., & Tsang, W. W. N. (2018, August 24). Falls prevention through physical and cognitive training (falls PACT) in older adults with mild cognitive impairment: A randomized controlled trial protocol. *BMC Geriatrics*, *18*(1), 1–12. doi:10.118612877-018-0868-2 PMID:30143002

Liu, J., Sohn, J., & Kim, S. (2017, November 26). Classification of daily activities for the elderly using wearable sensors. *Journal of Healthcare Engineering*, *8934816*, 1–7. Advance online publication. doi:10.1155/2017/8934816 PMID:29317996

Liu, Y., Qiu, B., Fan, X., Zhu, H., & Han, B. (2016, December). Review of smart home energy management systems. *Energy Procedia*, *104*, 504–508. doi:10.1016/j.egypro.2016.12.085

Lund, M. L., & Nygård, L. (2003, April 1). Incorporating or resisting assistive devices: Different approaches to achieving a desired occupational self-image. *OTJR (Thorofare, N.J.)*, *23*(2), 67–75. doi:10.1177/153944920302300204

Millán-Calenti, J. C., Maseda, A., Rochette, S., Vázquez, G. A., Sánchez, A., & Lorenzo, T. (2011, October 26). Mental and psychological conditions, medical comorbidity and functional limitation: Differential associations in older adults with cognitive impairment, depressive symptoms and co-existence of both. *International Journal of Geriatric Psychiatry*, *26*(10), 1071–1079. doi:10.1002/gps.2646 PMID:21905101

Moreland, J. D., Richardson, J. A., Goldsmith, C. H., & Clase, C. M. (2004, July). Muscle weakness and falls in older adults: A systematic review and meta-analysis. *Journal of the American Geriatrics Society*, *52*(7), 1121–1129. doi:10.1111/j.1532-5415.2004.52310.x PMID:15209650

Muir, S. W., Berg, K., Chesworth, B., Klar, N., & Speechley, M. (2010, April). Quantifying the magnitude of risk for balance impairment on falls in community-dwelling older adults: A systematic review and meta-analysis. *Journal of Clinical Epidemiology*, *63*(4), 389–406. doi:10.1016/j.jclinepi.2009.06.010 PMID:19744824

Onor, M. L., Trevisiol, M., Urciuoli, O., Misan, S., Bertossi, F., Tirone, G., Aguglia, E., & Pascolo-Fabrici, E. (2008, March). Effectiveness of telecare in elderly populations–a comparison of three settings. *Telemedicine Journal and e-Health*, *14*(2), 164–169. doi:10.1089/tmj.2007.0028 PMID:18361706

Pirzada, P., Wilde, A., Doherty, G. H., & Harris-Birtill, D. (2022, July 9). Ethics and acceptance of smart homes for older adults. *Informatics for Health & Social Care*, *47*(1), 10–37. doi:10.1080/17538157.2021.1923500 PMID:34240661

Reeder, B., Chung, J., Lyden, K., Winters, J., & Jankowski, C. M. (2020). Older women's perceptions of wearable and smart home activity sensors. *Informatics for Health & Social Care*, *45*(1), 96–109. doi:10.1080/17538157.2019.1582054 PMID:30919711

Shamsoshoara, A., Korenda, A., Afghah, F., & Zeadally, S. (2020, December 24). A survey on physical unclonable function (PUF)-based security solutions for Internet of Things. *Computer Networks*, *183*, 107593. doi:10.1016/j.comnet.2020.107593

Shirani, F., Groves, C., Henwood, K., Pidgeon, N., & Roberts, E. (2020, September). 'I'm the smart meter': Perceptions of smart technology amongst vulnerable consumers. *Energy Policy*, *144*, 111637. doi:10.1016/j.enpol.2020.111637

Thielke, S., Harniss, M., Thompson, H., Patel, S., Demiris, G., & Johnson, K. (2012). Maslow's hierarchy of human needs and the adoption of health-related technologies for older adults. *Ageing International*, *37*(4), 470–488. doi:10.100712126-011-9121-4

Tkatch, R., Musich, S., MacLeod, S., Alsgaard, K., Hawkins, K., & Yeh, C. S. (2016, September). Population health management for older adults: Review of interventions for promoting successful aging across the health continuum. *Gerontology & Geriatric Medicine*, *2*. doi:10.1177/2333721416667877 PMID:28680938

Touqeer, H., Zaman, S., Amin, R., Hussain, M., Al-Turjman, F., & Bilal, M. (2021, December). Smart home security: Challenges, issues and solutions at different IoT layers. *The Journal of Supercomputing*, *77*(12), 14053–14089. doi:10.100711227-021-03825-1

Vaidya, T., Zhang, Y., Sherr, M., & Shields, C. (2015). Cocaine noodles: Exploiting the gap between human and machine speech recognition. *Proceedings of the 9th USENIX Workshop on Offensive Technologies (WOOT 15).*

Wei, D., & Qiu, X. (2018, May). Status-based detection of malicious code in Internet of Things (IoT) devices. *2018 IEEE Conference on Communications and Network Security (CNS)*, 1-7. 10.1109/CNS.2018.8433183

Wilson, C., Hargreaves, T., & Hauxwell-Baldwin, R. (2017, April). Benefits and risks of smart home technologies. *Energy Policy*, *103*, 72–83. doi:10.1016/j.enpol.2016.12.047

Wong, J. K. W., & Leung, J. K. L. (2016, October 3). Modelling factors influencing the adoption of smart-home technologies. *Facilities*, *34*(13/14), 906–923. doi:10.1108/F-05-2016-0048

World Health Organization. (2019). *Global action plan on physical activity 2018-2030: more active people for a healthier world.* World Health Organization.

Yigitcanlar, T., Foth, M., & Kamruzzaman, M. (2019a). Towards post-anthropocentric cities: Reconceptualizing smart cities to evade urban ecocide. *Journal of Urban Technology*, *26*(2), 147–152. doi:10.1080/10630732.2018.1524249

Yigitcanlar, T., Han, H., Kamruzzaman, M., Ioppolo, G., & Sabatini-Marques, J. (2019b). The making of smart cities: Are Songdo, Masdar, Amsterdam, San Francisco and Brisbane the best we could build? *Land Use Policy*, *88*, 104187. doi:10.1016/j.landusepol.2019.104187

Yigitcanlar, T., & Kamruzzaman, M. (2019). Smart cities and mobility: Does the smartness of Australian cities lead to sustainable commuting patterns? *Journal of Urban Technology*, *26*(2), 21–46. doi:10.1080/10630732.2018.1476794

Zeng, E., Mare, S., & Roesner, F. (2017, July). End user security and privacy concerns with smart homes. *Thirteenth Symposium on Usable Privacy and Security (SOUPS 2017)*, 65-80.

Zhang, Q., Ren, L., & Shi, W. (2013, May). HONEY: A multimodality fall detection and telecare system. *Telemedicine Journal and e-Health*, *19*(5), 415–429. doi:10.1089/tmj.2012.0109 PMID:23537382

ADDITIONAL READING

Birchley, G., Huxtable, R., Murtagh, M., ter Meulen, R., Flach, P., & Gooberman-Hill, R. (2017). Smart homes, private homes? An empirical study of technology researchers' perceptions of ethical issues in developing smart-home health technologies. *BMC Medical Ethics*, *18*(1), 1–13. doi:10.118612910-017-0183-z PMID:28376811

Courtney, K. L. (2008, February). Privacy and senior willingness to adopt smart home information technology in residential care facilities. *Methods of Information in Medicine*, *47*(1), 76–81. doi:10.3414/ME9104 PMID:18213432

Hassija, V., Chamola, V., Saxena, V., Jain, D., Goyal, P., & Sikdar, B. (2019). A survey on IoT security: Application areas, security threats, and solution architectures. *IEEE Access: Practical Innovations, Open Solutions*, *7*, 82721–82743. doi:10.1109/ACCESS.2019.2924045

Portet, F., Vacher, M., Golanski, C., Roux, C., & Meillon, B. (2013). Design and evaluation of a smart home voice interface for the elderly: Acceptability and objection aspects. *Personal and Ubiquitous Computing*, *17*(1), 127–144. doi:10.100700779-011-0470-5

Ramesh, G., Krishnamurthi, I., & Kumar, K. S. S. (2014, May). An efficacious method for detecting phishing webpages through target domain identification. *Decision Support Systems*, *61*, 12–22. doi:10.1016/j.dss.2014.01.002

Sharma, P., Zawar, S., & Patil, S. B. (2016). Ransomware analysis: Internet of Things (Iot) security issues challenges and open problems inthe context of worldwide scenario of security of systems and malware attacks. *International Conference on Recent Innovation in Engineering and Management*, 2(3), 177-184.

Swamy, S. N., Jadhav, D., & Kulkarni, N. (2017, February). Security threats in the application layer in IOT applications. *2017 International Conference on i-SMAC (IoT in Social, Mobile, Analytics and Cloud)*, 477-480. 10.1109/I-SMAC.2017.8058395

Whittaker, C., Ryner, B., & Nazif, M. (2010, March 1). Large-scale automatic classification of phishing pages. *NDSS Symposium 2010.*

Xiang, G., Hong, J., Rose, C. P., & Cranor, L. (2011, September). Cantina+: A feature-rich machine learning framework for detecting phishing web sites. *ACM Transactions on Information and System Security*, *14*(2), 1–28. doi:10.1145/2019599.2019606

Zhao, W., Yang, S., & Luo, X. (2020, August). On threat analysis of IoT-based systems: A survey. *2020 IEEE International Conference on Smart Internet of Things (SmartIoT)*, 205-212. 10.1109/SmartIoT49966.2020.00038

Section 6

Moving Further Out of the Comfort Zone

Chapter 15
Ethics and Risks Associated With Self-Driving Automobiles

Tamara Phillips Fudge
https://orcid.org/0000-0002-8682-9711
Purdue University Global, USA

ABSTRACT

Autonomous vehicles are already on public roadways and may present hazards well beyond simply allowing a car or truck to drive itself. As the automobile industry prepares to add even more "smart" technologies, it is prudent to investigate the ethics behind what can be lethal machinery. Radar, LiDAR, sensors, cameras, and other technologies are briefly explored along with levels of automation and current standards that are supposed to apply to components. Central to the purpose of this investigation is the telling of individual stories in which drivers, passengers, and pedestrians have died or otherwise experienced life-changing events. Some positive effects are apparent, but whether machine learning can appropriately make life-and-death decisions is yet to be accepted by society. Other facets include economic and environmental complications, human behaviors, company responsibilities, and the need for legal guidance in a field that currently mostly operates on recommendations.

INTRODUCTION

Autonomous automobiles may be considered a technical wonder, but present an ethical dilemma: how can humans trust that operators, drivers in other vehicles, and pedestrians are truly safe from life-threatening accidents? This technology's basic concepts and description of specific incidents help in understanding the real-life implications, and lead to the exploration of risks that society must either accept, regulate, or maybe even renounce.

First, the idea of automating automobiles is not new. Research began as early as the 1920s with first trials conducted by the 1950s (Ondruš et al., 2020). In 1977, Tsukuba Mechanical of Japan developed a car that used "white street markers via two vehicle-mounted cameras" to set an unmanned vehicle in motion at about 20 miles an hour (Kopestinsky, 2022, Driverless Car Statistics section). By the 1980s the concept became a reality thanks to coordinating efforts of several universities and car manufactur-

DOI: 10.4018/978-1-6684-5892-1.ch015

ers (Ondruš et al., 2020). Cruise control, a component of self-driving cars, has been in use for decades; many cars today also have parallel parking assistance (Faulhaber et al., 2019). It is clear this is not a passing fancy; current statistics declare more than 80 companies are working on this technology with $54 billion invested in the global market as of 2022 (Kopestinsky, 2022).

Developers and proponents of autonomous vehicles claim they are safer than traditional cars, help to plan driving routes, reduce stops and waiting time for taxi services, share sensor data that is gathered with others on the road, and largely bypass human errors as the sensors do their job (Clayton & Kral, 2021). Traffic would be better regulated, and due to sensors, automobile theft might be reduced (Ondruš et al., 2020). A survey reported by Schneble and Shaw (2021) claims that emergency decisions may be faster since it does not involve the emotional reaction of a human driver – a term they call "autonowashing" (p. 2). Margan (2018) claims once this is widespread, "our roads will be safer as the driverless cars will always be attentive to the environment, obey the traffic rules, and never get tired, distracted, drunk or have fits of road rage" (p. 15). Another benefit may be lowered emissions due to smoother vehicle operation (Gurumurthy et al., 2021), although electric vehicles would produce an even more profound environmental impact. Also, those who are physically or cognitively unable to operate a motor vehicle on their own may benefit, and traffic may run more smoothly with electronic communication between vehicles (Faulhaber et al., 2019).

Mayer et al. (2021) states while human drivers are the predominant cause of accidents and the high number of road fatalities worldwide provides motivation for use of self-driving vehicles, it still must be considered that unpredictable pedestrians, animals, other drivers, and other factors make it impossible to expect zero misadventures. Part of the development of these vehicles, then, is to program them for dealing with events – and the question must be raised whether or not technology should be allowed to make moral decisions in life-or-death situations (Mayer et al., 2021). Ondruš et al., (2020) points out changes in weather conditions as a complication, and the reliance on technology may also result in generations of less experienced drivers, which can surely be problematic when the human must take back control of the vehicle.

Before a deeper exploration into this topic, it is helpful to consider what other technologies are already used on the road. Cameras capture speeding using radar detection, accidents can be detected and mapped on the internet so other drivers can check the web to avoid potential delays, and emergency services can receive priority through rerouting traffic (Stan et al., 2014). All-Electronic Tolling (AET) is employed on some toll roads, allowing funds to be collected by reading license plates or using a transponder instead of stopping to use cash (International Bridge, Tunnel and Turnpike Association, 2018). These and other measures are in place for safety and convenience as well as conserving fuel and potentially reducing air pollution (Stan et al., 2014). In a more automated future, there may be less need for roadway signage (Ondruš et al., 2020). Additionally, automation can be and is used in places other than public roads: in the agricultural community, an autonomous tractor is scheduled for launch in 2022; John Deere claims it will enhance efficiency due to the ability to deal with elevated landscapes, different soil types, moisture levels, and other issues ("John Deere," 2022).

TECHNOLOGY BASICS

Self-driving car technology is tremendously complex but a general understanding is helpful in contemplating the ethics of its use. Autonomous setup begin with an operating system so that algorithms can be run to gather and analyze data, and then connections made to the cloud for traffic and mapping/locational data. Convolutional neural networks (CNNs) deal with spatial orientation, capturing image and shape patterns, evaluating them, and making adjustments. There are Recurrent Neural Networks (RNNs), Partially Observable MDPs, Deep Reinforcement Learning (DRL) algorithms, and other methods involved (Barla, 2021).

In addition, Dedicated Short-Range Communication (DSRC) systems can be used for sending and receiving data, Inertial Navigation Systems (INS) to improve the precision of Global Positioning Systems (GPS), infrared sensors, and ultrasonic sensors (Vargas et al., 2021). Doppler frequencies must be considered whenever the vehicle is in motion, and weather conditions such as fog and dust may affect sensing (Shbat & Tuzlukov, 2014). Software updates must be possible and decisions made as to what collected data is appropriate to share and with whom (Dodig-Crnknovic et al., 2021). These technologies are mentioned to illustrate the systems' intricacies but further examination would take the reader beyond the scope intended here.

Complex systems are susceptible to complex problems, and both humans and technology – the latter created by and managed by humans – are prone to making errors. The following information will assist the reader in understanding errors made in real-life scenarios presented later.

Decision-Making

Decision-making and timely reactions are the crux of safe handling. Artificial Intelligence can recognize and react to stimuli because of what has been programmed into it; Machine Learning is a subcategory of AI in which the technology is not just *told* what it can do; it should *learn* from analysis of patterns and irregularities (Burnham, 2020). Barla (2021) lists four main decisions that must be made.

1. **Perception**: objects recognized using cameras, LiDAR (explained below), and radar.
2. **Localization**: determining objects' position relative to the vehicle using Visual Odometry (VO).
3. **Prediction**: choosing from a finite set of possibilities.
4. **Decision-making**: defining action by a set of complicated algorithms that include the Markov Decision Process (MDP) and Bayesian optimization (Barla, 2021).

In short, situations are predicted through modeling, and a wide variety of objects are intended to be detected by sensors and cameras.

Perception Tools: Cameras, Radar, and LiDAR

Cameras serve several purposes. Color can be detected via camera images and multiple cameras make it possible for the technology to see 3-dimensional representations (Vargas et al., 2021). Cameras tend to be inexpensive and can record incidents for later review.

There are three types of Radio Detection and Ranging (commonly known as radar) used to determine distance between the vehicle and objects. The table below is a general description and subject to some variance; some sources indicate a longer range for MRR or place Blind Spot Detection in the SRR category, for example.

Light Detection and Ranging (LiDAR) uses pulses of light instead of radio waves; setting the device on the top of the car roof allows for a full 360-degree viewing (Vargas et al., 2021). It is essentially a

Table 1. Radar deployments, ranges, and various uses

	Typical Deployments	Range	Some Uses
Short Range Radar (SRR)	Objects and movement detected by the side of the vehicle	.5 meter up to 20 meters	Cut-In Collision Warning
Medium Range Radar (MRR)	Objects and movement in front and in back of the vehicle with a rather wide arc	1 meter up to 60 meters	Blind Spot Detection Lane Change Assistance Rear-End Collision Warning Closing Vehicle Detection
Long Range Radar (LRR)	Objects and movement in front of the vehicle	10 meters up to 250 meters	Adaptive Cruise Control

Sources: Shbat & Tuzlukov (2014), Srinivasa (2018), "Frequency Ranges" (2019)

"spray of infrared dots" to measure and map distances and shapes, is used in some Augmented Reality applications and even iPhones. Although the naked eye cannot see the light bursts, it can be observed using night-vision cameras (Stein, 2022).

Other light sources can interfere with data accuracy and the cost of LiDAR is somewhat prohibitive (Vargas et al., 2021). That being said, LiDAR combined with radar is considered a suitable scheme (Barla, 2021).

Figure 1. A visual generalization of radar deployments
Source: Image by the author, content simplified from Shbat & Tuzlukov (2014),

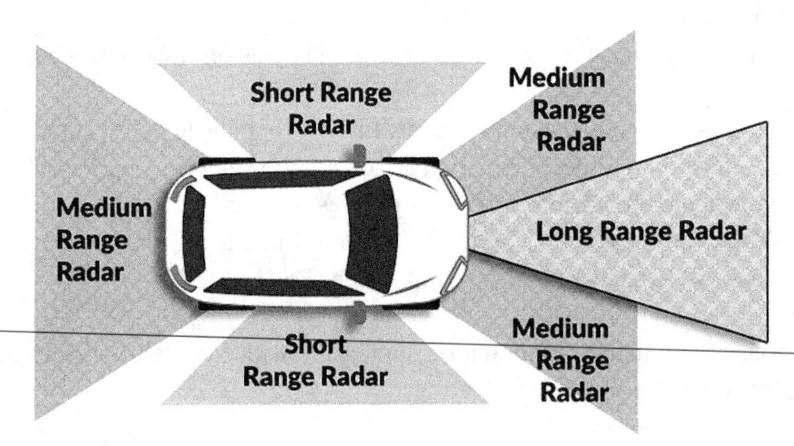

Inside the Vehicle

Obviously, there must be a computer interface inside the vehicle. The image below shows how an Open Pilot system adjusts to learn driving technique. The console may or may not have driver control, depending on the automation level explained in the next section. Drivers are expected to sit behind the wheel and be attentive enough to seize control of whatever feature is automated quickly.

Figure 2. Multi-unit LiDAR shown installed on a Chevrolet Bolt, 2017
Source: Dllu (2017).

ORGANIZING THE TECHNOLOGIES

Two systems assist the industry in organizing and identifying available tools: six SAE's Levels of Automation and ASIL's standards. Again, this information will be helpful in understanding the accident scenarios.

SAE: Six Levels of Automation

The Society of Automotive Engineers (2021) states there are three "actors" – the human, the driving automation system, and other vehicle mechanisms. There are six levels of implementation; full human

Figure 3. Comma Explorer drive annotation tool (Graphical User Interface on the dashboard)
Source: Jfrux (2019).

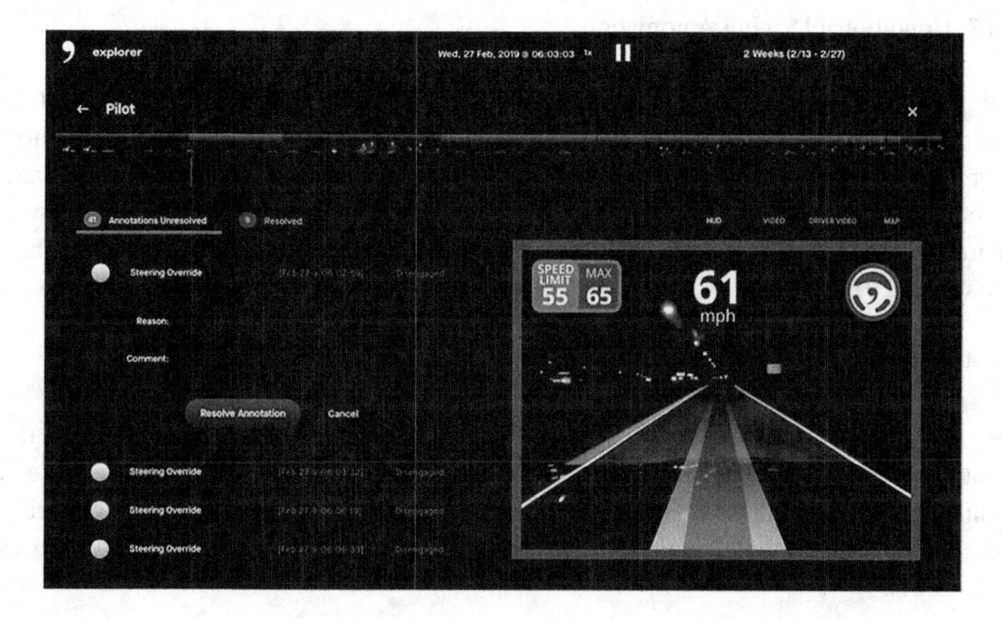

control is described as the lowest level, and the driver's involvement decreases as technology is added, ending with full technology control for the highest level. While each may point to specific technologies, any level can embrace features for crash avoidance (The Society of Automotive Engineers, 2021).

Level 0: No Automation

- Automobiles with full human control only are now rare, as cruise control (which places a car in Level 1) is a standard feature of modern vehicles (Faulhaber et al., 2019).

Level 1: Driver Assistance

- Some automation is employed at this level; it is typically either cruise control or may be adaptive cruise control, which assists in maintaining distance between the vehicle and others ("The 6 Levels," n.d.). With cruise control, the driver must still steer the vehicle and brake or turn off the feature when needed; at this level, the technology assists at a minimal level and the driver is still responsible for almost all control (Pearl, 2018). Another potential feature at this level is lane centering. Those assisting with parking might be designated as either Level 1 or Level 2 technologies (The Society of Automotive Engineers, 2021).

Level 2: Partial Driving Automation

- This level uses Advanced Driver Assistance Systems in which the technology controls both movement (acceleration, deceleration) and direction (steering). This still requires a human operator, who can take over with these functions when deemed necessary ("The 6 Levels," n.d.). Pearl (2018) claims at this level, a vehicle could work "safely without human supervision within certain parameters" but suggests the driver could text or otherwise be entertained until perceiving the need to resume control of the vehicle (p. 718).

Level 3: Conditional Driving Automation

- Although human intervention is still necessary for many circumstances, environmental detection is added at this level ("The 6 Levels," n.d.). Environmental detection involves the ability to perceive and classify objects and to understand positioning and movement. Range and distance in all directions is important, and can deal with what would be a "blind spot" for a human driver. Some environmental detection may also identify when the human operator is not fit to run the vehicle (Fässler, 2020).

Level 4: High Driving Automation

- Vehicles that meet this level can be driven mostly without human control, yet still allow the driver to intervene where deemed necessary ("The 6 Levels," n.d.). Geofencing – using these vehicles within a specific geographic region or set of regions – would likely reduce wait time for ride shares and control or at least better equalize access to transportation within urban sprawls (Gurumurthy et al., 2021).

Figure 4. ASIL variables and sub-classes
Source: Image by the author, content ideas from "What is ASIL?" (n.d.)

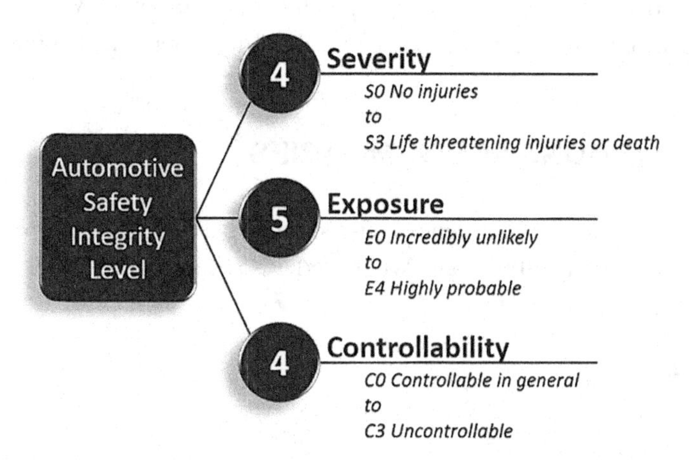

Level 5: Full Driving Automation

- As the name of this level suggests, these vehicles do not use any human intervention. These are still in development and testing ("The 6 Levels," n.d.), and some prototypes do not even include steering wheels, acceleration pedals, or brakes (Pearl, 2018).

Please see the Appendix for a chart prepared by the SAE with further details about each level. The concern at this time is focused on Levels 2 to 4.

ASIL: ISO Standard 26262

The Automotive Safety Integrity Level (ASIL) system is an ISO Standard meant to identify risks of separate electronic components in a vehicle. As self-driving innovations are developed, these standards still apply and must be considered. ASIL focuses on three variables, each with rankings:

- **Severity** describes driver and passenger injury risks.
  - Four sub-classes (S0-S3).
- **Exposure** relates to how frequently the dangers may be encountered.
  - Five sub-classes (E0-E4).
- **Controllability** is the ability of the vehicle operator to keep risks minimized.
  - Four sub-classes (C0-C3). ("What is ASIL?" n.d.).

Once variables and sub-classes have been identified, they are combined to provide a final classification: A, B, C, or D. Lowest at-risk results (S0, E0, and C0) result in an A grade component. As expected, the highest at-risk results (S3, E4, and C3) would thus result in a D grade. Anti-lock braking systems and air bags are examples of D-grade components; their quality must meet high risk activity ("What is ASIL?" n.d.).

While it is good to label component needs, a concern with this standard is vague rhetoric; words such as "usually" are not quantitative and so are open to interpretation ("What is ASIL?" n.d.). Should Level 5 self-driving cars become a reality, the "Controllability" factor will surely need to be revisited.

ACCIDENT DESCRIPTIONS AND SUMMARIES

The chronologies, post-accident consequences, and summarization of specific incidents are helpful in understanding roles and responsibilities, and where today's autonomous vehicles, humans, and regulations may fall short of moral expectations.

Incident Example No. 1

- Date: May 7, 2016
- Location: Williston, Florida
- Outcome: Autonomous car driver death

This was widely reported as the first self-driving car accident fatality. The incident involved a tractor-trailer traveling in front of a self-driving vehicle. The truck made a left turn but the Tesla Model S electric vehicle, set for autopilot, did not react and brakes did not activate (Deamer, 2016). The car driver, 40-year-old Joshua Brown, was killed when his vehicle's technology was unable to identify the truck directly in front of it – potentially due to overly sunny conditions and the white color of the 18-wheeler – and subsequently it drove at full speed into the truck (Yadron & Tynan, 2016). Even after initial impact under the truck's right side, the car did not stop until it had passed under the truck, off the road through a drainage ditch, through several fences, and ultimately hitting and breaking a utility pole before spinning counterclockwise and coming to a stop in the front yard of a residence (National Traffic Safety Board, 2017). Tesla issued a statement that in part attempted to excuse the technology as being "in a public beta [testing] phase" (Deamer, 2016, para. 4).

The National Highway Traffic Safety Commission (NHTSA) reported the Automatic Emergency Braking system failed to warn the driver but was "not designed to reliably perform in all crash modes, including crossing path collisions" (Pearl, 2018, p. 733). The NHTSA concluded the driver should have been able to discern the tractor-trailer's position for seven seconds or more prior to the crash, but apparently made no attempt to override the system and apply brakes; driver distraction and over-reliance on the technology was more at fault than the technology (Pearl, 2018).

The car in this incident would be considered a Level 2 vehicle. It was equipped with Tesla's Traffic-Aware Cruise Control and Autosteer systems (National Traffic Safety Board, 2017). The NTSB's report stated the Tesla's speed was 74 miles per hour at the time of impact. Neither driver was determined to be fatigued and both drivers were fully and lawfully licensed to drive – although the Tesla driver did have some speeding traffic violations on his record. No problems were noted with the road design or unusual weather conditions. Up to 10.4 seconds of time was available for line-of-sight acknowledgement and reaction for both drivers. The truck driver was subjected to a blood test following the event and tested positive for THC; it may not have been enough to indicate impairment, but he was cited for failure to yield to the Tesla (National Traffic Safety Board, 2017).

Importantly, the NTSB stated the Level 2 technologies should not be expected to "identify and respond to crossing vehicle traffic" and when testing these systems, they found no defects in how they were supposed to work (National Traffic Safety Board, 2017, p. 30). However, the car's manual explains these features should be used on divided roads such as interstates instead of two-lane roads with unlimited access. Since this accident, there have been some modifications to the technology including hands-off steering wheel alerts and the ability to identify if the road on which the car is traveling is appropriate, yet the use of automated systems remains in the hands of the drivers; there are no programmed restrictions. The recommendation of the NTSA was for manufacturers to develop safeguards so use of the systems is limited to appropriate conditions (National Traffic Safety Board, 2017).

A strikingly similar event occurred three years later on March 1, 2019 in Delray Beach, Florida. 50-year-old driver Tesla Model 3 Jimmy Banner turned on his car's autopilot feature and took his hands off the steering wheel. Ten seconds later, without any braking and at 68 miles per hour, his Tesla drove under a truck, killing Banner (Plungis, 2019).

Summary: These two incidents were caused by a combination of technology and human error. First, the Level 2 autonomous features failed to identify and react to a tractor-trailer in its path, and there were no safeguards against the use of these features on two-lane roads. Secondly, the drivers' inattentiveness to the situations and failure to override the Autopilot feature were critical errors in judgement.

Incident Example No. 2

- Date: March 18, 2018
- Location: Tempe, Arizona
- Outcome: Pedestrian death

The first pedestrian death caused by a self-driving car incident occurred in March 2018. An Uber-fleet Volvo XC90 helmed by Rafaela Vazquez, 44, was driving on Mill Avenue in the right lane when the it struck 49-year-old Elaine Herzberg (Kiley, 2018; Randazzo, 2019).

The speed limit at the crash site was 45 miles per hour (*Wood & Ziemann v. State of Arizona, City of Tempe*, 2019). The vehicle was traveling within the limit at 43 miles per hour but the LiDAR and cameras noted Herzberg's presence only 6 seconds before impact (Randazzo, 2019). Herzberg had been pushing her bicycle across the street but was not in a crosswalk at the time (Kiley, 2018). Herzberg died of her injuries shortly after being hit (*Wood & Ziemann v. State of Arizona, City of Tempe*, 2019).

While she was not in a crosswalk and it was nighttime, there were "brick-paved walkways apparently designed for pedestrian use" that traversed the road; these were not true crosswalks but emulated them and appeared to cause some confusion (*Wood & Ziemann v. State of Arizona, City of Tempe*, 2019, Count 3:88).

The car's autonomous systems had not identified Herzberg as a human at first; she was an "unknown object" until the technology realized she was a human with a bicycle. Reports also claim the technology could not discern the direction in which she was headed (Randazzo, 2019). The NTSB labeled this a "risk of automation complacency" and stated the lack of effective countermeasures contributed to the crash (Wamsley, 2020, para. 9).

The car itself made no attempt to brake, the Volvo's front end was significantly crushed by the impact, and Herzberg's bicycle was found nearby (Randazzo, 2019). Uber sent Vasquez's bosses to the scene. The driver expressed concern for Herzberg, but also for the self-driving car industry and her own future.

She was tested and determined not to have been impaired. Even at this early stage, the police suggested there could be criminal charges forthcoming (Randazzo, 2019).

The purpose for the drive that day was testing; no passengers were to be picked up. There were cameras in the Volvo and Vasquez's cell phone data was also confiscated (Randazzo, 2019). Video showed Vasquez looking down 166 times while the car was moving, and sometime after 9 pm, cell phone records identify "The Voice" playing on her phone (Randazzo, 2019). In the upcoming court case, prosecutors will introduce that she was reviewing Slack messages from the company at the time of the crash (Quach, 2022).

Important in this situation was that Uber disabled the emergency braking system ("City Safety Technology") in the Volvo. This meant only the vehicle operator could apply brakes (Randazzo, 2019). Vasquez did not apply brakes until after impact (Randazzo, 2019).

The National Transportation Safety Board investigated the accident as well as others in which self-driving cars had been unable to identify objects in the road. In a different incident, a bicycle lane post was hit and in another, a parked car was hit when the operator realized an approaching vehicle had not been clearly identified first and so swerved to avoid it (Shepardson, 2019). Uber's tampering with the software Volvo had provided was at least partially at fault, and the car also had a feature in which braking would be delayed by one second in an attempt to recalculate a suitable path. This 1-second delay feature has subsequently been removed (Shepardson, 2019).

Ultimately, Vasquez was charged with negligent homicide, and is currently on trial as of mid-2022 (Quach, 2022). Herzberg's daughter and husband sued the city of Tempe and the state of Arizona for allowing autonomous vehicles on the road without adequate testing, public notification, and resolution of other issues that led to the situation (*Wood & Ziemann v. State of Arizona, City of Tempe*, 2019).

Anuj Pradhan of the U-M Transportation Research Institute explained that self-driving cars must be able to "infer the intent of a pedestrian" (Kiley, 2018, para. 13). As human behavior is unpredictable, this poses major ethical problems. This event also brought up novel legal concerns: it is a new intersection of intellectual property, technology, and public safety (Kiley, 2018).

Business dealings also figure into situational analysis, as the history of Uber in Arizona is rife with controversy. Illegal in the state until 2015, ride services such as Uber and Lyft are now required to perform criminal background checks for drivers, carry $250k liability insurance, and undergo vehicle inspections (Randazzo, 2019). By the following year, Google's self-driving cars (called Waymo today) were the first to use public roads with autonomous vehicles. Uber, on the other hand, began testing on public roads without telling the public. Google accused Uber of stealing the technology (Randazzo, 2019). When an Uber car was caught running a light in California due to mapping errors and regulation was discussed among state legislators, nine cars from the San Francisco fleet were sent to Arizona; then after this particular crash, Uber's fleet was suspended and Waymo increased its presence in Arizona (Randazzo, 2019).

Summary: Many factors enter into this circumstance. Uber's disabling of an important braking function, the technology's inability to discern the person and/or bicycle in the car's path adequately, and the lack of a timely mechanism to alert a distracted driver to take control are notable, yet no culpability was assigned to the company. Whether the city or state were partially to blame has still to be determined, but the lawsuit brought by Herzberg's family demonstrates an interest in better regulation and legislation. In terms of human errors, the driver was distracted and the pedestrian was not in a true crosswalk. Lastly, while no mention was made of weather conditions, the accident occurred at night.

Incident Example No. 3

- Date: April 25, 2019
- Location: Monroe County, Florida
- Outcome: Bystander death, bystander injury

In April 2019, Dillon Angulo and Naibel Benavides Leon parked 40 feet from the roadway and stood near their vehicle in the vicinity of the intersection of Card Sound Road and Florida's County Road 905 in the Florida Keys (Shepard, 2019). It was after 9 pm and dark but there was a flashing red light (Boudette, 2021a). The intersection is in a "T" configuration, and the couple along with Argulo's Chevrolet Tahoe, while substantially away from the roadway, sat in the direct path of an approaching autonomous car (*Neima Benavides v. Tesla, Inc.*, 2021).

George McGee was running a 2019 Tesla Model S and using "Traffic Aware Cruise Control." He admitted to fumbling for a fallen phone, thereby having his hands off the wheel (Shepard, 2019). Ironically, the phone call he had made was to American Airlines to set up a flight for attending a funeral. The call had already been live for five minutes when he dropped the phone (Boudette, 2021a). When he looked up, it was too late to change course and the Tesla ran the stop sign and flashing light. Travelling at almost 70 miles per hour, the car then struck the Tahoe, which subsequently rotated and struck Angulo and Benavides Leon. There was no indication that brakes were applied (*Neima Benavides v. Tesla, Inc.*, 2021). Notably, automatic braking is considered a basic function of self-driving cars and has been in existence for quite some time (Boudette, 2021a). While one web article states McGee braked about 1 second before impact (Seamons, 2021), obviously, this was not enough time to avoid a severe impact due to the vehicle's speed.

Angulo, 27, was knocked unconscious but survived, suffering great bodily harm requiring a long hospital stay with traumatic brain injuries that have affected his memory (Shepard, 2019). A first responder spied a sandal under the Tahoe, which prompted the search for additional victims (Seamons, 2021). Paramedics then found 22-year-old Benavides Leon, who had been thrown about 75 feet into nearby bushes (Shepard, 2019; *Neima Benavides v. Tesla, Inc.*, 2021). She died after arrival at the hospital.

McGee was not issued a ticket and was not tested for drugs or alcohol at the scene (Shepard, 2019). He had been on the road for quite some time, as he was traveling from his Boca Raton office to his home in Key Largo, a trip of about 100 miles (Boudette, 2021a), yet no mention was noted of potential driver fatigue.

Benavides Leon's family filed a lawsuit in May 2021 charging Tesla for her wrongful death, citing design defects including "the failure to adequately monitor and determine driver-engagement" and failure to identify a stationary object in the vehicle's path (*Neima Benavides v. Tesla, Inc.*, 2021, Count I:23-24). The lawsuit did not cite McGee, claiming instead he "was driving the vehicle in a reasonably foreseeable and intended manner" whereas Tesla failed to provide failsafe mechanisms and sufficient warnings about the imperfect technologies (*Neima Benavides v. Tesla, Inc.*, 2021, Count II:33).

Summary: Again, there are many factors, although for this event, the technology issues may outweigh human factors. The autonomous Tesla in this incident did not identify a large object directly in front of it nor use automatic emergency braking. The ensuing lawsuit focuses on Tesla's failure to provide adequate self-driving features and to warn drivers of the limitations of the technology. While the driver was admittedly inattentive/distracted, it appears there was little investigation into any fatigue or impairment. Lastly, the speed in approaching a "T" intersection is of great concern.

Incident Example No. 4

- Date: August 24, 2019
- Location: Fremont, California
- Outcome: Passenger death, driver injury (in another vehicle)

Benjamin Maldonado Escudero and his 15-year-old son Jovani were returning to their California home in their Ford Explorer truck after a soccer tournament when a vehicle in front of them slowed down. This prompted the father (who was driving) to change lanes. His blinker flashed four times before the lane change was attempted, and the approaching Tesla was only noticed at the last moment. Suddenly, the Maldonados were hit from behind at the speed of about 70 miles per hour by a Tesla Model 3 on Autopilot (Boudette, 2021b). The driver of the Tesla, Romeo Lagman Yalung, had attempted to brake but did not have enough time to avoid the accident (*Escudero v. Tesla, Inc.*, 2021).

The Ford rolled over and crashed into a cement barrier, and both vehicles suffered tremendous damage. Benjamin Maldonado sustained injuries to his spine (*Escudero v. Tesla, Inc.*, 2021). The younger Maldonado was not wearing a seatbelt and was thrown from the truck's front passenger seat. He landed on the highway's shoulder and died at the scene (Boudette, 2021b). Neither Yalung nor his wife traveling with him were injured (Boudette, 2021b).

Raj Rajkumar of Carnegie Mellon University explained the monitoring technology can be tricked easily; this was backed by Consumer Reports trials in which test drivers slid out of their seats (and even into the back of the car) to see if Autopilot would notice, which it apparently does not (Boudette, 2021b). The Maldonado event occurred at about 6:35 pm (*Escudero v. Tesla, Inc.*, 2021); Rajkumar suggested the Tesla's cameras may have failed to identify problems due to sunny conditions or misperception of busy traffic, and a properly-employed radar could have helped avoid the crash (Boudette, 2021b).

The ensuing lawsuit brought by the Maldonado family claims Tesla sold the vehicle to Yalung's wife Vilma without inspection. Speed was a factor, including that the self-driving car was driving faster than traffic around it. Also, the suit claims that safety hazard instructions were inadequate (*Escudero v. Tesla, Inc.*, 2021, Second Cause:29).

Summary: A combination of problems precipitated this circumstance. The technologies in the autonomous vehicle did not appear to be sufficient to reduce speed based on adjacent traffic patterns or identify the Ford as a potential object to avoid. Lack of inspection and training is also a concern. Road conditions might have been a minor factor. The Tesla operator was inattentive to traffic speeds. It is unknown if seatbelt use could have prevented the young man's death, but may have been a contributing factor.

Incident Example No. 5

- Date: December 29, 2019
- Location: Gardena, CA
- Outcome: Deaths of another vehicle's driver and passenger

In this incident, a Honda Civic driven by Gilberto Alcazar Lopez had just entered the intersection of Vermont Avenue and Artesia Boulevard in Gardena, CA after the light turned green when it was hit by a 2016 Tesla Model S running on Autopilot. The Level 2 Tesla was entering the intersection from a freeway and ran its red light (Smith & Mitchell, 2022).

Lopez and his passenger Maria Guadalupe Nieves-Lopez perished at the scene. The driver of the limousine-service Tesla, Kevin George Aziz Riad, 27, and his female passenger were hospitalized but survived (Krisher & Dazio, 2022).

Charges have been filed by Los Angeles County against Riad for felony vehicular manslaughter – perhaps the first such charge for autonomous car situations (Smith & Mitchell, 2022). California Penal Code (§ 192(c)) identifies manslaughter charges as appropriate when a vehicle is operated unlawfully and with gross negligence; speeding and driving "without due caution" qualify this event for manslaughter on the condition that both can be proven.

Legal analyst Aron Solomon explains "the legal concept for negligence is 'situational'" (Burke, 2022, para. 8). For this situation, Lopez's family filed a civil suit claiming the limousine was travelling at an excessive speed (Burke, 2022). Both families claim Tesla is negligent because their vehicles can unexpectedly accelerate and a suitable emergency braking mechanism is not in place (Krisher & Dazio, 2022). While the criminal trial is not scheduled until 2023 (Krisher & Dazio, 2022), complainants could still receive damages in civil court ("Limousine Driver," 2022).

The lawsuits include the limousine company which is owned by Riad's father; as the Stanley Law Group states, "the owners of vehicles are legally responsible for damage caused by their vehicles" ("Limousine Driver," 2022, para. 3). State laws place responsibility on drivers for controlling their vehicles, per the NTSB (Krisher & Dazio, 2022).

Summary: As the lawsuits are still pending and more specific details are not yet published in news articles, a clear outcome is not yet known. That being said, it appears this incident may focus on human over-reliance on automation and a lack of failsafe technologies.

Other Incidents

There have been many other instances in which autonomous vehicles are put into question regarding safety. Some of these concerns:

- Cyberattacks are possible.
    - A test in 2015 demonstrated that hackers could take control of a self-driving car ("The dangers of driverless cars," 2021).
- Accidents can result in fires.
    - There have been several such incidents, including one in 2017 in which a self-driving car hit a house and set it on fire, and another that took an immense amount of water and four hours to extinguish ("The dangers of driverless cars," 2021).
- Approaching another mishap can result in further errors.
    - A 2018 incident near Tokyo, Japan began with a typical highway accident; those involved and helpers moved to the side of the road to await police response. Shortly thereafter, a car traveling in front of an autonomous vehicle switched lanes to avoid the roadside group of motorcycles, pedestrians, and a van. The self-driving car, however, accelerated and subsequently hit the group, killing Yoshihiro Umeda, 44. The self-driving car operator was said to have been "dozing," but Umeda's family sued Tesla for faulty design, as the technology did not assist in averting a tragedy (Murdock, 2020).
- Emergency vehicles can be impeded.
    - An accident in 2018 involved a firetruck hit by an autonomous car (Krisher & Dazio, 2022).

- "Stunts" for social media are a worry.
  - In 2021, a man was arrested in California for setting his car in motion and moving into the backseat. Even after his arrest, Param Sharma, 25, posted his exploits on Instagram, denies he broke any laws (even though he did), and claimed he will continue to ride in this fashion (Quintana, 2021).

MAJOR RISKS AND CONSIDERATIONS

The aforementioned scenarios identify unique situations, but some patterns are emerging. First, there are many stakeholders: manufacturers, car sales companies, repair shops, insurance companies, transportation companies, and other organizations, plus the individuals who might operate the cars or trucks or in any other way be affected by their use. The environment might also be considered a stakeholder as well as government entities which must contemplate lawmaking, licensing, enforcement, road management, and other facets. Because these vehicles are driven on public roads, the outreach and impact can be immense – and often unknown by those who may encounter such vehicles and their shortcomings.

Additional Technical Issues

Dodig-Crnknovic et al. (2021) include these major technical challenges: network availability and reliability, storing and appropriate sharing of collected data, potential algorithm vulnerabilities, basic security, software updates, and ensuring proper vehicle and systems maintenance. The availability of internet is reliant on many things, including meteorological conditions which cannot be controlled (Etienne, 2021) and driving through tunnels (Ondruš et al., 2020). Drones can extend the network range (Kopenstinsky, 2022) but present additional issues of airspace.

American Automobile Association (AAA) research indicates that there are many shortcomings of these technologies, claiming that problems arise in vehicles with Active Driving Assistance Systems as often as every eight miles (Edmonds, 2020, para. 1). Other potential road issues include that not all have adequate markings (Barla, 2021).

Road conditions can make a difference in how components react to traffic and objects in the car's path. Level 4 vehicles still need human intervention in the case of severe weather (Ondruš et al., 2020). Sun, glare, fog, rain, snow, dust storms, and other environments can impact what sensors see but can also affect how tires react to asphalt. According to Eliot (2020), run-flat tires tend to work decently for a short time after a puncture but could still "blow." As of 2020, less than one out of six new cars are actually outfitted with this kind of tire – and tire failures create nearly a quarter of all driving emergencies (Eliot, 2020). Drivers today are used to feeling and listening for flat tires and poor alignment; the AI in the higher-level autonomous vehicle of the future will need to be outfitted with the ability to deal with tire difficulties if humans relinquish that control.

According to Synposys, "today's cars have 100 million lines of code. Tomorrow's autonomous cars will have more than 300 million lines of code" ("What is an Autonomous Car?" n.d., para. 17). Along with this amount of coding is the real potential for hackers who may be able to commandeer a car despite a human operator's attempts to wrestle control back. Also, with this much code, there is great potential for miscalculations.

Certain biases may be inherent in programming. A *sample selection bias* occurs if the samples that teach the car about objects is based on samples "not representative of the population intended to be analyzed" (Srinivasan & Chander, 2021, p. 47). A *confounding bias* is one in which some important variable is omitted (Srinivasan & Chander, 2021). Cars' sensors must be programmed to understand what traffic cones are and how to avoid them, for example; they must correctly identify someone who is missing a limb, walking on crutches, is unusually tall, in a wheelchair, on a bicycle, or as learned from the lessons above, walking a bicycle in a roadway. Margan (2018) uses "goat on the road problem" as an illustration wherein an object (in this case, a goat) jumps into the road but the sensors have not been taught what the goat is so cannot understand it as a problem.

There may be some racial concerns as well. According to a 2018 study, error rates in facial recognition software are higher in some programs when the person is female and skin tone is dark (Metz, 2021). Whether an African-American woman in dark clothes walking across the street would be correctly identified as a human is a legitimate concern. As computer programmer Jacky Alciné said, "You have to be very intentional about what you put into it [the programming]. Otherwise, it is very difficult to undo" (Metz, 2021, para. 23).

System integrator and architect Maarten Vinkhuyzen (2022) provides an insider's view, claiming that the current problem in fully autonomous vehicle development is the lack of the technology knowing what of the "billions of miles of driving data" (para. 8) provides appropriate behavior examples and what does not; he says the data is based on human behaviors via Neural Net–based AI instead of rule-based automatons. The methods can be challenged.

Economic Ramifications

Vehicle autonomy would cause a disruption to current business models. Margan (2018) claims "every major automobile manufacturer is researching extensively" to develop and market autonomous cars (p. 15), which is in itself a heavy financial burden for these companies. While cameras and some of the other technologies previously mentioned are inexpensive, others such as LiDAR are not (Vargas et al., 2021).

The expense of research is typically passed down to consumers. Gurumurthy et al. (2021) estimate automation in a single vehicle could easily add $7,500-10,000 to its cost, which makes ownership an expensive proposition; due to this, studies suggest transportation fleets might be more likely to make the change first.

Liability insurance will still be necessary as more autonomous vehicles enter roadways. Injuries, property damage, and vehicle damage are still distinct possibilities (Kopestinksy, 2022), but while Margan (2018) suggests insurance premiums might drastically fall, the differences between level 2-to-4 vehicles would need assessment (which includes predictions and trends with collected historical data), and it may take years for the insurance industry to adapt.

Whether or not road maintenance costs are reduced as suggested by Margan (2018) would depend on the average driver being willing to use public autonomous transportation, thereby reducing the number of vehicles on the road.

Environmental Impact

Improved fuel economy and reduced gas emissions are perceived as positive environmental impacts (Clayton & Kral, 2021), although Hevelke and Nico-Rümelin (2015) suggest fully autonomous vehicles might encourage more travel. Even so, some sources indicate a potential reduction of emissions in urban areas of 80% ("What is an Autonomous Car?" n.d.). This does not take into consideration a larger change in emissions if a high number of new cars would be electric instead of gas-powered. Etienne (2021) claims electric vehicles might cost 2.3 times less to operate than gas-powered cars, eventually bringing some balance to initial high costs of vehicle purchase.

With new technologies there will also be the need to find new ways for recycling and sustainability (Dodig-Crnknovic et al., 2021). If there are indeed fewer cars in the future due to automation, it could be imagined some parking areas would no longer be needed and could instead be turned into more useful spaces ("What is an Autonomous Car?" n.d.,).

Choices: The Trolley Dilemma

The so-called trolley dilemma is often cited in the discussion of decision making: "trying to cause as little overall damage as possible" (Faulhaber et al., 2019, p. 399) yet realizing the trade-off is not likely ideal. The "trolley" alludes to a hypothetical circumstance in which a street car is careening towards pedestrians and a decision-maker must decide if switching tracks one way, which would kill one person, would be a better choice than switching another way, in which several people would be killed. Killing one to save many is still a difficult moral quandary, and in the case of autonomous cars, would involve decisions that would be made by programmers (Wolkenstein, 2018).

Meder et al. (2019) asserts the trolley dilemma gives a finite set of options; the choice is either one decision or another, but in real-life scenarios there may well be additional options. Choices for non-autonomous car drivers are guided by *default rules* wherein drivers follow standard procedures as a first option, such as braking instead of swerving (Meder et al., 2019), with the option to choose depending on perceived factors.

This can be extended to predicaments such as whether it is appropriate to change course to avoid hitting a pet versus a jaywalker, if the elderly or disabled might be more likely to be hurt than someone who is more "fit," or should passengers or pedestrians be protected first where there is an option – and again, whether or not an algorithm should be allowed to make these kinds of decisions (Etienne, 2021). This begs to ask if humans should relinquish control to technology, and indeed, Kopenstinsky (2022) reports that two-thirds of men and half of women surveyed do not think that choices between life and death "cannot be taught to any kind of vehicle" ("Self-Driving Car Safety Statistics" section).

A positive matter to note is that assistive technology allows for people with disabilities and the elderly to remain mobile (Faulhaber et al., 2019).

Human Behaviors and Intent

Millar (2014) argues while some claim technology is inherently *neutral* in ethics, there are technologies meant to control human behavior, such as alarms when drivers have not fastened their seatbelts or wheels that lock when a grocery cart is too far from its originating store. These are moral choices made for the consumer: buckle up, and do not steal grocery carts. Technology thus acts as a "moral proxy" much like healthcare proxies agreed upon by relatives for patients who are unable to make their own care decisions (Millar, 2014).

Slaby (2019) analyzed data from multiple sources in regard to human operator behaviors. While there were differences between urban and rural drivers and within different age groups, Slaby's statistics indicated a large percentage of drivers would text, phone, play games, or listen to music while letting a fully self-driving car transport them; some even would take the opportunity to nap. These behaviors would be in direct contrast to responsibilities expected on the road.

There is a pleasure of driving and controlling a vehicle that may be taken into account, plus the fact driving an autonomous car would mean rather strict surveillance (Etienne, 2021). A "more invasive digital surveillance age" is truly a concern, as the data gathered by the car could potentially be accessible by governmental agencies and pose unforeseen privacy issues (McLachlan, 2021, p. 16).

While some sources such as Margan (2018) suggest there could be a large reduction in driving under the influence and occurrences of road rage, this has yet to be proven. DUIs could still be appropriate if a driver is drunk and moves into the back seat, for example, and road rage is still possible with Level 2-4 cars since the human operator can still seize control of the vehicle when necessary or desired.

Potential terrorist use of unmanned vehicles for transporting and detonating explosives must be mentioned (Ondruš et al., 2020), as automation removes the need for perpetrator suicide. However, Lin (2014) points out the current high cost of self-driving vehicles today is a major deterrent to using them for car bombs, and GPS and other tracking capabilities of these machines would identify suspects quickly.

Licensure and Experience

Some legal considerations include vehicle registration, plus training, testing, and licensure of operators. Vehicle registration began in the years around the turn of the 20th century (McLachlan, 2020). A brief history is helpful in understanding the transformations of transport authorization: The earliest driver's licenses were focused on the collection of fees and did not require training or testing of skills. It did not take long, however – August 14, 1893 specifically – for France to take the first step into requiring driving tests (McLachlan, 2020). The United States left this discretion to municipal governments, with Chicago taking the lead in 1899. New York made it a state requirement two years later. Australia began to require testing in the 1910s after an automobile accident resulting in fatality; this included drivers of public transportation and taxis as well as those who owned their own vehicles (McLachlan, 2020). Current U.S. state laws require testing for a driver's license and these tests are usually preceded by driver's education training through schools or private companies.

In terms of vehicle registration, the United Kingdom's "Locomotives on Highways Act" of 1896 was one of the earliest attempts; it classified cars as "carriages" and regulated emissions, speed limits, and the use of lights and bells (McLachlan, 2020).

With autonomous vehicles, then, the question arises as to who and what to license, and if training and experience behind the wheel will be sufficient for newer drivers when they need to switch to manual control (Ondruš et al., 2020).

Company Responsibilities

Manufacturers of self-driving vehicles bear some responsibilities, of course. Tesla's experiments have not always ended in acceptance; in February of 2022, for example, the company had to recall 54,000 self-driving cars due to a feature allowing "rolling stops" at intersections with all-way stop signs. While movement was restricted to only 5.6 miles per hour, the NHTSA expressed concerns and this feature was subsequently removed (Krisher, 2021). There were also apprehensions about dashboard touch screens allowing video gaming, so this capability was eliminated in December of 2021 (Krisher, 2021).

Smith and Mitchell (2022) claim Tesla charges car owners $12k to test their new vehicles, whereas Waymo and other companies train their own testers and have developed strict rules. Waymo has been active in the field, managing driverless taxis in Arizona and partnering with J.B. Hunt to run Level 4 semi-tractor trailers in Texas ("Driverless car accidents," 2021).

The California DMV is investigating whether or not Tesla violates law via their advertising autonomous vehicles that do not have full-driving systems (Smith & Mitchell, 2022). Interestingly, Tesla's public relations department has been dissolved (Krisher & Dazio, 2022), making communication with the public another issue.

Tesla's Autopilot is said to be already in 765,000 U.S. vehicles (Krisher & Dazio, 2022). This might be applied on any kind of road, but General Motor's "Super Cruise" is only supposed to work when the vehicle is being driven on divided highways (Boudette, 2021b); limiting use in this fashion would help to avoid abuses on two-lane roads and residential areas.

Not all problems stem from manufacturers; they may occur after vehicle purchase. Uber at one point had disabled parts of the emergency braking system as mentioned in the Tempe incident (Randazzo, 2019).

On a more positive note, Comma, the maker of Open Pilot, exercises transparency on their website and makes use of the talents of the open-source community. Open Pilot uses a vehicle's existing radar data and supplements it with hardware (including a camera). Features include steering, acceleration and braking, Adaptive Cruise Control, and Automated Lane Centering. Its open-source coding can be found at GitHub and allows for review and improvements suggested by a large number of professional programmers. Comma's site lists compatible vehicles and the available features that can be used in each; their "Gold Standard" list is clearly marked for highway driving (maximum speed 84 miles per hour) and there are gentle warnings for use with automobiles on the other lists ("Open Pilot," n.d.).

Legal Progress

The U.S. government is working to be proactive with changes. In 2016, the National Highway Traffic Safety Administration issued a document titled, *Federal Automated Vehicles Policy*. Included in this document is a framework focusing development in areas such as location, weather conditions, time of day, speed, and similar parameters, plus differences between "normal" driving and operating under conditions that are less than optimal. Data recording and cybersecurity, the car's behavior after an accident, and testing are also covered. Importantly, documentation needs to be made of all processes and events. While this is provided by a federal organization, the document includes guidance for states in developing their own regulations and a section suggesting new regulatory tools (National Highway Traffic Safety Administration, 2016). Notably, the 116-page PDF is for guidance and following the recommendations therein is voluntary. It appears that it has not been updated as of mid-2022.

By 2019, the NHTSA announced that twenty auto manufacturers had agreed to begin installing low-speed Automatic Emergency Braking in new vehicles with the target date of September 2022. This technology forewarns and brakes to try to avoid front-end collisions (National Highway Traffic Safety Administration, 2019).

Governmental oversight has not always been robust, however; for example, early in the Trump administration, the NHTSA had no one at the helm for 15 months, followed by leadership that resulted in some rules being "rolled back." Transportation leadership since 2021 has refocused on safety problems (Hetzner, 2022).

In June of 2022, U.S. Senators Ed Markey of Massachusetts and Richard Blumenthal of Connecticut issued a press release voicing concerns about Tesla's potentially misleading marketing and they state, "every day that Tesla disregards safety rules … our roads become more dangerous" (Ed Markey, 2022, para. 2). Of course, this is not the only company with at least Level 2 vehicles on the road, but the sheer number of Teslas in use and perceived communication issues with the public, the concerns are real.

Also in June 2022, it was reported that the NHTSA was investigating Tesla's software problems and contemplating a mandatory, monitored recall which would affect 830,000 vehicles. This is in part due to incidents in which the cars relinquished control to the human driver less than one second prior to impact and the inquiry started with autonomous car failures when approaching an existing accident. Because the company's software is cloud-based, the updates could be administered via the cloud (Hetzner, 2022).

Laws for the future of autonomous vehicles will also need to involve human operators and vehicle owners. Evans et al. (2020) asserts the Ethical Valence Theory in relation to self-driving cars: "any and every road user in the vehicle's environment holds a certain claim on the vehicle's behavior, as a condition of their existence in the decision context" (p. 3289).

CONCLUSION

It is clear that while there are benefits, there are also many risks associated with autonomous vehicles and many unanswered questions. The currently known fatalities and injuries caused by Level 2 vehicles are cause for concern even if small in number; negligence from various parties, reliance on flawed technologies, and misunderstandings of the power of these machines can have dire consequences, and even one death or injury may be deemed as one too many if they could be prevented.

There is some rather recent guidance from organizations such as the National Highway Traffic Safety Administration, the National Traffic Safety Board, and the Society of Automotive Engineers International, plus well-organized ISO 26262 standards and ASIL risk classifications, but the promise of better safety and stringent regulation has not yet been delivered. There are already thousands of these vehicles on the road, many of which are being "tested" by untrained operators, and Level 5 vehicles are already in development. One must question the pace of development and lack of more specific laws that can protect all stakeholders, including the unsuspecting passenger or pedestrian.

Other questions must be asked, such as how or even if it is possible to replace the current worldwide fleet of cars, if companies will act ethically in development and deployment, if human nature will accept responsibilities even when reduced by the automation, and if the other issues noted can be controlled enough to meet societal human values. Lastly, responsibility and accountability for accidents must be defined, again to protect all stakeholders. Before a universal social investment in this technology is made and/or forced upon society, a clearer ethical path must be forged, and it may well need to begin within legal venues.

REFERENCES

Barla, N. (2021, August 25). *Self-driving cars with Convolutional Neural Networks.* Neptune AI. https://neptune.ai/blog/self-driving-cars-with-convolutional-neural-networks-cnn

Benavides, N. (PR) v. Tesla, Inc. (Fla. Miami-Dade County Court, 2021). https://www.plainsite.org/dockets/download.html?id=297527909&z=e1dd27ac

Boudette, N. E. (2021a, September 1). 'It happened so fast': Inside a fatal Tesla autopilot accident. *New York Times.* https://www.nytimes.com/2021/08/17/business/tesla-autopilot-accident.html

Boudette, N. E. (2021b, September 1). Tesla says autopilot makes its cars safer. Crash victims say it kills. *New York Times.* https://www.nytimes.com/2021/07/05/business/tesla-autopilot-lawsuits-safety.html

Burke, M. (2022, January 21). Tesla driver charged with manslaughter in deadly Autopilot crash raises new legal questions about automated driving tech. *NBC News.* https://www.nbcnews.com/news/us-news/tesla-driver-charged-manslaughter-deadly-autopilot-crash-raises-new-le-rcna12987

Burnham, K. (2020, May 6). *Artificial Intelligence (AI) vs. Machine Learning: What's the difference?* Northeastern University. https://www.northeastern.edu/graduate/blog/artificial-intelligence-vs-machine-learning-whats-the-difference

Cal. Penal Code § 192(c) (1872). https://leginfo.legislature.ca.gov/faces/codes_displaySection.xhtml?lawCode=PEN§ionNum=192

Clayton, E., & Kral, P. (2021, November). Autonomous driving algorithms and behaviors, sensing and computing technologies, and connected vehicle data in smart transportation networks. *Contemporary Readings in Law and Social Justice, 13*(2), 9–22. doi:10.22381/CRLSJ13220211

Deamer, K. (2016, July 1). What the first driverless fatality means for self-driving tech. *Scientific American.* https://www.scientificamerican.com/article/what-the-first-driverless-car-fatality-means-for-self-driving-tech/

Dllu. (2017, October 18). *Cruise Automation Bolt EV third generation in San Francisco* [Image]. Wikimedia. Creative Commons Share-Alike License 4.0: https://commons.wikimedia.org/w/index.php?curid=63450446

Dodig-Crnknovic, G., Holstein, T., & Pelliccione, P. (2021, July 16). *Future intelligent autonomous robots, ethical by design. Learning from autonomous cars ethics.* https://arxiv.org/ftp/arxiv/papers/2107/2107.08122.pdf

Driverless car accidents – who's at fault? (2021, June 25). *National Law Review, 11*(176). https://www.natlawreview.com/article/driverless-car-accidents-who-s-fault

Ed Markey, the United States Senator for Massachusetts. (2022, June 9). *Markey and Blumenthal statement on NHTSA's elevated investigation Into Tesla's autopilot system* [Press release]. https://www.markey.senate.gov/news/press-releases/markey-and-blumenthal-statement-on-nhtsas-elevated-investigation-into-teslas-autopilot-system

Edmonds, E. (2020, August 6). *AAA finds active driving assistance systems do less to assist drivers and more to interfere* [Press release]. AAA. https://newsroom.aaa.com/2020/08/aaa-finds-active-driving-assistance-systems-do-less-to-assist-drivers-and-more-to-interfere

Eliot, L. (2020, October 3). Tire blowouts could case self-driving cars to go astray. *Forbes.* https://www.forbes.com/sites/lanceeliot/2020/10/03/tire-blowouts-could-cause-self-driving-cars-to-go-astray

Escudero v. Tesla, Inc. (Cal. Alameda County Court, 2021). https://www.plainsite.org/dockets/download.html?id=296424442&z=a4a11de7

Etienne, H. (2021, May). The dark side of the 'Moral Machine' and the fallacy of computational ethical decision-making for autonomous vehicles. *Law, Innovation and Technology*, *13*(1), 85–107. doi:10.1080/17579961.2021.1898310

Evans, K., de Moura, N., Chauvier, S., Chatila, R., & Dogan, E. (2020). Ethical decision making in autonomous vehicles: The AV Ethics Project. *Science and Engineering Ethics*, *26*(6), 3285–3312. doi:10.100711948-020-00272-8 PMID:33048325

Fässler, G. (2020, September 8). *The evolution of environmental detection: What can it accomplish?* [Web video session]. Continental. https://www.continental.com/en/press/fairs-events/continental-commercial-vehicle-days/environmental-detection

Faulhaber, A. K., Dittmer, A., Blind, F., Wachter, M. A., Timm, S., Sütfeld, L. R., Stephan, A., Pipa, G., & König, P. (2019, April). Human decisions in moral dilemmas are largely described by utilitarianism: Virtual car driving study provides guidelines for autonomous driving vehicles. *Science and Engineering Ethics*, *25*(2), 399–418. doi:10.100711948-018-0020-x PMID:29357047

Frequency ranges for automotive radar technology. (2019, June 27). Cetecom. https://www.cetecom.com/en/news/frequency-ranges-for-automotive-radar-technology

Gurumurthy, K. M., Auld, J., & Kockelman, K. M. (2021, September 8). A system of shared autonomous vehicles for Chicago: Understanding the effects of geofencing the service. *Journal of Transport and Land Use*, *14*(1), 933–948. doi:10.5198/jtlu.2021.1926

Hetzner, C. (2022, June 10). *Elon Musk's regulatory woes mount as U.S. moves close to recalling Tesla's self-driving software*. Fortune. https://fortune.com/2022/06/10/elon-musk-tesla-nhtsa-investigation-traffic-safety-autonomous-fsd-fatal-probe/

Hevelke, A., & Nida-Rümelin, J. (2015, June). Responsibility for crashes of autonomous vehicles: An ethical analysis. *Science and Engineering Ethics*, *21*(3), 619–630. doi:10.100711948-014-9565-5 PMID:25027859

International Bridge, Tunnel and Turnpike Association. (2018, August). *Glossary of Terms*. https://www.ibtta.org/sites/default/files/documents/IBTTA%20Glossary.pdf

Jfrux. (2019, February 27). *Comma Explorer drive annotation tool* [Image]. Licensed by Creative Commons Attribution-Share Alike 4.0 International. No changes made to the image. https://commons.wikimedia.org/wiki/File:Comma_Explorer_Drive_Annotation_Tool.png

John Deere says fully autonomous tractor advances future of farming. (2022, March 16). *KWQC*. https://www.kwqc.com/2022/01/04/john-deere-announces-fully-autonomous-tractor-ces

Kiley, D. (2018, March 19). First death of a pedestrian struck by an autonomous vehicle may set tone for lawyers and liability. *Forbes*. https://www.forbes.com/sites/davidkiley5/2018/03/19/the-first-pedestrian-fatality-with-an-autonomous-vehicle-could-set-tone-for-lawyers-and-liability

Kopestinsky, A. (2022, March 5). *25 astonishing self-driving car statistics for 2022.* Policy Advice. https://policyadvice.net/insurance/insights/self-driving-car-statistics

Krisher, T. (2021, February 1). Tesla issues recall after Full Self-Driving software allows cars to run stop signs. *LA Times.* https://www.latimes.com/world-nation/story/2022-02-01/tesla-recall-self-driving-software-runs-stop-signs

Krisher, T., & Dazio, S. (2022, January 18). Felony charges are 1st in a fatal crash involving Autopilot. *AP News.* https://apnews.com/article/tesla-autopilot-fatal-crash-charges-91b4a0341e07244f3f03051b-5c2462ae

Limousine driver faces felony charges after causing fatal crash while relying on Tesla autopilot. (2022, February 23). The Stanley Law Group. https://www.thestanleylawgroup.com/limousine-driver-faces-felony-charges-after-causing-fatal-crash-while-relying-on-tesla-autopilot

Lin, P. (2014, August 17). *Don't fear the robot car bomb.* Bulletin of the Atomic Scientists. https://thebulletin.org/2014/08/dont-fear-the-robot-car-bomb

Margan, S. K. (2018, December 20). Autonomous vehicles and insurance. *BimaQuest: The Journal of Insurance & Management, 18*(3), 15–24.

Mayer, M. M., Bell, R., & Buchner, A. (2021, December 23). Self-protective and self-sacrificing preferences of pedestrians and passengers in moral dilemmas involving autonomous vehicles. *PLoS One, 16*(12), 1–25. doi:10.1371/journal.pone.0261673 PMID:34941936

McLachlan, S. (2020, February 22). *How might driver licensing and vehicle registration evolve if we adopt autonomous cars and digital identification?* https://arxiv.org/ftp/arxiv/papers/2202/2202.09861.pdf

Meder, B., Fleischhut, N., Krumnau, N.-C., & Waldmann, M. R. (2019, February). How should autonomous cars drive? A preference for defaults in moral judgments under risk and uncertainty. *Risk Analysis: An International Journal, 39*(2), 295–314. doi:10.1111/risa.13178 PMID:30157299

Metz, C. (2021, March 15). Who is making sure the A.I. machines aren't racist? *The New York Times.* https://www.nytimes.com/2021/03/15/technology/artificial-intelligence-google-bias.html

Millar, J. (2014). Technology as moral proxy: Autonomy and paternalism by design. In *Proceedings of the IEEE 2014 International Symposium on Ethics in Engineering, Science, and Technology (ETHICS '14).* IEEE Press. Article 23, 1–7.

Murdock, J. (2020, April 30). Tesla faces lawsuit after Model X on Autopilot with 'dozing driver' blamed for fatal crash. *Newsweek.* https://www.newsweek.com/tesla-lawsuit-model-x-autopilot-fatal-crash-japan-yoshihiro-umeda-1501114

National Highway Traffic Safety Administration. (2016, September). *Federal automated vehicles policy: Accelerating the next revolution in roadway safety.* https://www.transportation.gov/sites/dot.gov/files/docs/AV%20policy%20guidance%20PDF.pdf

National Highway Traffic Safety Administration. (2019, December 17). *NHTSA announces update to historic AEB commitment by 20 automakers.* https://www.nhtsa.gov/press-releases/nhtsa-announces-update-historic-aeb-commitment-20-automakers

National Traffic Safety Board. (2017). *Collision between a car operating with automated vehicle control systems and a tractor-semitrailer tuck near Williston, Florida, May 7, 2016.* Highway Accident Report NTSB/HAR-17/02. https://www.ntsb.gov/investigations/AccidentReports/Reports/HAR1702.pdf

Ondruš, J., Kolla, E., Vertal, P., & Šarić, Z. (2020). How do autonomous cars work? *Transportation Research Procedia, 44*, 226–23. doi:10.1016/j.trpro.2020.02.049

Open Pilot supports 150+ vehicles. (n.d.) Comma. https://comma.ai/vehicles

Pearl, T. H. (2018, Summer). Hands on the wheel: A call for greater regulation of semi-autonomous cars. *Indiana Law Journal (Indianapolis, Ind.), 93*(3), 713–756.

Plungis, J. (2019, May 16). *Tesla driver in fatal march crash was using autopilot, NTSB says.* Consumer Reports. https://www.consumerreports.org/car-safety/tesla-driver-in-fatal-march-crash-was-using-autopilot-ntsb-says

Quach, K. (2022, March 14). Driver in Uber's self-driving car death goes on trial, says she feels 'betrayed.' *The Register.* https://www.theregister.com/2022/03/14/in_brief_ai

Quintana, S. (2021, May 13). Backseat Tesla driver unapologetic after arrest for reckless driving. *NBC Bay Area.* https://www.nbcbayarea.com/news/local/backseat-tesla-driver-unapologetic-after-arrest-for-reckless-driving/2544875

Randazzo, R. (2019, March 18). Who was really at fault in fatal Uber crash? Here's the whole story. *AZ Central.* https://www.azcentral.com/story/news/local/tempe/2019/03/17/one-year-after-self-driving-uber-rafaela-vasquez-behind-wheel-crash-death-elaine-herzberg-tempe/1296676002/

Schneble, C. O., & Shaw, D. M. (2021, September). Driver's views on driverless vehicles: Public perspectives on defining and using autonomous cars. *Transportation Research Interdisciplinary Perspectives, 11*, 100446. doi:10.1016/j.trip.2021.100446

Seamons, K. (2021, August 21). *There was something different about this fatal crash.* Newser. https://www.newser.com/story/309951/there-was-something-different-about-this-fatal-tesla-crash.html

Shbat, M. S., & Tuzlukov, V. (2014). Radar sensor detectors for vehicle safety systems. In N. Bizon, L. Dascalescu, & N. Mahdavi Tabatabaei (Eds.), *Autonomous vehicles: Intelligent transport systems and smart technologies* (pp. 3–55). Nova Science Publishers. https://search-ebscohost-com.libauth.purdueglobal.edu/login.aspx?direct=true&db=nlebk&AN=809607&site=eds-live

Shepard, W. (2019, August 15). South Florida man wants answers after crash killed girlfriend. *NBC Miami.* https://www.nbcmiami.com/news/local/man-wants-answers-after-deadly-crash/124944

Shepardson, D. (2019, November 5). In review of fatal Arizona crash, U.S. agency says Uber software had flaws. *Reuters.* https://www.reuters.com/article/us-uber-crash/in-review-of-fatal-arizona-crash-u-s-agency-says-uber-software-had-flaws-idUSKBN1XF2HA

Slaby, C. (2019, November 15). Decision-making self-driving car control algorithms: Intelligent transportation systems, sensing and computing technologies, and connected autonomous vehicles. *Contemporary Readings in Law and Social Justice, 11*(2), 29–35. doi:10.22381/CRLSJ11220194

Smith, H., & Mitchell, R. (2022, January 19). A Tesla on autopilot killed two people in Gardena. Is the driver guilty of manslaughter? *Los Angeles Times.* https://www.latimes.com/california/story/2022-01-19/a-tesla-on-autopilot-killed-two-people-in-gardena-is-the-driver-guilty-of-manslaughter

Srinivasa, A. (2018, July 27). *Understanding Radar for automotive (ADAS) solutions.* Path Partner. https://www.pathpartnertech.com/understanding-radar-for-automotive-adas-solutions

Srinivasan, R., & Chander, A. (2021, August). Biases in AI systems. *Communications of the ACM, 64*(8), 44–49. https://dl.acm.org/doi/10.1145/3464903 doi:10.1145/3464903

Stan, V. A., Timnea, R. S., & Gheorghiu, R. A. (2014). Intelligent highway surveillance and safety systems. In N. Bizon, L. Dascalescu, & N. Mahdavi Tabatabaei (Eds.), *Autonomous vehicles: Intelligent transport systems and smart technologies* (pp. 147–184). Nova Science Publishers. https://search-ebscohost-com.libauth.purdueglobal.edu/login.aspx?direct=true&db=nlebk&AN=809607&site=eds-live

Stein, S. (2022, February 7). Lidar is one of the iPhone and iPad Pro's coolest tricks: Here's what else it can do. *CNet.* https://www.cnet.com/tech/mobile/lidar-is-one-of-the-iphone-ipad-coolest-tricks-its-only-getting-better

The 6 levels of vehicle autonomy explained. (n.d.). Synopsis. https://www.synopsys.com/automotive/autonomous-driving-levels.html

The dangers of driverless cars. (2021, May 5). *National Law Review, 11*(125). https://www.natlawreview.com/article/dangers-driverless-cars

The Society of Automotive Engineers International. (2021, April). *Surface vehicle recommended practice.* https://www.sae.org/standards/content/j3016_202104/preview

Vargas, J., Alsweiss, S., Toker, O., Razdan, R., & Santos, J. (2021, August 10). An overview of autonomous vehicles sensors and their vulnerability to weather conditions. *Sensors (Basel), 21*(16), 5397. doi:10.339021165397 PMID:34450839

Vinkhuyzen, M. (2022, May 26). *Tesla FSD training – Garbage in, garbage out.* Clean Technica. https://cleantechnica.com/2022/05/26/tesla-fsd-training-garbage-in-garbage-out

Wamsley, L. (2020, September 16). Backup driver of autonomous Uber SUV charged with negligent homicide in Arizona. *NPR.* https://www.npr.org/2020/09/16/913530100/backup-driver-of-autonomous-uber-suv-charged-with-negligent-homicide-in-arizona

What is an autonomous car? (n.d.) *Synopsis.* https://www.synopsys.com/automotive/what-is-autonomous-car.html

What is ASIL? (n.d.). *Synopsis.* https://www.synopsys.com/automotive/what-is-asil.html

Wolkenstein, A. (2018, June 8). What has the Trolley Dilemma ever done for us (and what will it do in the future)? On some recent debates about the ethics of self-driving cars. *Ethics and Information Technology, 20*(3), 163–173. doi:10.100710676-018-9456-6

Wood & Ziemann v. State of Arizona, City of Tempe. (Az. Maricopa County Court, 2019). https://bloximages.newyork1.vip.townnews.com/azfamily.com/content/tncms/assets/v3/editorial/5/e2/5e235f8a-4b6d-11e9-9b1c-6bd0b0191f91/5c92d61c84180.pdf.pdf

Yadron, D., & Tynan, D. (2016, June 30). Tesla driver dies in first fatal crash while using autopilot mode. *The Guardian*. https://www.theguardian.com/technology/2016/jun/30/tesla-autopilot-death-self-driving-car-elon-musk

APPENDIX

Figure 5. SAE International Levels
Source: © SAE International from SAE J3016™
Taxonomy and Definitions for Terms Related to Driving Automation Systems for On-Road Motor Vehicles (2021-04-30), https://saemobilus.sae.org/content/J3016_202104/
Used by kind permission.

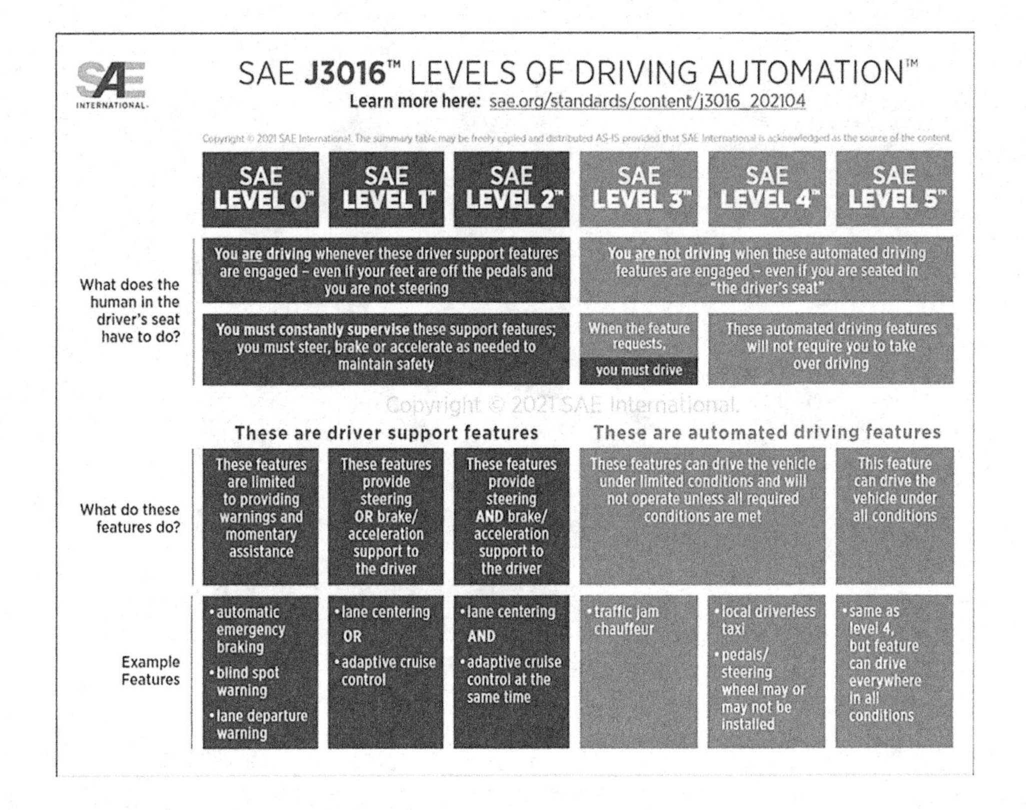

SAE **J3016**™ LEVELS OF DRIVING AUTOMATION™

Learn more here: sae.org/standards/content/j3016_202104

	SAE LEVEL 0™	SAE LEVEL 1™	SAE LEVEL 2™	SAE LEVEL 3™	SAE LEVEL 4™	SAE LEVEL 5™
What does the human in the driver's seat have to do?	**You are driving** whenever these driver support features are engaged – even if your feet are off the pedals and you are not steering			**You are not driving** when these automated driving features are engaged – even if you are seated in "the driver's seat"		
	You must constantly supervise these support features; you must steer, brake or accelerate as needed to maintain safety			When the feature requests, you must drive	These automated driving features will not require you to take over driving	
	These are driver support features			**These are automated driving features**		
What do these features do?	These features are limited to providing warnings and momentary assistance	These features provide steering **OR** brake/acceleration support to the driver	These features provide steering **AND** brake/acceleration support to the driver	These features can drive the vehicle under limited conditions and will not operate unless all required conditions are met		This feature can drive the vehicle under all conditions
Example Features	• automatic emergency braking • blind spot warning • lane departure warning	• lane centering **OR** • adaptive cruise control	• lane centering **AND** • adaptive cruise control at the same time	• traffic jam chauffeur	• local driverless taxi • pedals/steering wheel may or may not be installed	• same as level 4, but feature can drive everywhere in all conditions

Chapter 16
Metaverse:
The Ethical Dilemma

Susan Shepherd Ferebee
Purdue University Global, USA

ABSTRACT

Metaverse combines the virtual and physical, synthesizing the internet, the web, and extended reality into a world where digital and physical elements blend in differing degrees. The core purpose of the metaverse is social connection. This consists of users coexisting in an ecosystem that includes cultures, laws, economies, societies, and related obligations. Therefore, it will demand standards of how humans should behave and what is considered right and wrong. Ethics is based on these standards within any given universe. The ethical challenges faced in developing and deploying a metaverse are explored, and ethical design considerations of metaverse environments are defined. The metaverse ecosystem is described and critical technologies that support it are included. Examples of current metaverses are provided so that the reader can explore and experience a metaverse and further assess the ethics and benefits.

INTRODUCTION

Metaverse combines the virtual and physical, consolidating the web and internet with Extended Reality (Lee et al., 2021; Milgram et al., 1995). Extended Reality exists on a continuum where digital and physical blend to varying degrees (Milgram et al., 1995). According to Parisi (2021), the metaverse offers a 3D space for the convergence of 3D characters and digital information. The connection between the physical and virtual selves is their shared data (Lee et al., 2021). Social connection is the primary purpose of the metaverse. The first element of the metaverse is users expressing themselves through digital avatars to live a life representing a metaphor for their real world (Lee et al., 2021). The second element of the metaverse is the creation of ecosystems that encompass culture, laws, financial economies, and societal norms. Within the virtual ecosystems, products and intangible content can be produced, shared, bought, and sold across virtual domains (Vijoen, 2020). The metaverse is uncontrolled, governed only as needed for the greater good of the largest number (Parisi, 2021).

DOI: 10.4018/978-1-6684-5892-1.ch016

Modern technologies, over and above the existing internet, social media, and gaming environments, are required. Augmented and virtual reality, edge computing, blockchain, and artificial intelligence will be necessary constructs for the metaverse. With these technologies, the metaverse will offer a three-dimensional virtual space that is shared, persistent, and synchronous, coming together into a perceived virtual world (Lee et al., 2021). Metaverse makes possible for one human to have multiple digital existences simultaneously but never have to be physically present (Dow, 2022). Grider and Maximo (2021) place the metaverse into internet development: 1) Web 1.0 - Netscape succeeded in connecting people online, 2) Web 2.0 - social media like Facebook progressed to connecting people within virtual communities, and 3) Web 3.0 – metaverse connects people within a virtual community-owned space. The metaverse adds dimensions of virtual economics, organizational networks, blockchain infrastructure, and decentralized control (Grider & Maximo, 2021). The internet, through metaverse, offers 3 dimensional information and experience with real-time communication in 3 dimensional spaces (Parisi, 2021).

Hyperreality frames metaverse environments. Hyperreal means to display extraordinary vividness and incredible realism in terms of detail (Merriam Webster, n.d.).But at a more theoretical and philosophical level, as perceived in a metaverse, hyperreality is best described by Baudrillard (1994). In a metaverse, the users (citizens) create the world they live in – and no prior world existed before the one they made. Baudrillard (1994) describes this as "the generation by models of a real without origin or reality; a hyperreal. The territory no longer precedes the map, nor does it survive it. It is nevertheless the map that precedes the territory" (p. 1). For Baudrillard (1994), hyperreality lies beyond what is already known to be real.

In addition to hyperreal media is synthetic media which is entirely generated through artificial intelligence. Using artificial intelligence, very real-looking images of synthetic people can be easily generated. Artificial intelligence can be used to create synthetic video that are ultra-realistic and sometimes not easily recognized as not being real (Graham, 2022). Avatars may no longer be cartoonish-looking characters but images of synthetically generated people that are not easily distinguished from real people. Likewise, environments, while not being environments that exist in the real-world, may appear to have realistic objects that are not discernible from real-world objects.

The metaverse, comprised of users coexisting in an ecosystem which includes cultures, laws, economies, societies, and related obligations, will demand standards of how humans should behave and of what is considered right and wrong. Ethics is based on these standards within any given universe (Velasquez et al., 2010). Ethical standards exist outside of prescribed laws, religions, feelings, and what is socially acceptable. Ethics define what humans should do in terms of their rights, their obligations, what most benefits society, fairness, and particular virtues that are respected by a community (Velasquez et al, 2010). Challenges to future metaverse development include privacy, governance, and ethics. The questions arise as to how user behavior will be controlled in the metaverse, and what policies will emerge for user and platform governance (Fernandez et al., 2022). The important question is, what will the ethics of metaverse be and what are the dilemmas to determining those ethics?

This article explores the ethical challenges faced in developing and deploying a metaverse and discusses metaverse governance and defines ethical design considerations of metaverse environments. The metaverse ecosystem is described and critical technologies that support a metaverse are discussed. Examples of current metaverses are provided so that the reader can explore and experience a metaverse.

METAVERSE ETHICS

Spence (2008) defines a virtual world as a computer-mediated, continual environment where agents (represented by avatars) interact with each other and the environment. Avatars are a digital depiction of a person's body in a virtual world (Maloney et al., 2019). Spence's (2008) argument is that in a virtual world, an avatar acts as a representative (agent) of a person that created it in the virtual world with a goal and moral purpose in mind, which the person hopes to accomplish within the virtual world. Avatars, by purposively representing a person, have rights to well-being and freedom (Spence, 2008). Additionally, according to Gerwith (1978), since avatars, acting as agents with a purpose, have rights to liberty and well-being, they must be committed to the same rights of other avatars with which they interact. Spence (2008) states, "the rights of avatars in virtual worlds like the rights of their counterpart persons in the real world are universal rights that apply always and everywhere" (p. 11).

An interesting real-world example of Meta Platforms, the parent company of Facebook and Instagram, ethics challenge arose related to the Russia/Ukraine war. Meta (Meta Platform's metaverse) recently eased rules on hate speech to allow Ukraine users to make threats against Russian forces attacking their country. Meta Platforms has since modified the rule to state that posts calling for the assassination of Putin, Russia's president, are forbidden (Boorstin & Mangan, 2022). This is a clear example of blurred ethics and a lack of clear ethical guidelines in a metaverse environment. It also reflects the issue of who controls data in the metaverse. In this example, questions arise such as who has the authority to determine what can be stated, who is the authority monitoring actions in the metaverse, and how was that authority selected (Agerskov, 2021). Agerskov (2021) also asks what ethical code of conduct frames the metaverse, what is the legal system, and how is ethical behavior different for the human user, the human-generated avatar, and a machine generated and controlled avatar.

THE METAVERSE ECOSYSTEM

Lee, et al (2021) describe six factors of the metaverse ecosystem, each presenting unique ethical considerations. Each will be defined, and a summary of related ethical issues presented. The six factors of the metaverse ecosystem are a) avatar, b) created content, c) virtual finance, d) social acceptability, e) privacy and security, and f) accountability and trust.

Avatar

The original meaning of avatar derived from the Hindu term for a human or animal incarnation of a God. In virtual worlds, an avatar is a digital representation of a human user within a virtual environment (Lee et al., 2021). Park and Kim (2022) differentiate two types of avatars: digital twin and digitalMe. The digital twin avatar is a replication of the actual person in real life. In contrast, a digitalMe avatar represents an expression unique from the real self, projecting a self that cannot exist in real life. A digitalMe avatar can have a different gender, race, ethnicity, and age, and with the use of synthetic AI generation might appear exactly as a human image.

Regardless of how the avatar appears, ethical issues related to avatars have been discussed above: a) the rights of avatars, b) the obligation of ethical behavior between avatars, and differentiation of machine-driven avatars compared to human-driven avatars. An interesting concept in a metaverse environment is how the avatar characteristic (appearance, powers, attitude, rank) might influence human behavior and self-perception (Lee et al., 2021). Other avatar-related ethical issues include bias issues in avatar selection as well as privacy protection of the human behind the avatar as well as when there could be a need for the co-presence or revealing of the human behind the avatar (e.g., when the avatar is presenting material that requires them to be qualified to present.

Created Content

Content creation in the metaverse includes avatar creation, structure creation, environmental creation, and interaction creation (games, instructions for task achievement, virtual traveling, conferencing, live streams, social functions, schools, and stores) (Lee et al., 2021). Collaboration is a key element of content creation in a metaverse and as Lee et al. (2021) describe, "The metaverse could serve as a medium to knit the speakers (the primary actor of user-generated content) and the viewers virtually onto a unified landscape" (p. 31). Each form of content creation is an intellectual property to be protected. Censorship is another ethical issue that will arise.

Intellectual Property

Intellectual property refers to unique ideas, writings, inventions, and other non-tangible creations or ideas. They are protected by patents, copyrights, trade secret laws, and trademarks (Laas, 2021). This will not change in a metaverse, however, questions to answer are whether an avatar, in and of itself, can be the owner of content, or is only the creator of the avatar the owner. Because content creation in the metaverse environment relies on authoring tools for creating digital objects and these authoring tools rely on reuse of patterns, the originality of the work could be called into question. Multiple avatars can collaborate in content creation, just as multiple authors write articles, or produce movies in the world today. Understanding what constitutes original work in metaverse content creation, and which avatars on collaborative teams own the original design will raise ethical and potentially legal issues. An additional problem is raised when viewers are also asked to contribute to the content (Lee et al., 2021).

Another interesting question arises as to when an avatar, equipped with artificial intelligence and machine learning, co-authors a paper in the metaverse with a human metaverse user. Answers must be determined as to how the avatar represented as an author, and if a human-generated avatar with artificial intelligence and machine learning has the potential to plagiarize from internet content. If so, the next question is whether the human user or the human-generated avatar would be liable for the plagiarism. Lastly, if the writing is an automated process usable by the avatar, the legal and ethical implications related to plagiarism must be explored.

Censorship

According to the American Civil Liberties Union (n.d.), censorship occurs when one group of people stifles the expression of another group of people, imposing their own values, beliefs, and morality on another group. Censorship refers to the suppression of ideas, words, or images that are found to be unacceptable. When censorship occurs by the U.S. government, it is unconstitutional. With individuals, the First Amendment protects everyone's right to free speech (American Civil Liberties Union, n.d.). Expression can only be limited if the expression causes an immediate and imminent harm to a societal group (American Civil Liberties Union, n.d.). These same censorship principles and laws will apply in a metaverse. Censorship increased during the Trump administration, and users saw Twitter cancel many individuals' accounts, sometimes permanently, based solely on their content. With Elon Musk's purchase of the Twitter platform (Sy & Norris, 2022), it appears that these account terminations will be revisited as Musk has stated he will examine issues of free speech and extreme content (Sy & Norris, 2022).

How censorship will be overseen in the metaverse is an ethical question to continue to monitor. Minecraft, while not an entire metaverse, is seen as an advanced virtual world where avatars are free to generate content, meet, and share information (Lee et al., 2021). Minecraft has a repository of censored information called the "Uncensored Library" claiming it as a safe harbor for free written expression which is considered illegal (Lee et al., 2021). An ethical question that will arise in the metaverse is whether, if virtual content is destroyed in a virtual world, can the destroyed records be preserved (Lee et al., 2021).

Virtual Economics

The metaverse economy includes digital products and services as well as digital assets that can generate real-world value for individuals (Grider & Maximo, 2021). Non-fungible tokens are becoming a large part of this virtual metaverse economy. A non-fungible token refers to an absolutely unique digital item (e.g., music, artwork, video, book, cartoon). The purpose of NFTs is to give the buyer an asset that cannot be duplicated (Clark, 2022). New virtual businesses arising in the metaverse environments are NFT art galleries, virtual business spaces, digital marketing through digital billboards in the metaverse, digital metaverse concert venues, and spaces sponsored by real-world businesses (Grider & Maximo, 2021). These digital assets and business environments operate within a "crypto cloud economy" (Grider & Maximo, 2021, p. 12). Several types of cryptocurrencies are available, decentralized financial exchanges that provide loan opportunities, decentralized file storage options, and decentralized governance options where metaverse communities create their own rules for collectively owned virtual objects (Grider & Maximo, 2021).

Decentralization is a key concept in the metaverse referring to control being transferred from large corporations to the people inhabiting virtual communities (Canvesi, 2022). While this decentralized concept sounds like renewed opportunity and freedom for small players, reality does not bear this out. The concept of giving total economic governance to the metaverse users is not practical since cryptocurrency creation is hindered by a slow growth in money supply. Cryptocurrency does not have a built-in way to increase money supply easily, if at all. Cryptocurrency as it emerges in beginning stages attempts to reduce this barrier by continually producing new cryptocurrencies, but these new currencies are not always accepted and are not a reliable source of new money supply (Lee et al., 2021). What appears to be happening is large technical companies like Facebook or Google controlling the metaverse market share in what appears to be a shaping oligopolistic economy as opposed to a monopolistic economy, it is still far-removed from the idealistic decentralized world that metaverse developers strive for (Lee et al., 2021).

The ethics of the metaverse economy revolve around who is going to be the builder, and thereby controller, of the metaverse. Answers must include who will own the infrastructure and develop regulations, if the metaverse will be just another realm for current real-world businesses to make more money, and the consequences of real-world and metaverse data sources combined and now funneled through an imagined decentralized metaverse. It is just a shifting of current Facebook and Fortnite users, as examples, to a new and mis-defined decentralized world (Agerskov, 2021).

Social Acceptability

An avatar's social acceptability affects the avatar creator due to avatar attachment (Wolfendale, 2007). People become attached to possessions, other people, cultural artefacts and groups. Therefore, attachment to a self-created avatar is not unreasonable (Wolfendale, 2007). The avatar expresses identify for the creator, so if an avatar is not perceived as socially acceptable, the result can be cyberbullying with severe consequences for the creator (Wolfendale, 2007). Social acceptability elements of a metaverse include generational inclusion, avatar design options, and support for sustainability (Lee et al., 2021). Design elements that influence social acceptability are how diversity, inclusion, privacy, and fairness are demonstrated. Whether or not the metaverse design discourages or inhibits cyberbullying and user addiction also affects the social acceptability of a particular metaverse.

With regard to data privacy, questions to be asked include 1) will users have control over how their personal data is used and 2) will user-shared data be used for research? The former German chancellor, Angela Merkel, presented a unique privacy-related option that fits the decentralized model of metaverse. She suggested that users, themselves, have the option to sell their own data for monetary gain, a concept called privacy-trading (Lee et al., 2021). Some data-privacy ethical issues are removed when individuals have the right to manage and sell their own personal data.

Personalization of content offered to a user can be designed to support diversity and inclusion. By delivering only content that a particular user enjoys, has emotional connection to, and is inspired by, diversity and inclusion are represented within the metaverse. A problem for this approach to diversity and inclusion falls back to the algorithms used to determine what content matches a particular user (Lee et al., 2021). How fairness could be programmed is the first step, and what constitutes fairness (which may not be the same to everyone) determined. The ethical issue revolves around who is determining how the artificial intelligence (AI) is being programmed and who is judging the effectiveness of the results (Ruidong et al., 2021). Maxine Williams, the Chief Diversity Officer of Meta offers the following approaches being taken at Meta to insure diversity and inclusion in a three-dimensional virtual shared space:

- Diverse people must include not only metaverse users but metaverse designers and developers
- Language barriers must disappear in a metaverse. Translation tools must be an integral element of a metaverse to increase equal participation.
- True diversity and inclusion mean removing technological barriers and historical cultural barriers. Potential users must be asked to describe what inclusivity looks like to them in general and in a metaverse.
- Access to a metaverse must not be limited to particular devices.
- Avatar representation must demonstrate inclusivity and diversity at every level including disabilities. (Williams, 2022).

Privacy and Security

Fernandez et al. (2022) describes levels of privacy concerns in a metaverse: 1) user and bystander safety, 2) privacy violations at a sensory level, and 3) communication and action privacy. Data collected through sensory tracking such as eye movement in a virtual reality headset or recording of the user's real-world environment can provide highly personal data about the user, violating fundamental privacy rights (Fernandez et al., 2022). Conversations with other avatars in the metaverse are captured and could result in behavior manipulation by bad actors in the metaverse (Fernandez et al., 2022).

An interesting fact about existing metaverses is that while users can create avatars that are not in any way like their real existence, thereby creating an elevated level of personal data privacy, other elements such as the ability of other users to completely eavesdrop on and monitor the movements and locations of other avatars, removes a different type of privacy. A metaverse solution to this monitoring issue is for a person to create multiple avatars performing different behaviors in various places to confuse other users about which avatar is the primary avatar reflecting the real user (Leenes, 2007). A similar privacy protection technique is to create multiple metaverse spaces (homes, parks, streets) some of which are private. When a user is in a private space, monitoring cannot occur. Invisible avatars can also be created to perform functions undetected (Leenes, 2007). Clearly ethical, and potentially legal issues could arise with behaviors performed by invisible avatars – this could be considered the perfect way to commit virtual crimes. Another ethical issue that emerges from multiple avatars and invisible avatars is the lack of trust that would exist in an environment where a person could never rely on the authenticity with whome they are interacting (Lee et al., 2021). An unethical and unsafe result of the ability to create multiple avatars and objects is the ability to create inauthentic digital products that a user might buy. The ethical issues surrounding privacy and security are critical and demand that novel solutions to protect user identity be explored.

In terms of cybersecurity in the metaverse, the metaverse integrates a wide range and number of systems and therefore increases the attack surface. Improved methods of access control are required for the more complex system presented by the metaverse (Di Pietro, 2022). Social engineering attacks are simplified with the ability to create multiple avatars and masking one's true identity. Further research is needed to describe the architectural layers of the metaverse so that an improved and more complete security approach can be defined (Di Pietro, 2022).

Accountability and Trust

Trust

Users will not adopt a system that they cannot trust and where they cannot see a clear path to accountability for behaviors and events (Lee et al., 2021). According to Zhou and Lee (2022), the artificial intelligence that currently drives metaverse technology is not trustworthy and has the ability to cause harm due to biased and inaccurate decision-making from machine learning models. These same artificial intelligence algorithms will be making decisions on inclusion and diversity within a metaverse system. An example is Facebook, in 2019, was sued by the United States Department of Housing and Urban Development because their AI algorithms allowed and caused discrimination based on religion, race, and gender on Facebook's advertising platform (Zhou & Lee, 2022). To date, the algorithms used appear to still deliver biased results. Not only do the algorithms lead to bias but also the data collection and model designs. The ethical issue of determining what is fair arises in writing these algorithms, models, and processes. To achieve trustworthy AI, an algorithm must guarantee that results are neutral, but the metrics measuring fairness are complex (Zhou & Lee, 2022). Zhou and Lee (2022) suggest an approach that will improve fairness in AI:

- Because data acquires bias from the real world that it was collected from, data analysts and system operators must select data that reflects data diversity
- A system should ensure that use of the system should not be limited by gender, age, education, location, nationality, ethnicity, and socioeconomics.
- Access to the system must be fair and non-discriminatory.
- The system should ensure that users with poor internet and slower equipment are able to access and use the system. (Zhou & Lee, 2022)

Ammanath (2022) discusses elements that would create trust in a retail shopping experience in a virtual space using AI in a metaverse. Ammanath lists questions to be asked in answering if the AI is creating increased trust in the virtual space. These include: 1) if the sales agent is artificial intelligence, is that disclosed to the shopper, 2) can the user easily opt out of data-sharing, 3) what is the error level of the AI output in the complex metaverse environment, and 4) who is accountable for any ethical issue that arises? Ammanath (2022) emphasizes that employees, metaverse ecosystems, and procedures and processes must align to trust factors and states that ethics must be built into the design process.

Lee et al (2021) discuss the importance of dependable technology in a metaverse as an important contributor to trust. A metaverse user should be able to expect that the Extended Reality technology will use their data as they expect it to be used. Users also need to be educated in protecting themselves and removing themselves from untrustworthy situations in the metaverse. A sense of self-control contributes to trust. The IEEE Global Initiative on Ethics of Autonomous and Intelligent Systems provides recommendations for users that when they enter a virtual space, they should immediately receive instruction on how to quickly exit the virtual environment (Lee et al., 2021). Security awareness and self-protection knowledge will be an ethical requirement for building trust in the metaverse.

Accountability

Accountability is strongly linked to trust in the metaverse. A metaverse user must readily understand the methods for remedy regarding any aspect of the metaverse experience. An example would include content moderation policies. Avatars are user generated content, and the user must understand the policies and who they can interact with when there is a stated violation of a policy. Content moderation will need to be able to delineate between a human-created avatar and a machine-generated bot, as an example. Ethical and legal issues arise because a human-generated avatar will have freedom of expression (for the most part), but a machine-generated avatar would not (Lee et al., 2021).

One other aspect of accountability for both the metaverse system and the metaverse user centers around liability in Extended Reality scenarios where an injury might occur. If an injury occurs because an Extended Reality technology takes a user's total attention away, it needs to be determined if liability is to be placed on the device manufacturer, software developer, system designer, third-party organizations, or the user. These are similar issues that surround autonomous cars (Lee et al., 2021).

BLOCKCHAIN – FOUNDATION OF METAVERSE PLATFORM

Blockchain, as a distributed ledger system, is the foundation of cryptocurrencies but has expanded to storage of digital assets and smart contract capability. Economic transactions occur without a broker or intermediary (Tang et al., 2019). Blockchain technology alleviates some of the data privacy issues that emerge in a metaverse. Blockchain stores records and transactions as blocks that are linked together with cryptographic processes which guarantees that the ledger is sacrosanct and can share information securely even when the surrounding environment or system is not secure. Blockchain aligns with the metaverse concept because it is decentralized, having no centralized authority (such as a bank or financial institution). Blockchain uses a proof of work concept and for a metaverse environment, it facilitates enforcing accountability (Gadekallu et al., 2022). The use of blockchain in the metaverse is inevitable with the key purpose being to secure all digital content for the metaverse user (Gadekallu et al., 2022). In the metaverse, blockchain will both store content and provide an economic system that connects the real world to the metaverse (Gadekallu et al., 2022). A challenge to the effective use of blockchain in the metaverse is that the verification of the encrypted data is not as fast as what individuals are used to. The current challenge is to increase the speed and scalability of data access. Additionally, and an ethical concern of blockchain, is for public blockchains, data is available to all users leading to a privacy issue (Lee et al., 2021).

Blockchain has an interesting progression. Bitcoin is the first generation blockchain and it only decentralizes transaction records. The second generation blockchain, Ethereum, added smart contract functionality and asset management. In Ethereum, transactions trigger smart contract rules which perform actions like transferring funds or sending notifications. An insurance company can be united with hospitals and record and track patient health data in the blockchain and use a smart contract to pay the patient. Non-fungible tokens represent a smart contract variation (Gadekallu et al., 2022).

Blockchain for Metaverse AI

Challenges of using AI in a metaverse include 1) users knowing if they are interacting with an avatar generated by a human or one generated by a computer, 2) AI code can be used in the metaverse to perform illegal operations such as stealing or winning games, and 3) AI can make errors that result in users losing trust in the metaverse environment. Blockchain can provide solutions. Blockchain encryption gives metaverse users complete control over their information and allows them to authenticate themselves or their information in applications without sharing the information with the application. This occurs through blockchain's zero-knowledge proofs. Blockchain's audit trail provides full accountabilities for all transactions that occur within the metaverse. Blockchain allows AI to collect and use data for machine learning with no ability to exploit that gathered data (Gadekallu et al., 2022). However, to protect public blockchain data (Lee et al., 2021), standards and regulations for blockchain will be required. Overall, AI as well as Extended Reality are enhanced from the strong data privacy offered by blockchain access control, authentication, security, audits, and anonymity.

Blockchain for Metaverse Internet of Things (IoT)

With the massive amount of IoT devices that exist in a metaverse, data storage and security become a significant concern. This data is unstructured and often streamed real-time. A decentralized approach is needed for tracking for regulatory compliance as well as for safety. Blockchain allows IoT devices to use cross-chain networks to create archives of shared transactions that cannot be tampered. Every separate transaction is authenticated and archived. In addition, a blockchain that is enabled by IoT facilitates real-time data storage. Trust in the data is enhanced allowing users to make decisions and act quickly (Gadekallu et al., 2022).

Cross Chain Development

Cross-chain is seen as the primary solution that will allow interoperation between different blockchains facilitating users' ability to perform transactions between different blockchains. A user could make a transaction between a Bitcoin blockchain and an Ethereum blockchain. The ultimate goal is a decentralized system using cross-chain links. A further goal is seen as achieving Blockchain as a Service (Gadekallu et al., 2022).

Blockchain Ethics

While blockchain is growing rapidly and expanding into different areas of use, as it is adopted, there is little understanding of the overall consequences of blockchain use. This uncertainty raises ethical issues around blockchain use. According to Tang et al. (2019), there are four dimensions of blockchain, each with its own ethical issues: technology stack (the peer-to-peer decentralized ledger), cryptocurrencies, smart contracts, and decentralization.

The blockchain distributed shared ledger (the technology stack) is known for storing historical data immutably and for collective authentication. Ethical issues that apply to this blockchain ledger include data accuracy, data privacy, data accessibility, and data property (Tang et al., 2019).

Cryptocurrencies are untrusted as they play a role in disrupting national currencies and are seen to be Ponzi schemes and multilevel marketing schemes. However, cryptocurrencies are ethical as a payment mechanism. Ethical questions arise about how they will disrupt current payment processing institutions and how they will affect economies. Tax evasion is also an ethical issue (Tang et al., 2019).

Smart contracts are deployed in the blockchain network and perform actions based on a set of rules. These automated processes can be implemented outside of commercial agreements and an ethical issue exists as to how that automated changing and transferring of digital assets might be abused (Tang et al., 2019).

While cloud providers and cloud platforms have increased centralization, it raises ethical issues as to who owns and controls the data now in the cloud. New power has accumulated for cloud providers. Blockchain, on the other hand, decentralizes which provides both benefits and disadvantages. The combination of blockchain and cloud provides a balance (Tang et al., 2019).

METAVERSE GOVERNANCE

Governance is tied to ethics as it provides the framework for policies and regulations that helps ensure ethical design and ethical user conduct (Fernandez et al., 2022). Ethical design relates to how developers choose features, write code, and what interaction operability they add to the metaverse system. An example would be a developer choosing to include privacy bubbles, where users can limit access to interactions, thereby protecting their own privacy. Similarly, a developer and/or engineer can build in automated processes that allow banning of certain language in the chats (Fernandez et al., 2022).

Blockchain technology (described later in this article) includes features that allow autonomous decisions. The algorithms used for this autonomous decision-making exert incredible control in the metaverse and the average metaverse user will not understand the algorithms. A governance approach supporting only open-source development of algorithms is optimal, but still the user does often not understand the code. This reveals the need for internal automated auditing processes and systems. In the metaverse, more governance processes will be automated. Within blockchain technology exist decentralized autonomous organizations (DAOs) which allow every metaverse member to participate in decisions about changes to the metaverse platform (Fernandez et al., 2022).

Governance of youth in the metaverse focuses on shaping positive behaviors. Governance tools manage youth user misbehavior by encouraging and incentivizing positive behaviors. Community governance is a key element of metaverse governance with both youth and adults. Governance intends to develop awareness of decision-making that is for the good of society (Fernandez et al., 2022).

METAVERSE – ETHICAL DESIGN

The metaverse will expand the world and remodel society. New forms of interaction and creative expression emerge in limitless ways as a metaverse develops. Time, location, and language barriers dissolve. However, for metaverses to achieve the positive good to society desired, ethics must be built in at the design level. Fernandez et al. (2022) proposed a modular model of the Ethical Hierarchy of Needs to frame the metaverse design process. Modules include: 1) rule creation, secure software, behavior management, creation tools, creators, avatar privacy, sensory privacy, and communication privacy.

Creating Spaces, Avatars, and Creative Content

In a metaverse, opportunities are emerging for users/players to create an improved avatar or game item and sell for monetary gain. In addition, game players in a metaverse can earn money (digital currency) for their game achievement. A game player or metaverse member can create virtual structures and sell these as well. With this opportunity comes open doors for unethical scammers to sell unauthenticated items or poorly developed items. Design elements that can improve this situation include invite-only policies for selling items. Another design element is the inclusion of a reputation-based system where sales are made and sellers rated.AI can provide further moderation and punitive actions (Fernandez et al., 2022). Digital twin avatars should be protected through use of a blockchain ledger to authenticate the digital twin

Design Elements to Support the Societal Good

The metaverse should focus ethically on providing a diverse world, an equalizing world, an accessible world, a portal for cultural interaction and communication, and a trusted environment (Fernandez et al, 2022). While accessibility refers to everyone's ability to access and function within a metaverse, it also refers to new opportunities to expand for group gathering and collaboration. Examples that already exist are global concerts with unlimited virtual attendance and global education with unlimited attendance (Fernandez et al., 2022). Examples would include Berkeley holding virtual graduation within the Minecraft metaverse, Austin, Tx exploration, a Travis Scott concert and Ariana Grande concert in Fortnite, and virtual campuses in metaverse worlds.

Equalization occurs in a metaverse because individuals create an avatar to reflect how they want to portray themselves thus removing race, disability, gender, age, and socio-economic status. A person in a wheelchair can portray themselves as an athlete, gender identification is expressed through a self-created avatar, young individuals can portray as older, and the elderly can regain their youthful portrayal of themselves. A human can become a cat. The creative potential is unlimited as users consider the ability to create multiple avatars that might operate in different metaverse environments.

Trust is a social element critical to continued acceptance of the metaverse and the key element to be addressed for ethical design. A design element to create trust is incentives to reward authentic behavior and honesty and to participate and share in trust behaviors. Misleading information and false information are a critical issue to be addressed in metaverse design. Avatars, rather than actual people, contribute to the difficulty in trusting information being disseminated (Fernandez et al., 2022). Again, incentive systems should be built into the metaverse design to promote the sharing of accurate information (Fernandez et al., 2022). A design approach toward an ethical universe typically relies on positive incentive followed by punitive action for negative behaviors. However, the emphasis is typically on incentivization.

According to Fernandez et al. (2022), ethical design must address core individual rights, individual experience, and individual endeavor. In a metaverse, all people must be welcome and maintain their right to privacy and have the right to participate in all community decisions. Individual actions will be evaluated through a reputation-based system, reported to by all metaverse users. The right of individuals to design and create avatars, structures, communities, and environments will promote inclusion. There is an assumption in this ethical design framework that individuals seek societal good, but in the event where they may not, punitive actions are built in.

METAVERSE PLATFORMS TO VISIT

There are a number of metaverse platforms in various stages and with various purposes that people can explore today. Six of these are described:

- **The Sandbox** – The Sandbox is a metaverse in which an individual can purchase, sell, and trade pieces of property. It consists of virtual real estate, gaming, and a Disneyland-like amusement park. It is developed on the Ethereum Blockchain, meaning it includes the use of smart contracts. The Ethereum token used in Sandbox is called SAND. Sandbox maintains partnerships with real-world organizations like the South China Morning Post, Adidas, Warner Music Group, Atari, and Snoop Dogg. This metaverse platform is considered secure and stable.
- **Decentraland** – Decentraland has two tokens which each serve a different purpose. The LAND token is used to purchase virtual land. All historical data of land ownership is recorded on the blockchain and can be traced back to the original owner, similar to the U.S. Title service. MANA is a second token which allows the metaverse member to create a virtual reality land parcel. The token also allows purchase of the metaverse services. Virtual businesses can be created an operated as an online business. Decentraland has a sovereign state called Barbados and Barbados has its own embassy. The Decentraland bank branch called Onyx Lounge is owned by J. P. Morgan.

- **Roblox** – Roblox was strictly a gaming environment but has morphed into a metaverse platfor offering a variety of virtual environments and events for users. Roblox uses a token called Robux which is purchases with real-world cash and used to make purchases within Roblox. The Roblox organization has gone public and currently has over forty million players.
- **Metahero** – Metahero is unique in offering high quality 3D modeling and scanning tools. Hyperreal 3D avatars can be created by users as well as other 3D virtual objects. NFTs can be created from existing real-world artwork. The cryptocurrency used is called $HERO.
- **Second Life** – Second Life is one of the older metaverses. It was created by Linden Lab and is not considered a game, but rather an open-ended experience, different for each user. Users of Second Life design and build avatars, cities, spaces, and structures. This metaverse contains only user created content. The currency in Second Life is the Linden. Although Second Life is not considered a game, users can create games within Second Life. Second Life is not considered reliable or stable as it does experience outages
- **Horizon Worlds** – Horizon Worlds is the virtual social platform developed bv Meta, owned by Mark Zuckerberg. It is limited to individuals 18 years and older and only individuals from Canada and the U.S. and is only accessible through an Oculus system.

CONCLUSION

Metaverse ethics, while mirroring ethical issues that are faced in the real-world environment, also adds layers of complexity related to ethical behavior in non-human avatars that, through AI, can often act autonomously, and can be considered not just an extension of the human creator but a separate entity with completely separate characteristics, abilities, and behaviors. Adding to the complexity is the fact that one human being can create an endless number of avatars as well as an endless number of virtual environments within a metaverse. Differentiating between a human created avatar and a machine created avatar introduces additional ethical and safety issues.

The key to developing an ethical metaverse platform is to build ethics in at the design level. The goal is to have building blocks and code or technology modules that support ethical behavior for avatars. Incentives for authentic and honest avatar communication as well as incentive for trustworthy behaviors are core design elements as are reputation-based systems to control untrustworthy behavior. Ensuring data privacy and system accountability are also critical design elements to implement. The use of blockchain significantly improves data privacy, accountability, traceability, and transparency of data in a metaverse environment. However, blockchain introduces new ethical issues in four realms:: technology stack (the peer-to-peer decentralized ledger), cryptocurrencies, smart contracts, and decentralization. These are the issues that exist today with the current versions of blockchain. As blockchain continues to advance, new ethical issues will continue to emerge.

An ethical metaverse will offer a diverse, accessible, and equalizing environment that opens doors for innovative and advanced global communication, interaction, and collaboration. An environment that allows users to create a desired identity through an avatar equalizes bias based on age, gender, race, disability, and socio-economic status. Communication between people from other countries is empowered through the use of immediate translation capability. The barriers for interaction and communication in the real world are removed creating vibrant opportunities for social connection.

REFERENCES

Agerskov, S. (2021, November 2). *The metaverse – a dystopian future?* DataEthics. https://dataethics.eu/the-metaverse-a-dystopian-future

American Civil Liberties Union. (n.d.). *What is censorship*. https://www.aclu.org/other/what-censorship

Ammanath, B. (2022, March 21). An ethical metaverse takes trustworthy AI. *The Wall Street Journal*. https://deloitte.wsj.com/articles/an-ethical-metaverse-takes-trustworthy-ai-01647884164

Baudrillard, J. (1994). *Simulacra and simulation* (S. F. Glaser, Trans.). University of Michigan.

Boorstin, J., & Mangan, D. (2022, March 14). Meta now says you can't threaten to kill Russia's President Putin on Facebook because of the Ukraine war. *CNBC*. https://www.cnbc.com/2022/03/14/facebook-bans-death-threats-russia-leader-putin-despite-ukraine-war.html

Canvesi, B. (2022, February 2). *The foundation of the metaverse: centralization versus decentralization*. Association for Talent Development. https://www.td.org/atd-blog/the-foundation-of-the-metaverse-centralization-versus-decentralization

Clark, M. (2022, June 6). *NFTs, explained*. The Verge. https://www.theverge.com/22310188/nft-explainer-what-is-blockchain-crypto-art-faq

Di Pietro, R. (2022, April 6). *The metaverse: Technology, privacy, and security risks and the road ahead*. Modern Diplomacy. https://moderndiplomacy.eu/2022/04/06/the-metaverse-technology-privacy-and-security-risks-and-the-road-ahead

Dow, L. (2022, January 12). *The metaverse is coming. Are your ethics ready?* FastCompany. https://www.fastcompany.com/90712458/the-metaverse-is-coming-are-your-ethics-ready

Fernandez, C. B., & Hue, P. (2022, March 22). *Life, the metaverse and everything: An overview of privacy, ethics, and governance in metaverse*. Cornell University. doi:10.48550/arXiv.2204.01480

Gadekallu, T. R., Huynh-The, T., Wang, W., Yenduri, G., Ranaweera, P., Pham, Q.-V., da Costa, D. B., & Liyanage, M. (2022, March 21). *Blockchain for the Metaverse: A review*. Cornell University. doi:10.48550/arXiv.2203.09738

Gerwith, A. (1978). *Reason and morality*. University of Chicago Press.

Graham, T. (2022, January 17). *An introduction to the hyperreal metaverse*. Medium. https://medium.com/synthetic-media/an-introduction-to-hyperreal-synthetic-media-8a22e06509c5

Grider, D., & Maximo, M. (2021, November). *The metaverse: Web 3.0 virtual cloud economies*. Grayscale Investments, LLC. https://grayscale.com/wp-content/uploads/2021/11/Grayscale_Metaverse_Report_Nov2021.pdf

Laas, K. (2021, February 16). *Intellectual property*. Illinois Tech. https://guides.library.iit.edu/c.php?g=474695&p=3248753

Lee, L.-K., Braud, T., Zhou, P., Wang, L., Zu, D., Lin, Z., Kumar, A., Bermejo, C., & Hui, P. (2021, November). All one needs to know about Metaverse: A complete survey on technological singularity, virtual ecosystem, and research agenda. *Journal of Latex Class Files, 14*(8), 1–66. doi:10.48550/arXiv.2110.05352

Leenes, R. (2007). *Privacy in the metaverse. In IFIP International Summer School on the Future of Identity in the Information Society*. Springer. doi:10.1007/978-0-387-79026-8_7

Maloney, D., Rajasabeson, S., Moore, A., Caldwell, J., Archer, J., & Robb, A. (2019, March). Ethical concerns of the use of virtual avatars in consumer entertainment. *2019 IEEE Conference on Virtual Reality and 3D User Interfaces (VR)*. Osaka, Japan. 10.1109/VR.2019.8797926

Milgram, P., Takemura, H., Utsumi, A., & Kishino, F. (1995). Augmented reality: A class of displays on the reality-virtuality continuum. Telemanipulator and Telepresence Technologies, 2351, 282-292.

Parisi, T. (2021, October 22). *The seven rules of the metaverse*. Medium. https://medium.com/metaverses/the-seven-rules-of-the-metaverse-7d4e06fa864c

Park, S.-M., & Kim, Y.-G. (2022). A metaverse: Taxonomy, components, applications, and open challenges. *IEEE Access: Practical Innovations, Open Solutions, 10*, 4209–4251. doi:10.1109/ACCESS.2021.3140175

Ruidong, Y., Li, Y., Li, D., Yongcai, W., Yuqing, Z., & Weili, W. (2021, March). A stochastic algorithm based on reverse sampling technique to fight against the cyberbullying. *ACM Transactions on Knowledge Discovery from Data, 15*(4), 1–22. doi:10.1145/3441455

Spence, E. (2008). Meta ethics for the Metaverse: The ethics of virtual worlds. In A. R. Briggle, K. Waelbers, & P. Brey (Eds.), *Current Issues in Computing and Philosophy* (pp. 3–13). IOS Press.

Sy, S., & Norris, C. (2022, April 25). What Elon Musk's $44 billion purchase of Twitter may mean for the company and free speech. *PBS*. https://www.pbs.org/newshour/show/what-elon-musks-44-billion-purchase-of-twitter-may-mean-for-the-company-and-free-speech

Tang, Y., Xiong, J., Becerril-Arreola, R., & Iyer, L. (2019, June). Blockchain ethics research: A conceptual model. *SIGMIS-CPR '19: Proceedings of the 2019 on Computers and People Research Conference*, pp. 43–49. 10.1145/3322385.3322397

Velasquez, M., Andre, C., Shanks, T., & Meyer, M. (2010, January 1). *What is ethics?* Markkuka Center for Applied Ethics at Santa Clara University. https://www.scu.edu/ethics/ethics-resources/ethical-decision-making/what-is-ethics/

VijoenS. (2020, November 23). The promise and limits of lawfulness: Inequality, law, and the techlash. SSRN. https://papers.ssrn.com/sol3/papers.cfm?abstract_id=3725645

Williams, M. (2022, February 24). *Building the metaverse with diversity and inclusion from the start*. Meta. https://about.fb.com/news/2022/02/building-the-metaverse-with-diversity-and-inclusion-from-the-start/

Wolfendale, J. (2007, July). My avatar, my self: Virtual harm and attachment. *Ethics and Information Technology, 9*(2), 111–119. doi:10.100710676-006-9125-z

Zhou, P., & Lee, L. (2022, May 23). *The metaverse is coming but we still don't trust AI*. 360. https://360info.org/the-metaverse-is-coming-but-we-still-dont-trust-ai

Compilation of References

(ISC)². (n.d.). *Prepare for your (ISC)² exam day*. https://www.isc2.org/Exams/Exam-Day

15 tools and resources top tech leaders use every day. (2021, May 28). Forbes. https://www.forbes.com/sites/forbestechcouncil/2021/05/28/15-tools-and-resources-top-tech-leaders-use-every-day

15 ways leaders can encourage employees to take initiative. (2020, June 11). Small Biz Trends. https://smallbiztrends.com/2020/06/15-wasy-leaders-can-encourage-employees-to-take-initiative.html

33 Accessibility statistics you need to know in 2021. (2021, April 14). Monsido. https://monsido.com/blog/accessibility-statistics

4 tech tools every business leader should know about. (2020, February 4). Tanveer Naseer Leadership. https://tanveer-naseer.com/4-tech-tools-critical-for-organizational-success/

8 things to know about palm oil. (n.d.). World Wildlife Fund. https://www.wwf.org.uk/updates/8-things-know-about-palm-oil

Aasheim, C., Kaleta, J., & Rutner, P. (2021, March 13). Assessing IT students' intentions to commit unethical actions. *Journal of Computer Information Systems*, *61*(3), 219–228. doi:10.1080/08874417.2019.1584544

Abdulkader, B., Magni, D., Cillo, V., Papa, A., & Micera, R. (2020, August 28). Aligning firm's value system and open innovation: A new framework of business process management beyond the business model innovation. *Business Process Management Journal*, *26*(5), 999–1020. doi:10.1108/BPMJ-05-2020-0231

Abugabah, A., Nizamuddin, N., & Abuqabbeh, A. (2020). A review of challenges and barriers implementing RFID technology in the Healthcare sector. *Procedia Computer Science*, *170*, 1003–1010. doi:10.1016/j.procs.2020.03.094

Accessibility Guidelines Working Group (A.G.W.G.). (2018, June 5). *Web Content Accessibility Guidelines (WCAG) 2.1*. World Wide Web Consortium (W3C). https://www.w3.org/TR/WCAG21

Acharya, A. S., Prakash, A., Saxena, P., & Nigam, A. (2013, July-December). Sampling: Why and how of it. *Indian Journal of Medical Specialties*, *4*(2), 330–333.

Acien, A., Morales, A., Vera-Rodriguez, R., & Fierrez, J. (2020). Mobile active authentication based on multiple biometric and behavioral patterns. In T. Bourlai, P. Karampelas, & V. M. Patel (Eds.), *Securing Social Identity in Mobile Platforms. Advanced Sciences and Technologies for Security Applications* (1st ed.) pp. 161–177. Springer. doi:10.1007/978-3-030-39489-9_9

Acquisti, A., Adjerid, I., Balebako, R., Brandimarte, L., Cranor, L. F., Komanduri, S., & Wilson, S. (2017, August). Nudges for privacy and security: Understanding and assisting users' choices online. [CSUR]. *ACM Computing Surveys*, *50*(3), 1–41. https://doi.org/ doi:10.1145/3054926

Act to Increase Privacy and Security by Regulating the Use of Facial Surveillance Systems by Departments, Public Employees and Public Officials, Publ. L. No. 2021, c. 394, §1, 25 MRSA Pt. 14 (2021). https://legislature.maine.gov/legis/statutes/25/title25sec6001.html

Afifi, M., Kalra, D., Ghazal, T., & Mago, B. (2020, January). Information technology ethics and professional responsibilities. *International Journal of Advanced Science and Technology*, *29*(4), 11336–11343.

Agerskov, S. (2021, November 2). *The metaverse – a dystopian future?* DataEthics. https://dataethics.eu/the-metaverse-a-dystopian-future

AI Fairness 360 – Resources: Guidance on choosing metrics and mitigation. (n.d.). IBM Research Trusted AI. http://aif360.mybluemix.net/resources#guidance

AirSlate Legal Forms, Inc. (n.d.). *State law and legal definition.* U.S. Legal. https://definitions.uslegal.com/s/state-law

Akinyemi, A. (2021, December 10). *International web accessibility laws and policies.* Who Is Accessible. https://www.whoisaccessible.com/guidelines/international-web-accessibility-laws-and-policies

Ali, M., Sapiezynski, P., Bogan, M., Korolova, A., Mislove, A., & Rieke, A. (2019, November). Discrimination through optimization: How Facebook's ad delivery can lead to skewed outcomes. *Proceedings of the ACM on Human-Computer Interaction*, *2019*(3), 1–17. doi:10.1145/3359301

Allameh, E., Heidari Jozam, M., de Vries, B., Timmermans, H., Beetz, J., & Mozaffar, F. (2012, August). The role of Smart Home in smart real estate. *Journal of European Real Estate Research*, *5*(2), 156–170. doi:10.1108/17539261211250726

Álvarez López, Y., Franssen, J., Álvarez Narciandi, G., Pagnozzi, J., González-Pinto Arrillaga, I., & Las-Heras Andrés, F. (2018, August 13). RFID technology for management and tracking: e-Health applications. *Sensors (Basel)*, *18*(8), 2663. doi:10.339018082663 PMID:30104557

Alzheimer's Association. (2019). *Alzheimer's Disease Facts and Figures.* https://www.alz.org/media/Documents/alzheimers-facts-and-figures-2019-r.pdf

American Civil Liberties Union. (n.d.). *What is censorship.* https://www.aclu.org/other/what-censorship

Amirat, C., & Reeps, R. (2018, June-July). Continuous innovation through experimentation. *2018 IEEE Technology and Engineering Management Conference (TEMSCON).* doi:10.1109/TEMSCON.2018.8488399

Ammanath, B. (2021, November 9). Thinking through the ethics of new tech… before there's a problem. *Harvard Business Review.* https://hbr.org/2021/11/thinking-through-the-ethics-of-new-techbefore-theres-a-problem

Ammanath, B. (2022, March 21). An ethical metaverse takes trustworthy AI. *The Wall Street Journal.* https://deloitte.wsj.com/articles/an-ethical-metaverse-takes-trustworthy-ai-01647884164

Amsler, S. (2021, March). *RFID (radio frequency identification).* TechTarget: IoT agenda. https://internetofthingsagenda.techtarget.com/definition/RFID-radio-frequency-identification

Ancestry Guide for Law Enforcement. (n.d.). Ancestory.com. https://www.ancestry.com/c/legal/lawenforcement

Anderson, J. M., Rodrguez, A., & Chang, D. T. (2008, April). Foreign body reaction to biomaterials. *Seminars in Immunology*, *20*(2), 86–100. doi:10.1016/j.smim.2007.11.004 PMID:18162407

Andrews v. Blick Art Materials, LLC, 1:17-cv-00767 (EDNY, 2017). https://www.classaction.org/media/andrews-v-blick-art-materials.pdf

Apffel, C., Bernad, P., Gollenia, L. A., Lupo, C., Mijnarends, H., & Westland, J. (2020, December). *The future of technology leadership*. Spencer Stuart. https://www.spencerstuart.com/research-and-insight/the-future-of-technology-leadership

Apte, A., Ingole, V., Lele, P., Marsh, A., Bhattacharjee, T., Hirve, S., Campbell, H., Nair, H., Chan, S., & Juvekar, S. (2019, June). Ethical considerations in the use of GPS-based movement tracking in health research - lessons from a care-seeking study in rural west India. *Journal of Global Health, 9*(1), 010323. doi:10.7189/jogh.09.010323 PMID:31275566

Arenth, T. L. (2019). ADA web site accessibility claims on the rise: Practical strategies for defense. *Journal of Internet Law, 23*(4), 12–14.

Arians, H. (2017, September 10). *The impact of technology on leadership*. The People Development Magazine. https://peopledevelopmentmagazine.com/2017/09/10/technology-leadership

Arora, A., Gupta, S., & Sehgal, U. (2019, April). Reader collision and tag payoff matrix resulting in a Wal-Mart RFID. *International Journal of Computer Science and Network, 8*(2), 154–156.

Artegoni, A. (Ed.). (2021, January). *Evaluation of energy efficiency and flexibility in smart buildings*. Multidisciplinary Digital Publishing Institute. doi:10.3390/books978-3-03943-850-1

Artificial Intelligence in society. (2019, June 11). OECD. doi:10.1787/eedfee77-en

Assistive Technology Industry Association (ATIA). (n.d.). *What is AT?* https://www.atia.org/home/at-resources/what-is-at

Association of Bioethics Program Directors. (n.d.). *Graduate bioethics education programs results*. https://www.bioethicsdirectors.net/graduate-bioethics-education-programs-results

Auditing machine learning algorithms. (2020, November 24). Supreme Audit Institutions of Finland, Germany, the Netherlands, Norway, & the UK at Auditing Algorithms. https://www.auditingalgorithms.net

Autopilot and full self-driving capability. (n.d.). Tesla. https://www.tesla.com/en_AE/support/autopilot-and-full-self-driving-capability

Ayala, F. J. (2015, July 20). Cloning humans? Biological, ethical, and social considerations. *Proceedings of the National Academy of Sciences of the United States of America, 112*(29), 8879–8886. https://doi.org/10.1073/pnas.1501798112

Bacsa Palmer, Z., & Palmer, R. H. (2018). Legal and ethical implications of website accessibility. *Business and Professional Communication Quarterly, 81*(4), 399–420. https://doi.org/10.1177%2F2329490618802418

Bales, S., & van Rensburg, H. (2019). *Innovation wars: Driving successful corporate innovation programs*. Morgan James Publishing.

Balta-Ozkan, N., Davidson, R., Bicket, M., & Whitmarsh, L. (2013, December). Social barriers to the adoption of smart homes. *Energy Policy, 63*, 363–374. doi:10.1016/j.enpol.2013.08.043

Bammert, S., König, U. M., Roeglinger, M., & Wruck, T. (2020). Exploring potentials of digital nudging for business processes. *Business Process Management Journal, 26*(6), 1329–1347. https://doi.org/ doi:10.1108/BPMJ-07-2019-0281

Banafa, A. (2021, April 5). *Technology under your skin: 3 challenges of microchip implants*. BBVA Open Mind. https://www.bbvaopenmind.com/en/technology/innovation/technology-under-your-skin

Baraniuk, C. (2019, July 17). Can you trust FaceApp with your face? *BBC News*. https://www.bbc.com/news/technology-49018103

Barbaschow, A. (2020, December 9). *Australian committee calls for independent review of CovidSafe app*. ZDNet. https://www.zdnet.com/article/australian-committee-calls-for-independent-review-of-covidsafe-app

Barla, N. (2021, August 25). *Self-driving cars with Convolutional Neural Networks.* Neptune AI. https://neptune.ai/blog/self-driving-cars-with-convolutional-neural-networks-cnn

Barrot, J. S. (2018, August 24). Facebook as a learning environment for language teaching and learning: A critical analysis of the literature from 2010 to 2017. *Journal of Computer Assisted Learning*, *34*(6), 863–875. doi:10.1111/jcal.12295

Baudier, P., Ammi, C., & Deboeuf-Rouchon, M. (2020, April). Smart home: Highly-educated students' acceptance. *Technological Forecasting and Social Change*, *153*, 119355. doi:10.1016/j.techfore.2018.06.043

Baudrillard, J. (1994). *Simulacra and simulation* (S. F. Glaser, Trans.). University of Michigan.

Bayram, O. (2020). Importance of Blockchain use in cross-border payments and evaluation of the progress in this area. *Dogus University Journal, 21*(1), 171–189. https://dergipark.org.tr/en/pub/doujournal/issue/66682/1043235

Beever, J., Kuebler, S. M., & Collins, C. (2021, October). Where ethics is taught: An institutional epidemiology. *International Journal of Ethics Education*, *6*, 215–238. https://doi.org/10.1007/s40889-021-00121-7

Bellamy, R. K. E., Dey, K., Hind, M., Hoffman, S. C., Houde, S., Kannan, K., Lohia, P., Martino, J., Mehta, S., Mojsilovic, A., Nagar, S., Ramamurthy, K. N., Richards, J., Saha, D., Sattigeri, P., Singh, M., Varshney, K. R., & Zhang, Y. (2018). AI fairness 360: An extensible toolkit for detecting, understanding, and mitigating unwanted algorithmic bias. *IBM Journal of Research and Development*, *63*(4/5), 1–15. doi:10.48550/arxiv.1810.01943

Benavides, N. (PR) v. Tesla, Inc. (Fla. Miami-Dade County Court, 2021). https://www.plainsite.org/dockets/download.html?id=297527909&z=e1dd27ac

Bennett, W. J. (1980). The teacher, the curriculum, and values education. *New Directions for Higher Education, 1980*(31), 27–34. https://doi.org/10.1002/he.36919803106

Bergey, R., Movit, M., Baird, A. S., & Faria, A.-M. (2018). *Serving English language learners in higher education: Unlocking the potential.* American Institutes for Research. https://www.air.org/sites/default/files/downloads/report/Serving-English-Language-Learners-in-Higher-Education-2018.pdf

Bernard, L., & Sim, I. (2020). Ethical framework for assessing manual and digital contact tracing for COVID-19. *Medicine and Public Issues*, *174*(3), 395–400. doi:10.7326/M20-5834 PMID:33076694

Berryhill, J., Heang, K. K., Clogher, R., & McBride, K. (2019). Hello, World: Artificial Intelligence and its use in the public sector. *OECD Observatory of Public Sector Innovation, 36*, 1–148. doi:10.1787/19934351

Beym, J. (2014, March 18). *N.J. High School student claims she was punished for tweet about principal.* https://www.nj.com/camden/2014/03/nj_high_school_student_claims_she_was_punished_for_tweet_about_principal.html

Bezuidenhout, L., Quick, R., & Shanahan, H. (2020, February 17). "Ethics when you least expect it": A modular approach to short course data ethics instruction. *Science and Engineering Ethics*, *26*(4), 2189–2213. doi:10.100711948-020-00197-2 PMID:32067185

Bhargava, R., Deahl, E., Letouze, E., Noonan, A., Sangokoya, D., & Shoup, N. (2015, September). *Beyond data literacy: Reinventing community engagement and empowerment in the age of data.* Data-Pop Alliance (Harvard Humanitarian Initiative, MIT Media Lab and Overseas Development Institute) and Internews. https://dspace.mit.edu/bitstream/handle/1721.1/123471/Beyond%20Data%20Literacy%202015.pdf

Bhaskar Rao, J., Hrithik, P. S. S., Sushma, I., Karthikeya, M. S. S. G., Meghana, T., & Anurag, K. (2021, October). Toll collection system using image processing. *IUP Journal of Computer Sciences*, *15*(4), 34–44.

Bible Gateway. (n.d.a). *Passage lookup: Matthew 7:12.* https://www.biblegateway.com/passage/?search=Matthew+7%3A12%2C&version=NIV

Bible Gateway. (n.d.b). *Passage lookup: Matthew 22:39.* https://www.biblegateway.com/passage/?search=Matthew+22%3A39&version=NIV/

Billah, M. M., & Atbani, F. M. (2019, January-March). SWOT analysis of cryptocurrency an ethical thought. *Journal of Islamic Banking & Finance*, *36*(1), 22–27.

Billinger, M. S. (2009). A technoethical approach to the race problem in anthropology. In R. Luppicini & R. Adell (Eds.), Handbook of Research on Technoethics. IGI-Global., https://dx.doi.org/10.4018/9781605660226.ch004

Biometric Information Privacy Act, Ill Comp. Stat. Ann., ch. 740, § 14/1 *et seq.* (2018). https://www.akingump.com/a/web/101105/Biometric-Information-Privacy-Act-740-ILCS-14-1-et-seq.pdf

Bischoff, P. (2019, October 15). *Data privacy laws & government surveillance by country: Which countries best protect their citizens?* Comparitech. https://www.comparitech.com/blog/vpn-privacy/surveillance-states/

Boorstin, J., & Mangan, D. (2022, March 14). Meta now says you can't threaten to kill Russia's President Putin on Facebook because of the Ukraine war. *CNBC.* https://www.cnbc.com/2022/03/14/facebook-bans-death-threats-russia-leader-putin-despite-ukraine-war.html

Bostian, I. L. (2005). Cultural relativism in international war crimes prosecutions: The international criminal tribunal for Rwanda. *ILSA Journal of International & Comparative Law*, *12*(1), 1.

Boudette, N. E. (2021a, September 1). 'It happened so fast': Inside a fatal Tesla autopilot accident. *New York Times.* https://www.nytimes.com/2021/08/17/business/tesla-autopilot-accident.html

Boudette, N. E. (2021b, September 1). Tesla says autopilot makes its cars safer. Crash victims say it kills. *New York Times.* https://www.nytimes.com/2021/07/05/business/tesla-autopilot-lawsuits-safety.html

Bourekkache, S., & Kazar, O. (2020). Mobile and adaptive learning application for English language learning. *International Journal of Information and Communication Technology Education*, *16*(2), 36–46. doi:10.4018/IJICTE.2020040103

Boutet, A., & Gambs, S. (2019, November 3). Inspect what your location history reveals about you: Raising user awareness on privacy threats associated with disclosing his location data. *CIKM '19: Proceedings of the 28th ACM International Conference on Information and Knowledge Management*, Beijing, China. https://dl.acm.org/doi/abs/10.1145/3357384.3357837

Bowen, G. (2021, January). Digital leadership, ethics, and challenges. In H. Jahankhani, L. M. O'Dell, G. Bowen, D. Hagan, & A. Jamal (Eds.), Strategy (pp. 23–29). Leadership, and AI in the Cyber Ecosystem. https://doi.org/10.1016/B978-0-12-821442-8.00013-6

Boy, N., Jacobsen, E. K. U., & Lidén, K. (2018). *Societal ethics of biometric technologies* (PRIO Paper 2018). Peace Research Institute Oslo. https://www.prio.org/publications/11199

Boyette, C. (2013, May 27). N.Y. student suspended after controversial Twitter hashtag. *CNN.* https://www.cnn.com/2013/05/26/us/n-y-student-suspended-after-controversial-twitter-hashtag

Bradshaw, A. C. (2017). Critical pedagogy and educational technology. In A. D. Benson, R. Joseph, & J. L. Moore (Eds.), *Culture, learning, and technology: Research and practice* (pp. 8–27). Routledge. doi:10.4324/9781315681689-2

Brey, P. A. E. (2004). Ethical aspects of facial recognition systems in public spaces. *Information. Communication & Ethics in Society*, *2*(2), 97–109. doi:10.1108/14779960480000246

Brown, A., Croft, B., Dello Stritto, M. E., Heiser, R., McCarty, S., McNally, D., Nyland, R., Quick, J., Thomas, R., & Wilks, M. (2022, February 9). *Learning analytics from a systems perspective: Implications for practice.* EDUCAUSE. https://er.educause.edu/articles/2022/2/learning-analytics-from-a-systems-perspective-implications-for-practice

Budd, J. M. (2018). Teaching ethics: A framework for thought and action. *Journal of Education for Library and Information Science*, *59*(3), 53–66. doi:10.3138/jelis.59.3.2018-0022.06

Buenadicha Sánchez, C., Galdon Clavell, G., Hermosilla, M. P., Loewe, D., & Pombo, C. (2019, March). *La gestión ética de los datos.* Inter-American Development Bank. doi:10.18235/0001623

Bunge, M. (1977, January). Towards a technoethics. *The Monist*, *60*(1), 96–107. https://doi.org/10.5840/monist197760134

Buolamwini, J., & Gebru, T. (2018, February 9). *Gender shades: Intersectional accuracy disparities in commercial gender classification.* Gender Shades Project. http://proceedings.mlr.press/v81/buolamwini18a/buolamwini18a.pdf

Burke, M. (2022, January 21). Tesla driver charged with manslaughter in deadly Autopilot crash raises new legal questions about automated driving tech. *NBC News.* https://www.nbcnews.com/news/us-news/tesla-driver-charged-manslaughter-deadly-autopilot-crash-raises-new-le-rcna12987

Burnham, K. (2020, May 6). *Artificial Intelligence (AI) vs. Machine Learning: What's the difference?* Northeastern University. https://www.northeastern.edu/graduate/blog/artificial-intelligence-vs-machine-learning-whats-the-difference

Cal. Penal Code § 192(c) (1872). https://leginfo.legislature.ca.gov/faces/codes_displaySection.xhtml?lawCode=PEN§ionNum=192

California Consumer Privacy Act, Cal. Civ. Code, Title 1.81.5, ch. 55, § 3 (2018). https://leginfo.legislature.ca.gov/faces/codes_displayText.xhtml?division=3.&part=4.&lawCode=CIV&title=1.81.5

Cambridge University Press. (n.d.a). Algorithm. In *Cambridge English Dictionary online.* https://dictionary.cambridge.org/us/dictionary/english/algorithm

Cambridge University Press. (n.d.b). Privacy. In *Cambridge English Dictionary online.* https://dictionary.cambridge.org/us/dictionary/english/privacy

Campbell, L., Vasquez, M., Behnke, S., & Kinscherff, R. (2010). *APA Ethics Code commentary and case illustrations.* American Psychological Association. https://psycnet.apa.org/record/2009-08922-000

Canadian National Institute for the Blind (CNIB). (n.d.). *Accessibility at CNIB.* https://cnib.ca/en/accessibility-cnib?region=gta

Canvesi, B. (2022, February 2). *The foundation of the metaverse: centralization versus decentralization.* Association for Talent Development. https://www.td.org/atd-blog/the-foundation-of-the-metaverse-centralization-versus-decentralization

Carson, R. (1962). *Silent spring.* Houghton Mifflin.

Carvalho, A., Sambhara, C., & Young, P. (2020, January). What the history of Linux says about the future of cryptocurrencies. *Communications of the Association for Information Systems*, *46*, 18–29. doi:10.17705/1CAIS.04602

Carzo, R. (2010). Under the watchful eye: The highly intrusive nature of facial recognition technology. *The Review: A Journal of Undergraduate Student Research, 12*(2010), 1-5.

Castillo, C. (2021, January). *Guía de Auditoría Algorítmica.* Eticas. https://www.eticasconsulting.com/eticas-consulting-guia-de-auditoria-algoritmica-para-desarrollar-algoritmos-justos-y-eficaces/

Cath, C. (2018, October 15). Governing Artificial Intelligence: Ethical, legal and technical opportunities and challenges. *Philosophical Transactions - Royal Society. Mathematical, Physical, and Engineering Sciences, 376*(2133), 20180080. https://doi.org/ doi:10.1098/rsta.2018.0080 PMID:30322996

Center for the study of Ethics in the Professions. (n.d.). Illinois Institute of Technology. https://ethics.iit.edu/research

Chan, E. (2006). Teacher experiences of culture in the curricula. *Journal of Curriculum Studies, 38*(2), 349–360. doi:10.1080/00220270500391605

Chan, E. Y., & Saqib, N. U. (2021, June). Privacy concerns can explain unwillingness to download and use contact tracing apps when COVID-19 concerns are high. *Computers in Human Behavior, 119*, 106718. doi:10.1016/j.chb.2021.106718 PMID:33526957

Chang, S.-I., Huang, S.-Y., Yen, D. C., & Chen, Y.-J. (2008, September). The determinants of RFID adoption in the logistics industry - A supply chain. *Communications of the Association for Information Systems, 23*, 197–218. https://aisel.aisnet.org/cais/vol23/iss1/1 doi:10.17705/1CAIS.02312

Charlton, C. (2014, June 18). Honor student who was suspended for tweet joking that he'd made out with teacher sues his old school district. *Daily Mail Online*. https://www.dailymail.co.uk/news/article-2661496/Honor-student-suspended-tweet-joking-hed-teacher-sues-old-school-district.html

Checrallah, M., Sonnett, C., & Desgres, J. (2020). Evaluating cost, privacy, and data. In T. Trust (Ed.), *Teaching with Digital Tools and Apps*. EdTech Books. https://edtechbooks.org/digitaltoolsapps/evaluatingcostprivacydata

Chen, W., & Quan-Haase, A. (2020). Big data ethics and politics: Toward new understandings. *Social Science Computer Review, 38*(1), 3–9. doi:10.1177/0894439318810734

Chen, Y. (2020). Your face is commodity, fiercely contract accordingly: Regulating the capitalization of facial recognition technology through contract law. *Notre Dame Journal of Law, Ethics & Public Policy, 34*(2), 501–528.

Cherry, K. (2020, July 21). *What is empathy?* Very Well Mind. https://www.verywellmind.com/what-is-empathy-2795562

Chilson, N., & Barkley, T. (2021, December). The two faces of facial recognition technology. *IEEE Technology and Society Magazine, 40*(4), 87–99. doi:10.1109/MTS.2021.3123752

Choi, Y., Kwon, Y. H., & Kim, J. (2018, April). The effect of the social networks of the elderly on housing choice in Korea. *Habitat International, 74*, 1–8. doi:10.1016/j.habitatint.2018.02.003

Choksey, J. S. (2021, October 4). *What is Genesis Face Connect and how does it work?* J. D. Power. https://www.jdpower.com/cars/shopping-guides/what-is-genesis-face-connect-and-how-does-it-work

Chugh, R., & Ruhi, U. (2018, March). Social media in higher education: A literature review of Facebook. *Education and Information Technologies, 23*(2), 605–616. doi:10.100710639-017-9621-2

Cimperman, M., Brenčič, M. M., Trkman, P., & Stanonik, M. de L. (2013, September 30). Older adults' perceptions of home telehealth services. *Telemedicine Journal and e-Health, 19*(10), 786–790. doi:10.1089/tmj.2012.0272 PMID:23931702

Clark, M. (2022, June 6). *NFTs, explained*. The Verge. https://www.theverge.com/22310188/nft-explainer-what-is-blockchain-crypto-art-faq

Clark, M. I., & Driller, M. W. (2020, February). University students' perceptions of self-tracking devices, data privacy, and sharing digital data for research purposes. *Journal for the Measurement of Human Behaviour, 3*(2), 128–134. doi:10.1123/jmpb.2019-0034

Classen, J., Chen, J., Steinmetzer, D., Hollick, M., & Knightly, E. (2015, September 11). The spy next door: Eavesdropping on high throughput visible light communications. *Proceedings of the 2nd International Workshop on Visible Light Communications Systems*, 9-14. 10.1145/2801073.2801075

Clayton, E., & Kral, P. (2021, November). Autonomous driving algorithms and behaviors, sensing and computing technologies, and connected vehicle data in smart transportation networks. *Contemporary Readings in Law and Social Justice*, *13*(2), 9–22. doi:10.22381/CRLSJ13220211

Clearview AI's founder Hoan Ton-That speaks out [*Extended interview*]. (2020, March 6). [Video file]. YouTube. https://www.youtube.com/watch?v=q-1bR3P9RAw

Cloud Storage. (n.d.). *Amazon Web Services*. https://aws.amazon.com/what-is-cloud-storage

Comparing different types of RFID tags. (2018, November 12). Resource Label Group. https://www.resourcelabel.com/comparing-different-types-of-rfid-tags

Cook, D. J. (2012, March 30). How smart is your home? *Science*, *335*(6076), 1579–1581. doi:10.1126cience.1217640 PMID:22461596

Cortellazzo, L., Bruni, E., & Zampieri, R. (2019, August 29). The role of leadership in a digitalized world: A review. *Frontiers in Psychology, 10*(1938). doi:10.3389/fpsyg.2019.01938

Coskun, A., Kaner, G., & Bostan, İ. (2018, April 30). Is smart home a necessity or a fantasy for the mainstream user? A study on users' expectations of smart household appliances. *International Journal of Design*, *12*(1), 7–20.

Costache, M., Lazariou, A. M., Contoleco, A., Costache, D., George, S., Sajin, M. & Patrascu, O. M. (2014, September). Clinical or post-mortem? The importance of the autopsy: A retrospective study. *Maedica (Bucur), 9*(3), 261–265. https://www.ncbi.nlm.nih.gov/pmc/articles/PMC4305994/pdf/maed-09-261.pdf

Costanza-Chock, S. (2020). *Design justice: Community-led practices to build the worlds we need.* The MIT Press. doi:10.7551/mitpress/12255.001.0001

Craig, D. (2017, April 11). *How technology is making leaders more responsive and responsible.* Thomson Reuters blogs. https://blogs.thomsonreuters.com/answerson/disruptive-leadership-technology-making-leaders-more-responsive-responsible

Crolic, C., Thomaz, F., Hadi, R., & Stephen, A. T. (2022). Blame the bot: Anthropomorphism and anger in customer–chatbot interactions. *Journal of Marketing, 86*(1), 132–148. doi:10.1177/00222429211045687

Crook, A. (n.d.). *What are the different types of technology?* https://digitalizetrends.com/types-of-technology

Crookes, D. (2018, September 19-October 2). Microchip implants. *Web User,* (458), 38-39.

Crowell, C., Narvaez, D., & Gomberg, A. (2007, January). Moral psychology and information ethics. In L. A. Freeman & A. G. Peace (Eds.), *Information Ethics: Privacy and Intellectual Property* (pp. 19–37). Information Science Reference/IGI-Global. https://www.researchgate.net/publication/314456458_Moral_Psychology_and_Information_Ethics

Cyranoski, D. (2020, July 7). *China's massive effort to collect its people's DNA concerns scientists.* Nature. https://www.nature.com/articles/d41586-020-01984-4

Dalhousie University. (n.d.). *What is e-Leadership?* https://www.dal.ca/sites/celnet/about/eleadership.html

David, Y., Judd, T. M., & Zambuto, R. P. (2020). Introduction to medical technology management practices. In Clinical Engineering Handbook (2nd ed., pp. 166–177). Academic Press. https://doi.org/10.1016/B978-0-12-813467-2.00028-6

Davis, V. (2017, November 1). *What your students really need to know about digital citizenship: Ideas on how to guide students to the knowledge and experience they need to act responsibly online*. Edutopia. https://www.edutopia.org/blog/digital-citizenship-need-to-know-vicki-davis

Davis, F. D. (1989, September). Perceived usefulness, perceived ease of use, and user acceptance of information technology. *Management Information Systems Quarterly, 13*(3), 319–340. doi:10.2307/249008

de Groot, N. F., van Beers, B. C., & Meynen, G. (2021). Commercial DNA tests and police investigations: A broad bioethical perspective. *Journal of Medical Ethics, 47*(12), 788–795. doi:10.1136/medethics-2021-107568 PMID:34509983

Deamer, K. (2016, July 1). What the first driverless fatality means for self-driving tech. *Scientific American*. https://www.scientificamerican.com/article/what-the-first-driverless-car-fatality-means-for-self-driving-tech/

Deiser, R., & Newton, S. (2015, January 15). *Social technology and the changing context of leadership*. The Wharton School. https://leadershipcenter.wharton.upenn.edu/research/social-technology-changing-context-leadership

Della Corte, V., Del Gaudio, G., & Sepe, F. (2019, November 20). Leadership in the digital realm: What are the main challenges? In M. Franco (Ed.), *A New Leadership Style for the 21st Century*. IntechOpen. doi:10.5772/intechopen.89856

Demir, E., Köseoğlu, E., Sokullu, R., & Şeker, B. (2017). Smart home assistant for ambient assisted living of elderly people with dementia. *Procedia Computer Science, 113*, 609–614. doi:10.1016/j.procs.2017.08.302

Denis, G., Hermosilla, M. P., Claudio, A., Ávalos, R. S., Alarcón, N. G., & Pombo, C. (2021, August). *Uso responsable de IA para política pública: Manual de formulación de proyectos*. Inter-American Development Bank. doi:10.18235/0003631

Dennis, A. R., Yuan, L., Feng, X., Webb, E., & Hsieh, C. J. (2020, March 1). Digital nudging: Numeric and semantic priming in e-commerce. *Journal of Management Information Systems, 37*(1), 39–65. https://doi.org/ doi:10.1080/07421222.2019.1705505

Deshaies, B., & Hall, D. (2021, December 1). *Responsible use of automated decision systems in the federal government*. Statistics Canada. https://www.statcan.gc.ca/en/data-science/network/automated-systems

Dewey, J. (1897, January). My pedagogic creed. *School Journal, 54*, 77–80.

Dey, A., Tharmavaram, M., Pandey, G., Rawtani, D., & Hussain, C. M. (2020). Conventional and emerging biometrics techniques in forensic investigations. In D. Rawtani & C. M. Hussain (Eds.), *Technology in forensic science: Sampling, analysis, data and regulations* (1st ed.) pp. 177–389. Wiley Online Library. doi:10.1002/9783527827688.ch9

Di Pietro, R. (2022, April 6). *The metaverse: Technology, privacy, and security risks and the road ahead*. Modern Diplomacy. https://moderndiplomacy.eu/2022/04/06/the-metaverse-technology-privacy-and-security-risks-and-the-road-ahead

Dieffenbacher, S. F. (2022, March 4). *Value creation definition, model, principles, importance & steps*. Digital Leadership. https://digitalleadership.com/blog/value-creation

Dierksmeier, C., & Seele, P. (2018). Cryptocurrencies and business ethics. *Journal of Business Ethics, 152*(1), 1–14. doi:10.100710551-016-3298-0 PMID:30930508

DiGeronimo, J. J. (n.d.). *How to be a curious leader*. https://jjdigeronimo.com/how-to-be-a-curious-leader

Dllu. (2017, October 18). *Cruise Automation Bolt EV third generation in San Francisco* [Image]. Wikimedia. Creative Commons Share-Alike License 4.0: https://commons.wikimedia.org/w/index.php?curid=63450446

DNA Technology in Forensic Science. (1992). *National Academies Press*. https://www.ncbi.nlm.nih.gov/books/NBK234542

Dobson, J. (2007, November 12). Applying virtue ethics to business. The agent-based approach. *Electronic Journal of Business Ethics and Organizational Studies, 12*(2). http://ejbo.jyu.fi/articles/0901_3.html

Dodig-Crnknovic, G., Holstein, T., & Pelliccione, P. (2021, July 16). *Future intelligent autonomous robots, ethical by design. Learning from autonomous cars ethics.* https://arxiv.org/ftp/arxiv/papers/2107/2107.08122.pdf

Doi, T., Shimada, H., Park, H., Makizako, H., Tsutsumimoto, K., Uemura, K., Nakakubo, S., Hotta, R., & Suzuki, T. (2015, August). Cognitive function and falling among older adults with mild cognitive impairment and slow gait. *Geriatrics & Gerontology International, 15*(8), 1073–1078. doi:10.1111/ggi.12407 PMID:25363419

Doris, B., Wagner, D., & Winokur, H. (2022, Winter). Director's cut: How boards can help ensure the responsible use of AI: The question for boards and management is not whether to use Artificial Intelligence, but how to ensure it is used responsibly. *NACD Directorship, 48*(1), 38–43.

Dow, L. (2022, January 12). *The metaverse is coming. Are your ethics ready?* FastCompany. https://www.fastcompany.com/90712458/the-metaverse-is-coming-are-your-ethics-ready

Draper, A. (2019, January 19). *3 ways technology has affected today's leaders.* https://www.business2community.com/leadership/3-ways-technology-has-affected-todays-leaders-02160570

Driverless car accidents – who's at fault? (2021, June 25). *National Law Review, 11*(176). https://www.natlawreview.com/article/driverless-car-accidents-who-s-fault

Duff, C. (2021, November 9). *Everything you need to know about education technology "EdTech."* Owl Labs. https://resources.owllabs.com/blog/education-technology

Duncan, C. (2022, January 18). Communication tools in the workplace. https://www.alert-software.com/blog/internal-communication-tools

Dupuy, L., Froger, C., Consel, C., & Sauzéon, H. (2017, September 28). Everyday functioning benefits from an assisted living platform amongst frail older adults and their caregivers. *Frontiers in Aging Neuroscience, 9*, 302. doi:10.3389/fnagi.2017.00302 PMID:29033826

Ed Markey, the United States Senator for Massachusetts. (2022, June 9). *Markey and Blumenthal statement on NHTSA's elevated investigation Into Tesla's autopilot system* [Press release]. https://www.markey.senate.gov/news/press-releases/markey-and-blumenthal-statement-on-nhtsas-elevated-investigation-into-teslas-autopilot-system

Edmonds, E. (2020, August 6). *AAA finds active driving assistance systems do less to assist drivers and more to interfere* [Press release]. AAA. https://newsroom.aaa.com/2020/08/aaa-finds-active-driving-assistance-systems-do-less-to-assist-drivers-and-more-to-interfere

Education technology market size, share & trends analysis report, by sector (preschool, k-12, higher education), by end-user (business, consumer), by type, by deployment, by region, and segment forecasts, 2022 – 2030. (2022, April). Grand View Research Publishers. https://www.marketresearch.com/Grand-View-Research-v4060/Education-Technology-Size-Share-Trends-31517238

Educator toolkit for teacher and student privacy: A practical guide for protecting personal data. (2018, October). Parent Coalition for Student Privacy & the Badass Teachers Association. https://cdn.ymaws.com/www.a4l.org/resource/resmgr/files/sdpc-publicdocs/PCSP_BATS-Educator-Toolkit.pdf

Edu, J. S., Such, J. M., & Suarez-Tangil, G. (2020, August). Smart home personal assistants: A security and privacy review. *ACM Computing Surveys, 53*(6), 1–36. doi:10.1145/3412383

Egebark, J., & Ekström, M. (2016, March). Can indifference make the world greener? *Journal of Environmental Economics and Management, 76*, 1–13. https://doi.org/ doi:10.1016/j.jeem.2015.11.004

Eggert, E., & Abou-Zahra, S. (Eds.). (2022, January 17). *Complex images*. W3C Web Accessibility Initiative (WAI). https://www.w3.org/WAI/tutorials/images/complex/#approach-3-structurally-associating-the-image-and-its-adjacent-long-description-html5

Eid, M. (2014, January). Ethics, media, and reasoning: Systems and applications. In R. Luppicini (Ed.), Evolving Issues Surrounding Technoethics and Society in the Digital Age (pp. 188–197). IGI-Global. https://dx.doi.org/10.4018/978-1-4666-6122-6.ch012

Eisa, S., & Moreira, A. (2017). A behaviour monitoring system (BMS) for ambient assisted living. *Sensors (Basel), 17*(9), 1946. doi:10.339017091946 PMID:28837105

Eisenberg, N., Hofer, C., Sulik, M. J., & Liew, J. (2014). The development of prosocial moral reasoning and a prosocial orientation in young adulthood: Concurrent and longitudinal correlates. *Developmental Psychology, 50*(1), 58–70. doi:10.1037/a0032990 PMID:23731289

Eitel-Porter, R. (2020, September). Beyond the promise: Implementing ethical AI. *AI Ethics, 1*(1), 73–80. doi:10.100743681-020-00011-6

Eliaz, N. (2019). Corrosion of metallic biomaterials: A review. *Materials (Basel, Switzerland), 12*(3), 407. doi:10.3390/ma12030407 PMID:30696087

Eliot, L. (2020, October 3). Tire blowouts could case self-driving cars to go astray. *Forbes*. https://www.forbes.com/sites/lanceeliot/2020/10/03/tire-blowouts-could-cause-self-driving-cars-to-go-astray

Eliot, L. (2022, July 5). AI ethics perturbed by latest China devised AI party-loyalty mind-reading facial recognition attestation that might foreshadow oppressive autonomous systems. *Forbes*. https://www.forbes.com/sites/lanceeliot/2022/07/05/ai-ethics-perturbed-by-latest-china-devised-ai-party-loyalty-mind-reading-facial-recognition-attestation-that-might-foreshadow-oppressive-autonomous-systems

Elliott, D., & June, K. (2018, May 9). The evolution of ethics education 1980–2015. In E. Englehardt & M. Pritchard (Eds.), Ethics Across the Curriculum—Pedagogical Perspectives. Springer., https://doi.org/10.1007/978-3-319-78939-2_2

Elliott, K., Meng, J., & Hall, M. (2021, July 14). An integrated approach for predicting consumer acceptance of self-driving vehicles in the United States. *Journal of Marketing Development and Competitiveness, 15*(2), 10–20. doi:10.33423/jmdc.v15i2.4330

Enslin, P., Pendlebury, S., & Tjiattas, M. (2001). Deliberative democracy, diversity and challenges of citizenship education. *Journal of Philosophy of Education, 35*(1), 115–130. doi:10.1111/1467-9752.00213

Erwin, S. (2022, February 10). Space Force eager to invest in debris removal projects. *SpaceNews*. https://spacenews.com/space-force-eager-to-invest-in-debris-removal-projects

Escudero v. Tesla, Inc. (Cal. Alameda County Court, 2021). https://www.plainsite.org/dockets/download.html?id=296424442&z=a4a11de7

ESPN. (2018, September 18). *Stephen A., Jalen Rose get into heated debate over Barkley's super teams comments | Get Up! | ESPN* [Video] YouTube. https://www.youtube.com/watch?v=dNh_WMJsJDE

Ethical A. I. Advisory. (2020, June 5). *What are technology ethics?* https://aiadvisory.ai/2020/06/05/what-are-technology-ethics/

Etienne, H. (2021, May). The dark side of the 'Moral Machine' and the fallacy of computational ethical decision-making for autonomous vehicles. *Law, Innovation and Technology*, *13*(1), 85–107. doi:10.1080/17579961.2021.1898310

European Space Agency. (2019, September 12). *ESA commissions world's first space debris removal.* https://www.esa.int/Safety_Security/Clean_Space/ESA_commissions_world_s_first_space_debris_removal

Evans, K., de Moura, N., Chauvier, S., Chatila, R., & Dogan, E. (2020). Ethical decision making in autonomous vehicles: The AV Ethics Project. *Science and Engineering Ethics*, *26*(6), 3285–3312. doi:10.100711948-020-00272-8 PMID:33048325

Ezenwakwelu, C. A., Nwakoby, I. C., Egbo, O. P., Nwanmuoh, E. E., Duruzo, C. E., & Ihegboro, I. M. (2020, August). Business ethics and organizational sustainability. *International Journal of Entrepreneurship*, *24*(3), 1–14.

Facial recognition in insurance. (2020, December 16). HDI. https://www.hdi.global/en-za/infocenter/insights/2020/facial-recognition/

Fässler, G. (2020, September 8). *The evolution of environmental detection: What can it accomplish?* [Web video session]. Continental. https://www.continental.com/en/press/fairs-events/continental-commercial-vehicle-days/environmental-detection

Faulhaber, A. K., Dittmer, A., Blind, F., Wachter, M. A., Timm, S., Sütfeld, L. R., Stephan, A., Pipa, G., & König, P. (2019, April). Human decisions in moral dilemmas are largely described by utilitarianism: Virtual car driving study provides guidelines for autonomous driving vehicles. *Science and Engineering Ethics*, *25*(2), 399–418. doi:10.100711948-018-0020-x PMID:29357047

Fernandez, C. B., & Hue, P. (2022, March 22). *Life, the metaverse and everything: An overview of privacy, ethics, and governance in metaverse.* Cornell University. doi:10.48550/arXiv.2204.01480

Fields of concentration - 2021-22 Harvard College Student Handbook. (2021). https://handbook.college.harvard.edu/files/collegehandbook/files/fields_of_concentration_2021_2022.pdf

Folger, B. (2016, September). Microchips, ownership, & ethics. *Veterinary Team Brief*, *4*(8), 38–42.

Food and Agricultural Organization of the United Nations. (2020). *The state of the world's forests 2020.* https://www.fao.org/documents/card/en/c/ca8642en

Formosa, P., Wilson, M., & Richards, D. (2021, October). A principlist framework for cybersecurity ethics. *Computers & Security*, *109*, 102382. doi:10.1016/j.cose.2021.102382

Fowler, G. A. (2020, June 18). You downloaded FaceApp. Here's what you've just done to your privacy. *The Washington Post*. https://www.washingtonpost.com/technology/2019/07/17/you-downloaded-faceapp-heres-what-youve-just-done-your-privacy

Fram, B. R., Rivlin, M., & Beredjiklian, P. K. (2020). On emerging technology: What to know when your patient has a microchip in his hand. *The Journal of Hand Surgery*, *45*(7), 645–649. doi:10.1016/j.jhsa.2020.01.008 PMID:32164995

Frankfurt, T. (2021, December 13). Why all companies must explore the role of ethics in technology. *Forbes*. https://www.forbes.com/sites/forbestechcouncil/2021/12/13/why-all-companies-must-explore-the-role-of-ethics-in-technology

Frequency ranges for automotive radar technology. (2019, June 27). Cetecom. https://www.cetecom.com/en/news/frequency-ranges-for-automotive-radar-technology

Fudge, T. P. (2022a, March 15). *Chip removed from debit/credit card casing* [Image]. Academic Press.

Fudge, T. P. (2022b, April 28). *RFID from commercial packaging* [Image]. Academic Press.

Fuentes, P., Rogerson, A., Westgarth, T., Lida, K., Mbayo, H., Finotto, A., Rahim, S., & Petheram, A. (2022). *Government AI readiness index 2021.* Oxford Insights. https://static1.squarespace.com/static/58b2e92c1e5b6c828058484e/t/61ead0752e7529590e98d35f/1642778757117/Government_AI_Readiness_21.pdf

Gadekallu, T. R., Huynh-The, T., Wang, W., Yenduri, G., Ranaweera, P., Pham, Q.-V., da Costa, D. B., & Liyanage, M. (2022, March 21). *Blockchain for the Metaverse: A review.* Cornell University. doi:10.48550/arXiv.2203.09738

Gallagher, K., Magid, L., & Pruitt, K. (2017, May 4). *The educator's guide to student data privacy.* Connect Safely. https://www.connectsafely.org/wp-content/uploads/2016/05/Educators-Guide-Data-.pdf

Galston, W. A. (2022, March 9). *Is seeing still believing? The deepfake challenge to truth in Politics.* Brookings. https://www.brookings.edu/research/is-seeing-still-believing-the-deepfake-challenge-to-truth-in-politics

Garcia, M. (Ed.). (2021, May 26). *Space debris and human spacecraft.* NASA. https://www.nasa.gov/mission_pages/station/news/orbital_debris.html

Gearhart, D. (2012, October). Lack of ethics for eLearning: Two sides of the ethical coin. *International Journal of Technoethics (IJT), 3*(4), 33-40. doi:10.4018/jte.2012100103

Gemeente Amsterdam. (2019). *Standard clauses for procurement of trustworthy algorithmic systems.* https://www.amsterdam.nl/innovatie/digitalisering-technologie/algoritmen-ai/contractual-terms-for-algorithms

Gençer, M. S., & Samur, Y. (2016, August 19). Leadership styles and technology: Leadership competency level of educational leaders. *Procedia: Social and Behavioral Sciences, 229,* 226–233. https://doi.org/10.1016/j.sbspro.2016.07.132

Georgopoulos, M., Oldfield, J., Nicolaou, M. A., Yannis, P., & Pantic, M. (2021, May 15). Mitigating demographic bias in facial datasets with style-based multi-attribute transfer. *International Journal of Computer Vision, 129*(7), 2288–2307. doi:10.100711263-021-01448-w

Gerber, N., Reinheimer, B., & Volkamer, M. (2018, August). Home sweet home? Investigating users' awareness of smart home privacy threats. *Proceedings of An Interactive Workshop on the Human aspects of Smarthome Security and Privacy (WSSP).*

Gerdon, S., Katz, E., LeGrand, E., Morrison, G., & Torres Santeli, J. (2020, June). *AI procurement in a box: Workbook.* World Economic Forum. https://www3.weforum.org/docs/WEF_AI_Procurement_in_a_Box_Workbook_2020.pdf

Gerwith, A. (1978). *Reason and morality.* University of Chicago Press.

Ghafir, I., Prenosil, V., Alhejailan, A., & Hammoudeh, M. (2016, August). Social engineering attack strategies and defence approaches. *2016 IEEE 4th International Conference on Future Internet of Things and Cloud (FiCloud),* 145-149.

Ghosal, A. (2018, April 25). *Why we should collectively worry about Facebook and Google owning our data.* The Next Web. https://thenextweb.com/news/why-should-you-care-if-google-and-facebook-own-your-data

Gibson, B. (n.d.). *Systems theory.* Encyclopedia Britannica. https://www.britannica.com/topic/systems-theory

Goldstein, N. J., Cialdini, R. B., & Griskevicius, V. (2008, October). A room with a viewpoint: Using social norms to motivate environmental conservation in hotels. *The Journal of Consumer Research, 35*(3), 472–482. https://doi.org/doi:10.1086/586910

González, F., Ortiz, T., & Sánchez Avalos, R. (2020, October). *Uso responsable de la IA para las políticas públicas: Manual de ciencia de datos.* Inter-American Development Bank. doi:10.18235/0002876

Goodman, E. (2015, April 28). *Privacy in the classroom: What you need to know about educational software.* The International Association of Privacy Professionals. https://iapp.org/news/a/privacy-in-the-classroom-what-you-need-to-know-about-educational-software/

Goodwin-Jones, R. (2017). Smartphones and language learning. *Language Learning & Technology, 21*(2), 3–17.

Gotterbarn, D. W., Brinkman, B., Flick, C., Kirkpatrick, M. S., Miller, K., Vazansky, K., & Wolf, M. J. (2018). *ACM code of ethics and professional conduct.* ACM. https://www.acm.org/code-of-ethics

GOV.CO. (2021). *Dashboard - Seguimiento Marco Ético de IA.* https://inteligenciaartificial.gov.co/dashboard-IA/

GOV.UK. (2020a, September 16). *Data Ethics Framework.* https://www.gov.uk/government/publications/data-ethics-framework/data-ethics-framework-2020

GOV.UK. (2020b, June 8). *Guidelines for AI procurement.* https://www.gov.uk/government/publications/guidelines-for-ai-procurement/guidelines-for-ai-procurement

GOV.UK. (2021, May 13). *Ethics, transparency and accountability framework for automated decision-making.* https://www.gov.uk/government/publications/ethics-transparency-and-accountability-framework-for-automated-decision-making/ethics-transparency-and-accountability-framework-for-automated-decision-making

Government of Canada. (2021, April 1). *Directive on automated decision-making.* https://www.tbs-sct.canada.ca/pol/doc-eng.aspx?id=32592

Government of Canada. (2022, April 19). *Algorithmic impact assessment tool.* https://www.canada.ca/en/government/system/digital-government/digital-government-innovations/responsible-use-ai/algorithmic-impact-assessment.html

Graafstra, A. (2005, March 23). *RFID hand 2* [Image]. Licensed by Creative Commons Share-Alike 2.0 Generic. No changes made to the image. https://commons.wikimedia.org/wiki/File:RFID_hand_2.jpg

Graham, T. (2022, January 17). *An introduction to the hyperreal metaverse.* Medium. https://medium.com/synthetic-media/an-introduction-to-hyperreal-synthetic-media-8a22e06509c5

Gramm-Leach-Bliley Act (GLB Act). (n.d.). *EDUCAUSE.* https://library.educause.edu/topics/policy-and-law/gramm-leach-bliley-act-glb-act

Green, B. (2021). The contestation of tech ethics: A sociotechnical approach to technology ethics in practice. *Journal of Scientific Computing, 2*(3), 209–225. doi:10.48550/arXiv.2106.01784

Gregersen, E. (n.d.). *History of technology timeline.* Encyclopedia Britannica. https://www.britannica.com/story/history-of-technology-timeline

Grider, D., & Maximo, M. (2021, November). *The metaverse: Web 3.0 virtual cloud economies.* Grayscale Investments, LLC. https://grayscale.com/wp-content/uploads/2021/11/Grayscale_Metaverse_Report_Nov2021.pdf

Grika. (2006, March 30). *RFID tags* [Image]. Licensed by Creative Commons 2.5 Generic. Image was rotated but otherwise unchanged. https://commons.wikimedia.org/wiki/File:RFID_Tags.jpg

Grother, P., Ngan, M., & Hanaoka, K. (2019, December). *Face recognition vendor test (FRVT) Part 3: Demographic effects.* NIST. https://nvlpubs.nist.gov/nistpubs/ir/2019/NIST.IR.8280.pdf

Guio Español, A., Tamayo Uribe, E., Gómez Ayerbe, P., & Mujica, M. P. (2021). *Marco Ético para la Inteligencia Artificial en Colombia.* https://dapre.presidencia.gov.co/TD/MARCO-ETICO-PARA-LA-INTELIGENCIA-ARTIFICIAL-EN-COLOMBIA-2021.pdf

Guo, X., Zhang, X., & Sun, Y. (2016). The privacy–personalization paradox in mHealth services acceptance of different age groups. *Electronic Commerce Research and Applications, 16*, 55–65. doi:10.1016/j.elerap.2015.11.001

Gupta, P., & Kumar, D. (2018). Ethical behavior and the development paradigm. In I. Oncioiu (Ed.), Ethics and Decision-Making for Sustainable Business Practices (pp. 258–267). IGI-Global. https://dx.doi.org/10.4018/978-1-5225-3773-1.ch015

Gupta, B. B., Arachchilage, N. A. G., & Psannis, K. E. (2018). Defending against phishing attacks: Taxonomy of methods, current issues and future directions. *Telecommunication Systems, 67*(2), 247–267. doi:10.100711235-017-0334-z

Gurumurthy, K. M., Auld, J., & Kockelman, K. M. (2021, September 8). A system of shared autonomous vehicles for Chicago: Understanding the effects of geofencing the service. *Journal of Transport and Land Use, 14*(1), 933–948. doi:10.5198/jtlu.2021.1926

Gutierrez, C. I., & Marchant, G. E. (2021, May). A global perspective of soft law programs for the governance of Artificial Intelligence. SSRN *Electronic Journal.* doi:10.2139/ssrn.3897486

Gutmann, A. (1996). Challenges of multiculturalism in democratic education. In R. K. Fullinwider (Ed.), *Public education in a multicultural society: Policy, theory, critique* (pp. 156–179). Cambridge University Press. doi:10.1017/CBO9781139172899.008

Guynn, J. (2020, January 29). What you need to know before clicking 'I agree' on that terms of service agreement or privacy policy. *USA Today.* https://www.usatoday.com/story/tech/2020/01/28/not-reading-the-small-print-is-privacy-policy-fail/4565274002

Gwebu, K. L., Wang, J., & Wang, L. (2018, May 15). The role of corporate reputation and crisis response strategies in data breach management. *Journal of Management Information Systems, 35*(2), 683–714. doi:10.1080/07421222.2018.1451962

Haber, E. (2021, October). Racial recognition. *Cardozo Law Review, 43*(1), 71–134. http://cardozolawreview.com/wp-content/uploads/2021/12/2_Haber.43.1.1.pdf

Haddad, G. M. (2021). Confronting the biased algorithm: The danger of admitting facial technology results in the courtroom. *Vanderbilt Journal of Entertainment and Technology Law, 23*(4), 891–918.

Haindel, D. B. (n.d.). *Theodore Burghard Hurt Brameld: The prophet father of the coming world.* Southeastern Louisiana University. https://www2.southeastern.edu/Academics/Faculty/nadams/educ692/Brameld.html

Hamilton, C., Swart, W., & Stokes, G. M. (2021). Developing a measure of social, ethical, and legal content for intelligent cognitive assistants. *Journal of Strategic Innovation and Sustainability, 16*(3), 1–37.

Han, A. (2020, December 30). *Two Sundance docs sound the alarm on the dangers of modern AI: Is the tech industry... bad?* Mashable. https://mashable.com/article/coded-bias-social-dilemma-documentary-review

Hansen, L. (2022, July 20). *How CRM and ERP integration can benefit your business.* CIO Insight. https://www.cioinsight.com/enterprise-apps/crm-erp-integration

Hao, K. (2019, April 9). *Facebook's ad-serving algorithm discriminates by gender and race.* MIT Technology Review. https://www.technologyreview.com/2019/04/05/1175/facebook-algorithm-discriminates-ai-bias

Hao, G., Bishwajit, G., Tang, S., Nie, C., Ji, L., & Huang, R. (2017, June 23). Social participation and perceived depression among elderly population in South Africa. *Clinical Interventions in Aging, 12*, 971–976. doi:10.2147/CIA.S137993 PMID:28694690

Hasselbalch, G., Olsen, B. K., & Tranberg, P. (2020). *White paper on data ethics in public procurement of AI-based services and solutions.* Data Ethics EU. https://dataethics.eu/wp-content/uploads/dataethics-whitepaper-april-2020.pdf

Hauser, C. (2017, April 27). In Connecticut murder case, a Fitbit is a silent witness. *The New York Times*. https://www.nytimes.com/2017/04/27/nyregion/in-connecticut-murder-case-a-fitbit-is-a-silent-witness.html

Hausman, D. M., & Welch, B. (2010, January 8). Debate: To nudge or not to nudge. *Journal of Political Philosophy*, *18*(1), 123–136. https://doi.org/ doi:10.1111/j.1467-9760.2009.00351.x

Hayden, C. (2018, January 11). *Entertainment technologies*. International Studies Association and Oxford University Press. doi:10.1093/acrefore/9780190846626.013.386

Hayes, N. (2019, March 18). *Ethics and biometric identity - security info watch*. Security Info Watch. https://www.securityinfowatch.com/access-identity/biometrics/article/21072152/ethics-and-biometric-identity

Healey, D., Hegelheimer, V., Hubbard, P., Ioannou-Georgiou, D., Kessler, G., & Ware, P. (2008). *TESOL technology standards framework*. Teachers of English to Speakers of Other Languages, Inc. https://www.tesol.org/docs/default-source/books/bk_technologystandards_framework_721.pdf

He, D., Ye, R., Chan, S., Guizani, M., & Xu, Y. (2018, April 13). Privacy in the internet of things for smart healthcare. *IEEE Communications Magazine*, *56*(4), 38–44. doi:10.1109/MCOM.2018.1700809

Henry, S. L. (Ed.). (2022a, March 18). *WCAG 2 overview*. https://www.w3.org/WAI/standards-guidelines/wcag

Henry, S. L. (Ed.). (2022b, March 18). *What's new in WCAG 2.2 working draft*. https://www.w3.org/WAI/standards-guidelines/wcag/new-in-22/#introduction-timeline-comments

Hermannsdottir, A., Štangej, O., & Kristinsson, K. (2018). When being good is not enough: Towards contextual education of business leadership ethics. *Management*, *23*(2), 1–13. doi:10.30924/mjcmi/2018.23.2.1

Hetzner, C. (2022, June 10). *Elon Musk's regulatory woes mount as U.S. moves close to recalling Tesla's self-driving software*. Fortune. https://fortune.com/2022/06/10/elon-musk-tesla-nhtsa-investigation-traffic-safety-autonomous-fsd-fatal-probe/

Hevelke, A., & Nida-Rümelin, J. (2015, June). Responsibility for crashes of autonomous vehicles: An ethical analysis. *Science and Engineering Ethics*, *21*(3), 619–630. doi:10.100711948-014-9565-5 PMID:25027859

Hewson, C. (2015). Ethics issues in digital methods research. In H. Snee, C. Hine, Y. Morey, S. Roberts, & H. Watson, (Eds.) Digital methods for social science: An interdisciplinary guide to research innovation. Palgrave Macmillan.

Hill, K. (2022, June 30). Deleting your period tracker won't protect you. *The New York Times*. https://www.nytimes.com/2022/06/30/technology/period-tracker-privacy-abortion.html

Hlebowwitsh, P. (1999, Autumn). The burdens of the new curricularist. *Curriculum Inquiry*, *29*(3), 343–353. doi:10.1111/0362-6784.00131

Hlebowwitsh, P. (2010, September). Centripetal thinking in curricula studies. *Curriculum Inquiry*, *40*(4), 503–512. doi:10.1111/j.1467-873X.2010.00497.x

Hoehne, J. (2020, May 20). *Untitled*. [Photograph]. https://unsplash.com/photos/iggWDxHTAUQ

Hoffman, M. (2000). *Empathy and moral development: Implications for caring and justice*. Cambridge University Press. doi:10.1017/CBO9780511805851

Hogarty, S. (2021, November 1). *Five common leadership styles, and how to find your own*. We Work Ideas. https://www.wework.com/ideas/professional-development/management-leadership/five-common-leadership-styles-and-how-to-find-your-own

Holland, P., & Tham, T. L. (2020, April 27). Workplace biometrics: Protecting employee privacy one fingerprint at a time. *Economic and Industrial Democracy, 43*(2), 501–515. doi:10.1177/0143831X20917453

Holmes, J., Powell-Griner, E., Lethbridge-Cejku, M., & Heyman, K. (2009, July). *Aging differently: Physical limitations among adults aged 50 years and over: United States, 2001-2007* (NCHS Data Brief No. 20). U.S. Department of Health and Human Services, Centers for Disease Control and Prevention, National Center for Health Statistics. https://www.cdc.gov/nchs/data/databriefs/db20.pdf

Holthe, T., Halvorsrud, L., Karterud, D., Hoel, K. A., & Lund, A. (2018, May). Usability and acceptability of technology for community-dwelling older adults with mild cognitive impairment and dementia: A systematic literature review. *Clinical Interventions in Aging, 13*, 863–886. doi:10.2147/CIA.S154717 PMID:29765211

Hope, A. (2018, May). Creep: The growing surveillance of students' online activities. *Education and Society, 36*(1), 55–72. doi:10.7459/es/36.1.05

Horelick, J. (2021). Four web accessibility developments that shaped the first half of 2021. *Labor & Employment Law, 49*(2), 4–5.

Hostetler, K. (2005, August 1). What is "good" education research? *Educational Researcher, 34*(6), 16–21. doi:10.3102/0013189X034006016

Hougaard, R. (2019, March 5). The power of putting people first. *Forbes*. https://www.forbes.com/sites/rasmushougaard/2019/03/05/the-power-of-putting-people-first

How to protect your online privacy with threat modeling [Video]. (2017, November 15). Above the Noise. https://www.youtube.com/watch?v=VlYjtWg4Thw&ab_channel=AboveTheNoise

Hubert, M., Blut, M., Brock, C., Zhang, R. W., Koch, V., & Riedl, R. (2018). The influence of acceptance and adoption drivers on smart home usage. *European Journal of Marketing, 53*(6), 1073–1098. doi:10.1108/EJM-12-2016-0794

Hughes, D. J., Lee, A., Tian, A. W., Newman, A., & Legood, A. (2018, October). Leadership, creativity, and innovation: A critical review and practical recommendations. *The Leadership Quarterly, 29*(5), 549–569. https://doi.org/10.1016/j.leaqua.2018.03.001

Humility. (n.d.). https://www.dictionary.com/browse/humility

Huss, J. A. (2019). Middle level education aims for equity and inclusion, but do our school websites meet ADA compliance? *Middle Grades Review, 5*(1). https://scholarworks.uvm.edu/mgreview/vol5/iss1/4

Hutagalung, J. M., Tobing, C. I., Debastri, J., & Amanda, R. T. (2020, April). Space debris as environmental threat and the requirement of Indonesia's prevention regulation. *IOP Conference Series: Earth and Environmental Science, 465*(1), 1. doi:10.1088/1755-1315/456/1/012081

Hutchinson, B., Smart, A., Hanna, A., Denton, E., Greer, C., Kjartansson, O., Barnes, P., & Mitchell, M. (2021, March). Towards accountability for Machine Learning datasets: Practices from software engineering and infrastructure. *Proceedings of the 2021 ACM Conference on Fairness, Accountability, and Transparency*, 560–575. 10.1145/3442188.3445918

Hye-ran, K., & Tae-gyu, K. (2021, September 17). *Hyundai to feature facial recognition in electric vehicles*. UPI. https://www.upi.com/Top_News/World-News/2021/09/17/Hyundai-facial-recognition-Genesis/5491631899253

Ikeda, S. (2021, March 15). *Verkada data breach exposes feeds of 150,000 security cameras; targets include health care facilities, schools, police stations and a Tesla plant*. CPO. https://www.cpomagazine.com/cyber-security/verkada-data-breach-exposes-feeds-of-150000-security-cameras-targets-include-health-care-facilities-schools-police-stations-and-a-tesla-plant

International Bridge, Tunnel and Turnpike Association. (2018, August). *Glossary of Terms*. https://www.ibtta.org/sites/default/files/documents/IBTTA%20Glossary.pdf

International Institute for Management Development. (2022, August). *The 5 leadership styles you can use*. https://www.imd.org/imd-reflections/reflection-page/leadership-styles

International Organization for Standardization. (n.d.). *Online browsing platform [search for RFID]*. Retrieved April 30, 2022 from https://www.iso.org/obp/ui/#search

Iskat, G. J., & Liebowitz, J. (2021, June). What to do when employees resist change. *Super Vision, 82*(6), 7–9.

Islam, G. (2020). Psychology and business ethics: A multi-level research agenda. *Journal of Business Ethics, 165*(1), 1–13. https://doi.org/10.1007/s10551-019-04107-w

Ismail, A., & Kuppusamy, K. S. (2022). Web accessibility investigation and identification of major issues of higher education websites with statistical measures: A case study of college websites. *Journal of King Saud University - Computer and Information Sciences, 34*(3), 901-911. doi:10.1016/j.jksuci.2019.03.011

Israel, E., & Batalova, J. (2021, January 14). *International students in the United States*. Migration Policy Institute. https://www.migrationpolicy.org/article/international-students-united-states-2020

Ito, M., Antin, J., Finn, M., Law, A., Manion, A., Mitnick, S., Schlossberg, D., & Yardi, S. (2010). *Hanging out, messing around, and geeking out. Kids living and learning with new media*. The MIT Press.

Ivanova, E., & Borzunov, G. (2020). Optimization of machine learning algorithm of emotion recognition in terms of human facial expressions. *Procedia Computer Science, 169*, 244–248. doi:10.1016/j.procs.2020.02.143

Ivey, G., & Fisher, D. (2007). *Creating literacy rich schools for adolescents*. Association for Supervision and Curriculum Development (ASCD).

Janoski, T., & Wilson, J. (1995, September). Pathways to voluntarism: Family socialization and status transmission models. *Social Forces, 74*(1), 271–292. doi:10.2307/2580632

Jenkins-Scott, J. (2020, January 29). *Responsive leadership: Needed now more than ever*. Leadership Now. https://www.leadershipnow.com/leadingblog/2020/01/responsive_leadership_needed_n.html

Jesse, M., Jannach, D., & Gula, B. (2021, December 20). Explorations in digital nudging for online food choices. *Frontiers in Psychology, 12*, 1–12. https://doi.org/ doi:10.3389/fpsyg.2021.729589 PMID:34987443

Jfrux. (2019, February 27). *Comma Explorer drive annotation tool* [Image]. Licensed by Creative Commons Attribution-Share Alike 4.0 International. No changes made to the image. https://commons.wikimedia.org/wiki/File:Comma_Explorer_Drive_Annotation_Tool.png

Jobin, A., Ienca, M., & Vayena, E. (2019, September 2). The global landscape of AI ethics guidelines. *Nature Machine Intelligence, 1*(9), 389–399. doi:10.103842256-019-0088-2

Johansen, A. G. (2019, February 8). *Biometrics and biometric data: What is it and is it secure?* Norton. https://us.norton.com/internetsecurity-iot-biometrics-how-do-they-work-are-they-safe.html

John Deere says fully autonomous tractor advances future of farming. (2022, March 16). *KWQC*. https://www.kwqc.com/2022/01/04/john-deere-announces-fully-autonomous-tractor-ces

Jones, K. M. L. (2019, July 2). Learning analytics and higher education: A proposed model for establishing informed consent mechanisms to promote student privacy and autonomy. *International Journal of Educational Technology in Higher Education, 16*(24), 24. Advance online publication. doi:10.118641239-019-0155-0

Jonk, G., Anscombe, J., & Aurik, J. C. (2018, March 29). *How technology can transform leadership – for the good of employees*. World Economic Forum. https://www.weforum.org/agenda/2018/03/how-technology-can-transform-business-performance-for-human-good

Jung, A.-K., Stieglitz, S., Kissmer, T., Mirbabaie, M., & Kroll, T. (2022, June 29). Click me …! The influence of clickbait on user engagement in social media and the role of digital nudging. *PLoS One*, *17*(6), 1–22. https://doi.org/ doi:10.1371/journal.pone.0266743 PMID:35767538

Jupudi, S. (n.d.). *Technology is the enabler, not the driver, for business transformation*. Dallas Business Journal. https://www.bizjournals.com/dallas/news/2021/01/05/technology-is-the-enabler-not-the-driver-for-business-transformation.html

Kahn, S. (2014, January 31). Can positive duties be derived from Kant's formula of universal law? *Kantian Review*, *19*(1), 93–108. doi:10.1017/S1369415413000319

Kalantri, R., Parekar, A., Mohite, A., & Kankapurkar, R. (2014). RFID based toll collection system. *International Journal of Computer Science and Information Technologies*, *5*(2), 2582–2585.

Kallio, H., Pietilä, A.-M., Johnson, M., & Kangasniemi, M. (2016, January). Systematic methodological review: Developing a framework for a qualitative semi-structured interview guide. *Journal of Advanced Nursing*, *72*(12), 2954–2965. https://doi.org/ doi:10.1111/jan.13031 PMID:27221824

Kamal, A. H. M. (2013, July-December). Legal aspects of the security concerns surrounding Radio Frequency Identification (RFID) technology. *ASA University Review*, *7*(2), 91–110.

Kanade, V. (2022, March 11). *What Is super Artificial Intelligence (AI)? Definition, threats, and trends*. Spice Works. https://www.toolbox.com/tech/artificial-intelligence/articles/super-artificial-intelligence

Kanchanara. (2021, September 14). *All crypto coins are together in the dark* [Photograph]. https://unsplash.com/photos/fsSGgTBoX9Y

Kang, S. (2019, June). Sustainable influence of ethical leadership on work performance: Empirical study of multinational enterprise in South Korea. *Sustainability*, *11*(11), 3101. https://doi.org/10.3390/su11113101

Kantayya, S. (2021). *Coded bias* [Film; online video]. Independent Lens. https://www.codedbias.com

Karale, A. (2021, September). The challenges of IoT addressing security, ethics, privacy, and laws. *Internet of Things*, *15*, 1–20. doi:10.1016/j.iot.2021.100420

Kark, K., Briggs, B., & Tweardy, J. (2019, May 13). *Reimagining the role of technology*. Deloitte. https://www2.deloitte.com/us/en/insights/focus/cio-insider-business-insights/reimagining-role-of-technology-business-strategies.html

Kaspersky. (n.d.). *What is facial recognition: definition and explanation*. https://www.kaspersky.com/resource-center/definitions/what-is-facial-recognition

Kassim, A. J., & Lawless, L. (2021). The ADA and website accessibility post-Domino's: Detangling employers' and business owners' web and mobile accessibility obligations. *Tort Trial & Insurance Practice Law Journal*, *56*(1), 53–66.

Katsanis, S. H., Claes, P., Doerr, M., Cook-Deegan, R., Tenenbaum, J. D., Evans, B. J., Lee, M. K., Anderton, J., Weinberg, S. M., & Wagner, J. K. (2021, October 14). A survey of U.S. public perspectives on facial recognition technology and facial imaging data practices in health and research contexts. *PLoS One*, *16*(10), e0257923. doi:10.1371/journal.pone.0257923 PMID:34648520

Katznelson, G., & Gerke, S. (2021, March 3). The need for health AI ethics in medical school education. *Advances in Health Sciences Education: Theory and Practice*, *26*(4), 1447–1458. doi:10.100710459-021-10040-3 PMID:33655433

Kaufman, S. (2015, March 12). *They abandoned their wheelchairs and crawled up the Capitol steps.* ShareAmerica. https://share.america.gov/crawling-up-steps-demand-their-rights

Kaur, M., Sandhu, M., Mohan, N., & Sandhu, P. S. (2011, February). RFID technology principles, advantages, limitations & its applications. *International Journal of Computer and Electrical Engineering, 3*(1), 151–157. doi:10.7763/IJCEE.2011.V3.306

Kavathatzopoulos, I. (2017). Ethical leadership in business: The significance of Information and Communication Technology. In *Keynote speech at Japan Society for Information and Management 75th Annual Conference* (pp. 1-4). Tokyo: Japan Society for Information and Management.

Keener, E. B. (2022, April 25). Facial recognition: A new trend in state regulation. *Business Law Today.* https://businesslawtoday.org/2022/04/facial-recognition-new-trend-state-regulation

Kenna, D. (2021, July). *Using adversarial debiasing to remove bias from Word Embeddings.* doi:10.48550/arXiv.2107.10251

Kerdeman, D. (1998). Between Interlochen and Idaho: Hermeneutics and education for understanding. *Philosophy of Education,* 272-279.

Kethineni, S., & Cao, Y. (2020). The rise in popularity of cryptocurrency and associated criminal activity. *International Criminal Justice Review, 30*(3), 325–344. https://doi.org/10.1177/1057567719827051

Kharpal, A. (2022, June 3). Crypto firms say thousands of digital currencies will collapse, compare market to early dotcom days. *CNBC.* https://www.cnbc.com/2022/06/03/crypto-firms-say-thousands-of-digital-currencies-will-collapse.html

Kidd, D., Miner, J., Schein, M., Blauw, M., & Allen, D. (2020, December). Ethics across the curriculum: Detecting and describing emergent trends in ethics education. *Studies in Educational Evaluation, 67*, 100914. https://doi.org/10.1016/j.stueduc.2020.100914

Kiley, D. (2018, March 19). First death of a pedestrian struck by an autonomous vehicle may set tone for lawyers and liability. *Forbes.* https://www.forbes.com/sites/davidkiley5/2018/03/19/the-first-pedestrian-fatality-with-an-autonomous-vehicle-could-set-tone-for-lawyers-and-liability

Kim, S. E. (2021, August 25). Can the world's first space sweeper make a dent in orbiting debris? *Smithsonian Magazine.* https://www.smithsonianmag.com/science-nature/can-worlds-first-space-sweeper-make-dent-orbiting-debris-180978515

Kimbell, J. P., & Dos Santos, P. L. (2020). Business ethics case competitions: A fresh opportunity to teach business ethics. *Southern Journal of Business and Ethics, 12*, 40–62.

Klar, R., & Lanzerath, D. (2020). The ethics of COVID-19 tracking apps – challenges and voluntariness. *Research Ethics Review, 16*(3), 1–9. doi:10.1177/1747016120943622

Kluz, A., & Nowak, B. E. (2016, December 23). *The impact of technology on leadership.* The Oxford University Politics Blog. https://blog.politics.ox.ac.uk/impact-pressures-technology-leadership

Ko, E., & Kim, Y. (2020, May 20). Why do firms implement responsible innovation? The case of emerging technologies in South Korea. *Science and Engineering Ethics, 26*(5), 2663–2692. doi:10.100711948-020-00224-2 PMID:32436167

Kohlberg, L. (1984). Essays on moral development: Vol. II. *The psychology of moral development.* Harper and Row.

Kohlrieser, G., Orlick, A. L., Perrinjaquet, M., & Rossi, R. L. (n.d.). *Resilient leadership: Navigating the pressures of modern working life.* https://www.imd.org/research-knowledge/articles/resilient-leadership-navigating-the-pressures-of-modern-working-life

Kopestinsky, A. (2022, March 5). *25 astonishing self-driving car statistics for 2022*. Policy Advice. https://policyadvice.net/insurance/insights/self-driving-car-statistics

Korinchak, D. (2021, October 22). *RFID security vulnerabilities*. Cyber Experts. https://cyberexperts.com/rfid-security-vulnerabilites

Koss, E. (2022, August 8). *Education technology: What is Edtech? A Guide*. Built In. https://builtin.com/edtech

Kostka, G., Steinacker, L., & Meckel, M. (2021, March). Between security and convenience: Facial recognition technology in the eyes of citizens in China, Germany, the United Kingdom, and the United States. *Public Understanding of Science (Bristol, England)*, *30*(6), 671–690. doi:10.1177/09636625211001555 PMID:33769157

Kouroupis, K. (2021). Facial recognition: A challenge for Europe or a threat to human rights? *European Journal of Privacy Law & Technologies*, *2021*(1), 142–156. https://universitypress.unisob.na.it/ojs/index.php/ejplt/article/view/1265/667

Kretzschmar, M. E., Rozhnova, G., Bootsma, M. C. J., van Boven, M., van de Wijgert, J. H. H. M., & Bonten, M. J. M. (2020, July 16). Impact of delays on effectiveness of contact tracing strategies for COVID-19: A modelling study. *The Lancet. Public Health*, *5*(8), e452–e459. doi:10.1016/S2468-2667(20)30157-2 PMID:32682487

Krisher, T. (2021, February 1). Tesla issues recall after Full Self-Driving software allows cars to run stop signs. *LA Times*. https://www.latimes.com/world-nation/story/2022-02-01/tesla-recall-self-driving-software-runs-stop-signs

Krisher, T., & Dazio, S. (2022, January 18). Felony charges are 1st in a fatal crash involving Autopilot. *AP News*. https://apnews.com/article/tesla-autopilot-fatal-crash-charges-91b4a0341e07244f3f03051b5c2462ae

Kroll, T., & Stieglitz, S. (2021, January). Digital nudging and privacy: Improving decisions about self-disclosure in social networks. *Behaviour & Information Technology*, *40*(1), 1–19. https://doi.org/ doi:10.1080/0144929X.2019.1584644

Kruse, K. (2013, April 9). What is leadership? *Forbes*. https://www.forbes.com/sites/kevinkruse/2013/04/09/what-is-leadership/?sh=52a9c3345b90

Kuligowski, K. (2022, June 29). How to be an ethical leader: 7 tips for success. *Business News Daily*. https://www.businessnewsdaily.com/5537-how-to-be-ethical-leader.html

Kumar, S. G., Prince, S., & Shankar, B. M. (2021, May). Smart tracking and monitoring in supply chain systems using RFID and BLE. *2021 3rd International Conference on Signal Processing and Communication (ICPSC)*, pp. 757-760. https://ieeexplore.ieee.org/document/9451700

Kumar, V. (2021, March 11). *What will happen when a facial recognition firm is hacked?* Analytics Insight. https://www.analyticsinsight.net/what-will-happen-when-a-facial-recognition-firm-is-hacked

Kussainova, A., Rakhimberdinova, M., Denissova, O., Taspenova, G., & Konyrbekov, M. (2018, Winter). Improvement of technological modernization using behavioral economics. *Journal of Environmental Management and Tourism*, *7*(31), 1470–1478. https://doi.org/ doi:10.14505/jemt.v9.7(31).11

Laas, K. (2021, February 16). *Intellectual property*. Illinois Tech. https://guides.library.iit.edu/c.php?g=474695&p=3248753

Landt, J. (2001, October 1). *Shrouds of time: The history of RFID*. The Association for Automatic Identification and Data Capture Technologies (AIM). https://transcore.com/wp-content/uploads/2017/01/History-of-RFID-White-Paper.pdf

Lave, J., & Wenger, E. (1991). *Situated learning: Legitimate peripheral participation*. Cambridge University Press. doi:10.1017/CBO9780511815355

Lavoie, D. (2022, March 10). Virginia lawmakers ok lifting ban on facial technology use. *Associated Press*. https://www.msn.com/en-us/news/us/virginia-lawmakers-ok-lifting-ban-on-facial-technology-use/ar-AAUU3rK

Law Insider, Inc. (n.d.). *Federal or state law definition.* https://www.lawinsider.com/dictionary/federal-or-state-law

Le Calvez, S., Perron-Lepage, M. F., & Burnett, R. (2006, March 6). Subcutaneous microchip-associated tumours in B6C3F1 mice: A retrospective study to attempt to determine their histogenesis. *Experimental and Toxicologic Pathology, 57*(4), 255–265. doi:10.1016/j.etp.2005.10.007 PMID:16427258

Leadership ethics - Traits of an ethical leader. (n.d.). *Management Study Guide.* https://www.managementstudyguide.com/leadership-ethics.htm

Lee, D. (2018, February 3). Deepfakes porn has serious consequences. *BBC News.* https://www.bbc.com/news/technology-42912529

Lee, J., & Choi, S. J. (2021, July). Hospital productivity after data breaches: Difference-in-differences analysis. *Journal of Medical Internet Research, 23*(7), e26157. doi:10.2196/26157 PMID:34255672

Lee, L.-K., Braud, T., Zhou, P., Wang, L., Zu, D., Lin, Z., Kumar, A., Bermejo, C., & Hui, P. (2021, November). All one needs to know about Metaverse: A complete survey on technological singularity, virtual ecosystem, and research agenda. *Journal of Latex Class Files, 14*(8), 1–66. doi:10.48550/arXiv.2110.05352

Leenes, R. (2007). *Privacy in the metaverse. In IFIP International Summer School on the Future of Identity in the Information Society.* Springer. doi:10.1007/978-0-387-79026-8_7

Legallet, C., Mankin, K. T., Spaulding, K., & Mansell, J. (2017, July/August). Granulomatous inflammatory response to a microchip implanted in a dog for eight years. *Journal of the American Animal Hospital Association, 53*(4), 227–229. doi:10.5326/JAAHA-MS-6418 PMID:28535132

Lembke, T.-B., Engelbrecht, N., Brendel, A. B., & Kolbe, L. (2019, June). To nudge or not to nudge: Ethical considerations of digital nudging based on its behavioral economics roots. *European Conference on Information Systems.* Stoholm & Uppsala, Sweden. https://www.researchgate.net/publication/333421600_To_Nudge_or_Not_To_Nudge_Ethical_Considerations_of_Digital_Nudging_Based_on_Its_Behavioral_Economics_Roots

LePan, N. (2020, April 18). *Visualizing the length of the fine print, for 14 popular apps.* Visual Capitalist. https://www.visualcapitalist.com/terms-of-service-visualizing-the-length-of-internet-agreements

Lestari, N. D. I., & Subriadi, A. P. (2021, September). EdTech investment: Optimism, pessimism, and uncertainty. *2021 International Conference on Electrical and Information Technology (IEIT)*, 239-245. 10.1109/IEIT53149.2021.9587429

Lewin, K. (1947, June 1). Frontiers in group dynamics: Concept, method, and reality in social science; social equilibria and social change. *Human Relations, 1*(1), 5–40. doi:10.1177/001872674700100103

Lewis, D. (2021, February 26). Contact-tracing apps help reduce COVID infections, data suggest. *Nature, 591*(7848), 18–19. doi:10.1038/d41586-021-00451-y PMID:33623147

Li, C., & Yang, H. J. (2021). Bot-X: An AI-based virtual assistant for intelligent manufacturing. *Multiagent & Grid Systems, 17*(1), 1–14. doi:10.3233/MGS-210340

Lilly, J., Durr, D., Grogan, A., & Super, J. F. (2021, September-October). Wells Fargo: Administrative evil and the pressure to conform. *Business Horizons, 64*(5), 587–597. https://doi.org/10.1016/j.bushor.2021.02.028

Limousine driver faces felony charges after causing fatal crash while relying on Tesla autopilot. (2022, February 23). The Stanley Law Group. https://www.thestanleylawgroup.com/limousine-driver-faces-felony-charges-after-causing-fatal-crash-while-relying-on-tesla-autopilot

Lin, P. (2014, August 17). *Don't fear the robot car bomb.* Bulletin of the Atomic Scientists. https://thebulletin.org/2014/08/dont-fear-the-robot-car-bomb

Lipardo, D. S., & Tsang, W. W. N. (2018, August 24). Falls prevention through physical and cognitive training (falls PACT) in older adults with mild cognitive impairment: A randomized controlled trial protocol. *BMC Geriatrics, 18*(1), 1–12. doi:10.118612877-018-0868-2 PMID:30143002

Liu, B. (2021). *"Weak AI" is likely to never become "Strong AI", so what is its greatest value for us?* doi:10.48550/arXiv.2103.15294

Liu, J., Sohn, J., & Kim, S. (2017, November 26). Classification of daily activities for the elderly using wearable sensors. *Journal of Healthcare Engineering, 8934816*, 1–7. Advance online publication. doi:10.1155/2017/8934816 PMID:29317996

Liu, Y., Qiu, B., Fan, X., Zhu, H., & Han, B. (2016, December). Review of smart home energy management systems. *Energy Procedia, 104*, 504–508. doi:10.1016/j.egypro.2016.12.085

Li, X., Wang, H., Dai, H.-N., Wang, Y., & Zhao, Q. (2016). An analytical study on eavesdropping attacks in wireless nets of things. *Mobile Information Systems, 431475*, 1–10. Advance online publication. doi:10.1155/2016/4313475

Lovegrove, M. (2020) *Why we need to talk about ethics in technology.* Hello World. https://helloworld.raspberrypi.org/articles/HW06-why-we-need-to-talk-about-ethics-in-technology

Lund, M. L., & Nygård, L. (2003, April 1). Incorporating or resisting assistive devices: Different approaches to achieving a desired occupational self-image. *OTJR (Thorofare, N.J.), 23*(2), 67–75. doi:10.1177/153944920302300204

Lunter, J. (2020, October). Beating the bias in facial recognition technology. *Biometric Technology Today, 2020*(9), 5–7. doi:10.1016/S0969-4765(20)30122-3

Luppicini, R. (2008). The emerging field of technoethics. In R. Luppicini & R. Adell (Eds.), Handbook of research on technoethics (pp. 1–18). Information Science Reference/IGI-Global. https://doi.org/10.4018/978-1-60566-022-6.ch001

Luppicini, R. (2009). Technoethical inquiry: From technological systems to society. Global Media Journal - Canadian Edition, 2(1), 5-21.

Luppicini, R. (2010). *Technoethics and the evolving knowledge society: Ethical issues in technological design, research, development, and innovation.* IGI-Global. doi:10.4018/978-1-60566-952-6

Lupton, D. (2020, March 3). 'Honestly no, I've never looked at it': Teachers' understandings and practices related to students' personal data in digitised health and physical education. *Learning, Media and Technology, 46*(3), 281–291. doi:10.1080/17439884.2021.1896541

Lu, S., Chen, G., & Wang, K. (2020, November 23). Overt or covert? Effect of different digital nudging on consumers' customization choices. *Nankai Business Review International, 12*(1), 56–74. doi:10.1108/NBRI-12-2019-0073

Lwuozor, J. (2022, February 12). *Best facial recognition software for enterprises in 2022.* eSecurity Planet. https://www.esecurityplanet.com/products/facial-recognition-soft

Macgilchrist, F. (2019). Cruel optimism in edtech: When the digital data practices of educational technology providers inadvertently hinder educational equity. *Learning, Media and Technology, 44*(1), 77–86. doi:10.1080/17439884.2018.1556217

Machado, A. M., & Brandão, C. (2019). Leadership and technology: Concepts and questions. In *New Knowledge in Information Systems and Technologies. World CIST'19: Advances in Intelligent Systems and Computing* (pp. 764-773). Springer. doi:10.1007/978-3-030-16184-2_73

Machado, H., & Granja, R. (2020). DNA databases and Big Data. In Forensic Genetics in the Governance of Crime (pp. 57–70). Palgrave Pivot. doi:10.1007/978-981-15-2429-5_5

Mackenzie, L. (2020, January 21). *Surveillance state: How Gulf governments keep watch on us.* Wired. https://wired.me/technology/privacy/surveillance-gulf-states

Majumder, M. A., Guerrini, C. J., & McGuire, A. L. (2021, January). Direct-to-Consumer genetic testing: Value and risk. *Annual Review of Medicine*, *72*(1), 151–166. doi:10.1146/annurev-med-070119-114727 PMID:32735764

Maloney, D., Rajasabeson, S., Moore, A., Caldwell, J., Archer, J., & Robb, A. (2019, March). Ethical concerns of the use of virtual avatars in consumer entertainment. *2019 IEEE Conference on Virtual Reality and 3D User Interfaces (VR)*. Osaka, Japan. 10.1109/VR.2019.8797926

Marett v. Five Guys Enterprises LLC, 1:17-cv-00788 (SDNY, 2017). https://www.classaction.org/media/marett-v-five-guys.pdf

Margan, S. K. (2018, December 20). Autonomous vehicles and insurance. *BimaQuest: The Journal of Insurance & Management*, *18*(3), 15–24.

Marín, V. I., Carpenter, J. P., & Tur, G. (2021, September 20). Pre-service teachers' perceptions of social media data privacy policies. *British Journal of Educational Technology*, *52*(2), 519–535. doi:10.1111/bjet.13035

Martinez, A. M. (2009). Face recognition, overview. In S. Z. Li & A. Jain (Eds.), *Encyclopedia of Biometrics*. Springer. doi:10.1007/978-0-387-73003-5_84

Maschinenjunge. (2008, March 11). *RFID Chip 003* [Image]. Licensed by CC Share-Alike 3.0 Unported. No changes made to the image. https://commons.wikimedia.org/wiki/File:RFID_Chip_003.JPG

Mason, R. O. (1986, March). Four ethical issues of the information age. *Management Information Systems Quarterly*, *10*(1), 5–12. https://www.researchgate.net/publication/242705009_Four_Ethical_Issues_of_the_Information_Age

Masumba, D. (2019). *Leadership for innovation: Three essential skill sets for leading employee-driven innovation.* Morgan James Publishing.

Mattes. (2018, November 24). *MasterCard (transparent) 2018, Bank- N26* [Image]. Licensed under Creative Commons 2.0 Germany. No changes made to the image. https://commons.wikimedia.org/wiki/File:MasterCard_(transparent)_2018,_Bank-_N26.jpg

Mayer, M. M., Bell, R., & Buchner, A. (2021, December 23). Self-protective and self-sacrificing preferences of pedestrians and passengers in moral dilemmas involving autonomous vehicles. *PLoS One*, *16*(12), 1–25. doi:10.1371/journal.pone.0261673 PMID:34941936

May, T. (2018, August 2). Sociogenetic risks–ancestry DNA testing, third-party identity, and protection of privacy. *The New England Journal of Medicine*, *379*(5), 410–412. doi:10.1056/NEJMp1805870 PMID:29924688

McCartney, S., & Parent, R. (n.d.) *Ethics in law enforcement.* Pressbooks. https://ecampusontario.pressbooks.pub/ethicslawenforcement

McDevitt, P. (2020). *Anchoring cultural change and organizational change: Case study research evaluation project All Hallows College Dublin 1995-2015.* Information Age Publishing.

McLachlan, S. (2020, February 22). *How might driver licensing and vehicle registration evolve if we adopt autonomous cars and digital identification?* https://arxiv.org/ftp/arxiv/papers/2202/2202.09861.pdf

McLeod, S. (2021). *Pavlov's dogs study and Pavlovian conditioning explained.* Simply Psychology. https://www.simplypsychology.org/pavlov.html

McLoughlin, M., Sheler, J. L., & Witkin, G. (1987, February 23). A nation of liars? *U.S. News & World Report, 103*(20), 54–60.

Meder, B., Fleischhut, N., Krumnau, N.-C., & Waldmann, M. R. (2019, February). How should autonomous cars drive? A preference for defaults in moral judgments under risk and uncertainty. *Risk Analysis: An International Journal, 39*(2), 295–314. doi:10.1111/risa.13178 PMID:30157299

Merener, M. M. (2012, August). Theoretical results on de-anonymization via linkage attacks. *Transactions on Data Privacy, 5*(2), 377–402.

Merriam-Webster. (n.d.). *Curiosity.* https://www.merriam-webster.com/dictionary/curiosity

Merriam-Webster. (n.d.a). Bias. In *Merriam-Webster.com dictionary.* https://www.merriam-webster.com/dictionary/bias

Merriam-Webster. (n.d.b). Deepfake. In *Merriam-Webster.com dictionary.* https://www.merriam-webster.com/dictionary/deepfake

Merriam-Webster. (n.d.c). Discrimination. In *Merriam-Webster.com dictionary.* https://www.merriam-webster.com/dictionary/discrimination

Merrill, M. (2007, July 30). *Human-implantable RFID chips: Some ethical and privacy concerns.* Healthcare IT News. https://www.healthcareitnews.com/news/human-implantable-rfid-chips-some-ethical-and-privacy-concerns

Methods used to provide E&C training to boards globally 2018. (2022, July 6). Statista. https://www.statista.com/statistics/896596/methods-used-to-provide-ethics-and-compliance-training-to-boards

Metz, C. (2021, March 15). Who is making sure the A.I. machines aren't racist? *The New York Times.* https://www.nytimes.com/2021/03/15/technology/artificial-intelligence-google-bias.html

Metz, C., & Conger, K. (2020, December 7). Uber is giving self-driving car project to a start-up. *The New York Times.* https://www.nytimes.com/2020/12/07/technology/uber-self-driving-car-project.html

Metz, R. (2018, August 17). *This company embeds microchips in its employees, and they love it.* Technology Review. https://www.technologyreview.com/s/611884/this-company-embeds-microchips-in-its-employees-and-they-love-it

Michael, K., McNamee, A., & Michael, M. G. (2006, June 26). The emerging ethics of humancentric GPS tracking and monitoring. *2006 International Conference on Mobile Business.* Copenhagen, Denmark. https://ieeexplore.ieee.org/xpl/conhome/4124088/proceeding

Milgram, P., Takemura, H., Utsumi, A., & Kishino, F. (1995). Augmented reality: A class of displays on the reality-virtuality continuum. Telemanipulator and Telepresence Technologies, 2351, 282-292.

Millán-Calenti, J. C., Maseda, A., Rochette, S., Vázquez, G. A., Sánchez, A., & Lorenzo, T. (2011, October 26). Mental and psychological conditions, medical comorbidity and functional limitation: Differential associations in older adults with cognitive impairment, depressive symptoms and co-existence of both. *International Journal of Geriatric Psychiatry, 26*(10), 1071–1079. doi:10.1002/gps.2646 PMID:21905101

Millar, J. (2014). Technology as moral proxy: Autonomy and paternalism by design. In *Proceedings of the IEEE 2014 International Symposium on Ethics in Engineering, Science, and Technology (ETHICS '14).* IEEE Press. Article 23, 1–7.

Miller, B. E. (2021, July 23). *Leadership skills in the age of technology.* Inc. https://www.inc.com/inc-masters/leadership-skills-in-the-age-of-technology.html

Milutinović, M. (2018). Cryptocurrency. *Economics, 64*(1), 105–122. doi:10.5937/ekonomika1801105M

Mittelstadt, B. D., Allo, P., Taddeo, M., Wachter, S., & Floridi, L. (2016, December 1). The ethics of algorithms: Mapping the debate. *Big Data & Society, 3*(2). doi:10.1177/2053951716679679

Molla, R. (2020, February 7). *Law enforcement is now buying cellphone location data from marketers.* Vox. https://www.vox.com/recode/2020/2/7/21127911/ice-border-cellphone-data-tracking-department-homeland-security-immigration

Molnar, D., & Wagner, D. (2004, October 25). Privacy and security in library RFID: Issues, practices, and architectures. [Conference presentation] *Proceedings of the 11th ACM Conference on Computer and Communications Security.* Washington DC. 10.1145/1030083.1030112

Mora, F., Quito, R., & Macías, L. (2021). Reading comprehension and reading speed of university English language learners in Ecuador. *Journal of English Language Teaching and Applied Linguistics, 3*(11), 11–31. doi:10.32996/jeltal.2021.3.11.3

Moreland, J. D., Richardson, J. A., Goldsmith, C. H., & Clase, C. M. (2004, July). Muscle weakness and falls in older adults: A systematic review and meta-analysis. *Journal of the American Geriatrics Society, 52*(7), 1121–1129. doi:10.1111/j.1532-5415.2004.52310.x PMID:15209650

Moss, E., & Metcalf, J. (2020, October 9). High tech, high risk: Tech ethics lessons for the COVID-19 pandemic response. *Patterns, 1*(7), 1–8. doi:10.1016/j.patter.2020.100102 PMID:33073256

Most important objectives of ethics and compliance training globally 2018. (2022, July 6). Statista. https://www.statista.com/statistics/896556/metrics-for-measuring-effectiveness-of-compliance-programs

Muhammed-Shittu, A.-R. B. (2021, September 30). A study of philosophical theory and educational science of insights on ethics, values, characters, and morals rooted into the Islamic and contemporary western perspectives. *Tarih Kültür ve Sanat Arastirmalari Dergisi, 10*(3), 47–58. https://doi.org/10.7596/taksad.v10i3.3090

Muir, S. W., Berg, K., Chesworth, B., Klar, N., & Speechley, M. (2010, April). Quantifying the magnitude of risk for balance impairment on falls in community-dwelling older adults: A systematic review and meta-analysis. *Journal of Clinical Epidemiology, 63*(4), 389–406. doi:10.1016/j.jclinepi.2009.06.010 PMID:19744824

Mulgan, T. (2020). *Utilitarianism.* Cambridge University Press. doi:10.1017/9781108582643

Münscher, R., Vetter, M., & Scheuerle, T. (2016, December). A review and taxonomy of choice architecture techniques. *Journal of Behavioral Decision Making, 29*(5), 511–524. doi:10.1002/bdm.1897

Murdock, J. (2020, April 30). Tesla faces lawsuit after Model X on Autopilot with 'dozing driver' blamed for fatal crash. *Newsweek.* https://www.newsweek.com/tesla-lawsuit-model-x-autopilot-fatal-crash-japan-yoshihiro-umeda-1501114

Murukannaiah, P. K., Singh, M. P., Singh, M. P., & Murukannaiah, P. K. (2020, July 24). From machine ethics to internet ethics: Broadening the horizon. *IEEE Internet Computing, 24*(3), 51–57. https://doi.org/10.1109/MIC.2020.2989935

Myers, S. (2021, June 28). *Supreme Court sides with cheerleader suspended for off-campus social media tirade.* Sharyl Attkisson. https://sharylattkisson.com/2021/06/read-supreme-court-sides-with-cheerleader-suspended-for-off-campus-social-media-tirade/

Nader, R. (2004, May 1). Legislating corporate ethics. *Journal of Legislation, 30*(2), 193–204. https://scholarship.law.nd.edu/jleg/vol30/iss2/1

Naghdipour, B. (2017). 'Close your book and open your Facebook': A case for extending classroom collaborative activities online. *The Journal of Asia TEFL, 14*(1), 130–143. doi:10.18823/asiatefl.2017.14.1.9.130

Najibi, A. (2020, October 26). *Racial discrimination in face recognition technology.* Science in the News. https://sitn.hms.harvard.edu/flash/2020/racial-discrimination-in-face-recognition-technology

National Highway Traffic Safety Administration. (2016, September). *Federal automated vehicles policy: Accelerating the next revolution in roadway safety.* https://www.transportation.gov/sites/dot.gov/files/docs/AV%20policy%20guidance%20PDF.pdf

National Highway Traffic Safety Administration. (2019, December 17). *NHTSA announces update to historic AEB commitment by 20 automakers.* https://www.nhtsa.gov/press-releases/nhtsa-announces-update-historic-aeb-commitment-20-automakers

National Highway Traffic Safety Administration. (n.d.). *Automated vehicles for safety.* https://www.nhtsa.gov/technology-innovation/automated-vehicles-safety

National Institute of Food and Agriculture. (n.d.). *Agriculture technology.* U.S. Department of Agriculture. https://www.nifa.usda.gov/topics/agriculture-technology

National Institutes of Health. (2021, June 23). *Annual review of ethics (case studies).* https://oir.nih.gov/sourcebook/ethical-conduct/responsible-conduct-research-training/annual-review-ethics-case-studies

National Traffic Safety Board. (2017). *Collision between a car operating with automated vehicle control systems and a tractor-semitrailer tuck near Williston, Florida, May 7, 2016.* Highway Accident Report NTSB/HAR-17/02. https://www.ntsb.gov/investigations/AccidentReports/Reports/HAR1702.pdf

Naudé, W., & Dimitri, N. (2021). Public procurement and innovation for Human-Centered Artificial Intelligence. SSRN *Electronic Journal.* doi:10.2139/ssrn.3762891

Nellis, M. (2021, July 30). Towards predictivity? Immediacy and imminence in the electronic monitoring of offenders. In B. Arrigo & B. Sellers (Eds.), *The pre-crime society: Crime, culture and control in the ultramodern age* (pp. 341–364). Bristol University Press. doi:10.1332/policypress/9781529205251.003.0016

Nickel, S. (2015, March 2). *Patriot puppies, 731st AMS port dogs open doors to new passengers 150302-F-FT438-003* [Image]. In Public Domain. https://commons.wikimedia.org/wiki/File:Patriot_puppies,_731st_AMS_port_dogs_open_doors_to_new_passengers_150302-F-FT438-003.jpg

Nicola, S. (2018, October 22). Invasion of the body hackers. *Bloomberg Business Week,* (4589), 22-23.

No Biometric Barriers to Housing Act H.R. 4008, 116th Cong. (2019). https://www.congress.gov/bill/116th-congress/house-bill/4008

North-Samardzic, A. (2020). Biometric technology and ethics: Beyond security applications. *Journal of Business Ethics, 167*(3), 433–450. https://doi.org/10.1007/s10551-019-04143-6

O'Flaherty, K. (2020, June 19). FaceApp privacy: What you need to know about the viral Russian app. *Forbes.* https://www.forbes.com/sites/kateoflahertyuk/2020/06/19/faceapp-privacy-what-you-need-to-know-about-the-viral-russian-app

OECD. AI. (2019). *OECD AI Principles overview.* https://oecd.ai/en/ai-principles

Office of the Federal Chief of Sustainability Officer. (n.d.). https://www.sustainability.gov

Okeke, F., Sobolev, M., Bell, N., & Estrin, D. (2018, September). Good vibrations: Can a digital nudge reduce digital overload? *Proceedings of the 20th International Conference on Human-Computer Interaction with Mobile Devices and Services, 4,* 1-12. https://dl.acm.org/doi/10.1145/3229434.3229463

Ondruš, J., Kolla, E., Vertal, P., & Šarić, Z. (2020). How do autonomous cars work? *Transportation Research Procedia, 44*, 226–23. doi:10.1016/j.trpro.2020.02.049

Onor, M. L., Trevisiol, M., Urciuoli, O., Misan, S., Bertossi, F., Tirone, G., Aguglia, E., & Pascolo-Fabrici, E. (2008, March). Effectiveness of telecare in elderly populations–a comparison of three settings. *Telemedicine Journal and e-Health, 14*(2), 164–169. doi:10.1089/tmj.2007.0028 PMID:18361706

Open Pilot supports 150+ vehicles. (n.d.) Comma. https://comma.ai/vehicles

Operational technology (OT) – definitions and differences with IT. (n.d.). *i-SCOOP*. https://www.i-scoop.eu/industry-4-0/operational-technology-ot

Opilo, E. (2021, June 14). Baltimore city council approves moratorium on facial recognition technology; City police exempt from ban. *The Baltimore Sun*. https://www.baltimoresun.com/politics/bs-md-ci-baltimore-council-facial-recognition-20210614-xbooqalr6be7zhzljcnpeb3cqm-story.html

Orlowski, J. (2020). *The social dilemma* [Film; online video]. Exposure Labs. https://www.thesocialdilemma.com

Østergaard, E. K. (2016, March 27). *Responsive leadership - A guide*. Slide Share. https://www.slideshare.net/ErikKorsvikstergaard/responsive-leadership-a-guide

Østergaard, E. K. (2018, February 21). *The responsive leader: How to be a fantastic leader in a constantly changing world*. LID Publishing.

Østergaard, E. K. (n.d.). *What is the place for modern, responsive leadership in 2020?* https://www.vunela.com/what-is-the-place-for-modern-responsive-leadership-in-2020

Ove, J. (2014, October 17). *Georgia Court of Appeals: Parents can be held liable for kids' social media misdeeds*. Patch. https://patch.com/georgia/acworth/georgia-court-appeals-parents-can-be-held-liable-kids-social-media-misdeeds-0

Ozeran, N. (2018, August 16). Insight: A mid-year review of the current state of ADA website accessibility lawsuits. *Bloomberg Law*. https://news.bloomberglaw.com/daily-labor-report/insight-a-mid-year-review-of-the-current-state-of-ada-website-accessibility-lawsuits

Özsungur, F. (2019). The impact of ethical leadership on service innovation behavior: The mediating role of psychological capital. *Asia Pacific Journal of Innovation and Entrepreneurship, 13*(1), 73–88. doi:10.1108/APJIE-12-2018-0073

Paaske, S., Bauer, A., Moser, T., & Seckman, C. (2017, June). The benefits and barriers to RFID technology in healthcare. *On-Line Journal of Nursing Informatics, 21*(2), 10–11.

Page, M. (n.d.). *The impact of technology on executive leadership*. Michael Page. https://www.michaelpage.com/advice/management-advice/development-and-retention/impact-technology-executive-leadership

Palanski, M., Newman, A., Leroy, H., Moore, C., Hannah, S., & Den Hartog, D. (2021). Quantitative research on leadership and business ethics: Examining the state of the field and an agenda for future research. *Journal of Business Ethics, 168*(1), 109–119. https://doi.org/10.1007/s10551-019-04267-9

Paliktzoglou, V., Oyelere, S. S., Suhonen, J., & Mramba, N. R. (2021, Summer). Social media: Computing educational perspective in diverse educational contexts. *Journal of Information Systems Education, 32*(3), 160–165.

Pandit, C., Kothari, H., & Neuman, C. (2020, November 22). Privacy in time of a pandemic [Paper presentation]. *2020 13th CMI Conference on Cybersecurity and Privacy (CMI) - Digital Transformation - Potentials and Challenges*, Copenhagen, Denmark. https://ieeexplore.ieee.org/abstract/document/9322737

Papademetriou, C., Anastasiadou, S., Konetos, G., & Papalexandris, S. (2022, April). COVID-19 pandemic: The impact of the social media technology on higher education. *Education Sciences, 12*(4), 261. doi:10.3390/educsci12040261

Pardales, M. J. (2002). "So, how did you arrive at that decision?" Connecting moral imagination and moral judgment. *Journal of Moral Education, 31*(4), 423–437. doi:10.1080/0305724022000029653

Parisi, T. (2021, October 22). *The seven rules of the metaverse.* Medium. https://medium.com/meta-verses/the-seven-rules-of-the-metaverse-7d4e06fa864c

Parker, W. C. (1997, February). The art of deliberation. *Educational Leadership, 54*(5), 18–21.

Parker, W. C. (2003). *Teaching democracy: Unity and diversity in public life.* Teachers College Press.

Park, S.-M., & Kim, Y.-G. (2022). A metaverse: Taxonomy, components, applications, and open challenges. *IEEE Access: Practical Innovations, Open Solutions, 10*, 4209–4251. doi:10.1109/ACCESS.2021.3140175

Parks, R., Hsu, C.-H., & Xu, H. (2014, September 10). RFID privacy issues in healthcare: Exploring the roles of technologies and regulations. *Journal of Information Privacy and Security, 6*(3), 3–28. doi:10.1080/15536548.2010.10855891

Patel, A. (2022). *How technology can be used to empower leadership.* About Leaders. https://aboutleaders.com/technology-empower-leadership

Patrol: Tesla autopilot driver was watching movie, crashed. (2020, August 28). *ABC News.* https://abcnews.go.com/Technology/wireStory/patrol-tesla-autopilot-driver-watching-movie-crashed-72685378

Pearl, T. H. (2018, Summer). Hands on the wheel: A call for greater regulation of semi-autonomous cars. *Indiana Law Journal (Indianapolis, Ind.), 93*(3), 713–756.

Perrenet, J. C., Bouhuijs, P. A., & Smits, J. G. (2010). The suitability of problem-based learning for engineering education: Theory and practice. *Teaching in Higher Education, 5*(3), 345–358.

Perry, F. (2020). *The tracks we leave: Ethics and management dilemmas in healthcare* (3rd ed.). ACHE Management Series.

Pescher, C., Reichhart, P., & Spann, M. (2014, February 1). Consumer decision-making processes in mobile viral marketing campaigns. *Journal of Interactive Marketing, 28*(1), 43–54. doi:10.1016/j.intmar.2013.08.001

Peters, C., & Bradbard, D. (2010). Web accessibility: An introduction and ethical implications. *Journal of Information Communication and Ethics in Society, 8*(2), 206–232. https://doi.org/10.1108/14779961011041757

Picchi, A. (2019, June 28). Medtronic recalls insulin pumps because hackers could hijack device. *CBS News.* https://www.cbsnews.com/news/medtronic-insulin-pump-recall-fda-says-hackers-could-hijack-device

Pinar, W. F. (1978). The reconceptualization of curricula studies. *Journal of Curriculum Studies, 3*(10), 150–157.

Pirzada, P., Wilde, A., Doherty, G. H., & Harris-Birtill, D. (2022, July 9). Ethics and acceptance of smart homes for older adults. *Informatics for Health & Social Care, 47*(1), 10–37. doi:10.1080/17538157.2021.1923500 PMID:34240661

Ploug, T., Holm, S., & Brodersen, J. (2014). Scientific second-order 'nudging' or lobbying by interest groups: The battle over abdominal aortic aneurysm screening programmes. *Medicine, Health Care, and Philosophy, 17*(4), 641–650. doi:10.100711019-014-9566-9 PMID:24807744

Plungis, J. (2019, May 16). *Tesla driver in fatal march crash was using autopilot, NTSB says.* Consumer Reports. https://www.consumerreports.org/car-safety/tesla-driver-in-fatal-march-crash-was-using-autopilot-ntsb-says

Pope, A. (1711) *An essay on criticism.* [Quote] https://www.quotes.net/quote/38071

Porter, M. E. (2008, January). The five competitive forces that shape strategy. *Harvard Business Review, 86*(1), 78–93.

Prensky, M. (2001, October). Digital natives, digital immigrants. In *On the Horizon, 9* (Vol. 5). MCB University Press.

Prentice, W. (2004, January). Understanding leadership. *Harvard Business Review.* https://hbr.org/2004/01/understanding-leadership

Prentza, G. (2020, March 24). *Untitled* [Photograph]. https://unsplash.com/photos/SRFG7iwktDk

Price, M. S. (2020, January 20). Internet privacy, technology, and personal information. *Ethics and Information Technology, 22*(2), 163–173. https://doi.org/10.1007/s10676-019-09525-y

Price, T. L. (2018). A "critical leadership ethics" approach to the ethical leadership construct. *Leadership, 14*(6), 687–706. https://doi.org/10.1177/1742715017710646

Prichard, S. (2021, April 12). *Master the four fields of leadership*. Skip Prichard. https://www.skipprichard.com/master-the-four-fields-of-leadership/

Prisacariu, A., & Shah, M. (2016, August 2). Defining the quality of higher education around ethics and moral values. *Quality in Higher Education, 22*(2), 152–166. https://doi.org/10.1080/13538322.2016.1201931

Pritchard, M. (1996). *Reasonable children: Moral education and moral learning*. University Press of Kansas.

Protect pets. (n.d.) ChipMeNot. https://chipmenot.info

Pruitt, S. (2018, September 20). *Here are 6 things Albert Einstein never said.* History.com. https://www.history.com/news/here-are-6-things-albert-einstein-never-said

Prusak, L. (2011, January 28). 25 years after Challenger, Has NASA's judgment improved? *Harvard Business Review.* https://hbr.org/2011/01/25-years-after-challenger-has

Purdy, M., & Daugherty, P. (2016). *Why Artificial Intelligence is the future of growth.* Accenture. https://dl.icdst.org/pdfs/files2/2aea5d87070f0116f8aaa9f545530e47.pdf

Pyjamas in public. Chinese city apologises for 'shaming' residents. (2020, January 21). *British Broadcasting Corporation (BBC).* https://www.bbc.com/news/world-asia-china-51188669

QR code security: What are QR codes and are they safe to use? (n.d.). Kaspersky. https://usa.kaspersky.com/resource-center/definitions/what-is-a-qr-code-how-to-scan

Quach, K. (2022, March 14). Driver in Uber's self-driving car death goes on trial, says she feels 'betrayed.' *The Register.* https://www.theregister.com/2022/03/14/in_brief_ai

Quinn, M. (2013, September 3). California School District hires company to monitor students' social media. *Washington Examiner.* https://www.washingtonexaminer.com/red-alert-politics/california-school-district-hires-company-to-monitor-students-social-media

Quintana, S. (2021, May 13). Backseat Tesla driver unapologetic after arrest for reckless driving. *NBC Bay Area.* https://www.nbcbayarea.com/news/local/backseat-tesla-driver-unapologetic-after-arrest-for-reckless-driving/2544875

Raggett, D., Lam, J., Alexander, I. F., & Kmiec, M. (1997). *Raggett on HTML 4* (2nd ed.). Addison Wesley Longman. https://www.w3.org/People/Raggett/book4/ch02.html

Rahaman, A. (2019, June). Address the real reasons employees resist change. *HR News Magazine*, 18–21.

Rahman, T., Kim, Y. S., Noh, M., & Lee, C. K. (2021, March). A student on the determinants on social media based learning in higher education. *Educational Technology Research and Development, 69*(2), 1325–1351. doi:10.100711423-021-09987-2

Rainie, L., Funk, C., Anderson, M., & Tyson, A. (2022, March 17). *AI and human enhancements: Americans' openness is tempered by a range of concerns*. Pew Research Center. https://www.pewresearch.org/internet/2022/03/17/public-more-likely-to-see-facial-recognition-use-by-police-as-good-rather-than-bad-for-society

Rajasekar, S. J. S. (2021, January 18). An enhanced IoT based tracing and tracking module for COVID-19 cases. *SN Computer Science, 2*(1), 42. doi:10.1007/s42979-020-00400-y

Ramey, J. (2021, November 9). *Walmart is already using driverless trucks*. Autoweek. https://www.autoweek.com/news/technology/a38198243/walmart-autonomous-delivery-trucks-gatik/

Ranchordás, S. (2020). Nudging citizens through technology in smart cities. *International Review of Law Computers & Technology, 34*(3), 254–276. doi:10.1080/13600869.2019.1590928

Randazzo, R. (2019, March 18). Who was really at fault in fatal Uber crash? Here's the whole story. *AZ Central*. https://www.azcentral.com/story/news/local/tempe/2019/03/17/one-year-after-self-driving-uber-rafaela-vasquez-behind-wheel-crash-death-elaine-herzberg-tempe/1296676002/

Reeder, B., Chung, J., Lyden, K., Winters, J., & Jankowski, C. M. (2020). Older women's perceptions of wearable and smart home activity sensors. *Informatics for Health & Social Care, 45*(1), 96–109. doi:10.1080/17538157.2019.1582054 PMID:30919711

Regan, P. M., & Jesse, J. (2019, September 15). Ethical challenges of edtech, big data and personalized learning: Twenty-first century student sorting and tracking. *Ethics and Information Technology, 21*(3), 167–179. doi:10.100710676-018-9492-2

Rehak, K. (2022, May 29). *Should I Use This EdTech Tool? Decision-Making Guide for Instructors* [Image]. Academic Press.

Reich, J. (2020). *Failure to disrupt: Why technology alone can't transform education*. Harvard University Press. doi:10.4159/9780674249684

Reinraum. (2003, January 1). *134 2khz rfid animal tag* [Image]. In Public Domain. https://commons.wikimedia.org/wiki/File:134_2khz_rfid_animal_tag.jpg

Rejon-Parrilla, J. C., Espin, J., & Epstein, D. (2022, January 3). How innovation can be defined, evaluated and rewarded in health technology assessment. *Health Economics Review, 12*(1), 1–11. doi:10.118613561-021-00342-y PMID:34981266

Renaud, K., & Zimmermann, V. (2018, December). Ethical guidelines for nudging in information security & privacy. *International Journal of Human-Computer Studies, 120*, 22–35. doi:10.1016/j.ijhcs.2018.05.011

Resilience. (n.d.). *American Psychological Association*. https://www.apa.org/topics/resilience

Ribble, M., & Bailey, G. (2007). *Digital citizenship in schools*. International Society for Technology in Education.

Richardson, R. (2021). Best practices for government procurement of data-driven technologies: A short guidance for key stages of government technology procurement. SSRN *Electronic Journal*. doi:10.2139/ssrn.3855637

Richter, F.-J., & Sinha, G. (2020, August 21). Why do your employees resist new tech? *Harvard Business Review*. https://hbr.org/2020/08/why-do-your-employees-resist-new-tech

Richter, S., Rhode, J., Arado, T., & Parks, M. (2021, Fall). Principles for conducting a comprehensive LMS review. *The Community College Enterprise, 27*(2), 89–94.

Rieback, M. R., Crispo, B., & Tanenbaum, A. S. (2006, January-March). The evolution of RFID security. *IEEE Pervasive Computing, 5*(1), 62–69. https://ieeexplore.ieee.org/document/1593573

Rip, A. (1986). Controversies as informal technology assessment. *Knowledge (Beverly Hills, Calif.), 8*(2), 349–371.

Ritchie, E., & Landis, E. A. (2021, June 23). Industrial robotics in manufacturing. *Journal of Leadership, Accountability and Ethics, 18*(2), 110–116. doi:10.33423/jlae.v18i2.4258

Rome youth basketball has rest of season canceled. (2022, February 15). *Daily Sentinel.* https://romesentinel.com/stories/rome-youth-basketball-has-rest-of-season-canceled,129459

Rooksby, J. H. (2020, January 13). Consider impact of institution's tracking apps on privacy, best interest of students. *Campus Legal Advisor: Interpreting the Law for Higher Education Administrators, 20*(66), 1–3. doi:10.1002/cala.40173

Roque, A., Moreira, J. M., Dias Figueiredo, J., Albuquerque, R., & Gonçalves, H. (2020, July). Ethics beyond leadership: Can ethics survive bad leadership? *Journal of Global Responsibility, 11*(3), 275–294. https://doi.org/10.1108/JGR-06-2019-0065

Rorabaugh, P. (2012, August 6). *Occupy the digital: Critical pedagogy and new media.* Hybrid Pedagogy. https://hybridpedagogy.org/occupy-the-digital-critical-pedagogy-and-new-media

Rosales Torres, C. S., Buenadicha Sánchez, C., & Tetsuro, N. (2021, May 1). *Autoevaluación Ética de IA para actores del Ecosistema Emprendedor.* doi:10.18235/0003269

Rouse, M., & Shea, S. (2017, December 12). *RFID (radio frequency identification).* Internet of Things (Tech Target). https://internetofthingsagenda.techtarget.com/definition/RFID-radio-frequency-identification

Roussi, A. (2020, November 18). Resisting the rise of facial recognition. *Nature, 587*(7834), 350–353. doi:10.1038/d41586-020-03188-2 PMID:33208966

Ruidong, Y., Li, Y., Li, D., Yongcai, W., Yuqing, Z., & Weili, W. (2021, March). A stochastic algorithm based on reverse sampling technique to fight against the cyberbullying. *ACM Transactions on Knowledge Discovery from Data, 15*(4), 1–22. doi:10.1145/3441455

Ryan, M. (2020). The future of transportation: Ethical, legal, social and economic impacts of self-driving vehicles in the year 2025. *Science and Engineering Ethics, 26*(3), 1185–1208. doi:10.100711948-019-00130-2 PMID:31482471

Saeed, N., & Green, A. R. L. (2019, January 29). *RFID pet monitoring & identification system with RFID.* University of West London. https://core.ac.uk/download/pdf/195384891.pdf

Saleiro, P., Kuester, B., Hinkson, L., London, J., Stevens, A., Anisfeld, A., Rodolfa, K. T., & Ghani, R. (2018). *Aequitas: A bias and fairness audit toolkit.* doi:10.48550/arxiv.1811.05577

Schad, T. (2022, March 12). 'It seems to be more extreme': Violent sports fans are causing alarm at every level. *USA Today.* https://www.usatoday.com/story/sports/college/2022/03/12/sports-fans-more-violent-abusive-since-returning-after-worst-covid/6986397001

Schein, E. H., & Schein, P. A. (2018). *Humble leadership: The power of relationships, openness, and trust.* Berrett-Koehler Publishers.

Schmidt, A. T., & Engelen, B. (2020, February 27). The ethics of nudging: An overview. *Philosophy Compass, 15*(4). Advance online publication. doi:10.1111/phc3.12658

Schmidt, F., Dröge-Rothaar, A., & Rienow, A. (2021, August 28). Development of a web GIS for small-scale detection and analysis of COVID-19 (SARS-CoV-2) cases based on volunteered geographic information for the city of Cologne, Germany, in July/August 2020. *International Journal of Health Geographics*, *20*(1), 40. Advance online publication. doi:10.118612942-021-00290-0 PMID:34454536

Schneble, C. O., & Shaw, D. M. (2021, September). Driver's views on driverless vehicles: Public perspectives on defining and using autonomous cars. *Transportation Research Interdisciplinary Perspectives*, *11*, 100446. doi:10.1016/j.trip.2021.100446

Schneider, C., Weinmann, M., & Vom Brocke, J. (2018, June 25). Digital nudging. *Communications of the ACM*, *61*(7), 67–73. doi:10.1145/3213765

Schoenhofer, S. O., van Wynsberghe, A., & Boykin, A. (2019). Engaging robots as nursing partners in caring: Nursing as caring meets care-centered value-sensitive design. *International Journal for Human Caring*, *23*(2), 157–167. doi:10.20467/1091-5710.23.2.157

Schrage, M., Pring, B., Kiron, D., & Dickerson, D. (2021, January 26). Leadership's digital transformation: Leading purposefully in an era of context collapse. *MIT Sloan Management Review*. https://sloanreview.mit.edu/projects/leaderships-digital-transformation

Schwitzgebel, E., Cokelet, B., & Singer, P. (2020, October). Do ethics classes influence student behavior? Case study: Teaching the ethics of eating meat. *Cognition*, *203*, 104397. https://doi.org/10.1016/j.cognition.2020.104397

Scott, N. L., Hansen, B., LaDue, C. A., Lam, C., Lai, A., & Chan, L. (2016, August). Using an active Radio Frequency Identification Real-Time Location System to remotely monitor animal movement in zoos. *Animal Biotelemetry*, *4*(16). https://doi.org/10.1186/s40317-016-0108-5

Seamons, K. (2021, August 21). *There was something different about this fatal crash.* Newser. https://www.newser.com/story/309951/there-was-something-different-about-this-fatal-tesla-crash.html

Sears, D. (2018, March 8). The sightless visionary who invented cruise control. *Smithsonian Magazine*. https://www.smithsonianmag.com/innovation/sightless-visionary-who-invented-cruise-control-180968418

Seivold, G. (2022, March 13). Are RFID-blocking wallets necessary to prevent credit card theft? *Loss Prevention Magazine*. https://losspreventionmedia.com/are-rfid-blocking-wallets-necessary-to-prevent-credit-card-theft

Sendak, M., Elish, M. C., Gao, M., Futoma, J., Ratliff, W., Nichols, M., Bedoya, A., Balu, S., & O'Brien, C. (2020, January). "The human body is a black box": Supporting clinical decision-making with deep learning. *Proceedings of the 2020 Conference on Fairness, Accountability, and Transparency*, 99–109. 10.1145/3351095.3372827

Setiyaningrum, A., & Aryanto, V. D. W. (2016). Corporate ethics and corporate social responsibility in reinforcing consumers bonding: An empirical study in controversial industry. *International Journal of Technoethics*, *7*(1), 1–5. https://doi.org/10.4018/IJT.2016010101

Shamsoshoara, A., Korenda, A., Afghah, F., & Zeadally, S. (2020, December 24). A survey on physical unclonable function (PUF)-based security solutions for Internet of Things. *Computer Networks*, *183*, 107593. doi:10.1016/j.comnet.2020.107593

Sharma, K., Zhan, X., Nah, F. F.-H., Siau, K., & Cheng, M. X. (2021, October). Impact of digital nudging on information security behavior: An experimental study on framing and priming in cybersecurity. *Organizational Cybersecurity Journal: Practice, Process and People*, *1*(1), 69–91. doi:10.1108/OCJ-03-2021-0009

Shbat, M. S., & Tuzlukov, V. (2014). Radar sensor detectors for vehicle safety systems. In N. Bizon, L. Dascalescu, & N. Mahdavi Tabatabaei (Eds.), *Autonomous vehicles: Intelligent transport systems and smart technologies* (pp. 3–55). Nova Science Publishers. https://search-ebscohost-com.libauth.purdueglobal.edu/login.aspx?direct=true&db=nlebk&AN=809607&site=eds-live

Sheng, E. (2019, April 15). Employee privacy in the US is at stake as corporate surveillance technology monitors workers' every move. *CNBC.* https://www.cnbc.com/2019/04/15/employee-privacy-is-at-stake-as-surveillance-tech-monitors-workers.html

Shepard, W. (2019, August 15). South Florida man wants answers after crash killed girlfriend. *NBC Miami.* https://www.nbcmiami.com/news/local/man-wants-answers-after-deadly-crash/124944

Shepardson, D. (2019, November 5). In review of fatal Arizona crash, U.S. agency says Uber software had flaws. *Reuters.* https://www.reuters.com/article/us-uber-crash/in-review-of-fatal-arizona-crash-u-s-agency-says-uber-software-had-flaws-idUSKBN1XF2HA

Shilton, K., Heidenblad, D., Porter, A., Winter, S., & Kendig, M. (2020, July 1). Role-playing computer ethics: Designing and evaluating the privacy by design (PbD) simulation. *Science and Engineering Ethics, 26*(6), 2911–2926. https://doi.org/10.1007/s11948-020-00250-0

Shirani, F., Groves, C., Henwood, K., Pidgeon, N., & Roberts, E. (2020, September). 'I'm the smart meter': Perceptions of smart technology amongst vulnerable consumers. *Energy Policy, 144*, 111637. doi:10.1016/j.enpol.2020.111637

Shortt, M., Tilak, S., Kuznetcova, I., Martens, B., & Akinkuolie, B. (2021, July 5). Gamification in mobile-assisted language learning: A systematic review of Duolingo literature from public release of 2012 to early 2020. *Computer Assisted Language Learning*, 1–38. Advance online publication. doi:10.1080/09588221.2021.1933540

Simon, K. G. (2003). *Moral questions in the classroom: How to get kids to think deeply about real life and their schoolwork.* Yale University Press.

Sjödin, D., Parida, V., Jovanovic, M., & Visnjic, I. (2020). Value creation and value capture alignment in business model innovation: A process view on outcome-based business models. *Journal of Product Innovation Management, 37*(2), 158–183. doi:10.1111/jpim.12516

Slaby, C. (2019, November 15). Decision-making self-driving car control algorithms: Intelligent transportation systems, sensing and computing technologies, and connected autonomous vehicles. *Contemporary Readings in Law and Social Justice, 11*(2), 29–35. doi:10.22381/CRLSJ11220194

Sloan, D. (1979). The teaching of ethics in the American undergraduate curriculum, 1876-1976. *The Hastings Center Report, 9*(6), 21–41. https://doi.org/10.2307/3561673

Slobogin, C., & Hazel, J. W. (2021). "A world of difference"? Law enforcement, genetic data, and the Fourth Amendment. *Duke Law Journal, 70*(4), 705–774.

Smiley, S. (2016, June 14). *7 types of security attacks on RFID systems.* Atlas RFID Store. https://www.atlasrfidstore.com/rfid-insider/7-types-security-attacks-rfid-systems

Smith, D. R. (2012, May 30). *What does it mean to be a responsive leader.* Smart Business. https://sbnonline.com/article/what-does-it-mean-to-be-a-responsive-leader

Smith, H., & Mitchell, R. (2022, January 19). A Tesla on autopilot killed two people in Gardena. Is the driver guilty of manslaughter? *Los Angeles Times.* https://www.latimes.com/california/story/2022-01-19/a-tesla-on-autopilot-killed-two-people-in-gardena-is-the-driver-guilty-of-manslaughter

Smith, M., & Miller, S. (2021, December 11). The rise of biometric identification: Fingerprints and applied ethics. In *Biometric Identification, Law and Ethics* (pp. 1–19). Springer. https://link.springer.com/chapter/10.1007/978-3-030-90256-8_1

Smith, M., & Miller, S. (2021, April 13). The ethical application of biometric facial recognition technology. *AI & Society, 37*(1), 167–175. doi:10.100700146-021-01199-9 PMID:33867693

SobolevM. (2021, March 24). Digital nudging: Using technology to nudge for good. doi:10.2139/ssrn.3889831

Soni, R. G., & Soni, B. (2019). Evolution of supply chain management: Ethical issues for leaders. *Competition Forum, 17*(2), 240–247.

Sparks, S. D. (2013, January 4). *Social-emotional needs entwined with youth' learning, security*. Education Week. https://www.edweek.org/leadership/social-emotional-needs entwined-with-youth-learning-security/2013/01

Spekking, R. (2021, May 16). *Amazon Echo Dot (RS03QR) - case - RFID-0589* [Image]. Licensed by Creative Commons Share-Alike 4.0 International. No changes made to the image. https://commons.wikimedia.org/wiki/File:Amazon_Echo_Dot_(RS03QR)_-_case_-_RFID-0589.jpg

Spence, E. (2008). Meta ethics for the Metaverse: The ethics of virtual worlds. In A. R. Briggle, K. Waelbers, & P. Brey (Eds.), *Current Issues in Computing and Philosophy* (pp. 3–13). IOS Press.

Spinello, R. A. (2021, Spring). Corporate data breaches: A moral and legal analysis. *Journal of Information Ethics, 30*(1), 12–32.

Spivey, M. (2016, September 27). *How RFID tags can change wildlife conservation*. Gateway RFID Store. https://gatewayrfidstore.com/rfid-tags-can-change-wildlife-conservation

Srinivasa, A. (2018, July 27). *Understanding Radar for automotive (ADAS) solutions*. Path Partner. https://www.pathpartnertech.com/understanding-radar-for-automotive-adas-solutions

Srinivasan, R., & Chander, A. (2021, August). Biases in AI systems. *Communications of the ACM, 64*(8), 44–49. https://dl.acm.org/doi/10.1145/3464903 doi:10.1145/3464903

Stahl, B. C. (2021, March 18). Ethical issues of AI. *Artificial Intelligence for a better future: An ecosystem perspective on the ethics of AI and emerging digital technologies*, 35-53. doi:10.1007/978-3-030-69978-9_4

Stan, V. A., Timnea, R. S., & Gheorghiu, R. A. (2014). Intelligent highway surveillance and safety systems. In N. Bizon, L. Dascalescu, & N. Mahdavi Tabatabaei (Eds.), *Autonomous vehicles: Intelligent transport systems and smart technologies* (pp. 147–184). Nova Science Publishers. https://search-ebscohost-com.libauth.purdueglobal.edu/login.aspx?direct=true&db=nlebk&AN=809607&site=eds-live

Stauffer, D. C., & Maxwell, D. L. (2020, May 18). Transforming servant leadership, organizational culture, change, sustainability, and courageous leadership. *Journal of Leadership, Accountability and Ethics, 17*(1), 105–116. doi:10.33423/jlae.v17i1.2793

Steele, J. (2022, January 27). *The history of credit cards*. Credit Cards. https://www.creditcards.com/statistics/history-of-credit-cards

Stein, S. (2022, February 7). Lidar is one of the iPhone and iPad Pro's coolest tricks: Here's what else it can do. *CNet*. https://www.cnet.com/tech/mobile/lidar-is-one-of-the-iphone-ipad-coolest-tricks-its-only-getting-better

Stephenson, C. (2011, July/August). How leadership has changed. *IVEY Business Journal*. https://iveybusinessjournal.com/publication/how-leadership-has-changed

Sternlich, A. (2020, April 30). South Korea's widespread testing and contact tracing lead to first day with no new cases. *Forbes*. https://www.forbes.com/sites/alexandrasternlicht/2020/04/30/south-koreas-widespread-testing-and-contact-tracing-lead-to-first-day-with-no-new-cases

Sternrenette. (2014, January 7). *Ablesen eines Mikrotransponders bei einem Hund* [Image]. Licensed by Creative Commons Share-Alike 3.0 Unported. No changes made to the image. https://commons.wikimedia.org/wiki/File:Ablesen_eines_Mikrotransponders_bei_einem_Hund.JPG

Stith-Flood, C. (2018, May/June). It's not hard to be humble: The role of humility in leadership. *Family Practice Management, 25*(3), 25–27. https://www.aafp.org/fpm/2018/0500/p25.html

Stöber, T., Kotzian, P., & Weißenberger, B. E. (2019). Design matters: On the impact of compliance program design on corporate ethics. *Business Research, 12*(2), 383–424. doi:10.1007/s40685-018-0075-1

Stoeklé, H.-C., Mamzer-Bruneel, M.-F., Vogt, G., & Hervé, C. (2016, March 31). 23andMe: A new two-sided data-banking market model. *BMC Medical Ethics, 17*(1), 1–11. doi:10.118612910-016-0101-9 PMID:27059184

Stommel, J. (2014, November 17). *Critical digital pedagogy: A definition.* Hybrid Pedagogy. https://hybridpedagogy.org/critical-digital-pedagogy-definition

Strate, L. (2020, July-October). The ethics of innovation. *Etc.; a Review of General Semantics, 77*(3–4), 182.

Stryja, C., & Satzger, G. (2019, November/December). Digital nudging to overcome cognitive resistance in innovation adoption decisions. *Service Industries Journal, 39*(15/16), 1123–1139. doi:10.1080/02642069.2018.1534960

Student privacy pledge. (n.d.). *Student Privacy Compass.* https://studentprivacycompass.org/audiences/ed-tech

Subbian, V., Solomonides, A., Clarkson, M., Rahimzadeh, V. N., Petersen, C., Schreiber, R., DeMuro, P. R., Dua, P., Goodman, K. W., Kaplan, B., Koppel, R., Lehmann, C. U., Pan, E., & Senathirajah, Y. (2021, January 15). Ethics and informatics in the age of COVID-19: Challenges and recommendations for public health organization and public policy. *Journal of the American Medical Informatics Association: JAMIA, 28*(1), 184–189. doi:10.1093/jamia/ocaa188 PMID:32722749

Sullivan, J. (2021, October 18). *Fintech under fire: Benartzi responds to SEC's 'Digital Nudge' warning.* 401k Specialist Magazine. https://401kspecialistmag.com/fintech-under-fire-benartzi-responds-to-secs-digital-nudge-warning

Summa Linguae Technologies. (2021, October 13). *Facing the future: Innovative uses of facial recognition.* https://summalinguae.com/language-technology/facial-recognition-uses/

Sun, P. (2022). Prison IOT. In *Smart Prisons.* Springer Singapore. doi:10.1007/978-981-16-9657-2_3

Survivors of Laos' worst dam disaster still struggling two years later. (2020, July 22). *Radio Free Asia.* https://www.rfa.org/english/news/laos/xe-pian-xe-namnoi-two-year-07222020211103.html

Sy, S., & Norris, C. (2022, April 25). What Elon Musk's $44 billion purchase of Twitter may mean for the company and free speech. *PBS.* https://www.pbs.org/newshour/show/what-elon-musks-44-billion-purchase-of-twitter-may-mean-for-the-company-and-free-speech

Symanovich, S. (2019, August 18). *Biometric data breach: Database exposes fingerprints, facial recognition data of 1 million people.* Norton. https://us.norton.com/internetsecurity-emerging-threats-biometric-data-breach-database-exposes-fingerprints-and-facial-recognition-data.html

Taghaboni-Dutta, F., & Velthouse, B. (2006, November). RFID technology is revolutionary: Who should be involved in this game of tag? *The Academy of Management Perspectives, 20*(4), 65–78.

Tangermann, V. (2018, May 14). *All the rage in Sweden: Embedding microchips under your skin.* Futurism. https://futurism.com/sweden-microchip-trend

Tang, Y., Xiong, J., Becerril-Arreola, R., & Iyer, L. (2019, June). Blockchain ethics research: A conceptual model. *SIGMIS-CPR '19: Proceedings of the 2019 on Computers and People Research Conference*, pp. 43–49. 10.1145/3322385.3322397

Tannen, D. (1998). *The argument culture: Stopping America's war of words.* Random House.

Tannert, C. (2006, May). Thou shalt not clone. An ethical argument against the reproductive cloning of humans. *EMBO Reports*, *7*(3), 238–240. https://doi.org/10.1038/sj.embor.7400653

Tanton, R., Vidyattama, Y., Nepal, B., & McNamara, J. (2011, April 8). Small area estimation using a reweighting algorithm. *Journal of the Royal Statistical Society. Series A, (Statistics in Society)*, *174*(4), 931–951. doi:10.1111/j.1467-985X.2011.00690.x

Taylor, Z. W. (2019). Web (in)accessible: Supporting access to Texas higher education for students with disabilities. *Texas Education Review, 7*(2), 60-75. doi:10.26153/tsw/2285

Taylor, L. (2016, April 1). No place to hide? The ethics and analytics of tracking mobility using mobile phone data. *Environment and Planning. D, Society & Space*, *34*(2), 319–336. doi:10.1177/0263775815608851

Technical explanation for RFID systems. (2018, May 8). *Omron.* https://www.ia.omron.com/data_pdf/guide/47/rfid_tg_e_2_1.pdf

Technology ethics cases. (n.d.). Markkula Center for Applied Ethics, Santa Clara University. https://www.scu.edu/ethics/focus-areas/technology-ethics/resources/technology-ethics-cases

Terms-of-Service Labeling, Design, and Readability Act, S.B. 3501, 117th Cong. (2022). https://trahan.house.gov/uploadedfiles/tldr_act.pdf

Testud, G., Vergnes, A., Cordier, P., Labarraque, D., & Miaud, C. (2019, October 22). Automatic detection of small PIT-tagged animals using wildlife crossings. *Animal Biotelemetry*, *7*(1). https://doi.org/10.1186/s40317-019-0183-5

Thaler, R. H., & Sunstein, C. R. (2008). *Nudge: Improving decisions about health, wealth, and happiness.* Penguin.

The 6 levels of vehicle autonomy explained. (n.d.). Synopsis. https://www.synopsys.com/automotive/autonomous-driving-levels.html

The beginner's guide to RFID systems. (2019). Atlas RFID Store. https://rfid.atlasrfidstore.com/hs-fs/hub/300870/file-252314647-pdf/Content/basics-of-an-rfid-system-atlasrfidstore.pdf

The dangers of driverless cars. (2021, May 5). *National Law Review, 11*(125). https://www.natlawreview.com/article/dangers-driverless-cars

The Dyslexia Foundation. (n.d.). *The Dyslexia Foundation.* https://dyslexiafoundation.org

The Society of Automotive Engineers International. (2021, April). *Surface vehicle recommended practice.* https://www.sae.org/standards/content/j3016_202104/preview

The strategic and responsible use of Artificial Intelligence in the public sector of Latin America and the Caribbean. (2022, March 22). OECD & CAF. doi:10.1787/1f334543-en

The Strategy Institute. (2020, November 20). *How technology leadership can accelerate disruption in your business strategy.* https://www.thestrategyinstitute.org/insights/how-technology-leadership-can-accelerate-disruption-in-your-business-strategy

The Telegraph. (2020, September 29). *First presidential debate in full: Trump vs Biden | US Election 2020* [Video]. YouTube. https://www.youtube.com/watch?v=CweqW7Pzxz8

The View. (2020, June 9). *National debate over defunding the police* [Video]. YouTube. https://www.youtube.com/watch?v=uMu5UshGUPM

Theisen, T. (2021, December 26). Microchip implanted under skin could be your COVID vaccine passport. *Orlando Sentinel.* https://www.nny360.com/news/publicservicenews/microchip-implanted-under-skin-could-be-your-covid-vaccine-passport/article_21a09e0a-c63c-5f23-85b7-76f65a4a8932.html

Thielke, S., Harniss, M., Thompson, H., Patel, S., Demiris, G., & Johnson, K. (2012). Maslow's hierarchy of human needs and the adoption of health-related technologies for older adults. *Ageing International, 37*(4), 470–488. doi:10.100712126-011-9121-4

Thompson, S. A., & Warzel, C. (2019, December 19). *Twelve million phones, one dataset, zero privacy.* The New York Times. https://www.nytimes.com/interactive/2019/12/19/opinion/location-tracking-cell-phone.html

Thomson, A. (n.d.). *U.S. Naval Space Command space surveillance system.* FAS Space Policy Project. https://spp.fas.org/military/program/track/spasur_at.htm

Thornhill, C., Meeus, Q., Peperkamp, J., & Berendt, B. (2019, June 6). A digital nudge to counter confirmation bias. *Workshop Proceedings of the 13th International AAAI Conference on Web and Social Media.* 10.3389/fdata.2019.00011

Thornley, C., Murnane, S., McLoughlin, S., & Carcary, M. (2018, October). The role of ethics in developing professionalism within the global ICT community. *International Journal of Human Capital and Information Technology Professionals, 9*(4), 56–71. https://dx.doi.org/10.4018/IJHCITP.2018100104

Throwback. (2008, August 21). *The 1st Kennedy/Nixon Presidential Debate - Part 1/4 (1960)* [Video]. YouTube. https://www.youtube.com/watch?v=C6Xn4ipHiwE

Tkacová, H., Králik, R., Tvrdoň, M., Jenisová, Z., & García Martin, J. (2022, February 27). Credibility and involvement of social media in education: Recommendations for mitigating the negative effects of the pandemic among high school students. *International Journal of Environmental Research and Public Health, 19*(5), 2767. doi:10.3390/ijerph19052767 PMID:35270460

Tkatch, R., Musich, S., MacLeod, S., Alsgaard, K., Hawkins, K., & Yeh, C. S. (2016, September). Population health management for older adults: Review of interventions for promoting successful aging across the health continuum. *Gerontology & Geriatric Medicine, 2.* doi:10.1177/2333721416667877 PMID:28680938

Tomkarlo. (2009, August 9). *Back side of disposable RFID tag used for race timing* [Image]. In Public Domain. https://commons.wikimedia.org/wiki/File:Back_side_of_disposable_RFID_tag_used_for_race_timing.png

Top challenges with ethics and compliance training globally in 2018. (2022, July 6). Statista. https://www.statista.com/statistics/896563/metrics-for-measuring-effectiveness-of-compliance-programs

Torre, T., & Sarti, D. (2020, November 11). The "way" toward e-leadership: Some evidence from the field. *Frontiers in Psychology, 11*, 1–14. https://doi.org/10.3389/fpsyg.2020.554253

Touqeer, H., Zaman, S., Amin, R., Hussain, M., Al-Turjman, F., & Bilal, M. (2021, December). Smart home security: Challenges, issues and solutions at different IoT layers. *The Journal of Supercomputing, 77*(12), 14053–14089. doi:10.100711227-021-03825-1

Transcription services. (n.d.). *uiAccess.* http://www.uiaccess.com/transcripts/transcript_services.html

Troester, M. (2021, May 20). *Accessibility in application design: Ethical, inclusive and good for the business.* Progress. https://www.progress.com/blogs/accessibility-in-application-design-ethical-inclusive-and-good-for-the-business

Trusted-AI. (2022). *AI Fairness 360 (AIF360) - Examples.* Git Hub. https://github.com/Trusted-AI/AIF360/tree/master/examples

Tsamados, A., Aggarwal, N., Cowls, J., Morley, J., Roberts, H., Taddeo, M., & Floridi, L. (2022). The ethics of algorithms: Key problems and solutions. *AI & Society, 37*(1), 215–230. doi:10.100700146-021-01154-8

Tuazon, O. M. (2021, January-June). Universal forensic DNA databases: Acceptable or illegal under the European Court of Human Rights regime? *Journal of Law and the Biosciences, 8*(1), 1–24. doi:10.1093/jlb/lsab022 PMID:34188945

Tucker, S. (2021, August 2). *Tesla update adds Wi-Fi while driving, new streaming entertainment.* Kelley Blue Book. https://www.kbb.com/car-news/tesla-update-adds-wi-fi-while-driving-new-streaming-entertainment

Turner, W. (2020, March 31). Chipping away at workplace privacy: The implementation of RFID microchips and erosion of employee privacy. *Washington University Journal of Law and Policy, 61*(1), 275–297. https://openscholarship.wustl.edu/law_journal_law_policy/vol61/iss1/18

Turri, A. M., Smith, R. J., & Kopp, S. W. (2017, January 13). Privacy and RFID technology: A review of regulatory efforts. *The Journal of Consumer Affairs, 51*(2), 329–354. https://doi.org/10.1111/joca.12133

Tussyadiah, I., & Miller, G. (2019, September). Nudged by a robot: Responses to agency and feedback. *Annals of Tourism Research, 78*, 102752. doi:10.1016/j.annals.2019.102752

Twitter. (2022, June 10). *Twitter Terms of Service.* https://twitter.com/en/tos

Tyler, R. W. (1948). *Basic principles of curricula and instruction.* University Chicago Press.

U.S. Access Board. (2022, March). *I.T. accessibility laws and policies.* General Services Administration. https://www.section508.gov/manage/laws-and-policies

U.S. Access Board. (n.d.a). *About the ICT accessibility 508 standards and 255 guidelines.* General Services Administration. https://www.access-board.gov/ict/#

U.S. Access Board. (n.d.b). *About the U.S. Access Board.* General Services Administration. https://www.access-board.gov/about

U.S. Census Bureau. (2017). *American Community Survey: Disability characteristics.* https://data.census.gov/cedsci/table?q=disability

U.S. consumers and cybercrime - statistics & facts. (2022, July 6). Statista. https://www.statista.com/topics/2588/us-consumers-and-cyber-crime

U.S. Department of Education. (2020). *Protecting students with disabilities: Frequently asked questions about Section 504 and the education of children with disabilities.* https://www2.ed.gov/about/offices/list/ocr/504faq.html

U.S. Department of Homeland Security. (2022, April 7). *Enhanced drivers licenses: What are they?* https://www.dhs.gov/enhanced-drivers-licenses-what-are-they

U.S. Department of Justice. (2008). *Americans with Disabilities Act of 1990, as amended.* https://www.ada.gov/pubs/adastatute08.htm

U.S. Department of Justice. (2016, October 11). *Americans with Disabilities Act Title II regulations.* https://www.ada.gov/regs2010/titleII_2010/titleII_2010_regulations.htm#a35103

U.S. Department of Justice. (2020). *A guide to disability rights laws.* https://www.ada.gov/cguide.htm

U.S. Department of State. (n.d.). *Sustainability at the U.S. Department of State.* https://www.state.gov/sustainability-at-the-u-s-department-of-state

U.S. Equal Employment Opportunity Commission (EEOC). (1973, September 26). *Rehabilitation Act of 1973 (original text).* https://www.eeoc.gov/rehabilitation-act-1973-original-text#

U.S. Federal Bureau of Investigation National Crime Information Center. (2020). *2020 NCIC missing person and unidentified person statistics.* https://www.fbi.gov/file-repository/2020-ncic-missing-person-and-unidentified-person-statistics.pdf

UNESCO. (2021, June). Draft text of the Recommendation on the Ethics of Artificial Intelligence. *Intergovernmental Meeting of Experts (Category II) related to a Draft Recommendation on the Ethics of Artificial Intelligence.* https://unesdoc.unesco.org/ark:/48223/pf0000377897

United States Government Accountability Office. (2021, June). *Facial recognition technology: Federal law enforcement agencies should better assess privacy and other risks, document GAO-21-518.* https://www.gao.gov/assets/gao-21-518.pdf

Usagreencardcenter. (2007, July 20). *Bill O'Reilly and Geraldo Rivera angry fight Immigration* [Video]. YouTube. https://www.youtube.com/watch?v=Z3U9ENaTPLY

Vaidya, T., Zhang, Y., Sherr, M., & Shields, C. (2015). Cocaine noodles: Exploiting the gap between human and machine speech recognition. *Proceedings of the 9th USENIX Workshop on Offensive Technologies (WOOT 15).*

Valuetainment Short Clips. (2022, March 15). *Who should be held responsible for COVID? Patrick Bet-David podcast episode 132* [Video]. YouTube. https://www.youtube.com/watch?v=GolGGSY4GJ0

van Est, R., & Brom, F. (2012, December). Technology assessment, analytic and democratic practice. In R. Chadwick (Ed.), Encyclopedia of Applied Ethics (2nd ed.) pp. 306–320. https://doi.org/10.1016/B978-0-12-373932-2.00010-7

Van Noorden, R. (2020, November 18). The ethical questions that haunt facial-recognition research. *Nature, 587*(7834), 354–358. doi:10.1038/d41586-020-03187-3 PMID:33208967

Vargas, J., Alsweiss, S., Toker, O., Razdan, R., & Santos, J. (2021, August 10). An overview of autonomous vehicles sensors and their vulnerability to weather conditions. *Sensors (Basel), 21*(16), 5397. doi:10.339021165397 PMID:34450839

Velasquez, M., Andre, C., Shanks, T. J. S., & Meyer, M. J. (1988, January 1). *Ethics and Virtue.* Markulla Center for Applied Ethics. https://www.scu.edu/ethics/ethics-resources/ethical-decision-making/ethics-and-virtue

Velasquez, M., Andre, C., Shanks, T. J. S., & Meyer, M. J. (1992, August 1). *Ethical relativism.* Markkula Center for Applied Ethics. https://www.scu.edu/ethics/ethics-resources/ethical-decision-making/ethical-relativism

Velasquez, M., Andre, C., Shanks, T. J. S., & Meyer, M. J. (2010, January 1). *What is ethics?* Markkula Center for Applied Ethics at Santa Clara University. https://www.scu.edu/ethics/ethics-resources/ethical-decision-making/what-is-ethics

Velasquez, M., Andre, C., Shanks, T., & Meyer, M. (2010, January 1). *What is ethics?* Markkuka Center for Applied Ethics at Santa Clara University. https://www.scu.edu/ethics/ethics-resources/ethical-decision-making/what-is-ethics/

Verkada. (2022). *Summary: March 9, 2021 security incident report.* https://www.verkada.com/security-update/report

Vertical Flight Society. (n.d.). VTOL. https://vtol.org

VijoenS. (2020, November 23). The promise and limits of lawfulness: Inequality, law, and the techlash. SSRN. https://papers.ssrn.com/sol3/papers.cfm?abstract_id=3725645

Vinkhuyzen, M. (2022, May 26). *Tesla FSD training – Garbage in, garbage out.* Clean Technica. https://cleantechnica.com/2022/05/26/tesla-fsd-training-garbage-in-garbage-out

Vogt, F., Haire, B., Selvey, L., Katelaris, A. L., & Kaldor, J. (2022, February 4). Effectiveness evaluation of digital contact tracing for COVID-19 in New South Wales, Australia. *The Lancet. Public Health*, *7*(3), e250–e258. doi:10.1016/S2468-2667(22)00010-X PMID:35131045

W3C. (2005, February). *Introduction to web accessibility.* https://www.w3.org/WAI/fundamentals/accessibility-intro

W3C. (2008, December 11). *Web Content Accessibility Guidelines 2.0.* https://www.w3.org/TR/2008/REC-WCAG20-20081211/#keybrd-interfacedef

W3C. (2016). *H4: Creating a logical tab order through links, form controls, and objects.* https://www.w3.org/TR/WCAG20-TECHS/H4.html

W3C. (2019a, September). *Captions/subtitles.* https://www.w3.org/WAI/media/av/captions

W3C. (2019b, September). *Transcripts.* https://www.w3.org/WAI/media/av/transcripts

W3C. (n.d.a). *Facts about W3C.* https://www.w3.org/Consortium/facts

W3C. (n.d.b). *W3C mission.* https://www.w3.org/Consortium/mission.html

Wadood, S., Gharleghi, B., & Samadia, B. (2016). Influence of change in management in technological enterprises. *Procedia Economics and Finance*, *37*, 129–136. https://doi.org/10.1016/S2212-5671(16)30103-4

Wald, N., & Harland, T. (2022). Reconsidering Vygotsky's 'more capable peer' in terms of both personal and knowledge outcomes. *Teaching in Higher Education*, *27*(3), 417–423. https://doi.org/10.1080/13562517.2021.2007474

Walsh, J. (2016, December 16). Rogue One: The CGI resurrection of Peter Cushing is thrilling – but is it right? *The Guardian*. https://www.theguardian.com/film/filmblog/2016/dec/16/rogue-one-star-wars-cgi-resurrection-peter-cushing

Walsom, C. (2014, June 5). Microchipping: a brief history. *Companion Animal, 19*(6), 288–290. doi:10.12968/coan.2014.19.6.288

Wamsley, L. (2020, September 16). Backup driver of autonomous Uber SUV charged with negligent homicide in Arizona. *NPR.* https://www.npr.org/2020/09/16/913530100/backup-driver-of-autonomous-uber-suv-charged-with-negligent-homicide-in-arizona

Wang, J. (2018). *What's in your face? Discrimination in facial recognition technology* [Master's thesis, Georgetown University]. https://repository.library.georgetown.edu/bitstream/handle/10822/1050752/Wang_georgetown_0076M_14043.pdf

Wang, V. (2009). Traditional leadership in light of E-HRMS. In T. Torres-Coronas & M. Arias-Oliva (Eds.), Encyclopedia of Human Resources Information Systems: Challenges in e-HRM (pp. 849–854). IGI-Global. https://doi.org/10.4018/978-1-59904-883-3.ch125

Wang, J., Ranganathan, V., Lester, J., & Kumar, S. (2022, March). Ultra low-latency backscatter for fast-moving location tracking. *Proceedings of the ACM on Interactive, Mobile, Wearable and Ubiquitous Technologies*, *6*(1), 1–22. doi:10.1145/3517255

Wang, Q., & Yan, P. (2019, November 21). Development of ethics education in science and technology in technical universities in China: Commentary on "Ethics 'upfront': Generating an organizational framework for a new university of technology.". *Science and Engineering Ethics*, *25*(6), 1721–1733. https://doi.org/10.1007/s11948-019-00156-6

Wang, Y.-Y., Wang, Y.-S., & Wang, Y.-M. (2020, October). What drives students' internet ethical behavior: An integrated model of the theory of planned behavior, personality, and internet ethics education. *Behaviour & Information Technology, 41*(3), 588–610. https://doi.org/10.1080/0144929X.2020.1829053

WCusr. (2019, March 28). *X-Ray of RFID implant* [Image]. Licensed under the Creative Commons CC0 1.0 Universal Public Domain Dedication. https://commons.wikimedia.org/wiki/File:X-Ray_of_RFID_Implant.jpg

WebAIM. (n.d.). *United States laws: The Rehabilitation Act of 1973 (Sections 504 and 508).* https://webaim.org/articles/laws/usa/rehab

WebAIM. (n.d.a). *Creating accessible tables.* https://webaim.org/techniques/tables/data

WebAIM. (n.d.b). *Contrast and color accessibility.* https://webaim.org/articles/contrast

WebAIM. (n.d.c). *Links and hypertext.* https://webaim.org/techniques/hypertext/link_text

WebAIM. (n.d.d). *Typefaces and fonts.* https://webaim.org/techniques/fonts

Wei, D., & Qiu, X. (2018, May). Status-based detection of malicious code in Internet of Things (IoT) devices. *2018 IEEE Conference on Communications and Network Security (CNS)*, 1-7. 10.1109/CNS.2018.8433183

Weinmann, M., Schneider, C., & vom Brocke, J. (2016, December). Digital nudging. *Business & Information Systems Engineering*, *58*(6), 433–436. doi:10.100712599-016-0453-1

Wells, F., & Farthing, M. (Eds.). (2009). *Fraud and misconduct in biomedical research* (4th ed.). Taylor and Francis.

West, D. M., & Allen, J. R. (2018, April 24). *How artificial intelligence is transforming the world.* Brookings. https://www.brookings.edu/research/how-artificial-intelligence-is-transforming-the-world

Wexler, J., Pushkarna, M., Bolukbasi, T., Wattenberg, M., Viégas, F., & Wilson, J. (2020, January). The What-If Tool: Interactive probing of Machine Learning models. *IEEE Transactions on Visualization and Computer Graphics*, *26*(1), 56–65. doi:10.1109/TVCG.2019.2934619 PMID:31442996

What is a Local Area Network? LAN definition, history and examples. (n.d.). *CompTIA.* https://www.comptia.org/content/guides/what-is-a-local-area-network

What is an autonomous car? (n.d.) *Synopsis.* https://www.synopsys.com/automotive/what-is-autonomous-car.html

What is ASIL? (n.d.). *Synopsis.* https://www.synopsys.com/automotive/what-is-asil.html

What is blockchain technology? How does blockchain work? (2022, August 9). *Simplilearn.* https://www.simplilearn.com/tutorials/blockchain-tutorial/blockchain-technology

What is empathy? (n.d.). *Greater Good Magazine, Science Center.* https://greatergood.berkeley.edu/topic/empathy/definition

What is ethics? (n.d.). *The Ethics Centre.* https://ethics.org.au/about/what-is-ethics

What is human capital management (HCM)? (n.d.). *Oracle.* https://www.oracle.com/human-capital-management/what-is-hcm

What is information technology? (n.d.). *CompTIA.* https://www.comptia.org/content/articles/what-is-information-technology

What is lipreading? (2018). Hearing Link Services. https://www.hearinglink.org/living/lipreading-communicating/what-is-lipreading

What is medical technology? (n.d.). *MedTech Europe*. https://www.medtecheurope.org/about-the-industry/what-is-medical-technology

What is technology? (2019, September 4). *4HL Net*. https://4hlnet.extension.org/what-is-technology

White, M. (2013). *The manipulation of choice: Ethics and libertarian paternalism* (1st ed.). Palgrave Macmillan. doi:10.1057/9781137313577

Wienroth, M., Granja, R., Lipphardt, V., Amoako, E. N., & McCartney, C. (2021, November 24). Ethics as ived practice. Anticipatory capacity and ethical decision-making in forensic genetics. *Genes*, *12*(12), 1–17. doi:10.3390/genes12121868 PMID:34946816

Wiggers, K. (2021, March 5). *Study warns deepfakes can fool facial recognition*. VentureBeat. https://venturebeat.com/2021/03/05/study-warns-deepfakes-can-fool-facial-recognition

Wigmore, I. (2018, February). *Artificial superintelligence (ASI)*. https://www.techtarget.com/searchenterpriseai/definition/artificial-superintelligence-ASI

Williams, M. (2022, February 24). *Building the metaverse with diversity and inclusion from the start*. Meta. https://about.fb.com/news/2022/02/building-the-metaverse-with-diversity-and-inclusion-from-the-start/

Wilson, C., Hargreaves, T., & Hauxwell-Baldwin, R. (2017, April). Benefits and risks of smart home technologies. *Energy Policy*, *103*, 72–83. doi:10.1016/j.enpol.2016.12.047

Wirtz, B. W., Weyerer, J. C., & Geyer, C. (2019). Artificial Intelligence and the public sector—Applications and challenges. *International Journal of Public Administration*, *42*(7), 596–615. doi:10.1080/01900692.2018.1498103

Wolfendale, J. (2007, July). My avatar, my self: Virtual harm and attachment. *Ethics and Information Technology*, *9*(2), 111–119. doi:10.100710676-006-9125-z

Wolkenstein, A. (2018, June 8). What has the Trolley Dilemma ever done for us (and what will it do in the future)? On some recent debates about the ethics of self-driving cars. *Ethics and Information Technology*, *20*(3), 163–173. doi:10.100710676-018-9456-6

Wong, J. K. W., & Leung, J. K. L. (2016, October 3). Modelling factors influencing the adoption of smart-home technologies. *Facilities*, *34*(13/14), 906–923. doi:10.1108/F-05-2016-0048

Wood & Ziemann v. State of Arizona, City of Tempe. (Az. Maricopa County Court, 2019). https://bloximages.newyork1.vip.townnews.com/azfamily.com/content/tncms/assets/v3/editorial/5/e2/5e235f8a-4b6d-11e9-9b1c-6bd0b0191f91/5c92d61c84180.pdf.pdf

Woodrum Setser, M. M., Cantor, M. C., & Costa, J. H. C. (2020, August). A comprehensive evaluation of microchips to measure temperature in dairy calves. *Journal of Dairy Science*, *103*(10), 9290–9300. https://doi.org/10.3168/jds.2019-17999

World Health Organization. (2019). *Global action plan on physical activity 2018-2030: more active people for a healthier world*. World Health Organization.

Wotapka, D. (2016, August 9). *How to teach ethics when your syllabus is packed*. AICPA. https://us.aicpa.org/interestareas/accountingeducation/newsandpublications/how-to-teach-ethics

Wray, S. (2020, September 17). *Portland bans private companies from using facial recognition technology*. Cities Today. https://cities-today.com/portland-bans-private-companies-from-using-facial-recognition-technology

Yadron, D., & Tynan, D. (2016, June 30). Tesla driver dies in first fatal crash while using autopilot mode. *The Guardian*. https://www.theguardian.com/technology/2016/jun/30/tesla-autopilot-death-self-driving-car-elon-musk

Yagelski, R. P. (2009, October). A thousand writers writing: Seeking change through the radical practice of writing as a way of being. *English Education*, *42*(1), 6–28.

Yagelski, R. P., Elbow, P., Freire, P., & Murray, D. M. (2006, May). Review: "Radical to many in the educational establishment": The writing process movement after the hurricanes. *College English*, *68*(5), 531–544. doi:10.2307/25472169

Yaghmaei, E., & van de Poel, I. (2021). *Assessment of responsible innovation: Methods and practices*. Routledge.

Yang, M. X., Xiaolin, H., Demir, A., Poon, A., & Wong, P. H.-S. (2021, March 16). Intracellular detection and communication of a wireless chip in cell. *Scientific Reports*, *11*(1), 5967. Advance online publication. doi:10.103841598-021-85268-5 PMID:33727598

Yang, Y., Yin, D., Easa, S. M., & Liu, J. (2022). Attitudes toward applying facial recognition technology for red-light running by e-bikers: A case study in Fuzhou, China. *Applied Sciences (Basel, Switzerland)*, *12*(211), 211. doi:10.3390/app12010211

Yan, P. (2017). *Research on C. Mitcham's thoughts of engineering ethics*. Dalian University of Technology.

Yermak, S. O., & Lisnichenko, O. O. (2016). Studying the aspects of establishing the definition of "innovation activity" and its determining factors. *Bìznes Ìnform*, *3*, 49–55.

Yigitcanlar, T., Foth, M., & Kamruzzaman, M. (2019a). Towards post-anthropocentric cities: Reconceptualizing smart cities to evade urban ecocide. *Journal of Urban Technology*, *26*(2), 147–152. doi:10.1080/10630732.2018.1524249

Yigitcanlar, T., Han, H., Kamruzzaman, M., Ioppolo, G., & Sabatini-Marques, J. (2019b). The making of smart cities: Are Songdo, Masdar, Amsterdam, San Francisco and Brisbane the best we could build? *Land Use Policy*, *88*, 104187. doi:10.1016/j.landusepol.2019.104187

Yigitcanlar, T., & Kamruzzaman, M. (2019). Smart cities and mobility: Does the smartness of Australian cities lead to sustainable commuting patterns? *Journal of Urban Technology*, *26*(2), 21–46. doi:10.1080/10630732.2018.1476794

Young, J. R. (2019, June 4). *What is critical digital pedagogy, and why does higher ed need it?* EdSurge. https://www.edsurge.com/news/2019-06-04-what-is-critical-digital-pedagogy-and-why-does-higher-ed-need-it

Zalnieriute, M. (2021, September 1). Burning bridges: The automated facial recognition technology and public space surveillance in the modern state. *The Columbia Science and Technology Law Review*, *22*(2), 284–307. doi:10.52214tlr.v22i2.8666

Zedner, L. (2003, March). The concept of security: An agenda for comparative analysis. *Legal Studies*, *23*(1), 153–176. https://doi.org/10.1111/j.1748-121X.2003.tb00209.x

Zelenák, M. (2020). *Mobile application for attendance monitoring* [Master's thesis, Masaryk University]. ProQuest Dissertations and Theses Global. https://is.muni.cz/th/tpg1p/Mobile_Application_for_Attendance_Monitoring_Archive.pdf

Zeng, E., Mare, S., & Roesner, F. (2017, July). End user security and privacy concerns with smart homes. *Thirteenth Symposium on Usable Privacy and Security (SOUPS 2017)*, 65-80.

Zha, W. (2012, September 23). *The difference principle: Inconsistent in Rawlsian theory?* E-International Relations. https://www.e-ir.info/2012/09/23/the-difference-principle-inconsistency-in-rawlsian-theory/

Zhang, Q., Ren, L., & Shi, W. (2013, May). HONEY: A multimodality fall detection and telecare system. *Telemedicine Journal and e-Health*, *19*(5), 415–429. doi:10.1089/tmj.2012.0109 PMID:23537382

Zhou, P., & Lee, L. (2022, May 23). *The metaverse is coming but we still don't trust AI*. 360. https://360info.org/the-metaverse-is-coming-but-we-still-dont-trust-ai

Zhou, P., Zhou, S., Zhang, M., & Miao, S. (2022, May 12). Executive overconfidence, digital transformation and environmental innovation: The role of moderated mediator. *International Journal of Environmental Research and Public Health, 19*(10), 5990. Advance online publication. doi:10.3390/ijerph19105990 PMID:35627526

Zhu, X., & Yang, T. (2019, October 18). Do I own it?: US and Chinese college students' digital ownership perceptions. *Proceedings of the Association for Information Science and Technology, 56*(1), 346–355. doi:10.1002/pra2.28

Zimmermann, V., & Renaud, K. (2021, February). The nudge puzzle: Matching nudge interventions to cybersecurity decisions. *ACM Transactions on Computer-Human Interaction, 28*(1), 1–45. doi:10.1145/3429888

Žnidaršič, A., Baggia, A., Pavliеček, A., Fischer, J., Rostański, M., & Weber, B. (2021, December). Are we ready to use microchip implants? An international cross-sectional study. *Organizacija, 54*(4), 275–292.

Zorzoli, E. (2018, October 19). *Why a responsive management style is the future of leadership*. https://www.wearebeem.com/why-a-responsive-management-style-is-the-future-of-leadership/

About the Contributors

Tamara Phillips Fudge is a full-time professor in the graduate technology programs at Purdue University Global. She has won fellowships and awards for innovation and teaching, and has taught a wide variety of topics, including web development, human-computer interaction, systems analysis and design, and those featuring documentation, diagramming, problem-solving, and presentation. Her career started with music degrees from Indiana University and Florida State University. She sang opera, oratorio, and in recital; her compositions have been heard on Public Radio, at various universities, and a state choral convention. She taught vocal and choral music, piano, pedagogy, foreign language diction, opera production, music theory, composition, and related courses in the traditional college classroom for 20-odd years. For seven years, she was a weekend correspondent for the Quad-City Times (Davenport, IA). Following a brief stint as an agent/registered representative selling life and health insurance and variable products, she returned to school with a keen interest in technology and has since distinguished herself in online teaching, coordination of large projects, and curriculum development.

* * *

Harold Brayan Arteaga-Arteaga is an Electronic Engineer from Universidad Autónoma de Manizales. He has been a member of the Research Group on Bioinformatics and Artificial Intelligence since 2018. His academic work includes data science, machine learning, deep learning, and automation to solve multiple problems. He supported the projects: detection of COVID-19 using X-ray images using convolutional neural networks and the project convolutional neural networks for image steganalysis in the spatial domain. He worked as a young researcher and taught programming languages and machine learning.

Ibidayo Awosola is a Doctoral Candidate in Technology Leadership and Innovation at Purdue University, West Lafayette, Indiana. His research interests focus on leadership and innovation, emphasizing how leadership propels innovation, thus bridging the technology divide in developing nations and innovating frameworks and policies to aid technological development. Among other essential professional IT and Project Management qualifications, he holds a bachelor's degree in computer science from the Tai Solarin University of Education in Ogun State, Nigeria, and a master's degree in information technology from Purdue University, West Lafayette, Indiana. Ibidayo Awosola is highly enthusiastic about technology leadership and uses his optimistic attitude and unwavering enthusiasm to inspire others to work hard and succeed. He is currently engaged as a Project Manager at InfoSys, a global leader in next-generation digital services and consulting. Prior to joining InfoSys, he worked at a number of respected corporations, including STANLEY Security, a subsidiary of Stanley Black & Decker, Emerson, KPMG, Esko, a subsidiary of Danaher Corporation, and the United Nations, to name a few. Ibidayo Awosola is also the Founder/President of TechXtramile, an organization that aims to bridge the technology gap by providing young people with job-ready digital skills and assets required to flourish in a highly competitive and dynamic world. In addition to being a fellow of ULA (Ubuntu Leaders Academy); he is also a member of IEEE (Institute of Electrical and Electronics Engineers), PMI (Project Management Institute), WYA (World Youth Alliance), UNA-USA (United Nations Association of the United States of America), and the IS (Internet Society).

Joshua Bernal-Salcedo is a Biomedical Engineer from Universidad Autónoma de Manizales. He has been a member of the Research Group on Bioinformatics and Artificial Intelligence since 2021. He is researching the ethical aspects of using artificial intelligence-based systems for automated decision-making, mainly oriented to public policies. He has participated in ethical evaluation projects, such as the COVID-19 study.

Erin A. Brennan is a licensed attorney in both Pennsylvania and New York and maintained a full-time legal practice until 2016 when she joined Dutchess Community College as a tenure-track Assistant Professor in Business. In the Fall of 2018, she joined Penn State University as an Assistant Teaching Professor of Business. In addition to courses in the business program, she teaches a First Year Seminar course and a course examining issues in sustainability. She graduated magna cum laude from the King's College Honors Program with a joint degree in English and French. She continued her education at the Villanova University School of Law, graduating cum laude in 2001 and earning distinction for her studies in employment law. She prioritizes teaching legal concepts through the lens of sustainability and ethics. Her teaching emphasizes the impact of ethics in the legal system and encourages discussion about the evolution of well-developed laws as a means to help foster a sustainable future.

Andrew J. Campbell has been researching and teaching in Digital Mental Health, Cyberpsychology and Child, Adolescent and Family Mental Health for over 20 years. He is the Chair of Australia's first Cyberpsychology Research Group located within the research theme of Biomedical Informatics & Digital Health in School of Medical Sciences in the Faculty of Medicine & Health at The University of Sydney. He is the Inaugural Australasian Editor of the journal Cyberpsychology, Behavior and Social Networking and speaks regularly through media and community events about consumer technologies impact on health and wellbeing.

Pauline S. Cho is the CEO and president of Seojoong International Korea Inc., a construction materials company that imports, trades, and distributes in South Korea. Her main focus is on importing unique and quality materials that are sustainable as well as environmentally friendly to lower the carbon footprint of the industry. With carbon emissions being the highest in the construction industry, she dedicates her time in researching ways to move to net zero construction. She has recently graduated from Pennsylvania State University with a major in finance and a minor in economics. She has completed her degree subsequently while running her business because of her love for higher education.

Gina M. Dignazio is a Penn State World Campus student majoring in Business Administration. She has over 15 years of experience working in the food service and hospitality industry. In addition to her day job, Gina is also an exhibiting artist whose paintings have been on display in various parts of the United States and abroad. Previously, Gina worked as a general volunteer, board member, and painting instructor for the Hazleton Art League, a non-profit arts organization in Hazleton, PA. Currently, she serves on the gallery committee for The Exchange, a non-profit arts organization in Bloomsburg, PA. She holds a BFA in Studio Art from Kutztown University.

Lori S. Elias Reno is currently an Assistant Teaching Professor of Business and Coordinator of the Honors Program at The Pennsylvania State University (PSU) Hazleton campus. Before joining Penn State Hazleton, she served as Business faculty at PSU Brandywine and Mass Communications faculty at Bloomsburg University. She also taught several business and MIS classes for the Penn State Wilkes-Barre worksite education program. Before teaching full-time, she worked for nearly three decades as middle and senior management of marketing-related teams across industries like financial services, pharmaceuticals, retail, academia, and non-profits. Currently the sole proprietor of a part-time strategic planning consultancy, she primarily focuses on pro bono work for new entrepreneurs. She also volunteers as part of the PSU Hazleton LaunchBox, an initiative that aids entrepreneurs and students in the Greater Hazleton area. Recent scholarship activities include a publication on crowdsourcing in the Journal of Higher Education Theory and Practice, a work-in-progress about the virality of social media, and several conference panel presentations on the topic of client-based learning in marketing curricula.

Susan Shepherd Ferebee earned her Ph.D. in Information Systems at Nova Southeastern University in Ft. Lauderdale, Florida. She also has an Executive Juris Doctorate from Concord Law School, and has a Masters in Educational Psychology from Purdue University Global.. She is a faculty member in the School for Business and Information Technology at Purdue University Global and has also served as a consultant with more than 25 years of experience working directly with organizations and higher education institutions. Susan has published many peer-reviewed articles. Her current research in progress includes studies on smart technology use in home schooling, personal cybersecurity behaviors, and the influence of interpretive communities on persuasion, Susan also served as a guest editor for a special issue of International Journal of Conceptual Structures and Smart Applications. Susan serves as an Editorial Review Board member for International Journal of Cognitive Informatics and Natural Intelligence. She is an active presenter at international and national conferences. Susan has received numerous teaching and outstanding contributor awards and has been awarded several research grants.

Jennifer Fleming received her B.S. in Computer Science from Spelman College. In 2004, she earned an Executive MBA from Georgia State University and in 2010, she earned a Ph.D. in Applied Management and Decision Science from Walden University. Jennifer has worked for several companies these include AT&T, Compaq (Hewlett-Packard), Delta Airlines, and The Coca-Cola Company. She currently resides in Atlanta, Georgia, and enjoys teaching graduate courses in information technology, business administration, project management, and decision sciences.

Zergio Nicolás García-Arias is an Electronic Engineer graduated from the Universidad Autónoma de Manizales (UAM). He is 23 years old and lives in Guadalajara de Buga, Colombia. He is a member of the Research Group on Bioinformatics and Artificial Intelligence at UAM. He is researching the ethical aspects of using artificial intelligence-based systems for automated decision-making, mainly oriented to public policies. He is passionate about machine learning, deep learning, and artificial intelligence research.

Linnea Hall is an academic editor at Research Square. She earned a doctorate in law from California Western School of Law in San Diego, CA and a doctorate in computer science from Colorado Technical University in Colorado Springs, CO. Her research interests include artificial intelligence, computational neuroscience, and ethics.

Maria Paz Hermosilla is the Director of GobLab UAI – a public innovation lab at the Adolfo Ibanez University's School of Government – whose mission is to help transform the public sector through data science. She has a Master of Public Administration from New York University. She is the Academic director of a certificate program in Data Science and Public Policy and researcher and professor of Data Ethics for graduate programs in the School of Engineering, School of Business and School of Government. She is leading several research projects related to the use of data science to improve public policy, including predictive models in the social sector, the integration of data ethics into the public procurement process, and algorithmic transparency. She was appointed by the Chilean Ministry of Science in the expert advisory committee to create the national strategy for Artificial Intelligence, and in the national committee for Public Interest Data.

Thomas Huston, Ph.D., Team Lead for the ASC's Business Center at Purdue University Global, is a course developer, educational consultant, and instructor of business writing for JNC Online Programs. Recently, Dr. Huston served as a lecturer of ENGL 106, ENGL 106DIST, and ENGL 106LCOM at Purdue University in West Lafayette, IN. His current scholarship examines curriculum development for critical media literacy and the deliberative arts.

Jennie Lee Khun is an adjunct professor for Purdue University Global. Prior to completing her Doctorate in Strategic Leadership she worked for Shepherd University as an IT Business Analyst and later IT Operations and Project Manager. Jennie received her MBA from Shepherd and her undergraduate IT degree from Kaplan University. She currently works as an IT Project Manager with the United States Federal government and holds industry certifications including CISSP, PMP, PMI-ACP, and ITILv4. Jennie grew up a military brat and cites that upbringing for her love of travel. She enjoys swimming, baking, and playing board games.

Kathleen McCain received a bachelor's degree in psychology from Mississippi State University in Starkville, MS. She is currently pursuing a master's degree in industrial and organizational psychology. Her current research interests include mindfulness, corporate ethics, and worker motivation.

Ricardo Ortega-Bolaños received the degree in biomedical engineering from the Universidad Autónoma de Manizales. He has been a member of the Research Group on Bioinformatics and Artificial Intelligence since 2021. He is researching the ethical aspects of using artificial intelligence-based systems for automated decision-making, mainly oriented to public policies. He has worked on personal projects to classify sign language symbols and coffee rust using CNNs and transfer learning.

Roneeta Purkayastha is currently working as an Assistant Professor in the Department of Computer Science and Engineering, School of Engineering and Technology, Adamas University, Kolkata, India. She has 4+ publications in Scopus-indexed international conferences. Her most notable paper presentation is in 11th World Congress on Information and Communication Technologies held on December

2021. Her current research interests include Cloud Computing, Artificial Intelligence and UML based Ontology representation.

Ellen Marie Raineri earned a Ph.D. in Organization and Management and an M.B.A. in Information Systems. She earned a graduate certificate in Information Assurance and Security as well as a B.A. in Computer Science and a B.A. in English and Education Dr. Raineri has over 15 years of experience teaching undergraduate and graduate students. She is a full-time faculty member at Pennsylvania State University. Dr. Raineri has also worked in corporate for over 15 years in executive technical roles leading corporate initiatives for new products, services, and company divisions. Dr. Raineri made conference presentations and wrote journal articles, book chapters, and a book endorsed by Mark Victor Hansen (Chicken Soup for the Soul series) and Ken Blanchard (One Minute Manager). From these accomplishments, she was interviewed on Lifetime TV as well as other TV, radio, newspaper, and online media. Dr. Raineri is the recipient of the Excellence in Business award, Excellence in Entrepreneurship Teaching award, the Coleman US Association for Small Business and Entrepreneurship (USASBE) scholarship, and the Distinguished Service award. She was also nominated for the Faculty Scholar award, Innovative Entrepreneurship Education Course award, Innovative Pedagogy for Entrepreneurship Education award as well as local and state-wide Excellence in Teaching awards.

Kimberly M. Rehak is a doctor of education candidate in curriculum and instruction at the Indiana University of Pennsylvania. She works an instructional designer for the University of Pittsburgh College of General Studies. She has a master in public policy and management, a master of arts in applied linguistics, and a graduate-level certificate in Teaching English to Speakers of Other Languages (TESOL) from the University of Pittsburgh. She has been teaching the English language since 2002 and has worked in the USA, Austria, and the United Arab Emirates.

Gonzalo A. Ruz received the B.Sc. and M.Sc. degrees in electrical engineering from the Universidad de Chile, Santiago, Chile, in 2002 and 2003, respectively, the P.E. degree in electrical engineering from the Universidad de Chile, and the Ph.D. degree from Cardiff University, U.K., in 2008. Currently, he is a Professor with the Faculty of Engineering and Sciences, Universidad Adolfo IbaìnÞez, Santiago. His research interests include machine learning, evolutionary computation, data science, gene regulatory network modeling, and complex systems.

Audrey E. B. Ryder is currently a student majoring in Accounting at Pennsylvania State University, having previously earned degrees in Biochemistry and Music Performance from Ithaca College. During her prior work in academia, her research team participated in two research publications on both quorum sensing substrates and antagonizing bacterial communication. Ms. Ryder hopes to eventually pursue her CPA license following graduation and currently enjoys researching new technological developments within data analytics.

Paige R. Sharp has received a BS in Biology from the University of South Florida in St. Petersburg and an MS in Microbiology from the University of South Florida in Tampa. She is an Instructor of Biological Sciences at Pasco Hernando State College in Florida. Having both learned as a student and taught as an instructor for online courses, she is interested in focusing on making the online classroom more engaging and accessible.

Reinel Tabares-Soto is an Electronic Engineer from Universidad Nacional de Colombia, Systems Engineer from Universidad de Caldas, Master in Industrial Automation from Universidad Nacional de Colombia and PhD in Computer Engineering from Universidad Autónoma de Manizales. His lines of academic work are Machine Learning, Data Mining, High-Performance Computing, Bioinformatics, and Modeling-Simulation of Dynamic Systems. His master thesis was on parallel programming on heterogeneous architectures, and his Ph.D. thesis was on convolutional neural networks for image steganalysis in the spatial domain. Since 2017 he has worked at the Bioinformatics and Computational Biology (BIOS) research center serving as leader of the scientific computing area, HPC cluster administrator, and researcher on topics related to his lines of work. Since 2020 he has been working as the ICT and competitiveness secretary of Manizales, and in the company, People Contact as a consultant in digital transformation and artificial intelligence. Since 2014 he has been a teacher, researcher, and coordinator of the Electronics Engineering program at the Universidad Autónoma de Manizales. He leads the projects entitled: Towards understanding plant genomes of productive interest using bioinformatics techniques, artificial intelligence, and HPC, COVID-19 detection using X-ray images from convolutional neural networks and convolutional neural networks for image steganalysis in the spatial domain.

Po Man Tse holds a BA in Psychology from the University of British Columbia, MSc in Hotel and Tourism Management from The Hong Kong Polytechnic University. Research interests include consumer behavior, smart tourism, and hospitality education.

Kelly Wibbenmeyer currently works at Mercy as an Executive Director of IT Automation. She is actively working on Robotic Process Automation (RPA), Artificial Intelligence (AI), and Knowledge Management (KM) projects within her organization. She enjoys teaching and helping others with their educational journey. She has two boys and has been married for over 20 years. She has her Ph.D. in MIS from Northcentral University, MBA in Project Management from Wright State University, and BS in MIS from Wright State University. Both her MBA and BS degrees were taken in a mixed on-ground/ online environment and her Ph.D. was completely attained online. She has taught project management, innovation management, and business statistics for over five years at various universities. She also has attained her Six Sigma Black Belt certification, has been a certified project manager for over 17 years, and is an ITIL master as well.

Lynne Williams teaches at the graduate level in the School of Business and Information Technology for Purdue University Global which is the online branch of Purdue University. She was instrumental in developing the Masters of Science in Cybersecurity for Purdue University Global and has guest lectured internationally on the privacy concerns of digital footprints. Her areas of research encompass cybersecurity, networking technologies, and the effects of human interaction with technology.

Florence Wolfe Sharp holds a BA in English from the University of Dayton, MS in Educational Technology from Nova Southeastern University, and EdD in Educational Technology from the University of Florida. She is the Assistant Dean of Curriculum for Information Technology programs and served colleges and universities in Arizona, Florida, and Ohio before coming to Purdue University Global. Her interests include online presence, accessibility, and thoughtful technology integration.

Index

E

F

G

H

I

L

M

N

O

P

Q

R

S

T

U

V

W

Printed in the United States
by Baker & Taylor Publisher Services